Nanostructured Materials and Advanced Processes for Application in Water Purification

Nanostructured Materials and Advanced Processes for Application in Water Purification

Editor

Christos A. Aggelopoulos

MDPI

Basel • Beijing • Wuhan • Barcelona • Belgrade • Novi Sad • Cluj • Manchester

Editor
Christos A. Aggelopoulos
Institute of Chemical
Engineering Sciences
Foundation for Research and
Technology Hellas
Patras
Greece

Editorial Office
MDPI
St. Alban-Anlage 66
4052 Basel, Switzerland

This is a reprint of articles from the Special Issue published online in the open access journal *Nanomaterials* (ISSN 2079-4991) (available at: www.mdpi.com/journal/nanomaterials/special_issues/nano_processes_water).

For citation purposes, cite each article independently as indicated on the article page online and as indicated below:

Lastname, A.A.; Lastname, B.B. Article Title. *Journal Name* **Year**, *Volume Number*, Page Range.

ISBN 978-3-7258-0434-4 (Hbk)
ISBN 978-3-7258-0433-7 (PDF)
doi.org/10.3390/books978-3-7258-0433-7

Contents

About the Editor

Christos A. Aggelopoulos

Dr. Christos Aggelopoulos is a Principal Researcher at FORTH/ICE-HT and is the Head of the Laboratory of Cold Plasma and Advanced Techniques for Improving Environmental Systems (PlaNET). He has a BSc, an MSc, and a PhD in Physics (2007). Following postdoctoral research at Institut Francais du Petrole-IFP, France (2009–2010), he was subsequently continued as a Research Associate at FORTH/ICE-HT (2010–2012 and 2014–2018) and Université Pierre et Marie Curie—Ecole Nationale Supérieure de Chimie de Paris, France (2012–2014). He was also a member of the Adjunct Academic Staff of Hellenic Open University; from 2017 to 2020, he was a tutor in the postgraduate program (M.Sc.), known as the Environmental Design of Infrastructure Works.

His current research activities focus on (i) polluted soil–wastewater–waste treatment using cold atmospheric plasma, (ii) wastewater treatment using plasma catalysis, and (iii) the plasma-induced activation/regeneration of adsorbents capable of removing multi-species pollutants from wastewater. His recent research extends into the field of plasma treatment of cancer cells and the plasma-induced modification of materials' surfaces. Dr. Aggelopoulos has participated in 21 R&D projects (7 of which as PI/Co-PI), funded by the EU and national sources. For his research on plasma, he was awarded (2016) the 2nd Prize in the category "Applied Research" of the 3rd "Greece Innovates!" Applied Research & Innovation Competition. He is included in 2023 Stanford's list World Top 2% scientists.

He is co-author of 62 scientific articles (34 as a corresponding author and 2 monographs) in refereed scientific journals (Scopus h-index: 23, citations: ~1700; Google Scholar h-index: 26, citations: ~2150), 1 patent on soil remediation using non-thermal plasma, and about >70 publications in international/national conference proceedings. He is a reviewer for >30 international scientific journals and a member of scientific committees and associations.

Preface

Water contamination via heavy metal ions, organic pollutants, and pathogens is a severe threat worldwide, disturbing or threatening both human health and ecosystems even in very small concentrations. Thus, improving water quality is a crucial topic, with both social and economic benefits.

In recent years, the use of nanostructured materials to improve water purification systems has been increasingly considered. Nanostructured materials are promising candidates since their nano-scaled nature leads to important features. Nanomaterials have been tested for the treatment of water contaminated by different classes of pollutants, such as dyes, pharmaceuticals, pesticides, pathogens, as well as industrial and domestic wastes.

On the other hand, advanced oxidation processes, including photocatalysis and plasma, are considered advantageous pollutant destruction methods.

In this context, this Special Issue explores recent developments of nanostructured materials and advanced oxidation processes. It highlights cutting-edge research activities focusing on the synthesis, characterization, and implementation of both organic and inorganic nanostructured materials with applications related to water remediation.

Christos A. Aggelopoulos
Editor

nanomaterials

Editorial

Nanostructured Materials and Advanced Processes for Application in Water Purification

Christos A. Aggelopoulos

Laboratory of Cold Plasma and Advanced Techniques for Improving Environmental Systems, Institute of Chemical Engineering Sciences, Foundation for Research and Technology Hellas (FORTH/ICE-HT), 26 504 Patras, Greece; caggelop@iceht.forth.gr

Citation: Aggelopoulos, C.A. Nanostructured Materials and Advanced Processes for Application in Water Purification. *Nanomaterials* **2023**, *13*, 654. https://doi.org/10.3390/nano13040654

Received: 19 January 2023
Revised: 1 February 2023
Accepted: 6 February 2023
Published: 7 February 2023

Water pollution is a major environmental problem that has a significant impact on human and animal health and the ecosystem. Most pollution is caused by human activities, and pollutants can be categorized into inorganic, organic, biological and radioactive [1]. Thus, the critical issue of improving water quality has several social and economic benefits.

During the last decade, the field of nanotechnology has extensively developed, contributing to a significant impact on water purification. The use of nanomaterials has led to impressive findings in the field of water remediation, with a high efficiency for the removal of various pollutants, cost effectiveness and reusability. Thus, nanostructured materials, due to their unique physicochemical characteristics, such as catalytic activity; high physical, chemical and thermal stability; large specific surface area; high chemical reactivity; strong electron ability, etc., have gained the attention of many researchers. Different classes of nanomaterials can be used in water purification either as adsorbents, photocatalysts and/or antibacterial agents [2–5]. More specific, nanostructured materials used as adsorbents (nanosorbents), nanoparticles (e.g., Pd, Au, Cu, Fe_3O_4, TiO_2, ZnO, Ag, etc.) and nanocatalytic membrane systems are more efficient and less time-consuming than other treatment methods, environmentally friendly and consume a low amount of energy [4].

This Special Issue focuses on nanostructured materials and their applications in advanced water purification processes. This Special Issue contains fifteen articles and one communication that address the synthesis, modification and regeneration of novel nanomaterials for their applications to remediate water contaminated by various pollutants.

The first paper published in this Special Issue by Xiang et al. [6] illustrates the preparation of Fe_3O_4@C nanoparticles for the decolorization of high concentrations of methylene blue. According to the obtained results, an easy method for producing Fe_3O_4@C nanoparticles with excellent catalytic reactivity is presented, representing a promising approach for the industrial production of Fe_3O_4@C nanoparticles used to treat high concentrations of dyes in wastewater. The second paper by Zia et al. [7] presents the synthesis of silver/iron oxide nanocomposites (Ag/Fe_3O_4) for the efficient and specific removal of iodine anions from contaminated water. The findings presented in this study offer a novel method for desalinating radioiodine in various aqueous media. The next paper by Narath et al. [8] proposes a facile green synthetic method for the synthesis of zinc oxide nanoparticles (ZnO NPs) using the bio-template *Cinnamomum tamala* (*C. tamala*) leaves extract. According to the obtained results, the synthesized ZnO NPs exhibit excellent photocatalytic activity against dye molecules, through a green protocol. The fourth paper by Shrestha [9] illustrates the usage of wood dust of *Dalbergia sisoo* (Sisau) derived from activated carbon (AC) as an adsorbent material for the removal of rhodamine B dye from an aqueous solution. According to the author, this study successfully addressed a local problem of wastewater pollution from garment and textile industrial effluents using the locally available agro-waste of *Dalbergia sisoo*. Khan et al. [10] investigated the removal of nano-CuO from pure water and montmorillonite clay (MC) suspensions, using poly aluminum ferric chloride (PAFC) and cationic polyacrylamide (PAM) via the coagulation–flocculation–sedimentation (C/F/S) process. Moreover, PAFC and PAFC/PAM flocculation performance was investigated through a parametric analysis. The findings presented in this paper provide insights into enhanced flocculation and the coagulation of CuO in drinking water containing clay particles. Balarak et al. [11] found that activated porous carbon prepared from Azolla filiculoides (AF) (AFAC) could be a potential low-cost adsorbent of azithromycin (AZM) from aqueous solutions. In the next paper by Liou, Chen and Yang [12], amino-functionalized nanoporous Santa Barbara amorphous-15 (SBA-15) was fabricated by employing sodium silicate as a precursor and successfully applied as a high-efficiency adsorbent for the removal of tannic acid from aqueous media. Shalabayev et al. [13] reported the successful synthesis of CdS nanoparticles using cadmium acetate and sodium sulfide as Cd and S precursors, respectively. Moreover, the effect of using sodium thiosulfate as an additional sulfur precursor was also investigated (combined milling) as well as the antibacterial potential of mechanochemically prepared CdS nanocrystals on reference strains of *E. coli* and *S. aureus*. The results presented, demonstrated that solvent-free and sustainable mechanochemical synthesis can easily produce semiconductor nanocrystals in a multidisciplinary application. In the paper by Peñaranda et al. [14], three different magnetic microreactors based on torus geometries (i.e., one-loop, two-horizontal-loop, and two-vertical-loop microreactors) were designed, manufactured, and tested to increase the enzyme-based transformation of dyes by laccase bio-nanocomposites, improve particle suspension, and promote the interaction of reagents. Taken together, these results indicate that the novel microreactors introduced in this study have the potential to efficiently process wastewaters contaminated with dyes in a continuous mode, with potential for large-scale operations. Primo et al. [15] present two green synthesis routes that were used for the synthesis of Ag/ZnO nanoparticles, using cassava starch as a simple and low-cost effective fuel and aloe vera as a reducing and stabilizing agent. The results indicate that Ag/ZnO nanoparticles synthesized via green chemistry are a promising candidate for the treatment of wastewaters contaminated by bacteria (*E. coli*), due to their easy preparation, low-cost synthesis, and disinfection efficiency. Magro et al. [16] successfully developed sensors based on nanostructured thin films deposited on ceramic substrates with gold-interdigitated electrodes to detect azithromycin, clarithromycin, and erythromycin in concentrations ranging between 10^{-15} M and 10^{-5} M in mineral and river water matrices. Ramírez-Rodríguez et al. [17] proposed a novel method to fabricate a hybrid adsorbent membrane of whey protein isolate (WPI) and polycaprolactone (PCL) by electrospinning, finding promising results for the removal heavy metal ions (e.g., chromium ions) from water. Khan et al. [18] focus on the synthesis of a nanocomposite material (ZnO–CuO/g–C_3N_4) using a solution method for the successful extraction of arsenic (III), pointing out that a ZnO–CuO/g–C_3N_4 nanocomposite can be a potential candidate for the enhanced removal of arsenic from water reservoirs. Zhang et al. [19], prepared a new catalyst, copper oxide/graphene oxide–diatomaceous earth (CuO/GO-DE) using the ultrasonic impregnation method, which had excellent catalytic activity towards ciprofloxacin degradation. Tóth et al. [20] presented the effect of different phosphate sources on the synthesis and photocatalytic activity of Ag_3PO_4 for pollutants degradation. These authors concluded that Ag_3PO_4–based materials could be reliably used for the degradation of methyl orange (MO) as they mostly retain their photoactivity during the second recycling test. In the final paper by Giannoulia et al. [21], halloysite nanoclay (HNC) was examined as an adsorbent for the individual and simultaneous removal of antibiotic enrofloxacin (ENRO) and methylene blue (MB) from aqueous solutions, alongside its successful regeneration via cold atmospheric plasma (CAP) bubbling. In addition, CAP bubbling induced chemical modifications on the HNC surface, increasing its adsorption capacity when applied to new adsorption cycles.

Data Availability Statement: The data presented in this study are available on request from the corresponding author.

Acknowledgments: I am grateful to all the authors who contributed to this Special Issue. We also express our acknowledgments to the referees for reviewing the manuscripts.

Conflicts of Interest: The author declares no conflict of interest.

References

1. Schweitzer, L.; Noblet, J. Water contamination and pollution. In *Green Chemistry*; Elsevier: Amsterdam, The Netherlands, 2018; pp. 261–290.
2. Santhosh, C.; Velmurugan, V.; Jacob, G.; Jeong, S.K.; Grace, A.N.; Bhatnagar, A. Role of nanomaterials in water treatment applications: A review. *Chem. Eng. J.* **2016**, *306*, 1116–1137. [CrossRef]
3. Saleem, H.; Zaidi, S.J. Developments in the application of nanomaterials for water treatment and their impact on the environment. *Nanomaterials* **2020**, *10*, 1764. [CrossRef] [PubMed]
4. Nasrollahzadeh, M.; Sajjadi, M.; Iravani, S.; Varma, R.S. Green-synthesized nanocatalysts and nanomaterials for water treatment: Current challenges and future perspectives. *J. Hazard. Mater.* **2021**, *401*, 123401. [CrossRef] [PubMed]
5. Xu, C.; Nasrollahzadeh, M.; Sajjadi, M.; Maham, M.; Luque, R.; Puente-Santiago, A.R. Benign-by-design nature-inspired nanosystems in biofuels production and catalytic applications. *Renew. Sustain. Energy Rev.* **2019**, *112*, 195–252. [CrossRef]
6. Xiang, H.; Ren, G.; Zhong, Y.; Xu, D.; Zhang, Z.; Wang, X.; Yang, X. Fe_3O_4@C Nanoparticles Synthesized by In Situ Solid-Phase Method for Removal of Methylene Blue. *Nanomaterials* **2021**, *11*, 330. [CrossRef] [PubMed]
7. Zia, M.R.; Raza, M.A.; Park, S.H.; Irfan, N.; Ahmed, R.; Park, J.E.; Mushtaq, S. Removal of Radioactive Iodine Using Silver/Iron Oxide Composite Nanoadsorbents. *Nanomaterials* **2021**, *11*, 588. [CrossRef] [PubMed]
8. Narath, S.; Koroth, S.K.; Shankar, S.S.; George, B.; Mutta, V.; Wacławek, S.; Varma, R.S. Cinnamomum tamala leaf extract stabilized zinc oxide nanoparticles: A promising photocatalyst for methylene blue degradation. *Nanomaterials* **2021**, *11*, 1558. [CrossRef] [PubMed]
9. Shrestha, D. Efficiency of wood-dust of Dalbergia sisoo as low-cost adsorbent for rhodamine-B dye removal. *Nanomaterials* **2021**, *11*, 2217. [CrossRef] [PubMed]
10. Khan, R.; Inam, M.A.; Lee, K.H.; Channa, A.S.; Mallah, M.A.; Wie, Y.M.; Abbasi, M.N. Synergetic Effect of Organic Flocculant and Montmorillonite Clay on the Removal of Nano-CuO by Coagulation-Flocculation-Sedimentation Process. *Nanomaterials* **2021**, *11*, 2753. [CrossRef] [PubMed]
11. Balarak, D.; Mahvi, A.H.; Shahbaksh, S.; Wahab, M.A.; Abdala, A. Adsorptive removal of azithromycin antibiotic from aqueous solution by azolla filiculoides-based activated porous carbon. *Nanomaterials* **2021**, *11*, 3281. [CrossRef] [PubMed]
12. Liou, T.H.; Chen, G.W.; Yang, S. Preparation of Amino-Functionalized Mesoporous SBA-15 Nanoparticles and the Improved Adsorption of Tannic Acid in Wastewater. *Nanomaterials* **2022**, *12*, 791. [CrossRef] [PubMed]
13. Shalabayev, Z.; Baláž, M.; Khan, N.; Nurlan, Y.; Augustyniak, A.; Daneu, N.; Burkitbayev, M. Sustainable Synthesis of Cadmium Sulfide, with Applicability in Photocatalysis, Hydrogen Production, and as an Antibacterial Agent, Using Two Mechanochemical Protocols. *Nanomaterials* **2022**, *12*, 1250. [CrossRef] [PubMed]
14. Peñaranda, P.A.; Noguera, M.J.; Florez, S.L.; Husserl, J.; Ornelas-Soto, N.; Cruz, J.C.; Osma, J.F. Treatment of Wastewater, Phenols and Dyes Using Novel Magnetic Torus Microreactors and Laccase Immobilized on Magnetite Nanoparticles. *Nanomaterials* **2022**, *12*, 1688. [CrossRef] [PubMed]
15. Primo, J.D.O.; Horsth, D.F.; Correa, J.D.S.; Das, A.; Bittencourt, C.; Umek, P.; Anaissi, F.J. Synthesis and Characterization of Ag/ZnO Nanoparticles for Bacteria Disinfection in Water. *Nanomaterials* **2022**, *12*, 1764. [CrossRef] [PubMed]
16. Magro, C.; Moura, T.; Dionísio, J.; Ribeiro, P.A.; Raposo, M.; Sério, S. Nanostructured Metal Oxide Sensors for Antibiotic Monitoring in Mineral and River Water. *Nanomaterials* **2022**, *12*, 1858. [CrossRef] [PubMed]
17. Ramírez-Rodríguez, L.C.; Quintanilla-Carvajal, M.X.; Mendoza-Castillo, D.I.; Bonilla-Petriciolet, A.; Jiménez-Junca, C. Preparation and Characterization of an Electrospun Whey Protein/Polycaprolactone Nanofiber Membrane for Chromium Removal from Water. *Nanomaterials* **2022**, *12*, 2744. [CrossRef] [PubMed]
18. Khan, Q.U.; Begum, N.; Rehman, Z.U.; Khan, A.U.; Tahir, K.; Tag El Din, E.S.M.; Javed, M.S. Development of Efficient and Recyclable ZnO–CuO/g–C_3N_4 Nanocomposite for Enhanced Adsorption of Arsenic from Wastewater. *Nanomaterials* **2022**, *12*, 3984. [CrossRef] [PubMed]
19. Zhang, T.; Zhang, J.; Yu, Y.; Li, J.; Zhou, Z.; Li, C. Synthesis of CuO/GO-DE Catalyst and Its Catalytic Properties and Mechanism on Ciprofloxacin Degradation. *Nanomaterials* **2022**, *12*, 4305. [CrossRef] [PubMed]
20. Tóth, Z.R.; Debreczeni, D.; Gyulavári, T.; Székely, I.; Todea, M.; Kovács, G.; Hernadi, K. Rapid Synthesis Method of Ag_3PO_4 as Reusable Photocatalytically Active Semiconductor. *Nanomaterials* **2023**, *13*, 89. [CrossRef] [PubMed]
21. Giannoulia, S.; Triantaphyllidou, I.E.; Tekerlekopoulou, A.G.; Aggelopoulos, C.A. Mechanisms of Individual and Simultaneous Adsorption of Antibiotics and Dyes onto Halloysite Nanoclay and Regeneration of Saturated Adsorbent via Cold Plasma Bubbling. *Nanomaterials* **2023**, *13*, 341. [CrossRef] [PubMed]

nanomaterials

Article

Mechanisms of Individual and Simultaneous Adsorption of Antibiotics and Dyes onto Halloysite Nanoclay and Regeneration of Saturated Adsorbent via Cold Plasma Bubbling

Stefania Giannoulia [1,2], Irene-Eva Triantaphyllidou [1], Athanasia G. Tekerlekopoulou [2] and Christos A. Aggelopoulos [1,*]

[1] Laboratory of Cold Plasma and Advanced Techniques for Improving Environmental Systems, Institute of Chemical Engineering Sciences, Foundation for Research and Technology Hellas (FORTH/ICE-HT), 26504 Patras, Greece

[2] Department of Sustainable Agriculture, University of Patras, 2 G. Seferi St., 30100 Agrinio, Greece

* Correspondence: caggelop@iceht.forth.gr; Tel.: +30-2610965205

Abstract: Halloysite nanoclay (HNC) was examined as an adsorbent for the individual and simultaneous removal of antibiotic enrofloxacin (ENRO) and methylene blue (MB) from aqueous solutions, alongside its regeneration via cold atmospheric plasma (CAP) bubbling. Initially, batch kinetics and isotherm studies were carried out, while the effect of several parameters was evaluated. Both ENRO and MB adsorption onto HNC was better described by Langmuir model, with its maximum adsorption capacity being 34.80 and 27.66 mg/g, respectively. A Pseudo-second order model fitted the experimental data satisfactorily, suggesting chemisorption (through electrostatic interactions) as the prevailing adsorption mechanism, whereas adsorption was also controlled by film diffusion. In the binary system, the presence of MB seemed to act antagonistically to the adsorption of ENRO. The saturated adsorbent was regenerated inside a CAP microbubble reactor and its adsorption capacity was re-tested by applying new adsorption cycles. CAP bubbling was able to efficiently regenerate saturated HNC with low energy requirements (16.67 Wh/g-adsorbent) in contrast to Fenton oxidation. Most importantly, the enhanced adsorption capacity of the CAP-regenerated HNC (compared to raw HNC), when applied in new adsorption cycles, indicated its activation during the regeneration process. The present study provides a green, sustainable and highly effective alternative for water remediation where pharmaceutical and dyes co-exist.

Keywords: adsorption; nanoclays; halloysite; methylene blue; enrofloxacin; regeneration; cold atmospheric plasma

Citation: Giannoulia, S.; Triantaphyllidou, I.-E.; Tekerlekopoulou, A.G.; Aggelopoulos, C.A. Mechanisms of Individual and Simultaneous Adsorption of Antibiotics and Dyes onto Halloysite Nanoclay and Regeneration of Saturated Adsorbent via Cold Plasma Bubbling. *Nanomaterials* **2023**, *13*, 341. https://doi.org/10.3390/nano13020341

Academic Editor: Fernanda Cássio

Received: 19 December 2022
Revised: 6 January 2023
Accepted: 11 January 2023
Published: 13 January 2023

1. Introduction

Water pollution is regarded as a serious environmental problem with particularly negative impacts on modern societies [1]. Pharmaceuticals and dyes are usually present in aquatic systems and are responsible for adverse ecological and human health effects. In particular, the degradation-resistant properties of antibiotics lead to severe public health issues, including some types of cancer, skin problems, allergies, and other serious health problems [2]. Dyes cause non-esthetic pollution and eutrophication whilst at the same time reducing light penetration and photosynthetic activity. Thus, it is vital to remove both classes of pollutants from wastewater before disposal [3].

Many treatment techniques have been proposed for the simultaneous removal of pollutants, such as ion exchange, reverse osmosis, liquid membrane separation, adsorption, ozonation, photo-catalysis, coagulation, etc. [4–6]. Among these techniques, adsorption is one of the most efficient, easy-to-apply and cost-effective techniques. According to the literature, many adsorbents (e.g., clay minerals, biosorbents, polymers, etc.) have been already considered for the removal of dyes and antibiotics [7–13]. In the last decades, halloysite (HNC), a natural aluminosilicate nano-clay mineral ($Al_2(OH)_4Si_2O_5{\cdot}nH_2O$), has gained increasing attention due to its better adsorption performance compared to non-porous micron-sized kaolinite. This is mainly attributed to its morphology, chemical composition and the structural arrangement of its functional groups [14]. Indicatively for antibiotics, where the references are limited, tetracycline has been examined with various halloysite-based adsorption materials, with a maximum adsorption capacity ranging from 20.5 to 54.9 mg/g [15–17]; oxytetracycline has a maximum adsorption capacity of 54.9 mg/g [18] and ciprofloxacin a maximum adsorption capacity of 25.09 mg/g [19] and 21.7 mg/g [20]. The ultrahigh maximum adsorption capacity of 1297 mg/g for tetracycline and 1067.2 mg/g for chloramphenicol was achieved by in situ KOH activation of halloysite nanotubes [21]. In contrast to antibiotics, there is extensive literature on dye removal using halloysite and halloysite-based adsorbents. Regarding methylene blue, which is intensely studied as it is one of the most widely used dyes in many industries (e.g., paper, textile, chemical, pharmaceutical, etc.), the maximum adsorption capacity ranged from 29.33 to 689.66 mg/g [22].

Most studies to date focus on the adsorption of a single pollutant, and fewer have been published on the simultaneous adsorption of different classes of pollutants [23,24]. However, wastewater effluents are composed of mixtures of pollutants, and the challenge of simultaneously removing them remains. Furthermore, a key disadvantage of the adsorption method is the disposal of saturated adsorbent which can lead to secondary pollution. Conventional adsorbent regeneration methods are mainly achieved through chemical and thermal techniques, each having many disadvantages, such as the use of additional chemicals, increased energy consumption, and the requirement of high temperatures, etc. [25]. Recently, cold atmospheric plasma (CAP) is considered as an effective, environmentally friendly and low-energy regeneration process [26–30]. The effectiveness of CAP on adsorbent regeneration is attributed to the high oxidation potential of plasma-generated reactive oxygen and nitrogen species (RONS) such as $^1O_2$, $\cdot OH$, O, $\cdot O_2^-$, O_3, NO_2^-, NO_3^-, $ONOO^-$, H_2O_2, etc. [31,32].

In this study, the individual and simultaneous adsorption of pharmaceuticals and dyes onto HNC was thoroughly investigated; CAP bubbling was evaluated for the regeneration of saturated HNC, while CAP-regenerated HNC was compared with raw HNC for several adsorption cycles. Fluoroquinolone antibiotic enrofloxacin (ENRO) was selected as the model antibiotic contaminant, being mainly represented in veterinary studies [33,34] with limited assimilation by the organism and partial metabolism [33,35]. To the best of our knowledge, this is the first time that HNC has been used as an adsorbent for ENRO removal from water. Methylene blue (MB) was selected as the model dye, being the most frequently used industrial dye characterized as toxic, carcinogenic, and non-biodegradable. Batch experiments were conducted to evaluate the effect of adsorbent dosage, contact time, initial pollutants concentration and pH on the removal efficiency. HNC was thoroughly characterized by various techniques, such as the Brunauer–Emmett–Teller method (BET) and attenuated total reflectance-Fourier transform infrared spectroscopy (ATR-FTIR). The adsorption isotherms and kinetic parameters were determined, providing information on the adsorption mechanisms, whereas the potential reuse of HNC was evaluated through saturation experiments. Finally, the saturated adsorbent was regenerated inside a novel plasma microbubble reactor and its adsorption capacity was re-tested by applying new adsorption cycles.

2. Materials and Methods

2.1. Materials

Halloysite nanoclay (HNC, $Al_2Si_2O_5(OH)_4{\bullet}2H_2O$, M.W. = 294.19 g/mol), antibiotic enrofloxacin (ENRO, $C_{19}H_{22}FN_3O_3$, M.W. = 359.39 g/mol) and methylene blue (MB, $C_{16}H_{18}ClN_3S{\bullet}2H_2O$, M.W. = 373.90 g/mol) were purchased from Sigma-Aldrich (Saint Louis, MI, USA). The main properties of HNC, MB and ENRO are described in Table S1. Compressed dry air, used as plasma feeding gas during HNC regeneration, was supplied by

Linde (Athens, Greece). All chemicals used in the present study were of analytical grade and for the preparation of all solutions, triple distillation (3D) was used.

2.2. Batch Adsorption Experiments

Adsorption experiments were performed in tightly sealed glass bottles, placed on a rotor, at a speed of 18 rpm and a constant temperature of 28 °C within an incubator (Witeg, GmbH, Wertheim, Germany). A parametric study was carried out, and the effect of the adsorbent dosage, initial pH of the solution, contact time and initial pollutants' concentration on adsorption capacity of HNC was investigated. To that end, various adsorbent dosages (i.e., 0.5, 1, 2, and 3 g/L), pollutants' initial concentrations ranging from 10 to 150 mg/L, and pH values ranging from 2.0–11.0 were considered. Concerning the binary system, one pollutant was kept constant at 40 mg/L while the other was added at varying concentrations, i.e., 10, 20, 40, 80, 120 and 150 mg/L. Batch experiments were performed until the adsorbent's saturation.

At the end of each experiment, samples were collected, centrifuged for 1 min at 12,000 rpm, and analysed by a UV–Vis spectrophotometer (Shimadzu, UV-1900, Kyoto, Japan). For the UV/Vis analysis, the characteristic absorption bands of MB and ENRO at 663 and 272.5 nm, respectively were monitored for the quantification of the pollutants in the solutions.

The HNC adsorption capacity at equilibrium, q_e (mg/g), the adsorption capacity at a specific time t, q_t (mg/g) and the pollutants removal efficiency, R (%), were calculated according to Equations (1), (2) and (3), respectively:

$$q_e = \left(\frac{C_0 - C_e}{m}\right) \times V \tag{1}$$

$$q_t = \left(\frac{C_0 - C_t}{m}\right) \times V \tag{2}$$

$$R\ (\%) = \left(\frac{C_0 - C_e}{C_0}\right) \times 100 \tag{3}$$

where V is the volume of the artificially polluted water (L), m is the adsorbent mass (g) and C_o, C_e and C_t are the initial, equilibrium and at a specific time t concentration (mg/L), respectively.

2.3. Regeneration of Saturated HNC and Application on New Adsorption Cycles

The regeneration of saturated HNC was tested through different methods, i.e., Fenton oxidation, CAP bubbling and air bubbling. For the regeneration of the saturated raw HNC by Fenton oxidation, the saturated raw HNC (10 mg) was mixed with 25 mL of $FeCl_3/H_2O_2$ solution (C_{Fe} = 10^{-2} M and 3% H_2O_2) at room temperature for two days and then washed with distilled water.

Concerning the regeneration process based on CAP bubbling, the saturated raw HNC (30 mg) was inserted into a plasma reactor with 50 mL of 3D water. The CAP-regeneration system consisted of a specially designed plasma microbubble reactor driven by a nanosecond pulse generator (NPG-18/3500) and a discharge characterization system. The power supply was able to produce positive high voltage nanopulses of very short rising time (~4 ns). The detailed experimental setup along with a schematic diagram of the plasma microbubble reactor is presented in Figure 1. Briefly, the reactor consisted of a rod-like high voltage (HV) stainless-steel electrode placed between an inner quartz tube and an outer quartz tube. The plasma feeding was injected between the two dielectrics and entered the aqueous phase through 10 microholes located at the base of the outer dielectric tube, resulting in the production of underwater plasma bubbles inside the reaction tank [36]. The ground electrode was a stainless-steel mesh attached at the outer surface of the reaction tank. The air flow rate was controlled by a mass flow controller (Aalborg GFC17, Orangeburg, NY, USA) and kept constant during the adsorbent regeneration experiments at 3 L/min. The duration of adsorbent regeneration by plasma was 30 min, under constant pulse voltage (28 kV) and frequency (200 Hz).

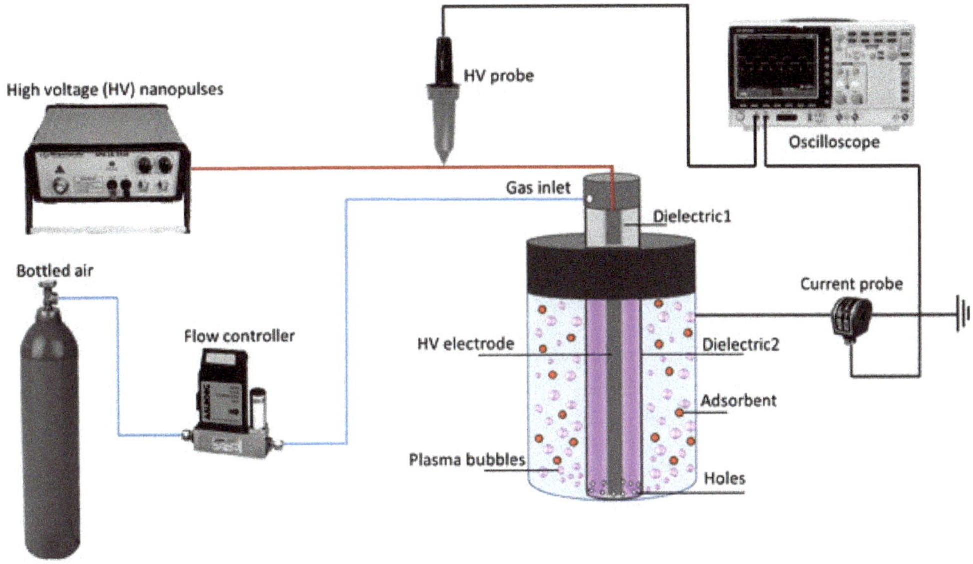

Figure 1. Schematic representation of the experimental setup used to regenerate HNC adsorbent by cold atmospheric plasma (CAP) bubbling.

Finally, in order to ensure that there was no desorption of the adsorbed pollutants through air bubbling, saturated raw HNC was inserted into the microbubble reactor and subjected to air bubbling (without plasma ignition) for 30 min.

According to the experimental results, the optimum regeneration method was selected (i.e., CAP bubbling) and the regenerated HCN (1st regenerated HNC) was subjected to new adsorption cycles until its saturation. After the saturation of the 1st regenerated HNC, a new regeneration treatment was applied, resulting in the second-time regenerated HNC (2nd regenerated HNC) and new adsorption cycles followed until saturation. All experiments were conducted in duplicate.

2.4. Characterization of the Adsorbent

The physicochemical properties of the raw adsorbent and the corresponding ones after adsorption experiments and/or regeneration were studied. For the determination of specific surface area, the Brunauer–Emmett–Teller (BET) method was employed. The ATR/FTIR spectra of the raw adsorbent and after the adsorption of MB and ENRO were recorded using a Bruker Optics ATR spectrometer (Alpha-P Diamond/Bruker Optics GmbH, Billerica, MA, USA). The ATR/FTIR spectra of ENRO and MB were also recorded for comparison and appropriate peak assignment. The point of zero charge (pH_{pzc}) was determined according to the solid addition method of Balistieri and Murray (1981) [37]. Briefly, the pH of 15 mL aqueous solutions was adjusted at a range 1.0–12.0 (pH_i) using HCl or NaOH 1 M, accordingly, following the addition of the adsorbent. The vials were then placed in the incubator (28 °C) and rotated (18 rpm) for 24 h until equilibrium (as confirmed by pH stabilization). After that point, samples were centrifuged, filtered, and subjected to pH measurement, indicating the final pH (pH_f). The pH_{pzc} was determined by the cross point of the curve plotting $\Delta pH = pH_i - pH_f$ versus pH_i. For all pH measurements, a multiparameter analyser was used (Consort C1020, Turnhout, Belgium).

3. Results and Discussion

3.1. Characterization of Halloysite

3.1.1. BET Analysis

The surface area of the adsorbents is indicative of their sorption capabilities due to the micropores and channels present in their structure, and their tiny particle size and surface roughness [38]. The specific surface area of the HNC was calculated by the BET method for the low relative vapor pressure N_2 adsorption isotherm. According to the results obtained, the HNC used in the present study has a relatively high specific surface area, 69 m^2/g, and a particle size of 30–70 nm (Figure S1), these are in accordance with those reported in the literature [39–41].

3.1.2. Point of Zero Charge

The surface of HNC is mainly covered by tetrahedral Si–O, while the inner and the edges are mainly covered by octahedral Al–OH. For pH values higher than the point of zero charge (pH_{pzc}), the charge balance is negative, while for pH values lower than pH_{pzc}, the charge balance is positive [40]. According to Bretti et al., (2016) [42], at pH 2.0, Si–OH and Al–OH are fully protonated. From this pH, the Si–OH are gradually deprotonated, and at pH of about 6.5 are totally deprotonated, and their outer surface reaches the maximum negative charge. The inner surface at pH 2.0 has the highest positive charge, and starting from pH 6.0 is gradually deprotonated. According to the results obtained in the present study, there are two cutting points at the curve of ΔpH versus pH_i (Figure S2). The first one is at pH = 2.5, mainly due to Si–O, and a second one at pH = 4.5, mainly due to Al–OH. Moreover, Sverjensky (1994) [43] has reported that the pH_{pzc} of Si–O and the $Al_2Si_2O_5(OH)_4$ crystal structure is 2.9 and 4.5, respectively. The pH of aqueous solutions of MB and ENRO is 6.2, so at this pH, the HNC outer surface is negatively charged, and the inner positively charged.

3.1.3. FTIR Analysis before/after Adsorption and for the Evaluation of the Regeneration Process

To identify any potential alterations in the surface chemistry of the materials during the adsorption/regeneration process, the adsorbents' spectroscopic characterisation was carried out using ATR-FTIR. The ATR/FTIR spectrum of HNC (Figure S3) exhibits all its characteristic peaks. In brief, the peaks at 3692 and 3626 cm^{-1} are attributed to the stretching vibrations of the inner-surface Al–OH groups, and the peaks at 752, 797, 907, 1004, 1118 and 1109 cm^{-1} are assigned to the stretching mode of apical Si–O [44,45].

In Figure 2, the ATR/FTIR spectra of HNC before and after the adsorption of ENRO (a) and MB (b) and after the regeneration of HNC with CAP are presented; the ATR/FTIR spectra of ENRO and MB are also presented for comparison. In both cases, the adsorption onto HNC is indicated in the spectra. More specifically, the HNC spectra before and after the adsorption of ENRO (Figure 2a) are similar, with the only difference an additional peak existing on the IR spectrum after the adsorption; this peak is one of the most intense peaks of the ENRO IR spectrum (i.e., 1506 cm^{-1} assigned to the C=C stretching vibration of the aromatic ring) [46]. The effectiveness of the CAP regeneration process is also verified, since this peak is disappeared from the spectrum after the regeneration of the HNC with CAP (grey highlighted area). In the case of MB (Figure 2b), the extended adsorption onto HNC is also evidenced through IR spectroscopy. In particular, after the adsorption, three additional peaks (i.e., 1595, 1392/1352sh/1332 cm^{-1} attributed to the CH=N group, and $-CH_2$ or $-CH_3$ stretching) are noticed [47]. It is interesting to note that after the CAP regeneration of HNC, the intensity of these peaks is significantly decreased (blue highlighted areas), but did not vanish, which is in accordance with the UV/Vis analysis of the evaluation study concerning regeneration with CAP (Section 3.7).

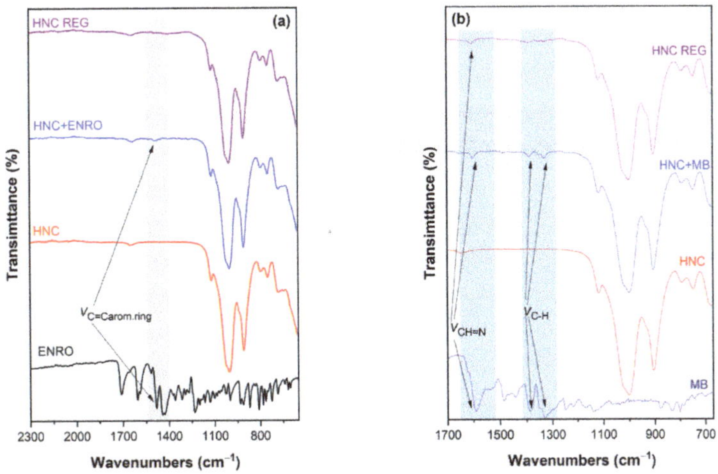

Figure 2. The ATR/FTIR spectra of (**a**) pristine HNC, ENRO, HNC after adsorption of ENRO and HNC after CAP regeneration, and (**b**) pristine HNC, MB, HNC after adsorption of MB and HNC after CAP regeneration in the selected spectral window. Grey and blue highlighted areas point out the differences in representative peaks for raw and regenerated HNC for ENRO and MB, respectively.

3.2. *Effect of Experimental Parameters*

3.2.1. Effect of Initial pH

The solution pH is considered a crucial parameter in the adsorption process, as it affects the charge of the adsorbent and thus its adsorption capacity. Concerning HNC, each nanotube consists of 15 to 20 aluminosilicate layers and in each layer; the Si–O and Al–OH groups are disposed on the external and internal surface, respectively. Because of that, the inner surface is charged positively and the outer negatively, according to the pH value [42]. In the present study, adsorption experiments were performed within a pH range of 2.0–11.0 for both pollutants (MB and ENRO) (Figure 3).

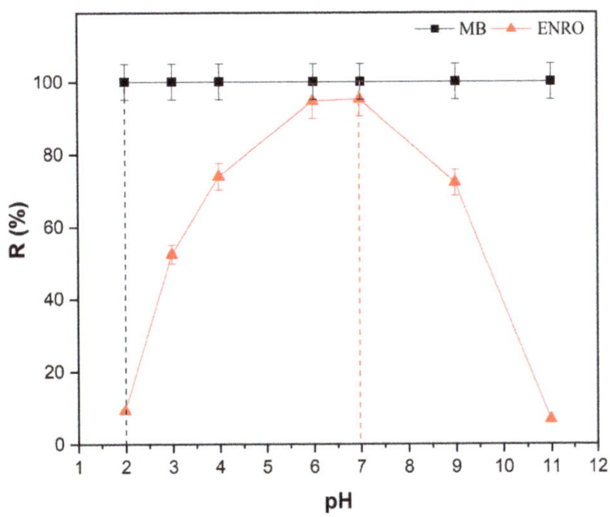

Figure 3. Effect of pH on MB and ENRO removal efficiency (C_{MB} = 40 mg/L, C_{ENRO} = 40 mg/L, C_{HNC} = 2 g/L, pH 2–11, T = 28 °C, contact time 1 h).

As may be seen in Figure 3, the removal efficiency of MB was not affected at all, remaining at 100% regardless of the pH value. This phenomenon can be attributed to the fact that since the halloysite nanotubes' external surface is negatively charged at a pH range 2–12 [48,49], strong electrostatic interactions are developed with the positively charged dye molecules, since MB is a cationic dye. Similar results have been reported by Zhao and Liu (2008) [50], using raw halloysite nanoclay, noting also that the adsorption equilibrium was achieved faster by increasing pH, since at higher pH values the outer negative charger is significantly increased.

In the case of ENRO, when the pH increased from 2 to 7, the removal efficiency intensely increased from 10% to 95% (at pH 6.2), and with a further increase of pH, the removal efficiency drastically decreased to lower than 10%. ENRO is a zwitterionic molecule, with pK_{a1} = 5.9 due to the carboxyl acid group and pK_{a2} = 7.7 due to the basic tertiary amine. When the pH of the solution is lower than pK_{a1}, the cationic form of ENRO ($ENRO^+$) is dominant, while at pH higher than pK_{a2}, a basic anionic form of ENRO ($ENRO^-$) dominates. When the pH value is between pK_{a1} and pK_{a2}, the intermediate zwitterion form ($ENRO^{\pm}$) dominates [51–53]. According to the results obtained (Figure 3), a gradual increase in ENRO adsorption on HNC was observed as the pH increased from 2 to 6. The $ENRO^+$ was able to form ionic interactions with the negatively charged external surface of HCN. Although the pK_{a1} value of ENRO is about 6.0, the amount of the absorbed ENRO on HCN reached its maximum at pH = 6.2 and remained up to pH = 7.0. Within this pH range, the $ENRO^+$ form decreases rapidly and zwitterionic form $ENRO^{\pm}$ becomes dominant in the solution, suggesting that the protonation of the piperazinyl group in $ENRO^{\pm}$ contributes to the adsorption mechanism of ENRO on HCN, maybe through the adsorption at the inner layer of the nanotube which is positively charged in this pH range. A significant decrease in ENRO adsorption on HNC was observed with further increase in the pH. Within this range of pH, the anionic form of ENRO ($ENRO^-$) is dominant, so ionic interactions cannot take place with the adsorbent external surface, while the inner layer of the halloysite nanotube bears no positive charge. At such high pH values, the low adsorption of $ENRO^-$ on HNC could possibly be attributed to the formation of cation bridging promoted by ions present [54–56]. To date, there are no research articles in the literature reporting the use of HNC as an adsorbent for removing ENRO. Nevertheless, the use of montmorillonite and kaolinite as adsorbents for ENRO has been reported, exhibiting common features with HNC such as structure, composition and genesis [57,58]. The results presented in the present paper are in accordance with the results reported in the literature, using montmorillonite and kaolinite as adsorbents and showing a common adsorption pH-dependent mechanism for ENRO [56,59,60]. Taking into consideration that the original pH of the aqueous solutions of the pollutants was 6.2 and the optimum pH value for the adsorption process was between 6 and 7, a pH value of 6.2 was selected for the subsequent experiments, since no pH adjustment was necessary.

3.2.2. Effect of HNC Dosage

In Figure 4, the removal efficiency of MB (C = 40 mg/L) and ENRO (C = 40 mg/L) in the single system, and of both pollutants (MB + ENRO, C_{MB} 40 + C_{ENRO} 40 mg/L) in the binary system, using different dosages of HNC (i.e., 0.5, 1.0, 2.0 and 3.0 g/L) is presented. It is obvious that for both pollutants and their mixture, the removal efficiency was increased by increasing the HNC dose from 0.5 to 3.0 g/L. In the case of MB in the single system, the removal efficiency increased from 37 to 99.8% when the HNC dosage increased from 0.5 to 2 g/L and reached 100% for 3 g/L of HNC (Figure 4a). Regarding ENRO in the single system, the removal efficiency was increased from about 10 to 93% by increasing the dose of HNC from 0.5 to 3 g/L (Figure 4b). This phenomenon can be attributed to the increase on the available active sites for adsorption in the same volume. The higher the HNC dosage, the greater the total available adsorbent area and therefore the higher the number of the adsorption active sites of HNC, either for MB or ENRO [61,62].

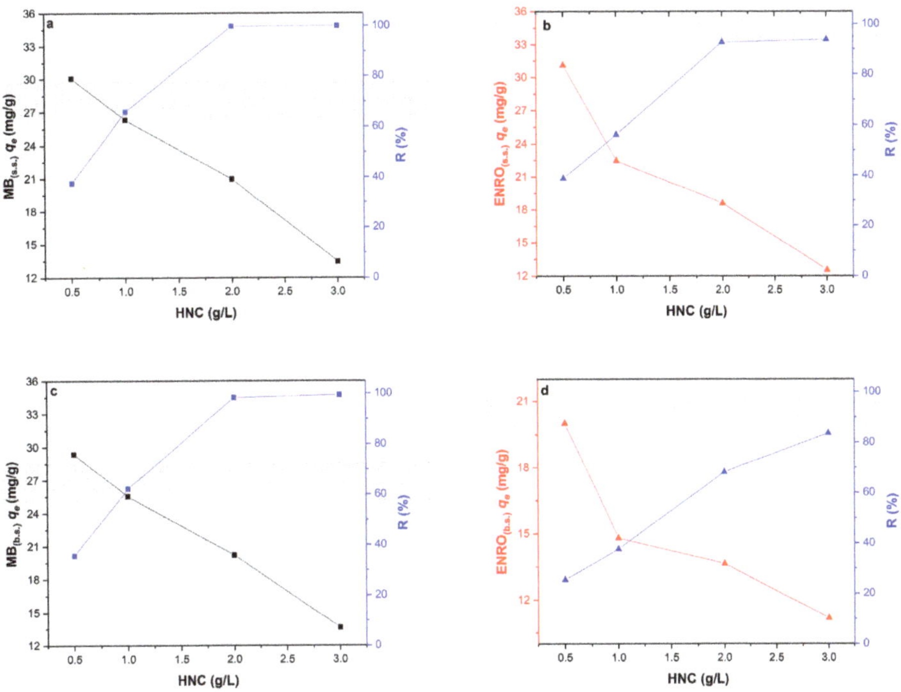

Figure 4. Effect of HNC dosage on removal efficiency and adsorption capacity for (**a**) $MB_{(s.s.)}$ (Co = 40 mg/L, pH = 6.2, contact time 1 h) (**b**) $ENRO_{(s.s.)}$ (C_0 = 40 mg/L, pH = 6.2, contact time 1 h, T = 28 °C) (**c**) $MB_{(b.s.)}$ (binary system, $C_{0\,MB}$ = 40 mg/L, $C_{0\,ENRO}$ = 40 mg/L, pH = 6.2, contact time 1 h, T = 28 °C) and (**d**) ENRO $_{(b.s.)}$ (binary system, $C_{0\,MB}$ = 40 mg/L, $C_{0\,ENRO}$ = 40 mg/L, pH = 6.2, contact time 1 h, T = 28 °C).

A similar trend was noticed also in the binary system. Nevertheless, it was observed that even though the removal efficiency of MB was not affected by the presence of ENRO in the solution, and reached up to 100% for 3 g/L of HCN (Figure 4c), for ENRO the presence of MB led to reduced removal efficiency, reaching up to 85% for 3 g/L of HCN (Figure 4d) instead of the 93% reached at the single system (Figure 4b). This can be attributed to the fact that during the adsorption process in a multicomponent system, the presence of some molecules can affect the adsorption of other molecules, either in an antagonistic or synergistic manner (see discussion below) [63].

Thus, in both single and binary systems, the removal efficiency increased with increasing adsorbent dosage, while the adsorption capacity of HNC was reduced (Figure 4). Similar results were reported by Liu et al. (2011) [61] who concluded that the optimum adsorbent dose was 2 g/L, and by Luo et al. (2011) [62], who reported that the optimum dose was 4 g/L. In the present study, an HNC dosage of 2 g/L was selected as the optimum value for further experimental procedure, since in this case, the adsorption is efficient for MB, ENRO and MB + ENRO without unnecessary use of excess adsorbent.

3.2.3. Effect of Contact Time

The effect of contact time on the adsorption capacity of MB, ENRO and their binary system onto HNC is presented in Figure 5. The adsorption capacity for both pollutants (Figure 5a) increased rapidly within the first 20 min and remained constant until 1 h of adsorption. The same trend was observed for both pollutants in the binary system (Figure 5b). In all cases, the equilibrium was reached after about 20–30 min of contact time. Similar results have been reported for MB adsorption onto HNC by Zhao and Liu (2008) [50], who reported that the adsorption reached equilibrium after 30 min of contact time. Concerning ENRO, there is no published research work on its adsorption onto HNC up to date. Nevertheless, it has been reported by Duan et al. (2018) [20] that ciprofloxacin, an antibiotic from the group of fluoroquinolones sharing a common structure with ENRO, exhibited high adsorption efficiency, using HNC as an adsorbent and reaching equilibrium at 60 min. Moreover, Wan M. et al. (2013) [56] reported that 93% of the maximum absorbed ENRO was achieved within 15 min for all the tested clay minerals, including montmorillonite.

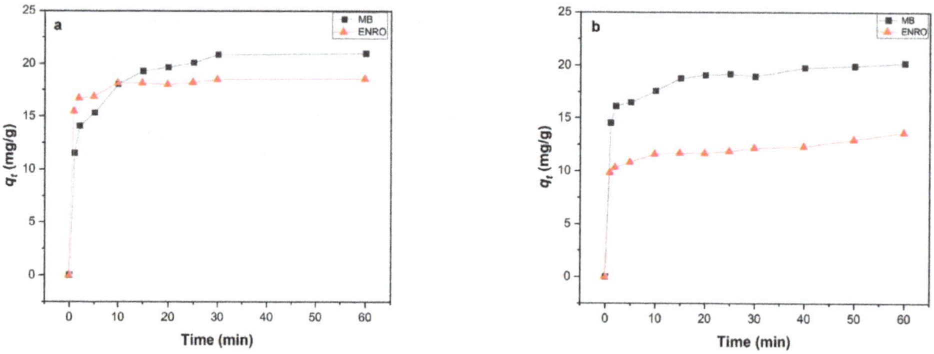

Figure 5. Effect of contact time on the amount adsorbed onto HNC of (**a**) single system, MB and ENRO (C_0 = 40 mg/L, pH = 6.2, C_{HNC} = 2 g/L, T = 28 °C) and (**b**) binary system ($C_{0\,MB}$ = 40 mg/L, $C_{0\,ENRO}$ = 40 mg/L pH = 6.2, C_{HNC} = 2 g/L, T = 28 °C).

3.2.4. Effect of Initial Pollutant Concentration

The initial pollutant concentration has a significant impact on the effectiveness of the adsorption process. The removal efficiency as a function of the initial pollutant concentration at single and binary systems is presented in Figure 6. In the single system (Figure 6a), the removal efficiency for both MB and ENRO decreased when the initial concentration of the pollutant increased. Specifically, experimental results for MB showed that removal efficiency decreased from 100% (for initial concentration ranging between 10–40 mg/L) to 36% (for initial concentration equal to 150 mg/L). The results were similar for ENRO, in which removal efficiency decreased from 97–92% (concertation range 10–40 mg/L) to 44% (concentration 150 mg/L). This can be attributed to the availability of adsorption sites on HNC for MB and ENRO. As the initial concentration of the pollutants increases, the available adsorption sites are limited and subsequently, the removal efficiency decreases [64].

The influence of different initial MB concentrations on removal efficiency, while the concentration of ENRO is kept stable at 40 mg/L, is presented in Figure 6b. From the results obtained, it was noticed that when the concentration of ENRO remained stable at 40 mg/L, the removal efficiency of MB was similar to that of the single system, mostly for the low MB concentrations. Specifically, MB removal efficiency was 100% for an initial concentration range of 10–40 mg/L and reached up to 25% for an initial concentration of 150 mg/L. ENRO removal efficiency was similar to the that achieved in the single system in the presence of low MB concentration (95% in the presence of 10 mg/L MB), but gradually decreased when the MB initial concentration increased, reaching up to 20% in the presence of 150 mg/L of MB. The negatively charged surface of HNC facilitates high adsorption of cationic molecules such as MB, $ENRO^+$ and the zwitterionic $ENRO^{\pm}$ that are present in the solution at experimental pH value; however, by increasing the concentration of MB, the available sites become limited and the absorption of MB is favored compared to $ENRO^+$ and $ENRO^{\pm}$.

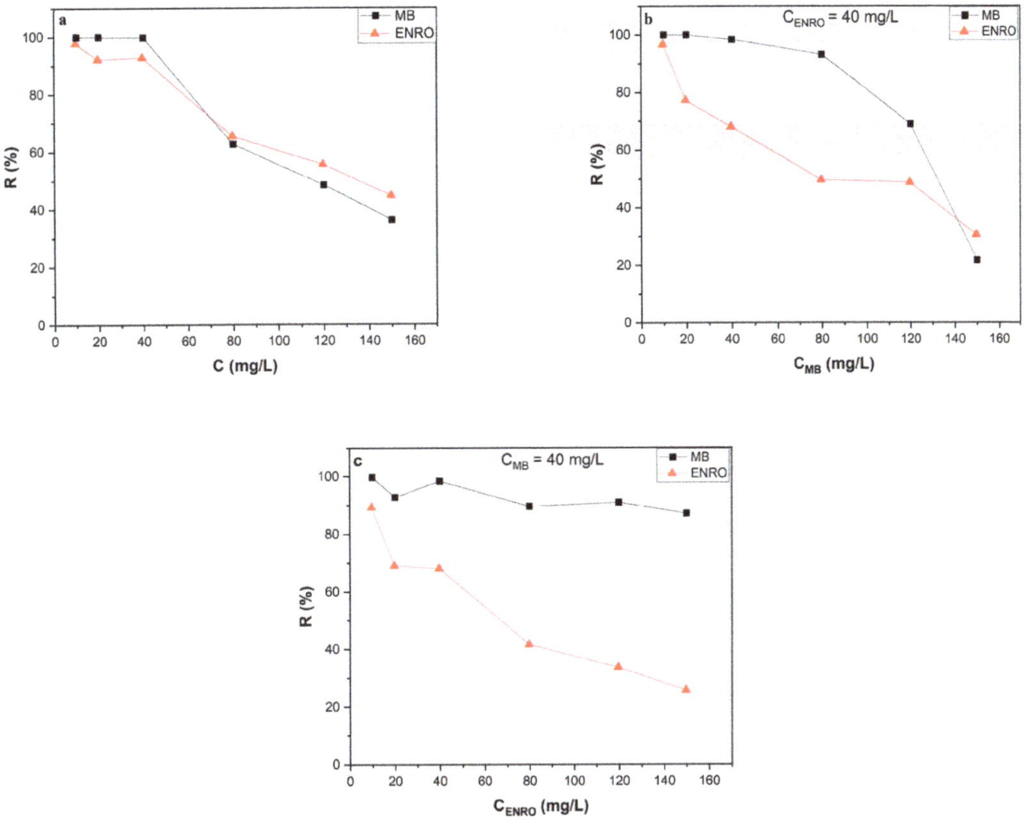

Figure 6. Effect of the initial concentration of MB and ENRO on the removal efficiency of the pollutants by HNC in single (**a**) and binary (**b**,**c**) systems. Conditions: (**a**) various concentrations of ENRO and MB, i.e., 10, 20, 40, 80, 120 and 150 mg/L (**b**) C_{ENRO} = 40 mg/L and various concentrations of MB, i.e., 10, 20, 40, 80, 120 and 150 mg/L and (**c**) C_{MB} = 40 mg/L and various concentrations of ENRO, i.e., 10, 20, 40, 80, 120 and 150 mg/L (pH = 6.2, C_{HNC} = 2 g/L, T = 28 °C).

In Figure 6c, the influence of different initial ENRO concentrations on removal efficiency while the concentration of MB is kept stable at 40 mg/L is presented. In this case, MB removal efficiency was similar to that of the single system, reaching up to 100%; it did not seem to be significantly influenced by the presence of high ENRO concentrations, since at 150 mg/L of ENRO it reached up to 90% removal efficiency. On the other hand, ENRO removal efficiency in the presence of MB decreased compared to the single system. According to these results, the electrostatic interactions between the negatively charged surface of HNC and the cationic dye MB are strong, and the adsorption is not influenced by the presence of ENRO, even at high concentrations, while ENRO removal efficiency is influenced by the presence of MB.

3.3. Adsorption Isotherms Models

The adsorption data of the single system were fitted to Freundlich and Langmuir isotherm models in order to elucidate the adsorption mechanism that takes place. Prior to applying the isotherm models, the HNC adsorption capacity at equilibrium, q_e (mg/g) was plotted versus Ce at equilibrium (mg/L) according to the obtained experimental data (Figure 7).

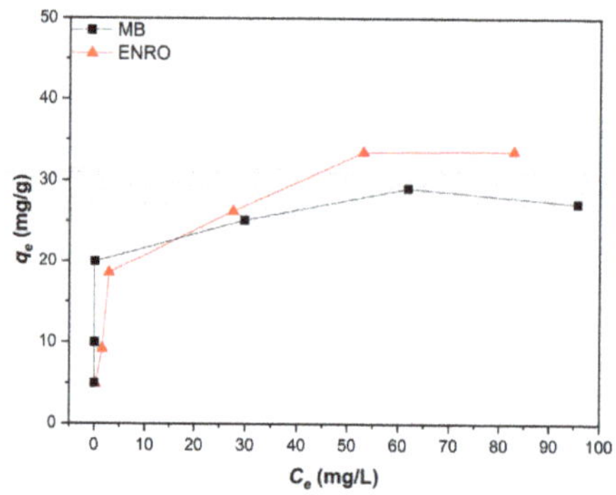

Figure 7. Adsorption isotherm of the single system for ENRO and MB (pH = 6.2, C_{HNC} = 2 g/L, T = 28 °C).

According to Figure 7, the amount of MB and ENRO that was adsorbed per adsorbent mass, q_e (mg/g), increased as a function of the initial concentrations of both pollutants. This tendency agrees with previous publications [65]. The q_e value increased rapidly within the first 5 min and reached its maximum value after ~30 min for both pollutants, being 33.57 and 27.18 mg/g for ENRO and MB, respectively. The increase in q_e as a function of the initial MB and ENRO concentration could be attributed to the higher concentration of MB and ENRO which increases the driving force for the mass transfer process, promoting the MB and ENRO molecules to be adsorbed onto the adsorbents' surface [65].

According to the Freundlich isotherm model, the adsorption occurs by multilayer sorption on a heterogeneous surface [66]. The linearized form of the Freundlich equation is expressed as follows [67]:

$$logq_e = logK + \frac{1}{n}logC_e \quad (4)$$

where K ((mg/g)(L/mg)$^{1/n}$) is the Freundlich adsorption constant related to the maximum adsorption capacity of the adsorbent and $\frac{1}{n}$ is a constant related to the adsorption intensity, varying with the heterogeneity of the adsorbent [68]. When $\frac{1}{n}$ values are within the range 0.1–1.0, the adsorption process could be considered favorable [69].

According to the Langmuir model, adsorption takes place only at specific homogenous sites with no lateral interaction between the adsorbed molecules, assuming a monolayer adsorption motif. The Langmuir model assumes that all active sites are identical and equally energetic, and interactions between molecules that have been adsorbed are prevented [70]. The linearized Langmuir equation is described by the following equation [11]:

$$\frac{C_e}{q_e} = \frac{1}{Qb} + \frac{C_e}{Q} \quad (5)$$

where q_e (mg/g) is the amount of pollutant adsorbed per unit mass of adsorbent at equilibrium, C_e (mg/L) is the equilibrium concentration of the pollutant in solution, Q (mg/g) is the maximum adsorption capacity of the adsorbent corresponding to monolayer coverage and b (L/mg) is the Langmuir adsorption constant which determines the affinities of binding sites and the sorption free energy. The dimensionless constant separation factor, R_L, is an indicator of the adsorption capacity:

$$R_L = \frac{1}{1 + bC_0} \quad (6)$$

where C_0 (mg/L) is the initial pollutant concentration in aqueous solution. The adsorption process is considered favorable when $0 < R_L < 1$ and unfavorable when $R_L > 1$.

In Table 1, the calculated Langmuir and Freundlich isotherm constants, along with the correlation coefficients (R^2) are presented. Based on the R^2 values, the Langmuir model fits almost perfectly to the experimental data for both pollutants (ENRO and MB), indicating that monolayer adsorption is probably taking place. The maximum monolayer adsorption capacities for ENRO and MB were found to be 34.80 and 27.66 mg/g, respectively, being very close to the experimental values for both pollutants. The R_L and $\frac{1}{n}$ values revealed the favorable character of the adsorption in both cases ($0 < R_L < 1$ and $\frac{1}{n} < 1$), indicating that the adsorption bond/interaction between the adsorbent and the adsorbate is strong. For MB, similar results have been published by Du and Zheng (2014) [71] and Jiang et al. (2014) [72], who reported that the Langmuir isotherm model fitted better to the experimental data when raw HNC was used as adsorbent, achieving a maximum adsorption capacity of 97.56 and 38.73 mg/L, respectively. Concerning ENRO, the isotherm model that perfectly fit the experimental results was Langmuir, and similar results have been reported in a study by Rivagli et al. (2014) [73], in which raw montmorillonite and kaolinite were used as adsorbates.

Table 1. Parameters of the isotherm models for MB and ENRO adsorption onto HNC, in single system, expressed in linear form.

Organic Pollutant	$q_{e,\,exp}$ (mg/g)	Langmuir				Freundlich		
		Q (mg/g)	b (L/mg)	R_L	R^2	K (mg/g) (L/mg)$^{1/n}$	$\frac{1}{n}$	R^2
MB	27.18	27.66	2.465	0.03–0.002	0.990	17.108	0.163	0.820
ENRO	33.57	34.80	0.268	0.27–0.02	0.993	9.067	0.324	0.937

3.4. Kinetic Adsorption Models

The adsorption kinetic study is regarded as a beneficial resource for adsorption studies since it contributes to the design of a suitable adsorption process, controlling the operation and any other practical aspect. In this study, four different adsorption kinetic models were investigated for both single and binary systems: pseudo-first-order (PFO), pseudo-second-order (PSO) and the Elovich and Weber–Morris intraparticle diffusion models.

The PFO equation is described by the following equation [74]:

$$\ln(q_e - q_t) = ln q_e - k_1 t \tag{7}$$

where k_1 (1/min) is the rate constant. According to the PFO kinetic model, the adsorbate is assumed to be adsorbed onto a single surface site at a time [74]. For both systems, single and binary, the PFO model did not satisfactorily fit the data either for MB or for ENRO (Table 2, Figures S4 and S8). In the case of the single system, the correlation coefficients (R^2) were found to be 0.914 and 0.836 for MB and ENRO, respectively, while the predicted adsorption capacity, $q_{e,cal}$, was not relatively close to the $q_{e,exp}$ for both pollutants. The same motif was observed for the binary system, where R^2 was found to be 0.949 and 0.924 for MB and ENRO, respectively, and the predicted adsorption capacity, $q_{e,cal}$, was found to be not so close to the $q_{e,exp}$ for both pollutants. Thus, the PFO kinetic model failed to describe the experimental data for both systems and for both pollutants; therefore, the PSO kinetic model was also tested, as described by the following equation [65]:

$$\frac{t}{q_t} = \frac{1}{k_2 q_e^2} + \frac{t}{q_e} \tag{8}$$

where k_2 (g/mg·min) is the rate constant. According to the PSO kinetic model, the adsorbate is adsorbed on two surface sites at time t, and the rate-limiting step is surface adsorption which involves chemisorption of the adsorbate on the adsorbent [75]. According to the results presented in Table 2 (Figures S5 and S9), in contrast to the PFO, the adsorption of MB and ENRO onto HNC is described perfectly by the PSO kinetic model for both systems (single and binary), indicating that chemisorption plays an important role in the process. R^2 was found to be 0.999 for MB and ENRO in the single system, while at the binary system, it was 0.999 and 0.994 for MB and ENRO, respectively. Moreover, the predicted adsorption capacities, $q_{e,cal}$, were close to the $q_{e,exp}$ for both pollutants in both systems.

Table 2. Kinetic parameters for the adsorption of MB and ENRO onto HNC for single and binary system.

		Pseudo-First-Order			Pseudo-Second-Order			Elovich			Intraparticle Diffusion Model		
	$q_{e,exp}$ (mg/g)	$q_{e,cal}$ (mg/g)	$k_1 \times 10^{-3}$ (1/min)	R^2	$q_{e,cal}$ (mg/g)	$k_2 \times 10^{-3}$ (g/mg·min)	R^2	α (mg/g·min)	1/b (mg/g)	R^2	K_{id} (mg/g·min$^{0.5}$)	C (mg/g)	R^2
Single system													
MB	20.96	10.81^{1h}	0.125^{1h}	0.914^{1h}	21.46^{1h}	31.2^{1h}	0.999^{1h}	322.22^{1h}	2.460^{1h}	0.968^{1h}	7.635^{2min} 1.688^{25min} 0.051^{1h}	3.451^{2min} 11.997^{25min} 20.564^{1h}	0.981^{2min} 0.959^{25min} 1^{1h}
ENRO	18.56	2.537^{1h}	104.76^{1h}	0.836^{1h}	18.65^{1h}	138.44^{1h}	0.999^{1h}	$1.25 \times 10^{-6\ 1h}$	0.747^{1h}	0.906^{1h}	5.004^{2min} 0.306^{1h}	9.868^{2min} 17.8921^{1h}	0.920^{2min} 0.951^{1h}
Binary system													
MB	20.20	4.672^{1h}	57.6^{1h}	0.949^{1h}	20.34^{1h}	43.16^{1h}	0.999^{1h}	322.22^{1h}	2.460^{1h}	0.968^{1h}	1.300^{5min} 0.370^{1h}	13.652^{5min} 17.297^{1h}	0.939^{5min} 0.902^{1h}
ENRO	13.64	3.37^{1h}	28.8^{1h}	0.924^{1h}	13.41^{1h}	42.25^{1h}	0.994^{1h}	$15.9 \times 10^{4\ 1h}$	0.791^{1h}	0.917^{1h}	0.478^{1h}	9.642^{1h}	0.950^{1h}

In order to further verify the significance of chemisorption in the process, the Elovich kinetic model was used, as it is regarded as one of the most useful models for describing chemical adsorption. The Elovich model is expressed by the following equation [76]:

$$q_t = \frac{1}{b}\ln(ab) + \frac{1}{b}\ln t \tag{9}$$

where a (mg/g min) is the initial adsorption rate and $1/b$ (mg/g) indicates the available sites for chemisorption. The Elovich model fitted the experimental data of MB and ENRO well enough for both systems (single and binary) (Table 2, Figures S6 and S10), indicating that a great number of active sites are available for interactions with HNC.

Finally, in order to evaluate the diffusion of MB and ENRO within the adsorbents' pores, the data were fitted by the Weber–Morris intraparticle diffusion model [77]:

$$q_t = K_{id}t^{1/2} + C \tag{10}$$

where q_t is the adsorption capacity (mg/min) of HNC for the pollutants at time t, k_{id} is the rate constant of intra-particle diffusion (mg/g·min$^{0.5}$) and C (mg/g) is a constant that is related to the thickness of the boundary layer. The experimental data of the single system showed that MB and ENRO curves were not linear, and could be divided into three linear regions for MB and two for ENRO (Table 2, Figure S7), meaning that the adsorption process is also controlled by film diffusion [78]. Therefore, a multilinear fitting was used to plot the data for three different linear regions. The first adsorption step is related to the transportation of the adsorbate from the solution to the external surface of the adsorbent, a process that is controlled by the film–liquid diffusion. The second step describes the diffusion of the adsorbate from the external surface to the pore structure (pore diffusion), and finally, the third one is attributed to the final equilibrium stage. In both pollutants (Figure S7), the linear fitting of the first stage did not pass through the origin ($C = 0$), which is an indication concerning the boundary layer resistance between adsorbent and adsorbate, while the deviation from this point is proportional to the boundary layer thickness. In the first region (0.5 min to 2 min), for both MB and ENRO, the experimental data were effectively fitted, and the plot q_t versus $t^{1/2}$ was linear, indicating that both pollutants were transferred from the solution to the halloysite's external surface through film diffusion. In the second region, which for MB is from 2 to 25 min and for ENRO from 2 min to 1 h, the K_{id} is significantly reduced. For MB, it dropped from 7.635 to 1.688 mg/g·min$^{1/2}$, and for ENRO from 5.004 to 0.306 mg/g·min$^{1/2}$, meaning that the diffusion of MB and ENRO inside the pores of halloysite is relatively low [11]. During these two stages, MB and ENRO molecules were diffused and adsorbed onto HNC through electrostatic interactions [79].

The third region for MB (25 min to 1 h) is attributed to the final equilibrium stage, and the K_{id} is even lower (0.051 mg/g·min$^{1/2}$). In the case of the binary system, the experimental data showed that the MB curve was not linear and could be divided into two linear regions, while for ENRO a single-step process occurred (Table 2, Figure S11).

3.5. Interaction Mechanism in the Binary System

In a multicomponent system that may contain two or more contaminants, the adsorbate molecules may interact in different ways. The mechanism is explained based on the ratio of the adsorption capacity of each component in the multicomponent system to the adsorption capacity of the adsorbent in the single system, $Q_{,multicomponent}:Q_{,single}$. The possible interactions among the adsorbate molecules are [80,81]:

The possible interactions among the adsorbate molecules are [80,81]:

(a) Antagonistic interaction: The adsorption capacity of an adsorbent decreases in a solution containing other components ($Q_{,multicomponent}:Q_{,single} < 1$).
(b) Synergistic interaction: The adsorption capacity of an adsorbent increases when it is in association with other components ($Q_{,multicomponent}:Q_{,single} > 1$).
(c) Non-interaction: The adsorption capacity is independent of the absence or presence of other components in a solution ($Q_{,multicomponent}:Q_{,single} = 1$).

In order to calculate the Q for each pollutant in the binary system, the extended Langmuir isotherm was applied, described by the following equation [80]:

$$q_{e1} = Q_1 \frac{b_1 C_{e1}}{1 + (b_1 C_{e1} + b_2 C_{e2})} \tag{11}$$

where Q (mg/g) is the maximum adsorption capacity and b (L/mg) the Langmuir constant related to the adsorption energy fitting parameters for each component.

Table 3 displays the interactive effects of MB and ENRO in the binary system. The calculated ratio for ENRO is less than one, indicating that its adsorption capacity in the binary system is decreased by the presence of MB. On the other hand, the calculated ratio for MB is slightly higher than 1, indicating that there is either synergistic or no interaction at all, and its adsorption capacity in the binary system is either enhanced or not influenced by the presence of ENRO. This could be attributed to the high negative charge on the external surface area of halloysite resulting in higher MB adsorption, since it has stronger cation character compared to ENRO at the investigated pH of the solution; however, further elucidation should be the subject of further research work.

Table 3. Interaction effect of MB and ENRO adsorption onto HNC in the binary system.

Pollutant	Adsorbent	$Q_{,binary}/Q_{,single}$	Interaction Effect
MB	HNC	1.08	Synergistic or non-interaction
ENRO	HNC	0.64	Antagonistic

3.6. Analysis of the Adsorption Mechanisms

The adsorption mechanism could be based either on chemisorption, physisorption, or both processes. Experimental data obtained in the present study, i.e., adsorption studies, ATR/FTIR and BET analysis contributed to the elucidation of the adsorption mechanism.

The HNC used in the present study has a specific surface area of 69 m^2/g, and particle size of 50 nm, indicating that the enclosure of a molecule inside the adsorbate is feasible through physisorption. Moreover, at the pH of the experimental procedure (6.2), the external surface of HNC poses a negative charge and the inner positive, while MB is positively charged and ENRO exists in zwitterionic form (ENRO$^{\pm}$), indicating that both pollutants can interact with the adsorbent through electrostatic interactions. The Langmuir isotherm and PSO kinetic model indicated that the main adsorption mechanism is based on monolayer chemisorption of the pollutants onto HNC driven by electrostatic interactions, while the Weber–Morris model revealed that the intraparticle diffusion was controlled by film diffusion. In Figure 8, the possible adsorption mechanism of MB and ENRO onto HNC is presented.

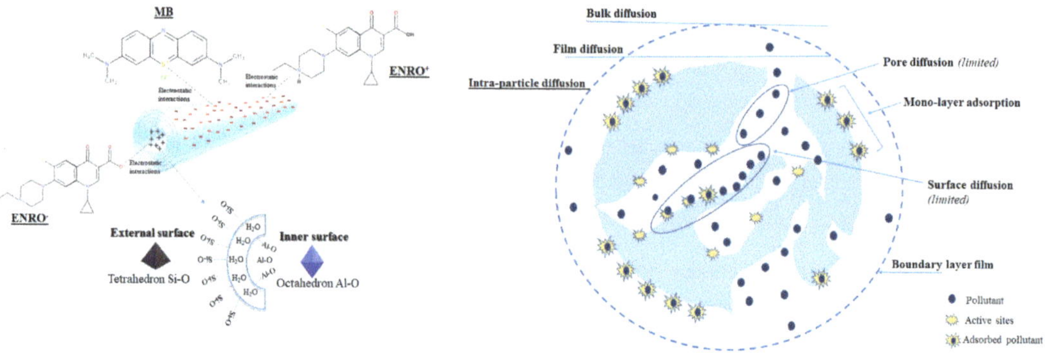

Figure 8. Possible adsorption mechanism of MB and ENRO onto HNC.

3.7. *Towards the Sustainability of the Process through the Regeneration of Adsorbent*

Recovery and sustainable management of used adsorbents are two of the most significant problems of the adsorption-treatment process. In order to be reused, several technologies have been investigated for their regeneration, including desorption, photodegradation, and biodegradation of the adsorbed molecules. Cold atmospheric plasma (CAP) has emerged as an interesting alternative over conventional regeneration methods due to exhibiting high regeneration efficiency and low energy consumption [29,82]. As reported in the literature, the CAP method has been effectively used for the regeneration of waste tea, which has been used for the adsorption of the methylene blue [29].

In the present study, CAP bubbling, Fenton oxidation and air bubbling were applied to the saturated raw HNC. When the saturated HNC was treated by CAP bubbling, its removal efficiency was almost completely restored for both pollutants (Figure 9). Specifically, the removal efficiency reached up to 88.5% and 81% for MB (Figure 9a) and ENRO (Figure 9b), respectively, revealing the potential of HNC to be reused. The removal efficiency of regenerated HNC through Fenton oxidation reached up to 32% and 47% for MB (Figure 9a) and ENRO (Figure 9b), respectively, which was significantly lower than the CAP-regenerated HNC. The application of air bubbling (without CAP) showed that the desorption of the pollutants was negligible. According to the results, it is obvious that CAP bubbling achieved almost complete restoration of the saturated HNC, reaching removal efficiency values, for both pollutants, similar to the raw HNC. This behavior was also verified by spectroscopic analysis (Section 3.1.3) of the HNC after CAP regeneration, where the MB and ENRO characteristic peaks that were noticed during adsorption were minimized or vanished.

After the first CAP regeneration of the saturated raw HNC, namely 1st CAP-regenerated HNC, new adsorption cycles were applied up to saturation, followed by new regeneration resulting in 2nd CAP-regenerated HNC and new adsorption cycles up to discern the removal efficiency. According to the results presented in Figure 10, it is impressive that the removal efficiencies for both pollutants for 1st CAP-regenerated HNC and 2nd CAP-regenerated HNC are significantly increased during the adsorption cycles compared to the raw HNC.

Indeed, for ENRO (Figure 10b), at the second adsorption cycle, raw HNC achieved a removal efficiency of up to 37%, whilst in the same cycle, 1st CAP-regenerated HNC reached up to 72%, and 2nd CAP-regenerated HNC 45%. The third cycle of 1st CAP-regenerated HNC was also impressive, achieving 50% of removal efficiency compared to the 10% of the raw HNC (Figure 10b). Similar results were obtained for MB (Figure 10a). In particular, the second cycle of adsorption for 1st CAP-regenerated and 2nd CAP-regenerated HNC reached up to 65%, while the corresponding value for raw HNC was 20%. Great enhancement was also noticed during the third and fourth adsorption cycles, where the raw HNC achieved removal efficiency of up to 5% and 1%, respectively, while for the 1st CAP-regenerated HNC was 38% and 20% and for 2nd CAP-regenerated HNC 45% and 30%, respectively. It is noteworthy that in the case of MB the removal efficiencies of the third and the fourth adsorption cycles of 2nd CAP-regenerated HNC were higher compared to the respective cycles of 1st CAP-regenerated HNC (Figure 10a). The enhancement of the removal efficiency of the CAP-regenerated HNC compared to the raw could be attributed to the chemical activity of plasma-generated species, e.g., $^1O_2$, $\cdot OH$, O, $\cdot O_2^-$, O_3, H_2O_2, etc., which effectively modified the adsorbent. Similar results have been reported [29,83,84], attributing the enhancement of the removal efficiency to the increment of the available adsorption sites through the surface oxidation induced by the plasma-generated species; the reactive species that are attached to the natural adsorption site could act beneficially, enhance the adsorption capacity of the adsorbent and be effectively re-used for several adsorption cycles.

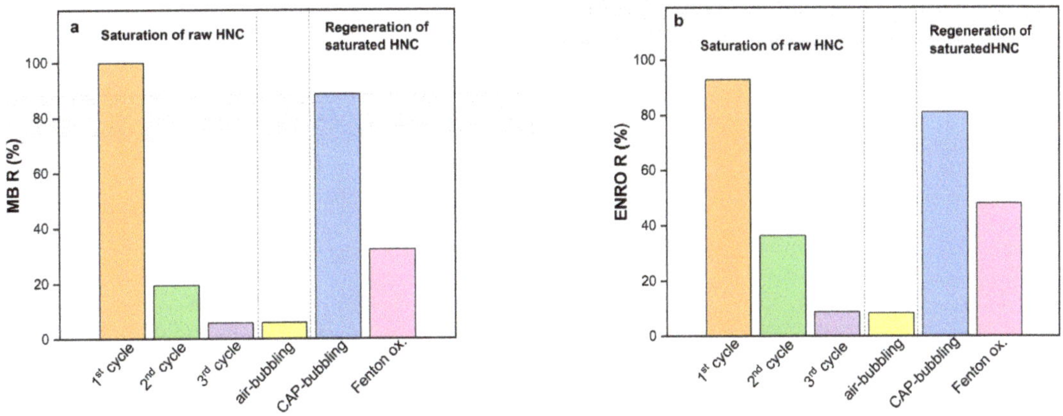

Figure 9. Removal efficiency for (**a**) MB and (**b**) ENRO after three adsorption cycles and regeneration of saturated HNC by applying air bubbling, CAP bubbling and Fenton oxidation (C_{ENRO} = 40 mg/L, C_{MB} = 40 mg/L, pH = 6.2, C_{HNC} = 2 g/L, T = 28 °C).

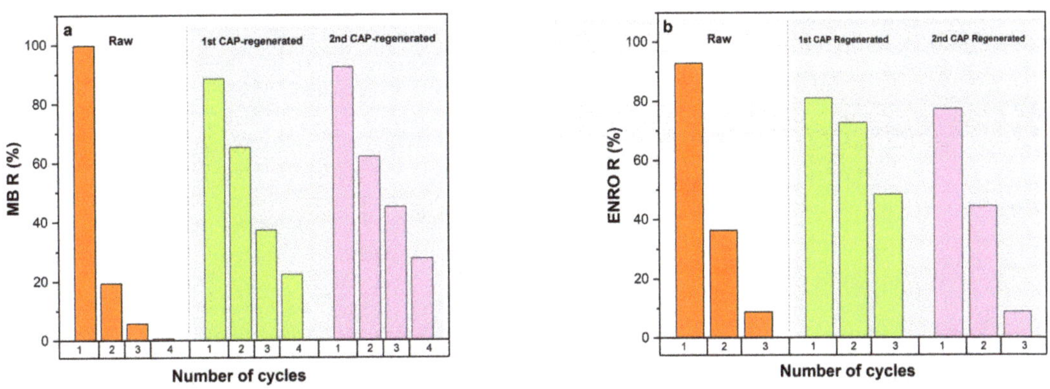

Figure 10. Adsorption cycles of raw, 1st CAP-regenerated HNC and 2nd CAP-regenerated HNC, for (**a**) MB and (**b**) ENRO (C_{ENRO} = 40 mg/L, C_{MB} = 40 mg/L, pH = 6.2, C_{HNC} = 2 g/L, T = 28 °C).

In conclusion, CAP bubbling not only effectively regenerated the adsorbent but also induced chemical modifications on its surface that were favorable for its adsorption capacity, even after several cycles; in contrast, the raw HNC's adsorption capacity decreased dramatically after the first adsorption cycle. At the same time, the ENRO and MB concentrations that might have been transferred from the solid phase to the aqueous solution during the regeneration experiments were completely eliminated by CAP bubbling; the latter was confirmed by UV-Vis measurements of the aqueous solution after regeneration experiments (data not shown). In contrast to chemical regeneration, in which the pollution is transferred from the adsorbent to the chemical solution, the present study presents a completely green process that does not result in any secondary pollution. Moreover, the power consumption during the regeneration of the saturated HNC was ~1 W, and thus the energy requirement is ~16.67 Wh/g of adsorbent, which is among the lowest compared to others in the literature [85], possibly due to the combination of HV nanopulses and plasma bubbles. Therefore, CAP bubbling provided a unique green solution, simultaneouslyexhibiting three different important functionalities: the regeneration of the adsorbent, its activation during the regeneration process and the elimination of the residual pollutant concentration in the water used for its regeneration.

4. Conclusions

The present study reports the efficiency of HNC adsorbent on the removal of the antibiotic ENRO and the cationic dye MB from artificially polluted water in a single and binary system. Several operating parameters were examined, and it was found that contact time, adsorbent dosage, pH and pollutant concentration affected the efficiency of the process. In the single system, both pollutants adsorbed onto HNC very quickly (10–20 min), whereas the same trend was observed for both pollutants in the binary system in which equilibrium was reached after ~20–30 min. Concerning pH, the natural pH of ENRO and MB solution (pH = 6.2) exhibited the maximum removal efficiency for both pollutants, as at this pH, the charges of the outer and inner surfaces of HNC and the pollutants favor the formation of electrostatic interactions. MB and ENRO adsorption onto HNC were better described by the Langmuir isotherm model and the PSO kinetic model for both pollutants in the single and the binary system, indicating that the process is controlled mainly by monolayer chemisorption through electrostatic forces. The maximum adsorption capacity Q was found to be 34.80 and 27.66 mg/g for ENRO and MB, respectively. Furthermore, the intraparticle diffusion model revealed that the adsorption of both pollutants onto HNC was mainly controlled by film diffusion. Regarding the simultaneous adsorption of pollutants, the presence of MB seems to have an antagonistic effect for ENRO adsorption onto HNC, while the presence of ENRO does not affect the adsorption of MB, revealing a synergism or non-interaction.Cold atmospheric plasma (CAP) bubbling regenerated the saturated HNC much more effectively compared to Fenton oxidation, with relative low energy cost (16.67 Wh/g-adsorbent). Moreover, the CAP-regenerated HNC was effectively applied to new adsorption cycles, achieving increased removal efficiencies for both pollutants compared to the raw HNC. Therefore, CAP bubbling induced chemical modifications on HNC surface that were favourable for its adsorption capacity even after several cycles in contrast to the raw HNC, whose adsorption capacity decreased dramatically after the first adsorption cycle. Finally, complete elimination of MB and ENRO, transferred from the solid phase to the aqueous solution during the regeneration process, was achieved. The present study could be considered a sustainable, green and cost-effective solution towards the remediation of wastewater in which pharmaceuticals and dyes are present. Our next steps include the elucidation of the interaction mechanism between the two pollutants as well as the combination of HNC adsorbent with CAP for the removal of emerging contaminants, such as perfluoroalkyl substances (PFAS), from water.

Supplementary Materials: The following supporting information can be downloaded at: https://www.mdpi.com/article/10.3390/nano13020341/s1, Figure S1: High-Resolution Transmission Electron Microscopy (HR-TEM) image for HNC; Figure S2. ΔpH versus pH_i for HNC; Figure S3. The ATR/FTIR spectrum of halloysite in the spectral region 500–4000 cm^{-1}; Figure S4. Pseudo-first order reaction kinetics for the adsorption of ENRO and MB on HNC in the single system; Figure S5. Pseudo-second order reaction kinetics for the adsorption of ENRO and MB on HNC in the single system; Figure S6. Elovich kinetic parameter for the adsorption of ENRO and MB on HNC in the single system; Figure S7. Intra-particle diffusion model curves for the adsorption of MB and ENRO on HNC, for the single system; Figure S8. Pseudo-first order reaction kinetics for the adsorption of ENRO and MB on HNC in the binary system; Figure S9. Pseudo-second order reaction kinetics for the adsorption of ENRO and MB on HNC in the binary system; Figure S10. Elovich kinetic parameter for the adsorption of ENRO and MB on HNC in the binary system; Figure S11. Intra-particle diffusion model curves for the adsorption of MB and ENRO on HNC, for the binary system; Table S1: Molecular structure and physicochemical properties of MB and ENRO.

Author Contributions: Conceptualization, C.A.A.; methodology, I.-E.T. and S.G.; investigation, S.G. and C.A.A.; resources, C.A.A.; writing—original draft preparation, I.-E.T. and C.A.A.; writing—review and editing, C.A.A. and A.G.T.; visualization, C.A.A. and S.G.; supervision, C.A.A.; funding acquisition, C.A.A. All authors have read and agreed to the published version of the manuscript.

Funding: This research was funded by European Union's Horizon 2020 research and innovation program, grant number No 101037509.

Data Availability Statement: Not applicable.

Acknowledgments: The authors thank Z. Lada, member of the Laboratory of Applied Molecular Spectroscopy at ICE-HT/FORTH (Head: G.A. Voyiatzis) for obtaining the ATR-FTIR spectra.

Conflicts of Interest: The authors declare no conflict of interest. The funders had no role in the design of the study; in the collection, analyses, or interpretation of data; in the writing of the manuscript; or in the decision to publish the results.

References

1. Schweitzer, L.; Noblet, J. Water Contamination and Pollution. In *Green Chemistry: An Inclusive Approach*; Elsevier Inc.: Amsterdam, The Netherlands, 2018; pp. 261–290.
2. Brusseau, M.L.; Artiola, J.F. Chemical Contaminants. In *Environmental and Pollution Science*; Elsevier: Amsterdam, The Netherlands, 2019; pp. 175–190.
3. Farhan Hanafi, M.; Sapawe, N. A Review on the Water Problem Associate with Organic Pollutants Derived from Phenol, Methyl Orange, and Remazol Brilliant Blue Dyes. *Mater. Today Proc.* **2020**, *31*, A141–A150. [CrossRef]
4. Yadav, A.; Yadav, P.; Labhasetwar, P.K.; Shahi, V.K. CNT Functionalized ZIF-8 Impregnated Poly(Vinylidene Fluoride-Co-Hexafluoropropylene) Mixed Matrix Membranes for Antibiotics Removal from Pharmaceutical Industry Wastewater by Vacuum Membrane Distillation. *J. Environ. Chem. Eng.* **2021**, *9*, 106560. [CrossRef]
5. Ajiboye, T.O.; Oyewo, O.A.; Onwudiwe, D.C. Simultaneous Removal of Organics and Heavy Metals from Industrial Wastewater: A Review. *Chemosphere* **2021**, *262*, 128379. [CrossRef] [PubMed]
6. Li, S.; Zhang, Y.; You, Q.; Wang, Q.; Liao, G.; Wang, D. Highly Efficient Removal of Antibiotics and Dyes from Water by the Modified Carbon Nanofibers Composites with Abundant Mesoporous Structure. *Colloids Surf. A Physicochem. Eng. Asp.* **2018**, *558*, 392–401. [CrossRef]
7. Phoon, B.L.; Ong, C.C.; Mohamed Saheed, M.S.; Show, P.L.; Chang, J.S.; Ling, T.C.; Lam, S.S.; Juan, J.C. Conventional and Emerging Technologies for Removal of Antibiotics from Wastewater. *J. Hazard. Mater.* **2020**, *400*, 122961. [CrossRef] [PubMed]
8. Rathi, B.S.; Kumar, P.S. Application of Adsorption Process for Effective Removal of Emerging Contaminants from Water and Wastewater. *Environ. Pollut.* **2021**, *280*, 116995. [CrossRef] [PubMed]
9. Senthil Kumar, P.; Janet Joshiba, G.; Femina, C.C.; Varshini, P.; Priyadharshini, S.; Arun Karthick, M.S.; Jothirani, R. A Critical Review on Recent Developments in the Low-Cost Adsorption of Dyes from Wastewater. *Desalin. Water Treat.* **2019**, *172*, 395–416. [CrossRef]
10. Yagub, M.T.; Sen, T.K.; Afroze, S.; Ang, H.M. Dye and Its Removal from Aqueous Solution by Adsorption: A Review. *Adv. Colloid Interface Sci.* **2014**, *209*, 172–184. [CrossRef]
11. Stavrinou, A.; Aggelopoulos, C.A.; Tsakiroglou, C.D. Exploring the Adsorption Mechanisms of Cationic and Anionic Dyes onto Agricultural Waste Peels of Banana, Cucumber and Potato: Adsorption Kinetics and Equilibrium Isotherms as a Tool. *J. Environ. Chem Eng.* **2018**, *6*, 6958–6970. [CrossRef]

12. Stavrinou, A.; Aggelopoulos, C.A.; Tsakiroglou, C.D. A Methodology to Estimate the Sorption Parameters from Batch and Column Tests: The Case Study of Methylene Blue Sorption onto Banana Peels. *Processes* **2020**, *8*, 1467. [CrossRef]
13. Makrygianni, M.; Lada, Z.G.; Manousou, A.; Aggelopoulos, C.A.; Deimede, V. Removal of Anionic Dyes from Aqueous Solution by Novel Pyrrolidinium-Based Polymeric Ionic Liquid (PIL) as Adsorbent: Investigation of the Adsorption Kinetics, Equilibrium Isotherms and the Adsorption Mechanisms Involved. *J. Environ. Chem Eng.* **2019**, *7*, 103163. [CrossRef]
14. Yuan, P.; Tan, D.; Annabi-Bergaya, F. Properties and Applications of Halloysite Nanotubes: Recent Research Advances and Future Prospects. *Appl. Clay Sci.* **2015**, *112–113*, 75–93. [CrossRef]
15. Zhang, W.; Wang, L.; Su, Y.; Liu, Z.; Du, C. Indium Oxide/Halloysite Composite as Highly Efficient Adsorbent for Tetracycline Removal: Key Roles of Hydroxyl Groups and Interfacial Interaction. *Appl. Surf. Sci.* **2021**, *566*, 150708. [CrossRef]
16. Erdem, S.; Öztekin, M.; Sağ Açıkel, Y. Investigation of Tetracycline Removal from Aqueous Solutions Using Halloysite/Chitosan Nanocomposites and Halloysite Nanotubes/Alginate Hydrogel Beads. *Environ. Nanotechnol. Monit. Manag.* **2021**, *16*, 100576. [CrossRef]
17. Dai, J.; Wei, X.; Cao, Z.; Zhou, Z.; Yu, P.; Pan, J.; Zou, T.; Li, C.; Yan, Y. Highly-Controllable Imprinted Polymer Nanoshell at the Surface of Magnetic Halloysite Nanotubes for Selective Recognition and Rapid Adsorption of Tetracycline. *RSC Adv.* **2014**, *4*, 7967–7978. [CrossRef]
18. Ramanayaka, S.; Sarkar, B.; Cooray, A.T.; Ok, Y.S.; Vithanage, M. Halloysite Nanoclay Supported Adsorptive Removal of Oxytetracycline Antibiotic from Aqueous Media. *J. Hazard. Mater.* **2020**, *384*, 121301. [CrossRef]
19. Cheng, R.; Li, H.; Liu, Z.; Du, C. Halloysite Nanotubes as an Effective and Recyclable Adsorbent for Removal of Low-Concentration Antibiotics Ciprofloxacin. *Minerals* **2018**, *8*, 387. [CrossRef]
20. Duan, W.; Wang, N.; Xiao, W.; Zhao, Y.; Zheng, Y. Ciprofloxacin Adsorption onto Different Micro-Structured Tourmaline, Halloysite and Biotite. *J. Mol. Liq.* **2018**, *269*, 874–881. [CrossRef]
21. Xie, A.; Dai, J.; Chen, X.; Ma, P.; He, J.; Li, C.; Zhou, Z.; Yan, Y. Ultrahigh Adsorption of Typical Antibiotics onto Novel Hierarchical Porous Carbons Derived from Renewable Lignin via Halloysite Nanotubes-Template and in-Situ Activation. *Chem. Eng. J.* **2016**, *304*, 609–620. [CrossRef]
22. Anastopoulos, I.; Mittal, A.; Usman, M.; Mittal, J.; Yu, G.; Núñez-Delgado, A.; Kornaros, M. A Review on Halloysite-Based Adsorbents to Remove Pollutants in Water and Wastewater. *J. Mol. Liq.* **2018**, *269*, 855–868. [CrossRef]
23. Yadav, A.; Bagotia, N.; Sharma, A.K.; Kumar, S. Simultaneous Adsorptive Removal of Conventional and Emerging Contaminants in Multi-Component Systems for Wastewater Remediation: A Critical Review. *Sci. the Total Environ.* **2021**, *799*, 149500. [CrossRef] [PubMed]
24. Girish, C.R. Simultaneous Adsorption of Pollutants onto the Adsorbent Review of Interaction Mechanism between the Pollutants and the Adsorbent. *Int. J. Eng. Technol.* **2018**, *7*, 3613–3622.
25. Baskar, A.V.; Bolan, N.; Hoang, S.A.; Sooriyakumar, P.; Kumar, M.; Singh, L.; Jasemizad, T.; Padhye, L.P.; Singh, G.; Vinu, A.; et al. Recovery, Regeneration and Sustainable Management of Spent Adsorbents from Wastewater Treatment Streams: A Review. *Sci. Total Environ.* **2022**, *822*, 153555. [CrossRef] [PubMed]
26. Kaliya Perumal Veerapandian, S.; de Geyter, N.; Giraudon, J.M.; Morin, J.C.; Esbah Tabaei, P.S.; de Weireld, G.; Laemont, A.; Leus, K.; van der Voort, P.; Lamonier, J.F.; et al. Effect of Non-Thermal Plasma in the Activation and Regeneration of 13X Zeolite for Enhanced VOC Elimination by Cycled Storage and Discharge Process. *J. Clean. Prod.* **2022**, *364*, 132687. [CrossRef]
27. Kalebić, B.; Škoro, N.; Kovač, J.; Rajić, N. Regeneration of the Ciprofloxacin-Loaded Clinoptilolite by Non-Thermal Atmospheric Plasma. *Appl. Surf. Sci.* **2022**, *593*, 153379. [CrossRef]
28. Jiang, N.; Qiu, C.; Guo, L.; Shang, K.; Lu, N.; Li, J.; Wu, Y. Post Plasma-Catalysis of Low Concentration VOC Over Alumina-Supported Silver Catalysts in a Surface/Packed-Bed Hybrid Discharge Reactor. *Water Air Soil Pollut.* **2017**, *228*, 1–11. [CrossRef]
29. Zhou, R.; Zhou, R.; Zhang, X.; Tu, S.; Yin, Y.; Yang, S.; Ye, L. An Efficient Bio-Adsorbent for the Removal of Dye: Adsorption Studies and Cold Atmospheric Plasma Regeneration. *J. Taiwan Inst. Chem. Eng.* **2016**, *68*, 372–378. [CrossRef]
30. Kaliya Perumal Veerapandian, S.; Giraudon, J.M.; de Geyter, N.; Onyshchenko, Y.; Krishnaraj, C.; Sonar, S.; Löfberg, A.; Leus, K.; van der Voort, P.; Lamonier, J.F.; et al. Regeneration of Hopcalite Used for the Adsorption Plasma Catalytic Removal of Toluene by Non-Thermal Plasma. *J. Hazard. Mater.* **2021**, *402*, 123877. [CrossRef]
31. Aggelopoulos, C.A. Recent Advances of Cold Plasma Technology for Water and Soil Remediation: A Critical Review. *Chem. Eng. J.* **2022**, *428*, 131657. [CrossRef]
32. Meropoulis, S.; Giannoulia, S.; Skandalis, S.; Rassias, G.; Aggelopoulos, C.A. Key-study on plasma-induced degradation of cephalosporins in water: Process optimization, assessment of degradation mechanisms and residual toxicity. *Sep. Purif. Techol.* **2022**, *298*, 121639. [CrossRef]
33. Grabowski, Ł.; Gaffke, L.; Pierzynowska, K.; Cyske, Z.; Choszcz, M.; Węgrzyn, G.; Węgrzyn, A. Enrofloxacin—The Ruthless Killer of Eukaryotic Cells or the Last Hope in the Fight against Bacterial Infections? *Int. J. Mol. Sci.* **2022**, *23*, 3648. [CrossRef] [PubMed]
34. Ezelarab, H.A.A.; Abbas, S.H.; Hassan, H.A.; Abuo-Rahma, G.E.D.A. Recent Updates of Fluoroquinolones as Antibacterial Agents. *Arch. Pharm.* **2018**, *351*, e1800141. [CrossRef] [PubMed]
35. Sodhi, K.K.; Singh, D.K. Insight into the Fluoroquinolone Resistance, Sources, Ecotoxicity, and Degradation with Special Emphasis on Ciprofloxacin. *J. Water Process. Eng.* **2021**, *43*, 102218. [CrossRef]

36. Zhang, T.; Zhou, R.; Wang, P.; Mai-Prochnow, A.; McConchie, R.; Li, W.; Zhou, R.; Thompson, E.W.; Ostrikov, K.; Cullen, P.J. Degradation of Cefixime Antibiotic in Water by Atmospheric Plasma Bubbles: Performance, Degradation Pathways and Toxicity Evaluation. *Chem. Eng. J.* **2021**, *421*, 127730. [CrossRef]
37. Balistrieri, L.S.; Murray, J.W. The surface chemistry of goethite (alpha FeOOH) in major ion seawater. *Am. J. Sci.* **1981**, *281*, 788–806. [CrossRef]
38. Bardestani, R.; Patience, G.S.; Kaliaguine, S. Experimental Methods in Chemical Engineering: Specific Surface Area and Pore Size Distribution Measurements—BET, BJH, and DFT. *Can. J. Chem. Eng.* **2019**, *97*, 2781–2791. [CrossRef]
39. Matusik, J.; Wścisło, A. Enhanced Heavy Metal Adsorption on Functionalized Nanotubular Halloysite Interlayer Grafted with Aminoalcohols. *Appl. Clay Sci.* **2014**, *100*, 50–59. [CrossRef]
40. Matusik, J. Halloysite for Adsorption and Pollution Remediation. In *Developments in Clay Science*; Elsevier: Amsterdam, The Netherlands, 2016; Volume 7, pp. 606–627.
41. Barrientos-Ramírez, S.; de Oca-Ramírez, G.M.; Ramos-Fernández, E.V.; Sepúlveda-Escribano, A.; Pastor-Blas, M.M.; González-Montiel, A. Surface modification of natural halloysite clay nanotubes with aminosilanes. Application as catalyst supports in the atom transfer radical polymerization of methyl methacrylate. *Appl. Catal. A: Gen.* **2011**, *406*, 22–33. [CrossRef]
42. Bretti, C.; Cataldo, S.; Gianguzza, A.; Lando, G.; Lazzara, G.; Pettignano, A.; Sammartano, S. Thermodynamics of Proton Binding of Halloysite Nanotubes. *J. Phys. Chem. C* **2016**, *120*, 7849–7859. [CrossRef]
43. Sverjensky, D.A. Zero-point-of-charge prediction from crystal chemistry and solvation theory. *Geochim. Et Cosmochim. Acta* **1994**, *58*, 3123–3129. [CrossRef]
44. Liu, Y.; Tang, Y.; Wang, P.; Zeng, H. Carbonaceous Halloysite Nanotubes for the Stabilization of Co, Ni, Cu and Zn in River Sediments. *Env. Sci Nano* **2019**, *6*, 2420–2428. [CrossRef]
45. Liu, H.; Xie, M.; Pan, B.; Li, N.; Zhang, J.; Lu, M.; Luo, J.; Wang, H. In-situ intercalated pyrolytic graphene/serpentine hybrid as an efficient lubricant additive in paraffin oil. *Colloids Surf. A Physicochem Eng Asp.* **2022**, *652*, 129929. [CrossRef]
46. El-Shwiniy, W.H.; El-Attar, M.S.; Sadeek, S.A. Metal Complexes of Enrofloxacin Part I: Preparation, Spectroscopic, Thermal Analyses Studies and Antimicrobial Evaluation. *J. Korean Chem. Soc.* **2013**, *57*, 52–62. [CrossRef]
47. Alshehri, A.A.; Malik, M.A. Biogenic Fabrication of ZnO Nanoparticles Using Trigonella Foenum-Graecum (Fenugreek) for Proficient Photocatalytic Degradation of Methylene Blue under UV Irradiation. *J. Mater. Sci. Mater. Electron.* **2019**, *30*, 16156–16173. [CrossRef]
48. Shchukin, D.G.; Sukhorukov, G.B.; Price, R.R.; Lvov, Y.M. Halloysite Nanotubes as Biomimetic Nanoreactors. *Small* **2005**, *1*, 510–513. [CrossRef]
49. Tarì, G.; Bobos, I.; Gomes, C.S.; Ferreira, J.M. Modification of surface charge properties during kaolinite to halloysite-7Å transformation. *J. Colloid Interface Sci.* **1999**, *210*, 360–366. [CrossRef] [PubMed]
50. Zhao, M.; Liu, P. Adsorption Behavior of Methylene Blue on Halloysite Nanotubes. *Microporous Mesoporous Mater.* **2008**, *112*, 419–424. [CrossRef]
51. Trouchon, T.; Lefebvre, S. A Review of Enrofloxacin for Veterinary Use. *Open J. Vet. Med.* **2016**, *06*, 40–58. [CrossRef]
52. Ötker, H.M.; Akmehmet-Balcioğlu, I. Adsorption and Degradation of Enrofloxacin, a Veterinary Antibiotic on Natural Zeolite. *J. Hazard. Mater.* **2005**, *122*, 251–258. [CrossRef]
53. Martínez-Mejía, M.J.; Sato, I.; Rath, S. Sorption Mechanism of Enrofloxacin on Humic Acids Extracted from Brazilian Soils. *Environ. Sci. Pollut. Res.* **2017**, *24*, 15995–16006. [CrossRef]
54. Batasheva, S.; Kryuchkova, M.; Fakhrullin, R.; Cavallaro, G.; Lazzara, G.; Akhatova, F.; Nigamatzyanova, L.; Evtugyn, V.; Rozhina, E.; Fakhrullin, R. Facile Fabrication of Natural Polyelectrolyte-Nanoclay Composites: Halloysite Nanotubes, Nucleotides and DNA Study. *Molecules* **2020**, *25*, 3557. [CrossRef] [PubMed]
55. Nowara, A.; Rgen Burhenne, J.; Spiteller, M. Binding of Fluoroquinolone Carboxylic Acid Derivatives to Clay Minerals. *J. Agric. Food Chem.* **1997**, *45*, 1459–1463. [CrossRef]
56. Wan, M.; Li, Z.; Hong, H.; Wu, Q. Enrofloxacin Uptake and Retention on Different Types of Clays. *J. Asian Earth Sci.* **2013**, *77*, 287–294. [CrossRef]
57. Deng, L.; Yuan, P.; Liu, D.; Annabi-Bergaya, F.; Zhou, J.; Chen, F.; Liu, Z. Effects of Microstructure of Clay Minerals, Montmorillonite, Kaolinite and Halloysite, on Their Benzene Adsorption Behaviors. *Appl. Clay Sci.* **2017**, *143*, 184–191. [CrossRef]
58. Lázaro, B.B. Halloysite and kaolinite: Two clay minerals with geological and technological importance. *Rev. La Acad. Cienc. Exactas Físicas Químicas Y Nat. Zaragoza* **2015**, *70*, 7–38.
59. Wu, Q.; Li, Z.; Hong, H.; Yin, K.; Tie, L. Adsorption and Intercalation of Ciprofloxacin on Montmorillonite. *Appl. Clay Sci.* **2010**, *50*, 204–211. [CrossRef]
60. Yan, W.; Hu, S.; Jing, C. Enrofloxacin Sorption on Smectite Clays: Effects of PH, Cations, and Humic Acid. *J. Colloid Interface Sci.* **2012**, *372*, 141–147. [CrossRef]
61. Liu, R.; Zhang, B.; Mei, D.; Zhang, H.; Liu, J. Adsorption of Methyl Violet from Aqueous Solution by Halloysite Nanotubes. *Desalination* **2011**, *268*, 111–116. [CrossRef]
62. Luo, P.; Zhang, B.; Zhao, Y.; Wang, J.; Zhang, H.; Liu, J. Removal of Methylene Blue from Aqueous Solutions by Adsorption onto Chemically Activated Hal.lloysite Nanotubes. *Korean J. Chem. Eng.* **2011**, *28*, 800–807. [CrossRef]

63. Adeyi, A.A.; Jamil, S.N.A.M.; Abdullah, L.C.; Choong, T.S.Y.; Lau, K.L.; Abdullah, M. Simultaneous Adsorption of Cationic Dyes from Binary Solutions by Thiourea-Modified Poly(Acrylonitrile-Co-Acrylic Acid): Detailed Isotherm and Kinetic Studies. *Materials* **2019**, *12*, 2903. [CrossRef]
64. Rápó, E.; Tonk, S. Factors Affecting Synthetic Dye Adsorption; Desorption Studies: A Review of Results from the Last Five Years (2017–2021). *Molecules* **2021**, *26*, 5419. [CrossRef] [PubMed]
65. Ho, Y.S.; McKay, G. Kinetic models for the sorption of dye from aqueous solution by wood. *Process. Saf. Environ. Prot.* **1998**, *76*, 183–191. [CrossRef]
66. Freundlich, H. Über die adsorption in lösungen. *Z. Für Phys. Chem.* **1907**, *57*, 385–470. [CrossRef]
67. Foo, K.Y.; Hameed, B.H. Preparation, Characterization and Evaluation of Adsorptive Properties of Orange Peel Based Activated Carbon via Microwave Induced K 2CO 3 Activation. *Bioresour Technol* **2012**, *104*, 679–686. [CrossRef] [PubMed]
68. Pathania, D.; Sharma, S.; Singh, P. Removal of Methylene Blue by Adsorption onto Activated Carbon Developed from Ficus Carica Bast. *Arab. J. Chem.* **2017**, *10*, S1445–S1451. [CrossRef]
69. Özcan, A.S.; Erdem, B.; Özcan, A. Adsorption of Acid Blue 193 from Aqueous Solutions onto Na-Bentonite and DTMA-Bentonite. *J. Colloid Interface Sci.* **2004**, *280*, 44–54. [CrossRef] [PubMed]
70. Foo, K.Y.; Hameed, B.H. Insights into the Modeling of Adsorption Isotherm Systems. *Chem. Eng. J.* **2010**, *156*, 2–10. [CrossRef]
71. Du, Y.; Zheng, P. Adsorption and Photodegradation of Methylene Blue on TiO2-Halloysite Adsorbents. *Korean J. Chem. Eng.* **2014**, *31*, 2051–2056. [CrossRef]
72. Jiang, L.; Zhang, C.; Wei, J.; Tjiu, W.; Pan, J.; Chen, Y.; Liu, T. Surface Modifications of Halloysite Nanotubes with Superparamagnetic Fe3O4 Nanoparticles and Carbonaceous Layers for Efficient Adsorption of Dyes in Water Treatment. *Chem. Res. Chin. Univ.* **2014**, *30*, 971–977. [CrossRef]
73. Rivagli, E.; Pastorello, A.; Sturini, M.; Maraschi, F.; Speltini, A.; Zampori, L.; Setti, M.; Malavasi, L.; Profumo, A. Clay Minerals for Adsorption of Veterinary FQs: Behavior and Modeling. *J. Env. Chem. Eng.* **2014**, *2*, 738–744. [CrossRef]
74. Lagergen, S. About the Theory of So-Called Adsorption of Soluble Substances. *Sven. Vetenskapsakad. Handingarl* **1898**, *24*, 1–39.
75. Blanchard, G.; Maunaye, M.; Martin, G. Removal of heavy metals from waters by means of natural zeolites. *Water Res.* **1984**, *18*, 1501–1507. [CrossRef]
76. Aharoni, C.; Tompkins, F.C. *Advances in Catalysis*; Eley, D.D., Pines, H., Weisz, P.B., Eds.; Academic Press: Cambridge, MA, USA, 1970; pp. 1–49.
77. Acelas, N.; Lopera, S.M.; Porras, J.; Torres-Palma, R.A. Evaluating the Removal of the Antibiotic Cephalexin from Aqueous Solutions Using an Adsorbent Obtained from Palm Oil Fiber. *Molecules* **2021**, *26*, 3340. [CrossRef]
78. Dawood, S.; Sen, T.K. Removal of Anionic Dye Congo Red from Aqueous Solution by Raw Pine and Acid-Treated Pine Cone Powder as Adsorbent: Equilibrium, Thermodynamic, Kinetics, Mechanism and Process Design. *Water Res* **2012**, *46*, 1933–1946. [CrossRef]
79. Ofomaja, A.E. Intraparticle Diffusion Process for Lead(II) Biosorption onto Mansonia Wood Sawdust. *Bioresour. Technol.* **2010**, *101*, 5868–5876. [CrossRef] [PubMed]
80. Laabd, M.; Chafai, H.; Essekri, A.; Elamine, M.; Al-Muhtaseb, S.A.; Lakhmiri, R.; Albourine, A. Single and Multi-Component Adsorption of Aromatic Acids Using an Eco-Friendly Polyaniline-Based Biocomposite. *Sustain. Mater. Technol.* **2017**, *12*, 35–43. [CrossRef]
81. Wang, F.; Pan, Y.; Cai, P.; Guo, T.; Xiao, H. Single and Binary Adsorption of Heavy Metal Ions from Aqueous Solutions Using Sugarcane Cellulose-Based Adsorbent. *Bioresour. Technol.* **2017**, *241*, 482–490. [CrossRef]
82. Chen, J.; Pan, X.; Chen, J. Regeneration of Activated Carbon Saturated with Odors by Non-Thermal Plasma. *Chemosphere* **2013**, *92*, 725–730. [CrossRef]
83. Qu, G.Z.; Lu, N.; Li, J.; Wu, Y.; Li, G.F.; Li, D. Simultaneous Pentachlorophenol Decomposition and Granular Activated Carbon Regeneration Assisted by Dielectric Barrier Discharge Plasma. *J. Hazard. Mater.* **2009**, *172*, 472–478. [CrossRef] [PubMed]
84. Sun, D.L.; Hong, R.Y.; Wang, F.; Liu, J.Y.; Rajesh Kumar, M. Synthesis and Modification of Carbon Nanomaterials via AC Arc and Dielectric Barrier Discharge Plasma. *Chem. Eng. J.* **2016**, *283*, 9–20. [CrossRef]
85. Kandel, D.R.; Kim, H.J.; Lim, J.M.; Poudel, M.B.; Cho, M.; Kim, H.W.; Oh, B.T.; Nah, C.; Lee, S.H.; Dahal, B.; et al. Cold Plasma-Assisted Regeneration of Biochar for Dye Adsorption. *Chemosphere* **2022**, *309*, 136638. [CrossRef] [PubMed]

nanomaterials

Article

Rapid Synthesis Method of Ag_3PO_4 as Reusable Photocatalytically Active Semiconductor

Zsejke-Réka Tóth [1,2,3,*], Diána Debreczeni [1], Tamás Gyulavári [1], István Székely [2,3], Milica Todea [2,4], Gábor Kovács [5], Monica Focșan [6], Klara Magyari [2], Lucian Baia [2,7,8], Zsolt Pap [1,2,8] and Klara Hernadi [1,9,*]

1 Department of Applied and Environmental Chemistry, Faculty of Science and Informatics, University of Szeged, Rerrich Béla sqr. 1, 6720 Szeged, Hungary
2 Nanostructured Materials and Bio-Nano-Interfaces Center, Interdisciplinary Research Institute on Bio-Nano-Sciences, Babeș-Bolyai University, Treboniu Laurian str. 42, 400271 Cluj-Napoca, Romania
3 Doctoral School in Physics, Faculty of Physics, Babes-Bolyai University, M. Kogălniceanu 1, 400084 Cluj-Napoca, Romania
4 Faculty of Medicine, Iuliu Hatieganu University of Medicine and Pharmacy, Victor Babeş 8, 400012 Cluj-Napoca, Romania
5 Department of Horticulture, Faculty of Technical and Human Sciences, Sapientia Hungarian University of Transylvania, Aleea Sighișoarei 1C, 530104 Târgu Mureș/Corunca, Romania
6 Nanobiophotonics and Laser Microspectroscopy Center, Interdisciplinary Research Institute on Bio-Nano-Sciences, Treboniu Laurian str. 42, Babes-Bolyai University, 400271 Cluj-Napoca, Romania
7 Faculty of Physics, Babeș-Bolyai University, M. Kogălniceanu str. 1, 400084 Cluj-Napoca, Romania
8 Institute of Research-Development-Innovation in Applied Natural Sciences, Babes-Bolyai University, Fântânele str. 30, 400294 Cluj-Napoca, Romania
9 Institute of Physical Metallurgy, Metal Forming and Nanotechnology, University of Miskolc, Miskolc-Egyetemváros, 3515 Miskolc, Hungary
* Correspondence: zsejke.toth@ubbcluj.ro (Z.-R.T.); klara.hernadi@uni-miskolc.hu (K.H.)

Abstract: The widespread use of Ag_3PO_4 is not surprising when considering its higher photostability compared to other silver-based materials. The present work deals with the facile precipitation method of silver phosphate. The effects of four different phosphate sources (H_3PO_4, NaH_2PO_4, Na_2HPO_4, $Na_3PO_4 \cdot 12\ H_2O$) and two different initial concentrations (0.1 M and 0.2 M) were investigated. As the basicity of different phosphate sources influences the purity of Ag_3PO_4, different products were obtained. Using H_3PO_4 did not lead to the formation of Ag_3PO_4, while applying NaH_2PO_4 resulted in Ag_3PO_4 and a low amount of pyrophosphate. The morphological and structural properties of the obtained samples were studied by X-ray diffractometry, diffuse reflectance spectroscopy, scanning electron microscopy, infrared spectroscopy, and X-ray photoelectron spectroscopy. The photocatalytic activity of the materials and the corresponding reaction kinetics were evaluated by the degradation of methyl orange (MO) under visible light. Their stability was investigated by reusability tests, photoluminescence measurements, and the recharacterization after degradation. The effect of as-deposited Ag nanoparticles was also highlighted on the photostability and the reusability of Ag_3PO_4. Although the deposited Ag nanoparticles suppressed the formation of holes and reduced the degradation of methyl orange, they did not reduce the performance of the photocatalyst.

Keywords: silver phosphate; reusability; precipitation method; photoluminescence; photocatalysis

Citation: Tóth, Z.-R.; Debreczeni, D.; Gyulavári, T.; Székely, I.; Todea, M.; Kovács, G.; Focșan, M.; Magyari, K.; Baia, L.; Pap, Z.; et al. Rapid Synthesis Method of Ag_3PO_4 as Reusable Photocatalytically Active Semiconductor. *Nanomaterials* **2023**, *13*, 89. https://doi.org/10.3390/nano13010089

Academic Editor: Christos A. Aggelopoulos

Received: 30 November 2022
Revised: 19 December 2022
Accepted: 20 December 2022
Published: 24 December 2022

1. Introduction

Since the discovery and first application attempts of photocatalysis, numerous materials, such as TiO_2-based semiconductors and composites, other metal oxides, and salts have been investigated regarding their degradation of organic pollutants [1]. However, even the TiO_2-based "flagships" (such as P25 [2] or other commercial titania) have severe limitations due to their relatively wide band gap, poor quantum efficiency, and/or fast recombination rate of photoexcited electron–hole pairs.

Silver-based semiconductors are promising candidates among the newer claimants to reach the throne of large-scale industrial applications due to their strong visible light response. One of their main benefits is that they can provide Ag nanoparticles that can be deposited in situ on the surface of catalysts. Depending on the light source applied during photocatalytic processes, deposited nanoparticles can act as charge separators (mostly under UV irradiation) or electron carriers (under visible light irradiation). Many photocatalysts (especially the Ag-based ones) are susceptible to photochemical corrosion, resulting in poor stability [3]. However, deposited nanoparticles can be effectively used to prevent the photocorrosion of semiconductors [4].

The application of Ag_3PO_4 as a photocatalytic material was first reported by Yi and coworkers [5]. Their investigation focused on water-splitting and degrading organic contaminants in wastewater by utilizing visible light. Ma et al. [6] focused on the structural peculiarities of Ag_3PO_4. They found that it can be characterized by a body-centered cubic structure (where all oxygen atoms are adjacent with three Ag and a P atom), a lattice parameter of 6 Å, and the space group of P4-3n [7,8]. Its band gap energy is relatively low ($E_g \approx 2.4$ eV), which can be mainly modified by fine-tuning its morpho-structural characteristics. During its synthesis, pH and the appearance of pyro- and polyphosphates are key parameters. It is well known that two phosphoric acid molecules can easily condensate, resulting in these phosphates. Since pH can strongly influence the formation of oxo-acid-based Ag salts, it is not surprising that the type of phosphate precursor can be decisive (as it can also strongly influence pH). Accordingly, several phosphate-containing precursors have already been investigated, focusing on the morphology of the target semiconductor, including NH_4^+ [9], Na^+ [10], and K^+ [11] based agents. Besides precursors, another decisive parameter is the applied synthesis pathway. Considering this aspect, the main methods applied to synthesize Ag_3PO_4 salts have been precipitation [9,10,12], ion-exchange [5,13], hydrothermal [14], ultrasonication [15], and other approaches [16]. It has been found that semiconductors prepared by ion-exchange reactions can yield electron–hole pairs with enhanced lifetime, resulting in higher photocatalytic efficiency [17]. Still, a deeper insight into the electronic structure of Ag_3PO_4 is necessary to understand the origin of the increased photocatalytic activity.

Thus, in this work, we focus on three main aspects related to the involvement of the Ag_3PO_4 semiconductor in photocatalytic approaches as follows:

- The effect of different phosphate sources on the synthesis and photocatalytic activity of Ag_3PO_4. For this purpose, H_3PO_4, NaH_2PO_4, Na_2HPO_4, and $Na_3PO_4 \cdot 12H_2O$ were used as phosphate sources. In this part, we propose a mechanism for the formation of Ag_3PO_4.
- Investigation of the stability of Ag_3PO_4 via recharacterization and reusability measurements.
- The effect of Ag nanoparticles on Ag_3PO_4, shown through the deposition mechanism of Ag nanoparticles. This aspect was demonstrated by photoluminescence measurements and their corresponding kinetic studies.

2. Materials and Methods

2.1. Materials

All chemicals were used as received: silver nitrate ($AgNO_3$, 99.8%, Penta industry; Prague, Czech Republic), methyl orange (MO, analytical reagent, Alfa Aesar; Tewksbury, MA, USA), Milli-Q (MQ; Budapest, Hungary) water, phosphoric acid (H_3PO_4, 85%, VWR Chemicals; Radnor, PA, USA), monosodium phosphate (NaH_2PO_4; >99.0%, Spektrum-3D; Debrecen, Hungary), disodium phosphate (Na_2HPO_4; >99.0%, Sigma-Aldrich; Schnelldorf, Germany), and trisodium phosphate dodecahydrate ($Na_3PO_4 \cdot 12H_2O$; analytical reagent, Sigma-Aldrich, Schnelldorf, Germany).

2.2. Methods

2.2.1. Synthesis of Ag_3PO_4 Semiconductors

A precipitation method was used [13,18] to synthesize Ag_3PO_4 microcrystals. Four phosphate sources (MPO_4: H_3PO_4, NaH_2PO_4, Na_2HPO_4, $Na_3PO_4 \cdot 12\ H_2O$) and $AgNO_3$ were used. The weight ratio of the precursors was: MPO_4:$AgNO_3$ = 3:2.

In each case, two different initial concentrations of phosphate sources were used (0.2 M and 0.1 M). The aqueous solution of phosphate sources was stirred for 5 min then 1.247 g of $AgNO_3$ was added. A yellow suspension formed from the colorless and transparent solution. After another 5 min of stirring, the suspensions were washed with 3 × 45 mL of MQ water (for 10 min at 4400 RPM) and dried overnight at 40 °C.

The sample abbreviations used in the manuscript were conceived as follows: Ag_3PO_4_source_concentration, where concentration is the initial concentration of the used phosphate precursor (example: Ag_3PO_4_Na_3PO_4_0.1M denotes the sample that was prepared using 0.1 M $Na_3PO_4 \cdot 12\ H_2O$ as the phosphate source). The word *"after"* was added to the names to designate samples investigated after photocatalytic tests (example: Ag_3PO_4_Na_3PO_4_0.1M_*after*).

2.2.2. Characterization and Instrumentation

The X-ray diffraction (XRD) measurements were performed on a Shimadzu 6000 X-ray diffractometer (Kyoto, Japan) at an accelerating voltage of 40 kV (30 mA), operated with CuKα radiation ($\lambda_{CuK\alpha}$ = 1.54 Å). The XRD patterns were recorded in 2θ range between 15 and 60°, with scan speed $1° \cdot min^{-1}$.

Fourier transform infrared spectroscopy (FT-IR) measurements were performed on a JASCO-6200 FT-IR spectrophotometer (Jasco, Tokyo, Japan) in the 4000–400 cm^{-1} wavelength range, with 4 cm^{-1} spectral resolution, using the well-known KBr pellet technique.

The morphology of the samples was identified with a Hitachi S-4700 Type II scanning electron microscope (SEM; Hitachi, Tokyo, Japan) equipped with an Everhart—Thornley detector using an electron beam with an acceleration voltage of 10 kV.

The band structure of the semiconductors was investigated by diffuse reflectance spectroscopy (DRS). The spectra were recorded in the 250–800 nm range with a JASCO-V650 spectrophotometer (equipped with an ILV-724 integration sphere; Jasco, Wien, Austria) using $BaSO_4$ as a reference. The band gap energy values were calculated based on the Kubelka-Munk theory [19].

X-ray photoelectron spectroscopy (XPS) measurements were carried out using a Specs Phoibos 150 MCD system (SPECS Surface Nano Analysis GmbH, Berlin, Germany) equipped with Al-Kα source (1486.6 eV) at 14 kV and 20 mA, a hemispherical analyzer, and a charge neutralization device. Care was taken to completely cover the double-sided carbon tapes with the silver—phosphate samples.

Fluorescence measurements were carried out using a Jasco LP-6500 spectrofluorometer (Jasco, Japan; PL) equipped with a Xenon lamp (excitation source) coupled to an epifluorescence accessory (EFA 383 module). Fluorescence spectra were collected with a 1 nm spectral resolution in the 350–600 nm wavelength range using a fixed excitation wavelength of 325 nm. Bandwidths of 1 nm or 10 nm were employed during excitation and emission.

2.2.3. Photocatalytic Activity

The photocatalytic investigation of the samples was carried out in a double-walled photoreactor, where MO (C_{MO} = 125 μM) was the model pollutant. The reactor was surrounded by 6 × 15 W visible light-emitting lamps (λ > 400 nm). The system was kept at a constant temperature (25 °C), and the suspension ($C_{suspension} = 1\ g \cdot L^{-1}$) was continuously stirred and purged by air at constant flow ($40\ L \cdot h^{-1}$). The concentration change of MO was followed with a JASCO-V650 UV-Vis spectrophotometer (UV-Vis; Jasco, Wien, Austria) at λ_{max} = 513 nm (using a 1 mm optical path length quartz cell). The suspension was kept in the dark for 10 min to reach the adsorption–desorption equilibrium. The experiments were conducted for 2 h. Samples were taken every 10 min in the first hour and every 20 min in the second hour. Last, the samples were centrifugated and filtrated before quantitative analysis.

The conversion of MO was calculated by the following equation:

$$H = (100 - (C_{120}/C_0 \times 100),$$

where C_{120} is the concentration of MO after 120 min and C_0 is the initial concentration of MO.

The reaction order (n) and apparent rate constants (k_1, k_2) were calculated to investigate the kinetics of MO degradation. The apparent rate constants (k_1, k_2) were determined by plotting the MO concentration vs. the irradiation time, where the slope was considered the apparent rate constants. For k_1 values, we took the first hour (0–60 min) of the degradation process into account, whereas for k_2 values, we considered the second (60–120 min) hour.

The adsorption of MO on the samples was also investigated. The photocatalysts (50 mg) and MO (50 mL; C_{MO} = 125 μM) were added to a beaker. The beaker was stirred (500 RPR) and covered with aluminum foil to eliminate all light sources. Samples were taken every 5 min in the first 30 min, every 10 min between 30–60 min, and every 20 min between 1–2 h. Then, they were centrifuged and filtered, and the adsorption of MO was determined with a JASCO-V650 spectrophotometer.

To investigate reusability, we used the same setup for the photoactivity and adsorption measurements, but sampling times were changed to 30, 60, and 120 min. Samples collected between two cycles were washed with distilled water three times and dried at 40 °C for 12 h.

3. Results and Discussion

3.1. Characterization of Ag_3PO_4

Synthesizing Ag_3PO_4 by using different MPO_4 sources is a rather complicated process. MPO_4 (used as different phosphate sources; Figure 1) has different disproportional rates in aqueous media, which resulted in different pH values (Table 1). Moreover, pH can also indirectly affect the samples' morphological and structural properties. Using H_3PO_4 did not lead to the formation of Ag_3PO_4. The lack of Ag_3PO_4 can be explained by the acidic environment set by H_3PO_4, hindering the precipitation of Ag_3PO_4 (Figure 1). Moreover, the presence of H_3PO_4 can facilitate the formation of pyrophosphate. Thus, this sample was omitted from all the experimental work presented here.

Table 1. pH, band gap energy, absorption band maxima, average particles sizes, and conversion values of the obtained Ag_3PO_4 samples (n.a.—no available).

	pH MPO_4	Band Gap Energy (eV)	λ_{max} (nm)	d_{SEM} (μm)	Conversion −2 h-(%)
Ag_3PO_4_NaH_2PO_4_0.2M	4.22	2.3	507	n.a.	38.98
Ag_3PO_4_NaH_2PO_4_0.1M	4.27	2.3	504	n.a.	49.53
Ag_3PO_4_Na_2HPO_4_0.2M	9.14	2.33	506	0.92	89.72
Ag_3PO_4_Na_2HPO_4_0.1M	9.24	2.34	506	0.97	88.80
Ag_3PO_4_Na_3PO_4_0.2M	11.46	2.27	498	0.58	95.94
Ag_3PO_4_Na_3PO_4_0.1M	11.71	2.22	495	0.33	94.53

The colors of the samples are the same as those used in the entire article.

XRD patterns were recorded to elucidate the effect of the other three phosphate sources on the formation of Ag_3PO_4 crystals. Cubic Ag_3PO_4 was identified (COD 00-101-0324) in all cases: the reflections of the Ag_3PO_4 were located at $2\theta° \approx 21.1°$, $\approx 29.8°$, $\approx 33.3°$, $\approx 36.7°$, $\approx 42.6°$, $\approx 47.9°$, $\approx 52.8°$, $\approx 55.2°$, and $\approx 57.3°$, which were assigned to (110), (200), (210),

(211), (220), (310), (222), (320), and (321) crystallographic planes (Figure 2a). Additional reflections were also observed in the Ag_3PO_4_NaH_2PO_4 sample series located at $2\theta^\circ \approx 27.0^\circ$, $\approx 30.6^\circ$ and $\approx 32.1^\circ$ (Figure 2a), which could be associated with the typical reflections of $Ag_4P_2O_7$ [20]. These observations are consistent with the presumed mechanism (Figure 1). It must be emphasized that the pyrophosphate formation is much higher for NaH_2PO_4 than for the other two samples. In this case, crystallized pyrophosphate was formed since the materials have higher proton concentrations, and the possibility of condensation is much higher than in other cases. Interestingly, diffraction peaks of other Ag species, such as Ag nanoparticles or Ag_xO particles, were not found in the XRD patterns (Figure 2a).

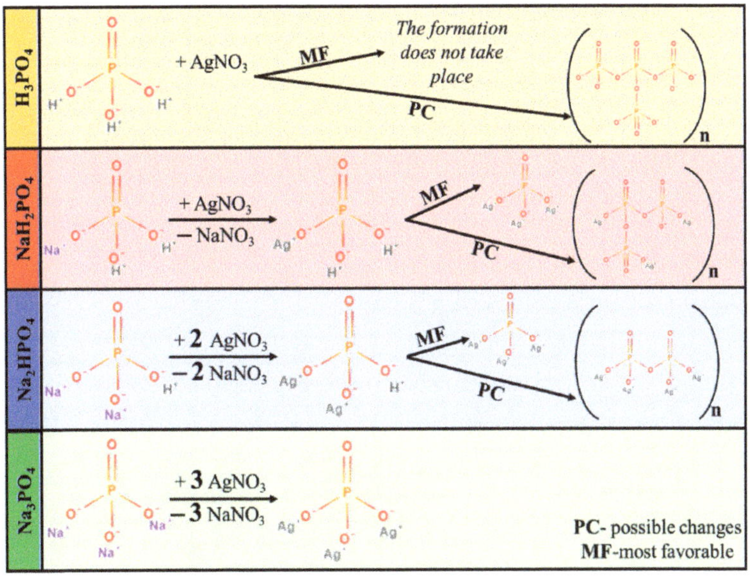

Figure 1. Proposed mechanism of Ag_3PO_4 formation.

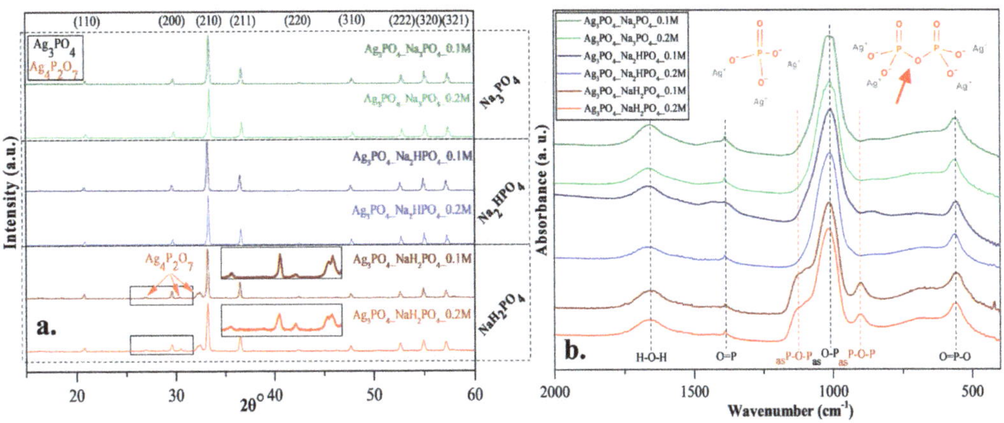

Figure 2. (**a**) X-ray diffraction patterns and (**b**) infrared spectra of Ag_3PO_4 samples synthesized by using different types (Na_3PO_4, Na_2HPO_4 and NaH_2PO_4) and concentrations of phosphate sources (C = 0.1 and 0.2 M).

FT-IR measurements (Figure 2b) were conducted to investigate the presence of pyrophosphate. The typical bands of asymmetrical vibrations of P–O–P bonds [20] were identified at ≈902 and ≈1116 cm^{-1}, confirming the presence of silver pyrophosphate (observed only in the Ag_3PO_4_NaH_2PO_4 sample series; Figure 2b). Additional bands were also observed [17]: at ≈554 cm^{-1} (O=P–O); ≈1007 cm^{-1} ($_{as}$O–P), ≈1389 cm^{-1} (O=P), and H–O–H (≈1655 cm^{-1}). Thereby, it can be concluded that the formation of $Ag_4P_2O_7$ depends on the nature of the applied phosphate source (Figure 2b).

When NaH_2PO_4 was used as the phosphate source, the formation of Ag_3PO_4 was incomplete, while the formation of $Ag_4P_2O_7$ was detected (Figure 1). The reason for this could be that the formation of HNO_3 was not favored. This could be because, during the synthesis, Na^+ exchange is more favored than that of H^+. This makes the free formation of $NaNO_3$ favorable because the electropositivity of Na^+ is higher than that of H^+. Against the Ag_3PO_4_NaH_2PO_4 samples, the $Ag_4P_2O_7$ was not detected using Na_2HPO_4 or Na_3PO_4 as a phosphate source.

The morphological properties of the samples were analyzed using SEM. A correlation was found with the XRD measurements (Figure 2a). Two differently shaped and sized particles—*spherical-like* structure with 1.5 μm diameter and *plates* with 0.2 μm height—were obtained in the Ag_3PO_4_NaH_2PO_4 sample series (Figure 3). The particles could not be distinguished based on their elemental composition. This can be attributed to the co-presence of phosphate and pyrophosphate according to XRD patterns (Figure 2a). The samples prepared by using the other two phosphate sources had a much higher monodispersity (Figure S1) and a more defined shape (Figure 3). Their average particle size was ≈0.9 μm regardless of the used concentration. The particle size distribution was lower than for the Na_3PO_4 samples series; moreover, lower concentration resulted in smaller particles. The Ag_3PO_4_Na_2HPO_4 sample series contained more polyhedral particles compared to the Ag_3PO_4_Na_3PO_4 sample series.

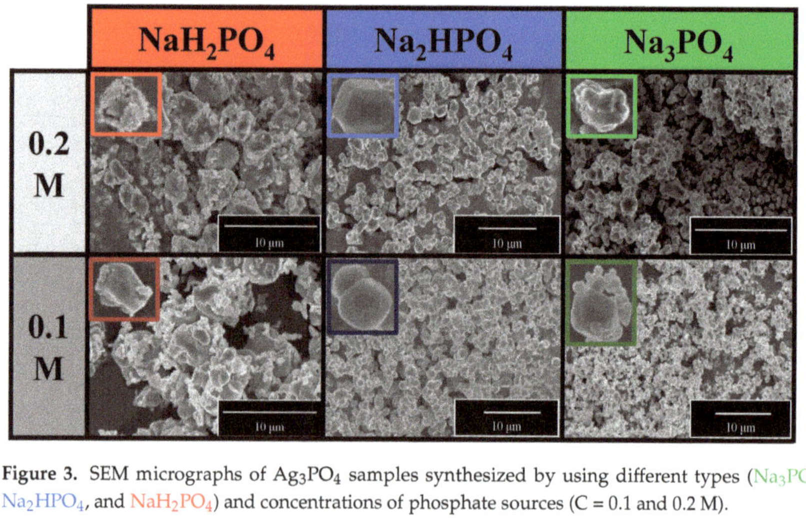

Figure 3. SEM micrographs of Ag_3PO_4 samples synthesized by using different types (Na_3PO_4, Na_2HPO_4, and NaH_2PO_4) and concentrations of phosphate sources (C = 0.1 and 0.2 M).

Diffuse reflectance spectroscopy was used to analyze the optical properties of the materials and to understand the light absorption properties of Ag_3PO_4_NaH_2PO_4 (which contained $Ag_4P_2O_7$, a stand-alone photocatalyst [20]). As shown in Figure 4a, using different phosphate sources significantly influenced the visible light absorption properties of the samples, whose reflectance intensities decreased in the following order:

Ag_3PO_4_Na_2HPO_4 > Ag_3PO_4_NaH_2PO_4 > Ag_3PO_4_Na_3PO_4. These differences may result in different photocatalytic performances. On the other hand, no specific plasmon resonance band of Ag was detected (which could be found at ≈320 nm [21]). These observations agreement with the XRD results, where the reflection of Ag could not be detected.

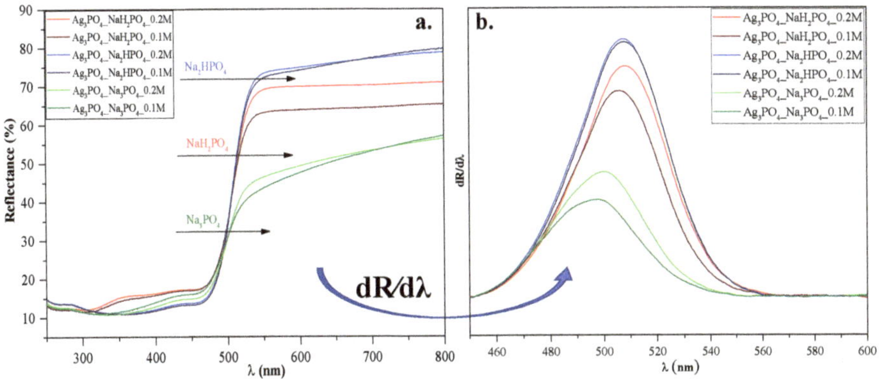

Figure 4. (**a**) Diffuse reflectance spectra and (**b**) their first derivative order of Ag_3PO_4 samples synthesized by using different types (Na_3PO_4, Na_2HPO_4, and NaH_2PO_4) and concentrations of phosphate sources (C = 0.1 and 0.2 M).

The Kubelka-Munk theory was used to calculate the indirect band gap energies of the samples. No significant changes could be observed between them ($E_g \approx 2.22$–2.34 eV; Table 1). Hence no clear conclusions could be drawn. To understand the relationship between the samples' light absorption and photocatalytic activity, we analyzed their first derivative spectra (Figure 4b). Still, the λ_{max} values were almost identical ($\lambda_{max} \approx 495$–$507$ nm; Figure 4). Since $Ag_4P_2O_7$ can be photoactive as well [20], the derivative DRS of sample Ag_3PO_4_NaH_2PO_4 should result in two specific electron transition peaks: (i) one corresponding to $Ag_4P_2O_7$ (observed at ≈300 nm [20]) and (ii) one corresponding to Ag_3PO_4, (observed at ≈548 nm (2.26 eV [11])).

The lack of the Ag nanoparticles was also demonstrated with high-resolution XPS (Figure 5). Symmetrical peaks were found in the Ag3d spectra (Ag $3d_{5/2}$ and $3d_{3/2}$ of Ag_3PO_4 corresponding to the peaks 373.67 and 367.67 eV, respectively [22]), which could be associated with Ag^+ from Ag_3PO_4.

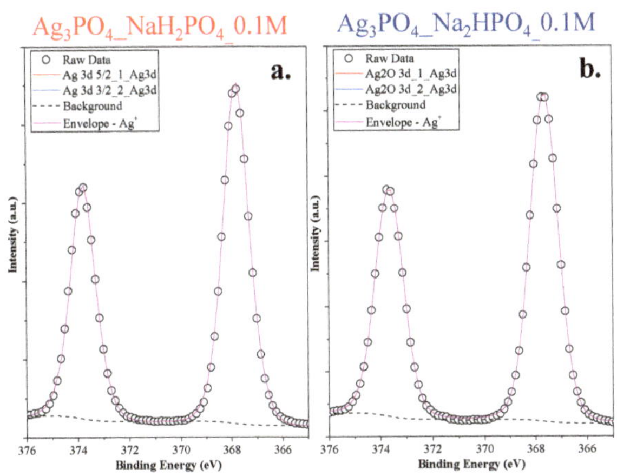

Figure 5. Ag3d XPS spectra of the Ag 3d (**a**) Ag_3PO_4_NaH_2PO_4_0.1M and (**b**) Ag_3PO_4_Na_2HPO_4_0.1M samples.

3.2. Photocatalytic Degradation

MO was employed as a model pollutant to investigate photoactivity. As shown in Figure 6, the photocatalytic activity of the samples can be correlated with the phosphate source used in the synthesis process. Regardless of the concentration of the phosphate source, the order was as follows: Ag_3PO_4_Na_3PO_4 > Ag_3PO_4_Na_2HPO_4 > Ag_3PO_4_NaH_2PO_4 (Figure 6). Because the samples had different absorption properties, similar photocatalytic activation was also expected to differ. According to our assumptions, samples containing $Ag_4P_2O_7$, a stand-alone photocatalyst, should have higher photocatalytic activity. Moreover, a photojunction between $Ag_4P_2O_7$ and Ag_3PO_4 could also form [23]. $Ag_4P_2O_7$ did not show any spectral features, that is, its characteristic secondary bands were not found in the DR and first-order derivate spectra (Figure 4). It could inhibit the photocatalyst and absorb electrons, which in turn could not be used in the photocatalytic process. As a result, a photojunction did not form. Similar to the DRS measurements, regardless of the used phosphate source, different photoactivity was observed (Figure 6). Using different phosphate sources resulted in different optical and structural properties, which influenced the degradation of MO. To clarify the degradation of MO, we evaluated the MO conversion values presented in Section 2.2.3. After two hours, 95.94 and 94.53% conversions were obtained for Ag_3PO_4_Na_3PO_4_0.2M and Ag_3PO_4_Na_3PO_4_0.1M, respectively (Table 1). It could also be noticed that the most photoactive materials (Ag_3PO_4_Na_3PO_4; Figure 6) had the highest degradation yield of MO degradation after one hour. On the other hand, the second-highest MO decolorization was achieved by using Na_2HPO_4 as a phosphate source. Based on these findings, we ascertained that the proposed mechanism of Ag_3PO_4 formation (Figure 1) is in good agreement with the photocatalytic performance. The more complete the transformation, the higher the photoactivity because when Na_3PO_4 was used during the synthesis, no $Ag_4P_2O_7$ or Ag nanoparticles could be observed.

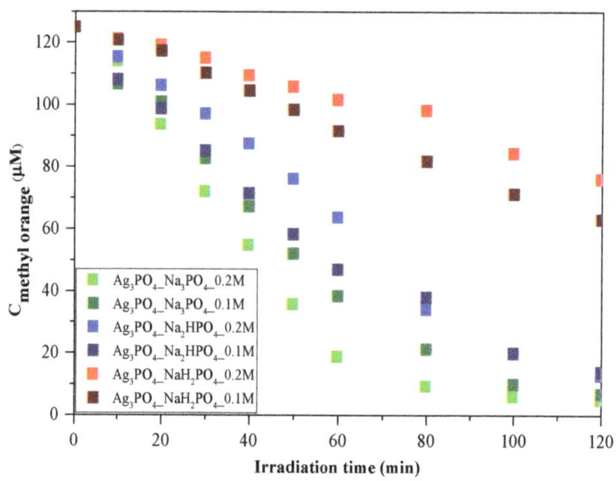

Figure 6. Photocatalytic investigation of Ag_3PO_4 samples using C = 125 μM of MO as a model pollutant and visible light (λ > 400 nm) as a light source.

As a parallel measurement, the adsorption capacity of the two most efficient photocatalysts was also investigated without using any light source (Figure S2). Adsorption did not occur throughout the process. Thus, it could be concluded that the photocatalytic degradation of MO could indeed be performed using these Ag-based materials.

Amornpitoksuk et al. [10] investigate the effect of different phosphates on the photocatalytic degradation of methylene blue. They found that using Na_2HPO_4 resulted in the highest photocatalytic activity. They also mentioned that the samples synthesized from Na_3PO_4 contained not only Ag_3PO_4 but Ag_2O as well, which inhibited photoactivity.

Similarly, in our case, samples that did not contain Ag_2O had the highest photocatalytic activity. However, it cannot be conclusively declared which phosphate source is the most suitable for achieving high photoactivity. In our case, the samples containing pyrophosphate proved to be less effective. Thus, it can be concluded that synthesizing pure Ag_3PO_4 is the best approach since both Ag_2O and $Ag_4P_2O_7$ are stand-alone photocatalysts that do not improve the efficiency of Ag_3PO_4.

Regardless of the employed phosphate source, after one hour (Figure 6), a change in the photocatalytic reaction was observed during MO removal. A kinetic study of the MO degradation curves was carried out.

Regarding the kinetics of MO degradation, a two-step mechanism was proposed (Table 2). The first step is the zero-order decolorization of azo dye bonds (R_1–N=N–R_2), probably by direct hole oxidation (0–60 min) since this is the thermodynamic and electrochemical facile pathway [24]. The second step (k_2) is the first-order mineralization (60–120 min) of the intermediates (benzenesulfonic acid; 4-hydrazinylaniline; phenyl diazene) by reactive radicals (•OH; $O_2^{\bullet-}$) [25]. The k_1 values are in excellent agreement with the correlation coefficients (R^2) regarding the MO conversion values after 1 h of visible light irradiation. The proposed mechanism can be applied to most samples; however, for Ag_3PO_4-Na_3PO_4-0.2M and Ag_3PO_4-Na_3PO_4-0.1M (Table 2), the degradation of intermediates follows second-order kinetics [26]. This could mean that the relatively fast mineralization of intermediates could also occur by direct hole oxidation and not by reactive radicals [27,28].

Table 2. Kinetics of methyl orange degradation of the Ag_3PO_4 samples.

	First Hour → 0 Order			Second Hour → 1 or 2 Order		
	NaH_2PO_4//Na_2HPO_4//Na_3PO_4 →0th Order			NaH_2PO_4//Na_2HPO_4 →1st order Na_3PO_4 → 2nd Order		
Sample	**Reaction Order (n)**	**k_1 (μM·min^{-1})**	**R^2**	**Reaction Order (n)**	**k_2 (s^{-1})**	**R^2**
Ag_3PO_4_NaH_2PO_4_0.1M	0	0.3945	0.990	1	0.0063	0.988
Ag_3PO_4-NaH_2PO_4-0.2M	0	0.5626	0.991	1	0.0065	0.998
Ag_3PO_4_Na_2HPO_4_0.1M	0	1.0027	0.996	1	0.0244	0.996
Ag_3PO_4_Na_2HPO_4_0.2M	0	1.2889	0.997	1	0.0250	0.974
Ag_3PO_4_Na_3PO_4_0.1M	0	1.8342	0.996	2	0.0023	0.97
Ag_3PO_4_Na_3PO_4_0.2M	0	1.4366	0.993	2	0.0025	0.998
Sample	**Reaction order (n)**	**k_1 (μM·min^{-1})**	**R^2**	**Reaction order (n)**	**k_2 (s^{-1}/M^{-1})**	**R^2**

3.3. *Stability Investigation of Ag_3PO_4*

3.3.1. Recharacterization of the Investigated Ag_3PO_4

After the photocatalytic tests, we once again investigated the structural and optical parameters of the samples to examine the stability of the photocatalysts. After the photocatalytic activity, a slight modification was observed in the XRD patterns (Figure 7a) because not only the typical Ag_3PO_4 reflections (presented in Figure 2a), but a specific reflection of Ag (nano)particles was also observed at 38.1° (COD-00-110-0136). Based on our previous measurements [29,30], a silver-based material might lose photocatalytic activity due to the presence of Ag nanoparticles on the surface. Still, several publications in the literature state that Ag-based materials can be used quasi-unlimited times in (photo)catalytic processes [10]. Besides the deposited Ag nanoparticles, the typical reflection of $Ag_4P_2O_7$ was also observed after the photocatalytic degradation, which was noticed before the degradation. Moreover, the optical parameters (Figure 7b–c) were also evaluated, and the decreasing intensity of the absorption band of Ag_3PO_4 was observed. The typical plasmon resonance band could not be clearly identified. Thus, it can be concluded that the amount of Ag nanoparticles was lower than the detection limit of the device applied. The appearance of Ag_xO is also possible because their reflection is close to that of Ag, and they do not have plasmon resonance bands in the reflectance spectra.

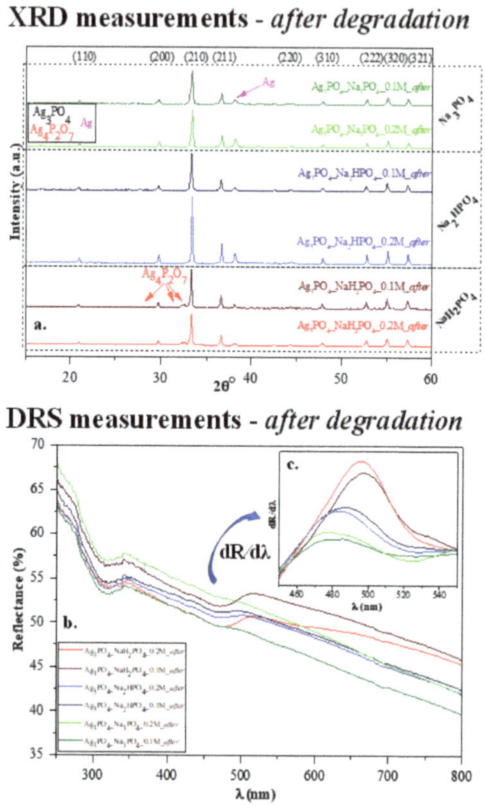

Figure 7. Stability measurements carried out after photocatalytic degradation: (**a**) XRD patterns and (**b**) DR spectra and (**c**) insert figure: their first derivative order. The term "*after*" was used in the sample names to indicate that the results were obtained after MO degradation.

3.3.2. Reusability of Ag_3PO_4 Samples: Recycling Tests and Photoluminescence Measurements

To investigate reusability, we have chosen the Ag_3PO_4_Na_2HPO_4_0.2M and Ag_3PO_4_Na_3PO_4_0.2M samples because these were the most efficient photocatalysts with a degradation of 89% and 95%, respectively (Table 1). The samples had photocatalytic activity after the second round as well (Figure 8a); however, a decrease in the degradation of MO was observed. The reason for this decrease was investigated by photoluminescence measurements (Figure 8b) to examine the recombination properties of Ag_3PO_4. Similar to Mohammadreza Batvandi et al. [31], two of the observed emission bands in our samples were in the violet and blue-cyan wavelength regions. The emission band observed in the violet region corresponds to charge transfer and self-trapping [32]. The blue-cyan wavelength can be assigned to surface oxygen vacancies and defects. After degradation, the intensity of these bands decreased, and in parallel, the band in the violet region increased slightly. The reason for the decreasing bands may be the disappearance of surface oxygen vacancies and defects, which may result from the deposition of Ag nanoparticles on the semiconductor surface. Based on our interpretation, the disappearance of oxygens vacancies and significant disappearances of defects resulted in a decrease in photocatalytic activity.

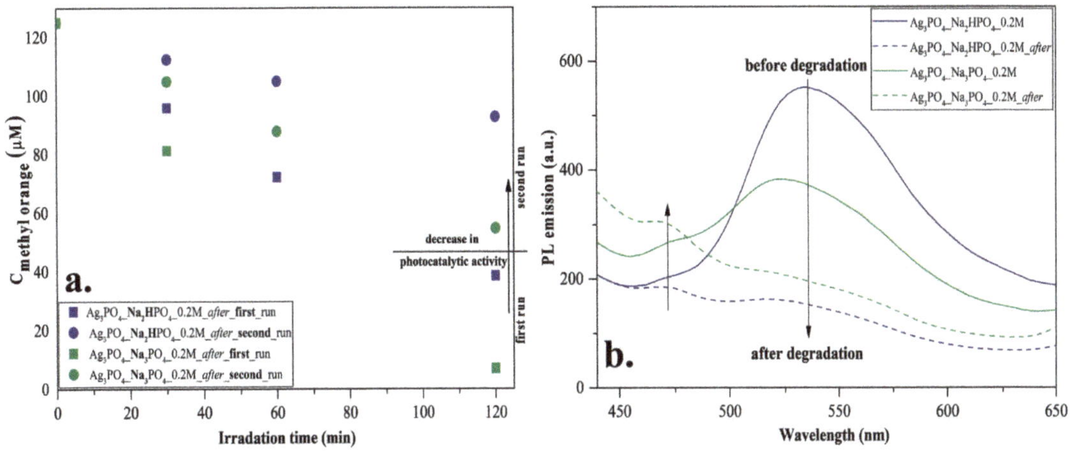

Figure 8. Stability measurements for Ag_3PO_4_Na_2HPO_4_0.2M and Ag_3PO_4_Na_3PO_4_0.2M: (**a**) reusability test with MO degradation under visible light irradiation and (**b**) photoluminescence spectra at 325 nm excitation.

Considering all the results presented above, we proposed a possible charge transfer mechanism (Figure 9). Before elaborating on the mechanism, we summarized the main conclusions that are indispensable for understanding the mechanism as follows:

- Besides the Ag_3PO_4, only $Ag_4P_2O_7$ was observed after the synthesis. Typical bands for surface oxygen vacancies and defects were also observed in the PL spectra.
- Two different pathways were observed using MO as a model pollutant, and the kinetics parameters changed after the first hour.
- The deposition of Ag nanoparticles was observed after MO degradation, which resulted in the lack of surface oxygen vacancies and defects.

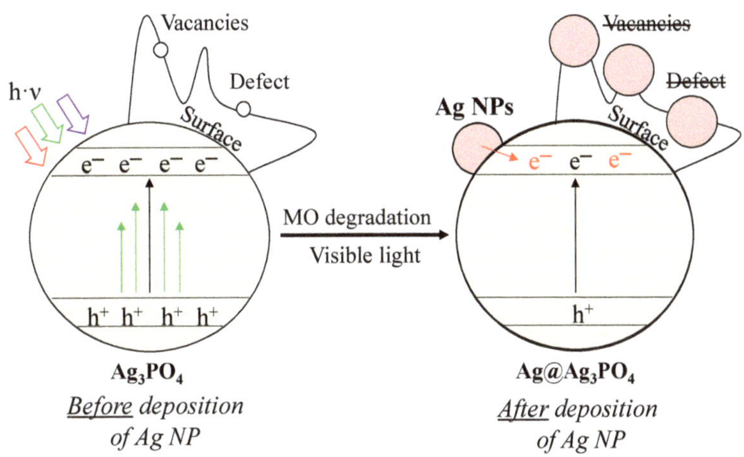

Figure 9. Proposed degradation mechanism of MO with Ag_3PO_4 photocatalysts.

Defects and vacancies are essential for the degradation of MO. The degradation of MO occurs by reactive radicals/h^+ formed in the valence bands of materials. After irradiation, structural changes were observed in our materials. We assume that Ag nanoparticles (Figure 9) overlap with or replace vacancies and defects (*demonstrated by PL measurements;* Figure 8b). These nanoparticles change the reaction mechanism, which was confirmed by the change in the kinetic parameters (*from 0th→1st and 2nd order reaction;* Table 2). This assumption was confirmed by the typical band observed in the PL spectra, attributed to charge transfer and self-trapping (Figure 8b). This also caused fewer holes to form. Thus, the charge transfer between Ag nanoparticles and Ag_3PO_4 occurs by utilizing their localized surface plasmon resonance effect (Figure 9). Thereby, it can be deduced that the formed Ag nanoparticles do not modify the photocatalysts; they only change the mechanism of MO degradation by promoting the transfer of electrons to the semiconductor conduction band. Consequently, the formation of holes is no longer favored. The lack of charge carrier holes also results in a lower photocatalytic efficiency, supported by the MO degradation rate slowing down during the 2-h-long experiment. This is also supported by the lower MO degradation in the second run of the reusability tests. In addition, the formation of Ag NPs could also degrade the Ag_3PO_4 and could damage the semiconductor via Ag^+ ions being reduced into Ag NPs. These are observable by the change in the DR spectra of the samples before (Figure 4a) and after (Figure 7b) the degradation.

Although Ag nanoparticles were formed on the surface, they participated in electron transfer (transferring electrons to the conduction band of the semiconductor) processes, which did not promote the degradation of MO.

4. Conclusions

This paper examined the effect of different phosphate sources on the synthesis and photocatalytic activity of Ag_3PO_4. We proved that the formation of $Ag_4P_2O_7$ depends on the nature of the phosphate source. The type of phosphate sources influenced the light absorption properties and photocatalytic activity of the samples. We concluded that $Ag_4P_2O_7$ inhibits the photocatalytic activity of Ag_3PO_4. In addition to other similar publications in the literature, we also investigated the stability and reusability of Ag_3PO_4. We concluded that Ag species were formed on the Ag_3PO_4, which resulted in a slightly lower methyl orange degradation during the reusability processes. The difference could be attributed to the localized surface plasmon resonance of Ag nanoparticles, promoting the transfer of electrons within the semiconductor and preventing hole formation. This fact was supported by PL measurements. Considering the characterization results obtained before and after the photocatalytic tests, we concluded that Ag_3PO_4-based materials could be reliably used for the degradation of MO as they mostly retain their photoactivity during second recycling test.

Supplementary Materials: The following supporting information can be downloaded at: https://www.mdpi.com/article/10.3390/nano13010089/s1, Figure S1: The particle size distribution of Ag_3PO_4 materials; Figure S2: The adsorption test of the Ag_3PO_4_Na_3PO_4_0.2M and Ag_3PO_4_Na_3PO_4_0.1M on MO (C = 125 μM).

Author Contributions: Conceptualization, Z.-R.T., G.K., Z.P. and D.D.; methodology, D.D., T.G. and K.M.; formal analysis, G.K.; investigation, M.T. and M.F.; resources, L.B. and K.H.; writing—original draft preparation, D.D., Z.-R.T. and G.K.; writing—review and editing L.B., K.H., I.S., G.K., Z.P., K.M. and T.G.; visualization, Z.-R.T.; supervision, K.H. and Z.P.; project administration, K.H., G.K., K.M. and Z.P. All authors have read and agreed to the published version of the manuscript.

Funding: K. Magyari: Zs. Pap and G. Kovács acknowledge the financial support of the Bolyai János fellowship.

Institutional Review Board Statement: Not relevant.

Informed Consent Statement: Not applicable.

Data Availability Statement: Not applicable.

Conflicts of Interest: The authors declare no conflict of interest.

References

1. al Kausor, M.; sen Gupta, S.; Chakrabortty, D. Ag_3PO_4-Based Nanocomposites and Their Applications in Photodegradation of Toxic Organic Dye Contaminated Wastewater: Review on Material Design to Performance Enhancement. *J. Saudi Chem. Soc.* **2020**, *24*, 20–41. [CrossRef]
2. Ohtani, B.; Prieto-Mahaney, O.O.; Li, D.; Abe, R. What Is Degussa (Evonic) P25? Crystalline Composition Analysis, Reconstruction from Isolated Pure Particles and Photocatalytic Activity Test. *J. Photochem. Photobiol. A Chem.* **2010**, *216*, 179–182. [CrossRef]
3. Feng, C.; Li, G.; Ren, P.; Wang, Y.; Huang, X.; Li, D. Effect of Photo-Corrosion of Ag_2CO_3 on Visible Light Photocatalytic Activity of Two Kinds of Ag_2CO_3/TiO_2 Prepared from Different Precursors. *Appl. Catal. B* **2014**, *158–159*, 224–232. [CrossRef]
4. Li, J.; Fang, W.; Yu, C.; Zhou, W.; Zhu, L.; Xie, Y. Ag-Based Semiconductor Photocatalysts in Environmental Purification. *Appl. Surf. Sci.* **2015**, *358*, 46–56. [CrossRef]
5. Yi, Z.; Ye, J.; Kikugawa, N.; Kako, T.; Ouyang, S.; Stuart-Williams, H.; Yang, H.; Cao, J.; Luo, W.; Li, Z.; et al. An Orthophosphate Semiconductor with Photooxidation Properties under Visible-Light Irradiation. *Nat. Mater.* **2010**, *9*, 559–564. [CrossRef]
6. Ma, X.; Lu, B.; Li, D.; Shi, R.; Pan, C.; Zhu, Y. Origin of Photocatalytic Activation of Silver Orthophosphate from First-Principles. *J. Phys. Chem. C* **2011**, *115*, 4680–4687. [CrossRef]
7. Huang, G.F.; Ma, Z.L.; Huang, W.Q.; Tian, Y.; Jiao, C.; Yang, Z.M.; Wan, Z.; Pan, A. Ag3PO4 Semiconductor Photocatalyst: Possibilities and Challenges. *J. Nanomater.* **2013**, *2013*, 371356. [CrossRef]
8. Ng, H.N.; Calvo, C.; Faggiani, R. A New Investigation of the Structure of Silver Orthophosphate. *Acta Cryst.* **1978**, *B34*, 898–899. [CrossRef]
9. Song, X.; Li, R.; Xiang, M.; Hong, S.; Yao, K.; Huang, Y. Morphology and Photodegradation Performance of Ag_3PO_4 Prepared by $(NH_4)_3PO_4$, $(NH_4)_2HPO_4$ and $NH_4H_2PO_4$. *Ceram. Int.* **2017**, *43*, 4692–4701. [CrossRef]
10. Amornpitoksuk, P.; Intarasuwan, K.; Suwanboon, S.; Baltrusaitis, J. Effect of Phosphate Salts (Na_3PO_4, Na_2HPO_4, and NaH_2PO_4) on Ag_3PO_4 Morphology for Photocatalytic Dye Degradation under Visible Light and Toxicity of the Degraded Dye Products. *Ind. Eng. Chem. Res.* **2013**, *52*, 17369–17375. [CrossRef]
11. Qin, L.; Tao, P.; Zhou, X.; Pang, Q.; Liang, C.; Liu, K.; Luo, X. Synthesis and Characterization of High Efficiency and Stable Spherical Ag_3PO_4 Visible Light Photocatalyst for the Degradation of Methylene Blue Solutions. *J. Nanomater.* **2015**, *2015*, 258342. [CrossRef]
12. Yang, Z.M.; Tian, Y.; Huang, G.F.; Huang, W.Q.; Liu, Y.Y.; Jiao, C.; Wan, Z.; Yan, X.G.; Pan, A. Novel 3D Flower-like Ag_3PO_4 Microspheres with Highly Enhanced Visible Light Photocatalytic Activity. *Mater. Lett.* **2014**, *116*, 209–211. [CrossRef]
13. Raudoniene, J.; Skaudzius, R.; Zarkov, A.; Selskis, A.; Karlsson, O.; Kareiva, A.; Garskaite, E. Wet-Chemistry Synthesis of Shape-Controlled Ag_3PO_4 Crystals and Their 3D Surface Reconstruction from SEM Imagery. *Powder Technol.* **2019**, *345*, 26–34. [CrossRef]
14. Krungchanuchat, S.; Ekthammathat, N.; Phuruangrat, A.; Thongtem, S.; Thongtem, T. High UV-Visible Photocatalytic Activity of Ag_3PO_4 Dodecahedral Particles Synthesized by a Simple Hydrothermal Method. *Mater. Lett.* **2017**, *201*, 58–61. [CrossRef]
15. Dong, P.; Wang, Y.; Li, H.; Li, H.; Ma, X.; Han, L. Shape-Controllable Synthesis and Morphology-Dependent Photocatalytic Properties of Ag_3PO_4 Crystals. *J. Mater. Chem. A Mater.* **2013**, *1*, 4651–4656. [CrossRef]
16. Bi, Y.; Hu, H.; Jiao, Z.; Yu, H.; Lu, G.; Ye, J. Two-Dimensional Dendritic Ag_3PO_4 Nanostructures and Their Photocatalytic Properties. *Phys. Chem. Chem. Phys.* **2012**, *14*, 14486–14488. [CrossRef]
17. Dhanabal, R.; Chithambararaj, A.; Velmathi, S.; Bose, A.C. Visible Light Driven Degradation of Methylene Blue Dye Using Ag_3PO_4. *J. Environ. Chem. Eng.* **2015**, *3*, 1872–1881. [CrossRef]
18. Tóth, Z.R.; Feraru, A.; Debreczeni, D.; Todea, M.; Popescu, R.A.; Gyulavári, T.; Sesarman, A.; Negrea, G.; Vodnar, D.C.; Hernadi, K.; et al. Influence of Different Silver Species on the Structure of Bioactive Silicate Glasses. *J. Non Cryst. Solids* **2022**, *583*, 121498. [CrossRef]
19. Tauc, J. Optical Properties and Electronic Structure of Amorphous Ge and Si. *Mat. Res. Bull.* **1968**, *3*, 37–46. [CrossRef]
20. da Pereira, W.S.; Gozzo, C.B.; Longo, E.; Leite, E.R.; Sczancoski, J.C. Investigation on the Photocatalytic Performance of $Ag_4P_2O_7$ Microcrystals for the Degradation of Organic Pollutants. *Appl. Surf. Sci.* **2019**, *493*, 1195–1204. [CrossRef]
21. Huang, K.; Lv, Y.; Zhang, W.; Sun, S.; Yang, B.; Chi, F.; Ran, S.; Liu, X. One-Step Synthesis of Ag_3PO_4/Ag Photocatalyst with Visible-Light Photocatalytic Activity. *Mater. Res.* **2015**, *18*, 939–945. [CrossRef]
22. Zhang, W.; Zhou, L.; Shi, J.; Deng, H. Synthesis of $Ag_3PO_4/g\text{-}C_3N_4$ Composite with Enhanced Photocatalytic Performance for the Photodegradation of Diclofenac under Visible Light Irradiation. *Catalysts* **2018**, *8*, 45. [CrossRef]
23. Zhang, M.; Li, Y.; Shi, Y.; Wang, Q.; Wang, A.; Gao, S. Effect of Blending Manner on Composition and Photocatalytic Performance of Ag/Ag3PO4/Ag4P2O7 Composites via an in-Situ Reduction-Precipitation Route. *Inorg. Chem. Commun.* **2021**, *130*, 108675. [CrossRef]
24. Nie, C.; Dong, J.; Sun, P.; Yan, C.; Wu, H.; Wang, B. An Efficient Strategy for Full Mineralization of an Azo Dye in Wastewater: A Synergistic Combination of Solar Thermo- and Electrochemistry plus Photocatalysis. *RSC Adv.* **2017**, *7*, 36246–36255. [CrossRef]
25. Putri, R.A.; Safni, S.; Jamarun, N.; Septiani, U. Kinetics Study and Degradation Pathway of Methyl Orange Photodegradation in the Presence of C-N-Codoped TiO_2 Catalyst. *Egypt. J. Chem.* **2020**, *63*, 563–575. [CrossRef]

26. Rodríguez-Cabo, B.; Rodríguez-Palmeiro, I.; Corchero, R.; Rodil, R.; Rodil, E.; Arce, A.; Soto, A. Photocatalytic Degradation of Methyl Orange, Methylene Blue and Rhodamine B with AgCl Nanocatalyst Synthesised from Its Bulk Material in the Ionic Liquid [P6 6 6 14]Cl. *Water Sci. Technol.* **2017**, *75*, 128–140. [CrossRef]
27. Hou, M.; Li, F.; Liu, X.; Wang, X.; Wan, H. The Effect of Substituent Groups on the Reductive Degradation of Azo Dyes by Zerovalent Iron. *J. Hazard. Mater.* **2007**, *145*, 305–314. [CrossRef]
28. Székely, I.; Baia, M.; Magyari, K.; Boga, B.; Pap, Z. The Effect of the PH Adjustment upon the WO3-WO3·0.33H_2O-TiO_2 Ternary Composite Systems' Photocatalytic Activity. *Appl. Surf. Sci.* **2019**, *490*, 469–480. [CrossRef]
29. Tóth, Z.R.; Hernadi, K.; Baia, L.; Kovács, G.; Pap, Z. Controlled Formation of Ag-AgxO Nanoparticles on the Surface of Commercial TiO_2 Based Composites for Enhanced Photocatalytic Degradation of Oxalic Acid and Phenol. *Catal. Today* **2020**. [CrossRef]
30. Tóth, Z.R.; Pap, Z.; Kiss, J.; Baia, L.; Gyulavári, T.; Czekes, Z.; Todea, M.; Magyari, K.; Kovács, G.; Hernadi, K. Shape Tailoring of AgBr Microstructures: Effect of the Cations of Different Bromide Sources and Applied Surfactants. *RSC Adv.* **2021**, *11*, 9709–9720. [CrossRef]
31. Batvandi, M.; Haghighatzadeh, A.; Mazinani, B. Synthesis of Ag_3PO_4 Microstructures with Morphology-Dependent Optical and Photocatalytic Behaviors. *Appl. Phys. A Mater. Sci. Process.* **2020**, *126*, 571. [CrossRef]
32. Botelho, G.; Andres, J.; Gracia, L.; Matos, L.S.; Longo, E. Photoluminescence and Photocatalytic Properties of Ag_3PO_4 Microcrystals: An Experimental and Theoretical Investigation. *Chempluschem* **2016**, *81*, 202–212. [CrossRef] [PubMed]

nanomaterials

Article

Synthesis of CuO/GO-DE Catalyst and Its Catalytic Properties and Mechanism on Ciprofloxacin Degradation

Ting Zhang *, Jingjing Zhang, Yinghao Yu, Jinxu Li, Zhifang Zhou and Chunlei Li *

School of Petrochemical Engineering, Lanzhou University of Technology, Lanzhou 730050, China
* Correspondence: zhangting@lut.edu.cn (T.Z.); licl@lut.edu.cn (C.L.)

Abstract: A new catalyst, copper oxide/graphene oxide–diatomaceous earth (CuO/GO-DE), was prepared by the ultrasonic impregnation method. The optimal conditions for catalyst preparation were explored, and its structure and morphology were characterized by BET, XRD, SEM, TEM, FTIR, Raman and XPS. By taking ciprofloxacin as the target pollutant, the performance and reusability of CuO/GO-DE to degrade antibiotic wastewater was evaluated, and the optimal operating conditions were obtained. The main oxidizing substances in the catalytic system under different pH conditions were analyzed, as well as the synergistic catalytic oxidation mechanism. The intermediate products of ciprofloxacin degradation were identified by LC-MS, and the possible degradation process of ciprofloxacin was proposed.

Keywords: diatomaceous earth; graphene oxide; copper oxide; ciprofloxacin; catalytic mechanism

Citation: Zhang, T.; Zhang, J.; Yu, Y.; Li, J.; Zhou, Z.; Li, C. Synthesis of CuO/GO-DE Catalyst and Its Catalytic Properties and Mechanism on Ciprofloxacin Degradation. *Nanomaterials* **2022**, *12*, 4305. https://doi.org/10.3390/nano12234305

Academic Editor: Zoltán Kónya

Received: 31 October 2022
Accepted: 2 December 2022
Published: 4 December 2022

1. Introduction

Ciprofloxacin (CIP), belonging to fluoroquinolones, is a common antibiotic [1,2], which is usually used to treat human or animal diseases, such as inhibiting the development of malignant cells or preventing humans or animals from being infected by bacteria [3–6]. However, due to the abuse of antibiotics [7] and the incomplete metabolism of antibiotics [8], CIP with a high concentration of about 10 μg/L has been detected in groundwater systems [9]. In effluents of hospitals and drug production facilities, the concentration of ciprofloxacin wastewater is as high as 150 μg/L, even more than 10 mg/L [10]. Antibiotic wastewater pollution has become very serious [11], which may bring significant harm to human health and the ecosystem [12], so it is urgent to explore efficient methods to treat ciprofloxacin wastewater [13]. At present, many treatment technologies have been developed, such as adsorption [14], biodegradation [15], advanced oxidation [16], etc. Heterogeneous catalysis (one of the advanced oxidation processes), as an efficient and environmentally friendly technology, has attracted much attention. Heterogeneous catalysts can catalyze H_2O_2 or H_2O and generate radicals with strong oxidation ability, such as ·OH, $\cdot O_2^-$, and oxidize and decompose refractory organics to simple organics, even carbon oxide [5,9,11,16].

Diatomaceous earth (DE) is a biogenic siliceous sedimentary rock with active silica, light weight, and porous characteristics. The pores are regularly distributed, and the pore size ranges from more than ten to a hundred nanometers. As an important nonmetallic mineral [17,18], diatomite has become a good adsorbent or catalyst carrier for functional materials [19–21] due to its large specific surface area, strong adsorption capacity, low price and rich reserves. Shen [22] et al. applied diatomite modified with NaOH and $MnCl_2$ to adsorb heavy metal ion Cd(II) in wastewater, and the adsorption ratio of Cd(II) reached 98.69%. Xiong [23] et al. prepared Ag_3PO_4/Fe_3O_4/DE composites by one-step hydrothermal method and studied their photocatalytic properties, and the results showed that when the content of Fe_3O_4 was 8%, the degradation ratio of pollutants reached 98%. Sun [24] et al. prepared porous $MnFe_2O_4$/diatomite (MFD) material by solvothermal method and explored the degradation effects of tetracycline hydrochloride (TC-HCl) by

MFD, and the results showed that the material had a good catalytic effect on TC degradation at pH ranging from 3 to 11. Dang [25] et al. prepared a diatomite composite coated with amorphous manganese oxide (MnO_2) in acidic conditions, which showed high efficiency in the degradation of methylene blue and methyl orange. Wang [26] et al. mixed g-C_3N_4 with diatomite to prepare CN/DE-10 composite and explored its degradation ability on Rhodamine B (RhB), which can be effectively degraded within 50 min. Daniel [27] et al. prepared spherical diatomite Fenton catalyst, and the degradation ratio of clindamycin reached 89.7% at pH 3.

Graphene oxide (GO), first discovered in 1859, is an ideal single-layer amorphous two-dimensional material [28,29]. As a derivative of graphene, it retains the carbon skeleton and the same properties of graphene [30], while it is easier to prepare than graphene [31]. Therefore, GO is widely used in many fields. It has been proven to have practical and efficient applications in the fields of environmental catalysis [32], functional materials [33] and conductivity [34,35]. Hassan [36] et al. prepared RGO/SZ material by loading silica and zirconia on reduced graphene oxide and explored its degradation ability of bisphenol A (BPA). The results showed that under optimal conditions, the degradation ratio for BPA reached 88%. Liu [37] et al. prepared GO-SH/DE composite adsorption material through the combination of diatomite (DE) and thiolated graphene oxide (GO-SH), and the adsorption capacity of the material for patulin (PAT) in apple juice was explored. Under the best conditions, the adsorption capacity for PAT reached 90%. Metal oxides (Fe_2O_3, CeO_2, CuO, etc.) can be employed as active components coated on catalyst supports because they have similar catalytic effects to Fe^{2+} or Cu^{2+} [38,39]; moreover, they are not easy to lose and can be used in a wide pH range. In this study, diatomaceous earth (DE) was modified by graphene oxide (GO) to improve the specific surface area and enhance surface capacity, and then the modified diatomite carrier was loaded with different active metal salts separately to screen the best active component (metal oxide) for CIP degradation. An optimized CuO/GO-DE catalyst was prepared, and the degradation performance and mechanism of ciprofloxacin (CIP) were studied.

2. Experimental Section

2.1. Materials

Diatomite was purchased from Aladdin Company (Los Angeles, CA, USA), with a silica content of 90%, alumina content of 3%, and magnesium oxide and calcium oxide content of 0.5%; Nano graphite powder and ciprofloxacin were also purchased from Aladdin company. Iron nitrate, copper nitrate, nickel nitrate, cerium nitrate, silver nitrate and sodium nitrate were purchased from Tianjin Kaixin Chemical Co., Ltd. (Tianjin, China) Potassium permanganate, concentrated hydrochloric acid, concentrated sulfuric acid and hydrogen peroxide were purchased from Shanghai Sinopharm Chemical Reagent Co., Ltd. (Shanghai, China). All chemical reagents were analytically pure and had not been further purified before use. In this study, the experimental water is distilled water, which is self-made by our laboratory.

2.2. Synthesis of Metal-Loaded GO-DE Composites

Graphene oxide (GO) was prepared by the Hummers method [38,40]. A total of 1 g graphite powder and 1 g sodium nitrate were added into 60 mL concentrated sulfuric acid and stirred vigorously in a low-temperature water bath, then 6 g potassium permanganate was slowly added and stirred for 3 h. The reaction mixture was put into a water bath with a temperature of 35 °C and continuously stirred for 3 h, then 140 mL of deionized water was added and stirred at room temperature for 12 h. A total of 200 mL of deionized water was added into the mixed solution, then 20 mL of hydrogen peroxide was slowly added, stirred for 1 h and stood for 5 h. After the product was settled, it was washed with dilute hydrochloric acid, then washed with deionized water to neutral, and ultrasonic for 2 h at 100 Hz to obtain a viscous graphene oxide solution.

Some distilled water was added to the viscous GO solution, and GO dispersion liquid was obtained by 100 Hz ultrasound. Different dosages of purified DE were added to the above GO dispersion liquid; the mixture was magnetically stirred and then centrifuged. After pouring out the supernatant, the pasty clay was fully stirred again and then dried in an oven. The sample was then ground and sieved by a 100-mesh sieve, and a GO-DE sample was obtained.

Metal-loaded GO-DE composite was prepared by ultrasonic dipping method. Five different metal salts (copper nitrate, iron nitrate, nickel nitrate, cerium nitrate and silver nitrate) were used as predecessors. They were dissolved in distilled water, respectively, and the prepared GO-DE was put into a different solution and ultrasonic impregnated for 30 min with 0.1 M NaOH solution added. Then the sample was filtered, separated, washed with distilled water 5 times, and dried at 110 °C for 2 h to obtain five different metal-loaded GO-DE composites, which were used as heterogeneous catalysts in our study.

2.3. Characterization

The specific surface area, pore volume and pore diameter of the material were measured at −196 °C with ASAP 2010 specific surface area meter produced by American Micromitrics Company (Atlanta, GA, USA). The surface morphology of the samples was observed by JEOL JSM-6701F scanning electron microscope (SEM) with an accelerating voltage of 20 kV. The microstructures were investigated by JEOL JEM-1200EX (JEOL Ltd., Tokyo, Japan) transmission electron microscope (TEM). Before SEM and TEM tests, the samples need to be ground and ultrasonically dispersed in an ethanol solution. The content and mapping of the main elements of the materials were tested and analyzed by energy spectrometer (EDS) from Brooke Technology Co., Ltd., (Beijing, China) Nicolet AVTAR 360 FT-IR infrared spectrometer (Thermo Electron Co., Waltham, MA, USA) was used to test the chemical structure, including the changes in the compositional or functional group of samples, with the scanning range from 4500 cm^{-1} to 400 cm^{-1}. The molecular structures of the samples were tested by a PERS-SR532 Raman spectrometer (Persetech Company, Xiamen, China). Crystal structures of the samples were identified by Panalytical X'Pert PRO X-ray diffractometer. The working voltage and current were 40 kV and 150 mA, the diffraction angle was 5° to 80°, and the scanning step was 0.02°/s. The PHI5702 X-ray photoelectron spectrometer (American Physical Electronics Company, Chanhassen, MN, USA) was used to quantitatively and qualitatively analyze the elements on the material surface and the chemical valence state and valence electron state of the elements. The X-ray emission sources are Mg, Al double anode target and Al monochromator target.

2.4. CIP Degradation

A total of 250 mg/L CIP stock solution was prepared for later use. Additionally, 20 mL CIP stock solution was taken in a conical flask and 80 mL distilled water was added to obtain 100 mL 50 mg/L CIP solution. The pH value of the solution was adjusted with sulfuric acid (1:9) or NaOH (0.1 mol/L), and a certain amount of catalyst and hydrogen peroxide were added into the conical flask. The conical flask was then placed in a thermostatic oscillator with a water bath at the speed of 120 rpm. The supernatant was taken at regular intervals to measure the absorbance by UV-1900 ultraviolet-visible spectrophotometer at λ of 277 nm, at which CIP has the maximum absorption peak. The degradation ratio of the CIP solution can be calculated by the following formula: $\eta = (C_0 - C_t)/C_0 \times 100\%$, in which C_0 is the initial concentration of CIP solution, and C_t is the concentration of CIP solution with the reaction time t.

2.5. Measurement of ·OH Concentration and ·OH Scavenging Experiment

·OH reacts with salicylic acid to produce 2,3-dihydroxybenzoic acid [39], which has the maximum absorption peak at the wavelength of 510 nm. ·OH concentration can be measured by the following methods: 100 mL salicylic acid solution with a certain concentration was put into a 250 mL conical flask, and 1 g/L catalyst and 0.1 mL hydrogen peroxide were added into it. The conical flask was put into a water bath at 60 °C. With the reaction happening, the ·OH concentration of the solution can be measured at regular intervals by UV-1900 ultraviolet-visible spectrophotometer at 510 nm.

Tert-butyl alcohol (TBA) is a strong ·OH scavenger [41], which can effectively inhibit the occurrence of the Fenton reaction. The CIP solutions (100 mL, 50 mg/L) were prepared under pH 4, 7 and 10 separately, and the catalyst, hydrogen peroxide and excessive TBA were also added to the solution. The control group was set, the conical flask was put into a water bath at 60 °C for reaction, and the degradation ratios of CIP were tested, respectively, when the reactions were all balanced.

2.6. *Capture of Cu(III) in Alkaline Condition*

According to the literature reports, when periodate is added to the alkaline reaction system, it can form Cu(III)—periodate complex to stabilize the unstable Cu(III) in the reaction system [42,43]. Based on the experimental scheme of Li et al. [43], periodate was added to the alkaline catalytic reaction system containing CuO/GO-DE compound material, hydrogen peroxide and NaOH, and the control group was set. After reacting in a water bath thermostatic oscillator for 5 min, the water sample was taken out and filtered through a 0.22 µm organic filter membrane, and the absorbance of the filtered solution was measured by UV-1900 ultraviolet-visible spectrophotometer at 415 nm.

3. Result and Discussion

3.1. *Optimization of Catalyst Preparation*

Diatomite (DE) does not have a high specific surface area and good adsorption capacity as activated carbon, and it cannot be used as an absorbent directly. In order to take good advantage of DE, graphene oxide (GO) was introduced into diatomite to improve its adsorption property. The removal efficiencies of CIP by diatomite modified with different GO proportions were compared, and the experimental conditions were as follows: catalyst dosage of 2 g/L, pH value of 7, reaction temperature of 60 °C, CIP initial concentration of 50 mg/L, the results were shown in Figure 1a. It can be found that the removal ratios of CIP rise when GO proportions increase from 3% to 10%. Therefore, the more GO is introduced into DE, the higher the adsorption efficiency of the material. As we know, The preparation of GO is relatively complex, costly and difficult to produce in large quantities; considering the low cost and time saving, the proportions of GO around 7~10% integrated with DE was better for this composite support's preparation.

The activities of different metal oxides loaded on GO-DE and their catalytic properties were also studied. CuO, Fe_2O_3, NiO_2, Ag_2O and CeO_2, which are five types of metal oxides (metal nitrates were used as the precursors), were selected as active substances and coated on GO-DE support; the CIP degradation capacities of these catalytic composite materials can be seen in Figure 1b. According to Figure 1b, the GO-DE support loaded with CuO or Ag_2O have the best degradation abilities of CIP, while Ag or Ag salt are all not cheap. From economic considerations, copper nitrate is the best one for GO-DE-based catalyst preparation. Not only the type of metal oxides but also the loading amount of it on the catalyst surface influences the catalytic efficiency of the catalyst. Different concentrations of precursor $Cu(NO_3)_2$ solutions were used to prepare the catalysts, which were applied for CIP degradation. It can be seen from Figure 1c that the different precursor concentration for the catalyst preparation has different catalytic performance on CIP degradation, and when using 5% $Cu(NO_3)_2$ precursor concentration to prepare the catalyst, the CIP degradation ratio is the highest.

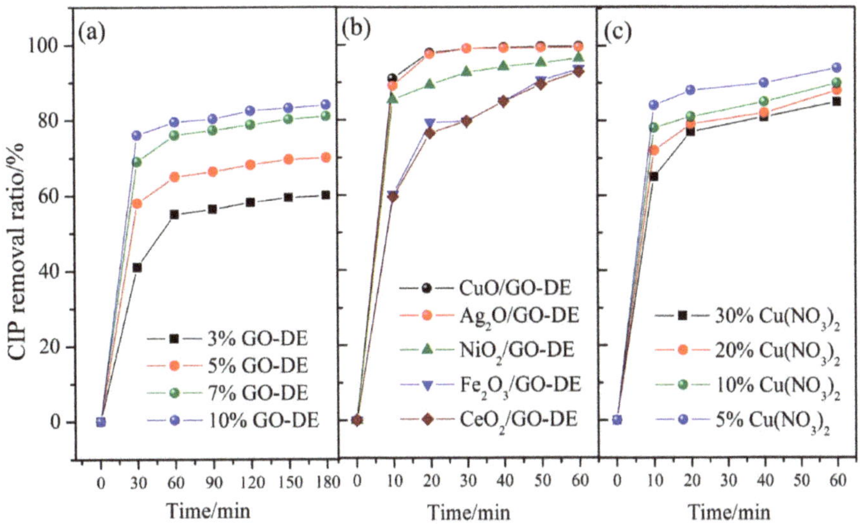

Figure 1. (**a**) Influence of proportions of GO integrated with DE on CIP removal ratio; (**b**) influence of metal type loaded on GO-DE on CIP removal ratio; (**c**) influence of precursor concentration on CIP removal ratio.

3.2. Characterization of CuO/GO-DE

In order to observe the surface morphology of the catalyst and verify the successful loading of copper oxide on GO-DE support, BET test, SEM, XRD, FT-IR and XPS were used to analyze it.

Table 1 shows the comparison of BET-specific surface area measurement results of DE, GO-DE and CuO/GO-DE. It can be seen from Table 1 that DE does not have a high specific surface area, while adding GO into DE can greatly improve the specific surface area of DE, nearly double that of DE. The specific surface area of CuO/GO-DE is reduced a little due to the metal loading. Total pore volume and average pore diameter have changed but not too much when introducing GO and CuO into DE.

Table 1. BET measurement results of DE, GO-DE and Cu/GO-DE.

Samples	BET Specific Surface Area (m^2/g)	Total Pore Volume (cm^3/g)	Average Pore Diameter (μm)
DE	10.7239	0.021179	0.483013
GO-DE	18.7833	0.027155	0.553611
CuO/GO-DE	15.4166	0.025061	0.526992

Figure 2a–c shows the SEM images of DE, GO-DE and CuO/GO-DE, respectively. From Figure 2a, holes with diameters from 0.05 to 0.6 μm on the diatomite surface can be seen clearly. Figure 2b shows that the surface of DE is evenly covered by folded thin-film GO with the holes preserved, and the diameter of the holes is even higher than before (0.05–0.75 μm), which can effectively improve the specific surface area of diatomite. After being coated by CuO, the holes become smaller than before (Figure 2c). Figure 2d shows the TEM diagram of CuO/GO-DE; it can be clearly seen that columnar CuO is embedded in GO. In addition, Figure 2e shows EDS mapping images of CuO/GO-DE composites. It can be seen that the composites contain Si, C, O and Cu elements. Si comes from DE, C, O from GO and DE, and Cu from loaded metal copper salt. It showed that the Cu element was loaded on the composite material GO-DE successfully and uniformly, which plays an important part in the catalytic performance of the prepared catalyst.

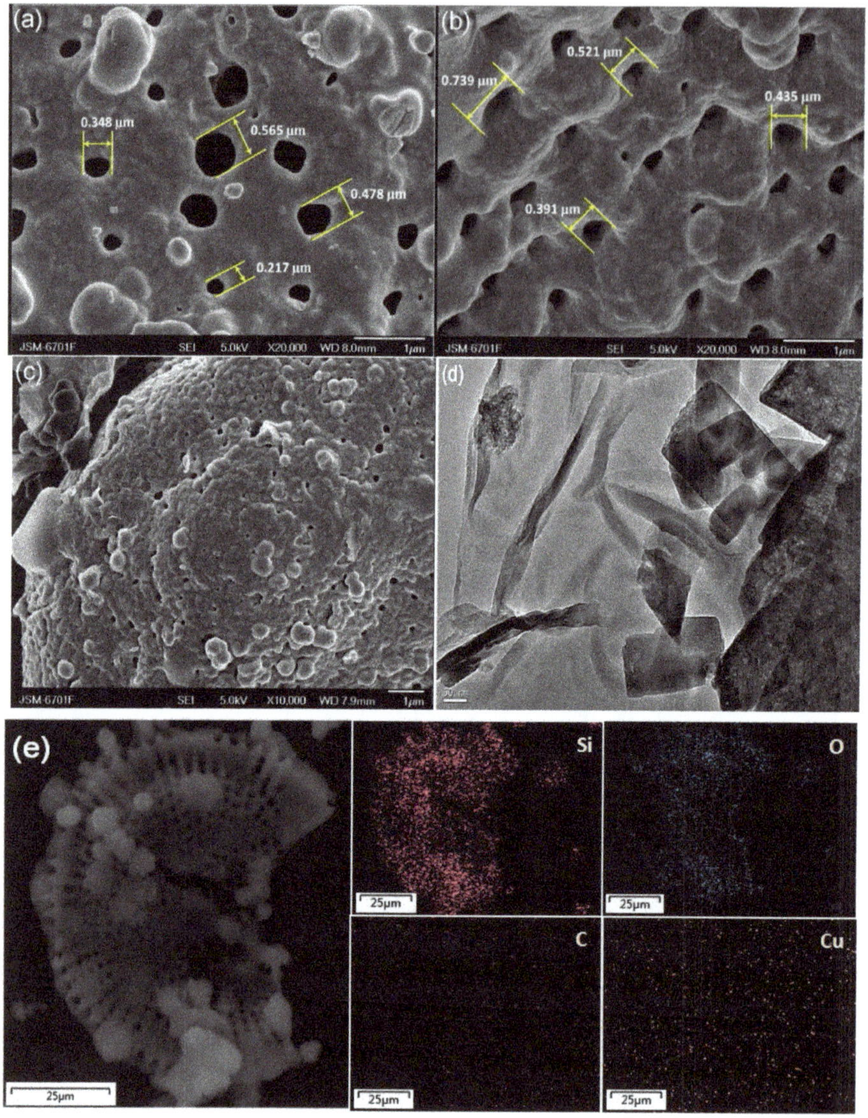

Figure 2. SEM micrographs of (**a**) DE, (**b**) GO-DE, and (**c**) CuO/GO-DE; (**d**) TEM micrograph of CuO/GO-DE; (**e**) EDS mapping images of CuO/GO-DE composites.

The FTIR spectra of GO, DE, GO-DE and CuO/GO-DE are shown in Figure 3a. In the spectra of GO and DE, the broad band absorption peaks at 3440 cm^{-1} are related to the surface hydroxyl stretching vibration of GO and DE, respectively [41,43]. The peak at 1623 cm^{-1} is attributed to the stretching vibration of the C=C bond on GO, and the peak at 1087 cm^{-1} is the stretching vibration of the C-O-C bond on GO [44]. The peak at 1094 cm^{-1} is attributed to the symmetric stretching vibration of the Si-O-Si bond of DE; The peak at 794 cm^{-1} is attributed to the vibration of the Si-OH bond of DE [23,45]. GO-DE and CuO/GO-DE have a C=C bond and C-O-O bond as GO has, which proves the existence of GO in composite materials. In CuO/GO-DE spectrum, there is a weak absorption peak at 620 cm^{-1}, related to the vibration of the Cu(II)-O bond [46].

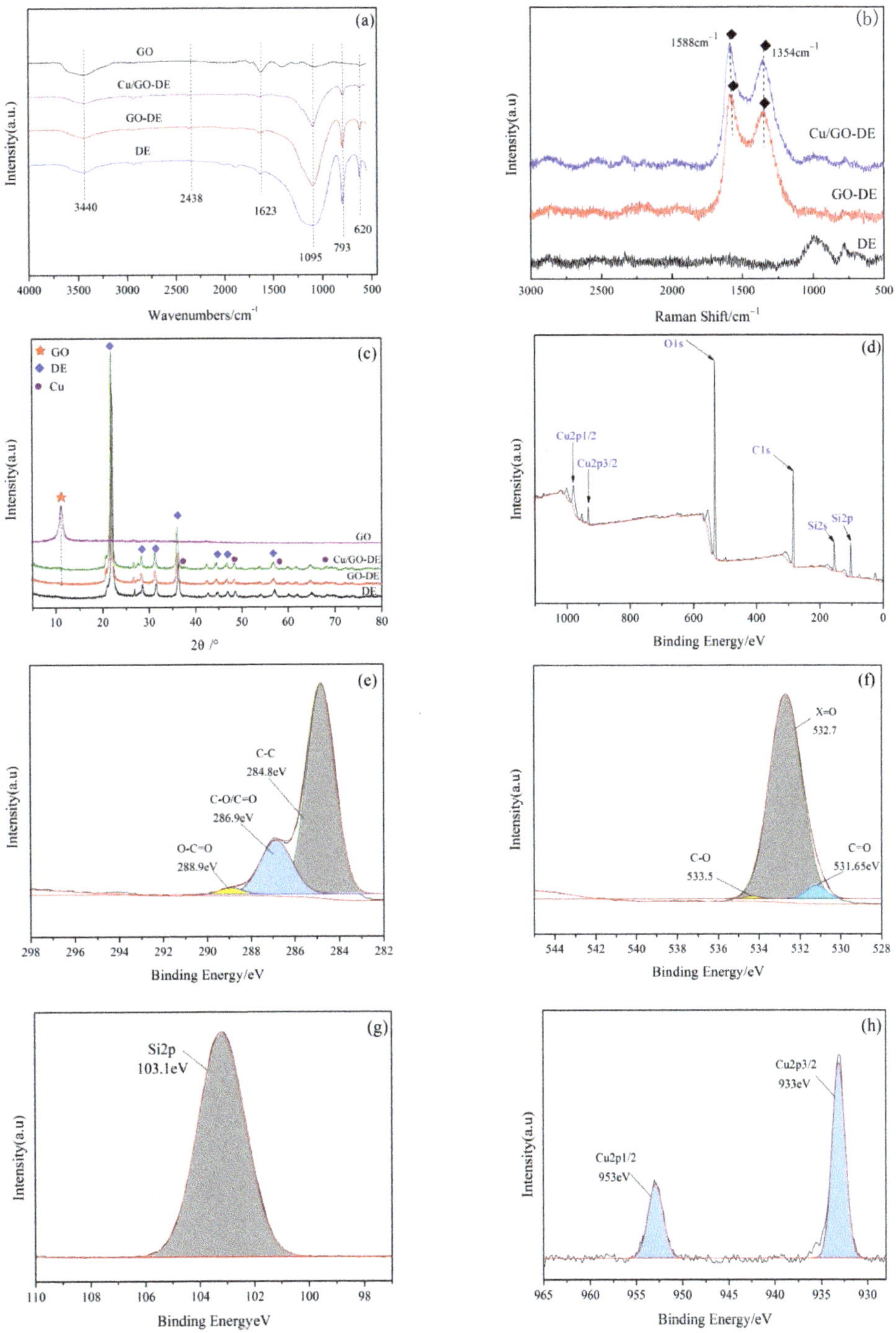

Figure 3. (**a**) FTIR spectra of DE, GO, GO-DE and CuO/GO-DE; (**b**) Raman spectra of DE, GO-DE and CuO/GO-DE; (**c**) XRD patterns of DE, GO, GO-DE and CuO/GO-DE; XPS spectra of CuO/GO-DE: (**d**) total spectrum, (**e**) C 1 s, (**f**) O 1 s, (**g**) Si 2 p and (**h**) Cu 2 p.

Raman spectroscopy can be used to characterize the defects of GO. In general, the Raman spectrum of GO has two characteristic peaks: G and D peaks. The G peak, caused by the first-order scattering of E_{2g} lattice vibration by sp^2 carbon atom hybrid vibration, is around 1580 cm^{-1}; the D peak, caused by sp^3 hybrid carbon atoms, is about 1350 cm^{-1} [47,48]. Figure 3b shows the Raman characterization of DE, GO-DE and CuO/GO-DE, from which it can be clearly seen that there are two characteristic peaks (the D peak at 1354 cm^{-1} and the G peak at 1588 cm^{-1}) of GO on the GO-DE support and CuO/GO-DE composite, GO is successfully integrated with DE.

The XRD patterns of GO, DE, GO-DE and CuO/GO-DE are shown in Figure 3c. The diffraction peaks at $2\theta = 21.80°, 26.04°, 29.01°, 32.56°, 46.78°$ and $57.23°$ correspond to the crystal structure of DE [49,50]. It can be seen that the preparation of CuO/GO-DE did not change the structure of DE. Jade6 XRD software was used to analyze weak diffraction peaks at $2\theta = 39.28°, 41.34°, 58.62°$ and $68.36°$ of CuO/GO-DE sample, which corresponds to 45-0937 card of the standard sample, indicating those peaks are the characteristic diffraction peaks of CuO [40,51]. It was confirmed that CuO was successfully loaded on the GO-DE carrier. GO has a strong characteristic peak at $2\theta = 11.42°$ [52], but the peaks are very weak in the spectra of GO-DE and CuO/GO-DE, which may be due to the small proportion of GO in the composite.

In order to make sure the chemical state of various elements in the sample, the CuO/GO-DE was tested by XPS. The total spectrum and the characteristic spectrums of each element are shown in Figure 3d–h. In the total spectrum (Figure 3d), it can be seen that the peaks near the binding energies of 103 eV and 150 eV correspond to the characteristic peaks of Si 2 p and Si 2 s: the peak with binding energy around 285 eV belongs to the characteristic peak of C 1 s; the peak of binding energy near 532 eV is the characteristic peak of O 1 s; the peaks of binding energy near 933 eV and 953 eV are the characteristic peaks of Cu 2 p3/2 and Cu 2 p1/2, respectively, indicating the existence of divalent copper (CuO) in the catalyst. Figure 3e shows the characteristic peaks at the binding energies of 284.8 eV, 286.9 eV and 288.9 eV corresponding to C-C, C-O/C=O and O-C=O [47,51], respectively. Figure 3f shows the spectrum of O 1 s with the peaks at 531.65 eV and 532.7 eV representing C=O of GO and the peaks at 533.5 eV representing C-O [38,46]. In Figure 3g, the peak at the binding energy of 103.1 eV corresponds to the characteristic peak of Si 2 p, indicating the main components of DE. The spectrum of Cu 2 p in the sample (Figure 3h) has the characteristic peaks of Cu 2 p3/2 and Cu 2 p1/2 at the binding energies of 933 eV and 953 eV, respectively [52], indicating that CuO exists on the surface of the catalyst and the copper is loaded on the material in the form of CuO.

3.3. Degradation of CIP

In order to determine the optimum operating conditions during CIP degradation, a series of single-factor experiments were carried out at different pH values, initial CIP concentration, catalyst dosage, H_2O_2 concentration and reaction temperature.

pH value is an important factor affecting the catalytic performance of CuO/GO-DE. Figure 4a shows the degradation capacity of the catalyst under different pH conditions. It can be seen that the catalytic system achieved high degradation efficiencies at pH ranging from 4 to 10, and at pH of 7, it can reach the highest CIP degradation ratio (99.9% after 240 min of reaction). Under acidic conditions, ·OH generated from catalyzing H_2O_2 by CuO/GO-DE is the main oxidant to oxidize the CIP under neutral and alkaline conditions, although the amount of ·OH reduced due to H_2O_2 decomposing to H_2O and O_2, Cu(I) will be oxidized by H_2O_2 and O_2 to high valence Cu(III) with strong oxidizability [53–56], which plays an important role as well as ·OH does in the degradation of CIP.

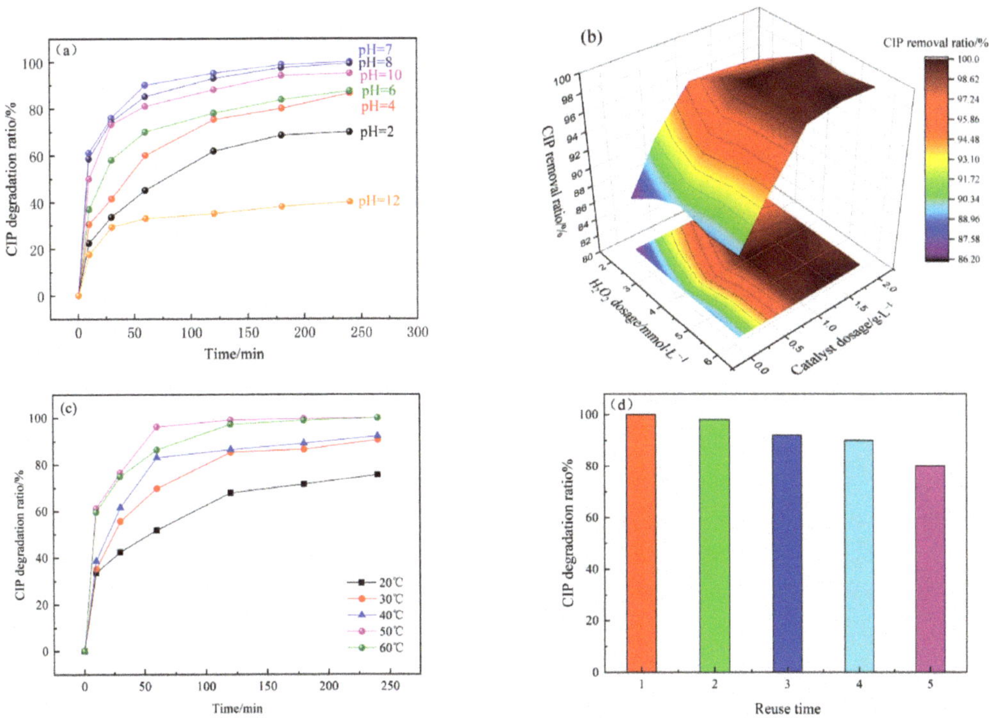

Figure 4. Effects of pH value (**a**), catalyst dosage and H_2O_2 concentration (**b**); temperature (**c**) on the degradation of CIP by CuO/GO-DE catalyst; the reusability of CuO/GO-DE catalyst (**d**).

The results fully showed that the catalyst could degrade ciprofloxacin in a wide pH range with H_2O_2. The traditional Fenton oxidation method is applicable to a very narrow pH range of 2.5~3.5. However, the pH of most organic wastewater is near neutral. The CuO/GO-DE catalyst has great advantages over the traditional Fenton catalyst.

When the concentration of pollutants is certain, the relationship between catalyst dosage and H_2O_2 dosage is mutual restriction. It can be seen from Figure 4b of the experimental results that it is not true; the more catalyst or H_2O_2, the better the catalytic effect. There should be a certain proportional relationship between the two factors. In other words, the active site (coated CuO) of the catalyst and H_2O_2 should be in a certain proportion in order to achieve the optimal catalytic effect. The experimental and calculation results showed that the catalytic effect is the best when the mass ratio of the catalyst to H_2O_2 is 0.133.

As can be seen from Figure 4c, with the increase in reaction temperature, the degradation ratio of ciprofloxacin gradually increased, and the degradation rate and efficiency reached the maximum at 50 °C. When the reaction temperature was increased to 60 °C, the degradation efficiency did not rise, but the degradation rate decreased. The reason is that with the increase in the temperature, the reaction activation energy increases, which accelerates the catalytic reaction speed and increases the amount of ·OH and Cu(III), thus improving the efficiency of the catalytic system for the degradation of CIP. The too-high temperature accelerates the decomposition of H_2O_2, so 50 °C is the best temperature for the catalytic reaction.

The recyclability of the CuO/GO-DE catalyst was studied through continuous reusability experiments. The catalyst was reused for the next run after washing it with distilled water without further treatment. The experimental results are shown in Figure 4d. After five cycles, the degradation ratio of ciprofloxacin decreased from 99.9% to 80%, which showed a certain stability of the catalyst. The reduction in degradation ratio may be because of the degradation product residue on the catalyst surface, which reduces the surface catalytic active sites.

3.4. Degradation Mechanism of Ciprofloxacin

In order to analyze the CIP degradation process and mechanism of the catalytic system, comparative experiments of CIP treatment in different systems, ·OH concentration, Cu(III) measurements of different systems and ·OH extinction experiment were carried out.

The comparative experiments were performed when (1) only H_2O_2 existed, (2) only CuO/GO-DE composite existed or (3) both CuO/GO-DE and H_2O_2 existed. The results are shown in Figure 5a. When there was only H_2O_2 in the system, just part of CIP was oxidized and degraded, and the removal ratio was low. When there was only CuO/GO-DE in the system, it can be seen that the removal ratio of CIP can reach about 85%. This is because, in the case of the absence of H_2O_2, CuO/GO-DE may catalyze O_2 to $\cdot O_2^-$, which also has an effect on the oxidative degradation of CIP. When there were both CuO/GO-DE and H_2O_2 in the system, the degradation rate of CIP jumped to 99.9%, indicating that CuO/GO-DE/H_2O_2 system has stronger oxidation than other systems.

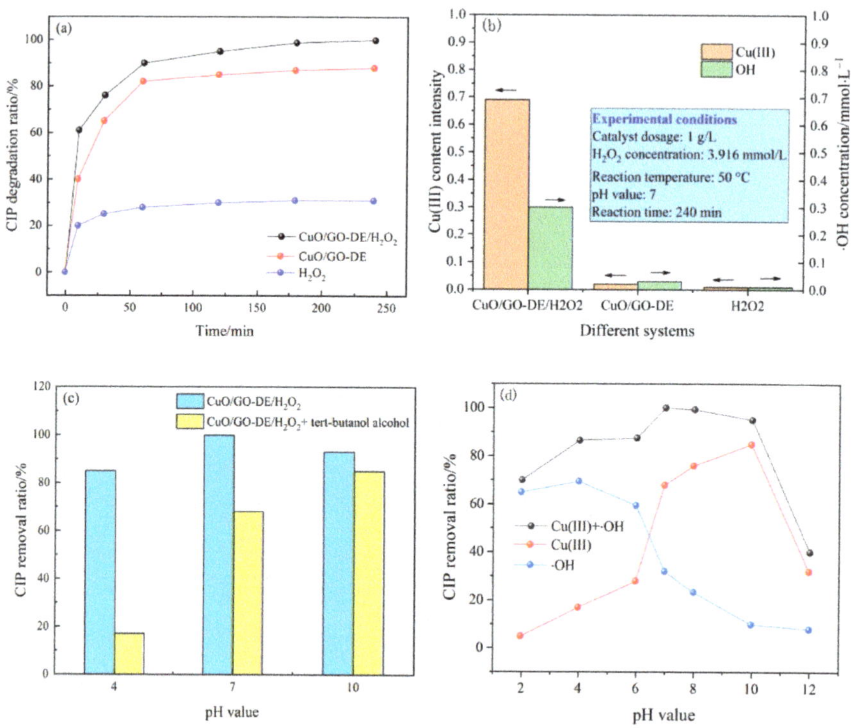

Figure 5. (**a**) Removal of CIP by different systems; (**b**) ·OH concentration and Cu(III) measurement in different systems; (**c**) ·OH scavenging experiments; (**d**) contribution of ·OH and Cu(III) on CIP removal ratios in different pH value.

In order to explore the main oxidizing substances in the reaction system, the concentration of ·OH and the content of Cu(III) in the system were determined. The results are shown in Figure 5b. It can be seen that under the conditions of catalyst dosage 1 g/L, H_2O_2 concentration 3.916 mmol/L, reaction temperature 50 °C, pH value 7 and reaction time 240 min, there were a lot of Cu(III) and ·OH produced in CuO/GO-DE/H_2O_2 system, while there were almost no Cu(III) and ·OH produced in CuO/GO-DE and H_2O_2 systems separately.

In order to further determine the contributions of Cu(III) and ·OH in the degradation of CIP in this heterogeneous system, the ·OH scavenging experiments were carried out, and the results are shown in Figure 5c. It can be seen that when the pH value was 4, the addition of tert-butyl alcohol significantly inhibited the produced ·OH as well as the CIP degradation ratio, which dropped from 86.5% to 17.6%. Under acidic conditions, copper existed in the form of Cu(I) or Cu(II) and catalyzed H_2O_2 to generate a large amount of ·OH. At this time, the reaction system mainly relied on ·OH to oxidize and decompose the target pollutants. At a pH of 7, the degradation ratio of CIP decreased from 99.9% to 68.3% with tert-butyl alcohol. The inhibition of tert-butyl alcohol on ·OH and CIP degradation ratio was significantly weakened, indicating that in addition to ·OH, there were other strong oxidizing substances, such as Cu(III), that play the role of oxidative decomposition to target pollutants at this time. At pH of 10, the degradation ratio of CIP decreased from 93.5% to 85.7% with tert-butyl alcohol, and the inhibition of tertbutyl alcohol on the degradation ratio was further weakened, indicating that at this time, there was little ·OH obtained from the decomposition of H_2O_2 in the solution, and Cu(III) was the main oxidant for oxidative decomposition of target pollutant. From the experimental results, it can be seen that under medium and alkaline conditions, Cu(II) is oxidized to higher valence Cu(III), and as the pH value rises, the number of Cu(III) is significantly increased, reflecting that Cu(III) is strongly dependent on pH value [41,57–59]. In summary, we can infer that CuO/GO-DE materials have different degradation modes of CIP at different pH values: when the solution is acidic, the degradation of CIP mainly depends on the oxidation of ·OH; when in neutral, it depends on the joint action of Cu(III) and ·OH; when in alkaline, it mainly depends on the oxidation of Cu(III). This explains why the catalyst has excellent degradation ability for CIP in a wide pH range due to the synergistic effect of ·OH and Cu(III). CIP removal ratios at pH values of from 2 to 12 were also tested in CuO/GO-DE/H_2O_2 system and CuO/GO-DE/H_2O_2 + tert-butyl alcohol system, and the contribution curves of Cu(III) + ·OH and Cu(III) to CIP removal ratios varied with pH was obtained, and the contribution curve of ·OH can also be deduced, as can be seen in Figure 5d.

According to the morphology analysis and mechanical analysis, the preparation process of Cu/GO-DE and degradation mechanisms of CIP by Cu/GO-DE/H_2O_2 system in different pH ranges were shown in Figure 6a,b separately.

In addition, the intermediate products of CIP degradation were described by LC-MS analysis, and the possible degradation pathway of the CIP molecule was speculated, as shown in Figure 7. The intermediate P1 (m/z = 362) was caused by the attacking of active oxides in the reaction system to the piperazine ring on the CIP molecule; P2 (m/z = 308) is diethylene CIP, which was produced by losing two C = O bonds of P1, and P3 (m/z = 263) is aniline, which was formed due to the loss of C_2H_5N of P2 [38]. P3 may also be formed due to the piperazine ring directly falling off caused by the active oxide's attack on the CIP molecule [60]; P3 then lost the cyclopropyl and amino group to obtain P4 (m/z = 156) [61], and the fluorine atom on P4 was replaced by -OH and the quinolone ring was cracked to form P5 (m/z = 154). In addition, the intermediate product P6 (m/z = 290) was formed by ring cracking due to the oxidation of the cyclopropyl group on the CIP molecule. The active oxides in the reaction system then continued to attack the piperazine ring and quinolone ring on the CIP molecule, resulting in its cracking and the formation of P8 (m/z = 152). P5 and P8 may be further transformed into low molecular organics and eventually mineralized into water and carbon dioxide.

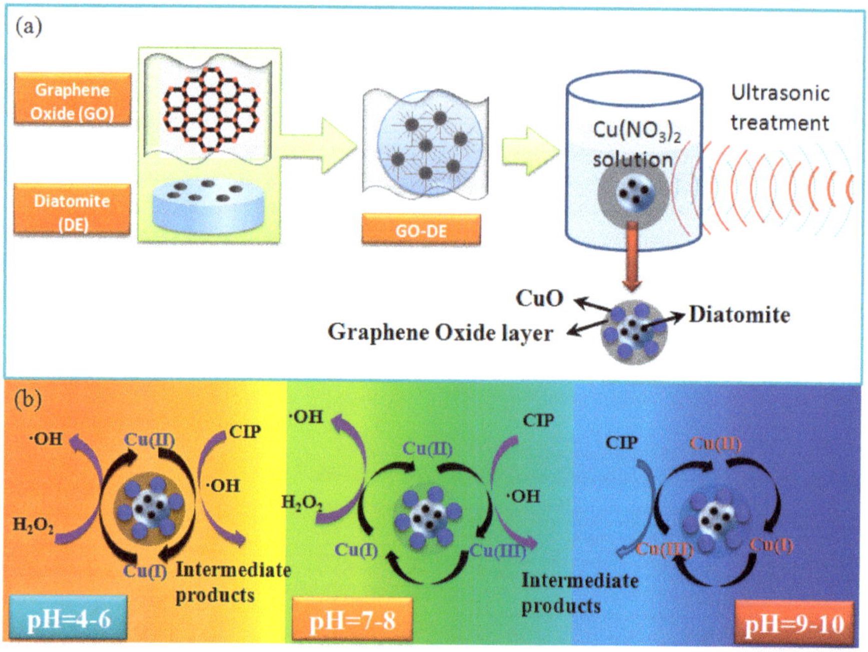

Figure 6. Preparation process of Cu/GO-DE (**a**) and degradation mechanism of CIP by Cu/GO-DE/H_2O_2 in different pH ranges (**b**).

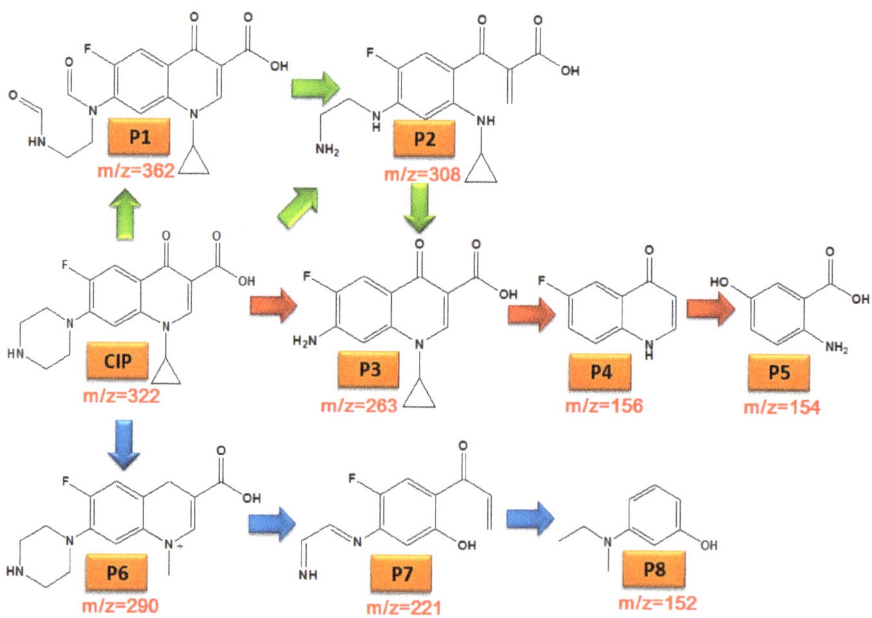

Figure 7. Possible degradation pathway of CIP.

4. Conclusions

In this study, a copper-loaded graphene–diatomaceous earth (CuO/GO-DE) catalyst was prepared. By changing the loading ratio of graphene oxide and copper, the optimal conditions for the preparation of the catalyst were explored. The catalyst was characterized by BET, SEM, TEM, Raman, FTIR, XRD and XPS, which showed that introducing GO into DE optimized the surface morphology and microscopic properties of DE and confirmed that copper oxide was successfully and evenly coated on GO-DE support. The effects of pH value, CIP initial concentration, reaction temperature, and catalyst and H_2O_2 dosages on CIP degradation by CuO/GO-DE catalytic system were studied. The results showed that CuO/GO-DE composites had excellent catalytic degradation activity in a wide range of pH. Under the best degradation conditions (pH of 7, CIP initial concentration of 50 mg/L, mass ratio of catalyst to H_2O_2 of 0.133, reaction temperature of 50 °C), the degradation ratio of CIP can reach 99%. The catalyst could be used repeatedly, and the degradation catalytic efficiency did not decrease significantly. Through the determination or quenching experiments of ·OH, Cu(III), etc., the degradation mechanisms of CuO/GO-DE catalyst for CIP at different pH values were proposed: under acidic conditions, the CIP degradation mainly depends on the oxidation of ·OH; under neutral conditions, it depends on the synergistic oxidation of Cu(III) and ·OH; under alkaline conditions, it mainly depends on the oxidation of Cu(III). The intermediate products of CIP degradation were identified by LC-MS. According to the main identified products, the possible degradation process of CIP in the catalytic system was proposed. This study provided a simple and effective method for the degradation of antibiotics in wastewater.

Author Contributions: Conceptualization, T.Z. and Y.Y.; methodology, T.Z. and J.Z.; validation, Y.Y. and J.Z.; formal analysis, T.Z.; investigation, Y.Y. and J.Z.; resources, T.Z.; data curation, J.Z.; writing—original draft preparation, T.Z., J.Z. and Y.Y.; writing—review and editing, T.Z. and Z.Z.; visualization, J.L.; supervision, C.L.; project administration, T.Z.; funding acquisition, C.L. All authors have read and agreed to the published version of the manuscript.

Funding: This work was financially supported by Major Science and Technology Projects in the Gansu Province of China [grant number 21ZD2JA001].

Data Availability Statement: Data available on request from the authors.

Conflicts of Interest: The authors declare no conflict of interest.

References

1. Przybylska, N.; Śliwińska-Bartkowiak, M.; Kościński, M.; Rotnicki, K.; Bartkowiak, M.; Jurga, S. Confined effect of water solution of ciprofloxacin in carbon nanotubes studied by Raman and Fourier Transform Infrared Spectroscopy methods. *J. Mol. Liq.* **2021**, *336*, 115938. [CrossRef]
2. Du, Y.; Dai, M.; Naz, I.; Hao, X.; Wei, X.; Rong, R.; Peng, C.; Ali, I. Carbothermal reduction synthesis of zero-valent iron and its application as a persulfate activator for ciprofloxacin degradation. *Sep. Purif. Technol.* **2021**, *275*, 119201. [CrossRef]
3. HGull, H.; Karadag, C.; Senger, B.; Sorg, R.V.; Möller, P.; Mellert, K.; Steiger, H.; Hänggi, D.; Cornelius, J.F. Ciprofloxacin enhances phototoxicity of 5-aminolevulinic acid mediated photodynamic treatment for chordoma cell lines. *J. Photodiagnosis Photodyn. Ther.* **2021**, *35*, 102346. [CrossRef]
4. Zhou, M.; Li, C.; Zhao, L.; Ning, J.; Pan, X.; Cai, G.; Zhu, G. Synergetic effect of nano zero-valent iron and activated carbon on high-level ciprofloxacin removal in hydrolysis-acidogenesis of anaerobic digestion. *J. Sci. Total. Environ.* **2021**, *752*, 142261. [CrossRef] [PubMed]
5. Sar, A.B.; Shabani, E.G.; Haghighi, M.; Shabani, M. Synergistic catalytic degradation of ciprofloxacin using magnetic carbon nanomaterial/$NiFe_2O_4$ promoted cold atmospheric pressure plasma jet: Influence of charcoal, multi walled carbon nanotubes and walnut shell. *J. Taiwan Inst. Chem. Eng.* **2022**, *132*, 104131. [CrossRef]
6. Ude, Z.; Flothkötter, N.; Sheehan, G.; Brennan, M.; Kavanagh, K.; Marmion, C.J. Multi-targeted metallo-ciprofloxacin derivatives rationally designed and developed to overcome antimicrobial resistance. *Int. J. Antimicrob. Agents* **2021**, *58*, 106449. [CrossRef]
7. Mahmoudpour, M.; Kholafazad-Kordasht, H.; Dolatabadi, J.E.N.; Hasanzaden, M.; Rad, A.H.; Torbati, M. Sensitive aptasensing of ciprofloxacin residues in raw milk samples using reduced graphene oxide and nanogold-functionalized poly (amidoamine) dendrimer: An innovative apta-platform towards electroanalysis of antibiotics. *Anal. Chim. Acta* **2021**, *1174*, 338736. [CrossRef]
8. Yu, F.; Yang, Z.; Zhang, X.; Yang, P.; Ma, J. Lanthanum modification κ-carrageenan/sodium alginate dual-network aerogels for efficient adsorption of ciprofloxacin hydrochloride. *Environ. Technol. Innov.* **2021**, *24*, 102052. [CrossRef]

9. Krishnan, S.; Shriwastav, A. Chlorophyll sensitized and salicylic acid functionalized TiO_2 nanoparticles as a stable and efficient catalyst for the photocatalytic degradation of ciprofloxacin with visible light. *Environ. Res.* **2023**, *216*, 114568. [CrossRef]
10. Li, X.; Wang, W.; Dou, J.; Gao, J.; Chen, S.; Quan, X.; Zhao, H. Dynamic adsorption of ciprofloxacin on carbon nanofibers: Quantitative measurement by in situ fluorescence. *J. Water Process. Eng.* **2016**, *9*, 14–20. [CrossRef]
11. Huang, Y.; Nengzi, L.C.; Zhang, X.; Gou, J.; Gao, Y.; Zhu, G.; Cheng, Q.; Cheng, X. Catalytic degradation of ciprofloxacin by magnetic $CuS/Fe_2O_3/Mn_2O_3$ nanocomposite activated peroxymonosulfate: Influence factors, degradation pathways and reaction mechanism. *Chem. Eng. J.* **2020**, *388*, 124274. [CrossRef]
12. Yu, F.; Sun, S.; Han, S.; Zheng, J.; Ma, J. Adsorption removal of ciprofloxacin by multi-walled carbon nanotubes with different oxygen contents from aqueous solutions. *Chem. Eng. J.* **2016**, *285*, 588–595. [CrossRef]
13. Alshaikh, H.; Shawky, A.; Mohamed, R.M.; Knight, J.G.; Selva, R.L. Solution-based synthesis of Co_3O_4/ZnO pn heterojunctions for rapid visible-light-driven oxidation of ciprofloxacin. *J. Mol. Liq.* **2021**, *334*, 116092. [CrossRef]
14. Chafyq, E.H.; Legrouri, K.; Aghrouch, M.; Oumam, M.; Mansouri, S.; Khouya, E.H.; Hannache, H. Adsorption of ciprofloxacin antibiotic on materials prepared from Moroccan oil shales. *Chem. Phys. Lett.* **2021**, *778*, 138707. [CrossRef]
15. Devi, M.G.; Dutta, S.; Al Hinai, A.T.; Feroz, S. Nano engineered biodegradable capsules for the encapsulation and kinetic release studies of ciprofloxacin hydrochloride. *J. Indian Chem. Soc.* **2021**, *98*, 100109. [CrossRef]
16. Nie, X.; Li, G.Y.; Li, S.S.; Luo, Y.M.; Wan, Q.; An, T.C. Highly efficient adsorption and catalytic degradation of ciprofloxacin by a novel heterogeneous Fenton catalyst of hexapod-like pyrite nanosheets mineral clusters. *Appl. Catal. B Environ.* **2022**, *300*, 120734. [CrossRef]
17. van Garderen, N.; Clemens, F.J.; Kaufmann, J.; Urbanek, M.; Binkowski, M.; Graule, T.; Aneziris, C.G. Pore analyses of highly porous diatomite and clay based materials for fluidized bed reactors. *Microporous Mesoporous Mater.* **2012**, *151*, 255–263. [CrossRef]
18. Lv, Z.Q.; Jiang, A.N.; Jin, J.X. Influence of ultrafine diatomite on cracking behavior of concrete: An acoustic emission analysis. *Constr. Build. Mater.* **2012**, *308*, 124993. [CrossRef]
19. Zhang, Y.T.; Wei, W.; Wang, Y.; Ni, B.J. Enhancing methane production from algae anaerobic digestion using diatomite. *J. Clean. Prod.* **2021**, *315*, 128138. [CrossRef]
20. Taoukil, D.; Meski, Y.E.; Lahlaouti, M.L.; Djedjig, R.; Bouardi, A.E. Effect of the use of diatomite as partial replacement of sand on thermal and mechanical properties of mortars. *J. Build. Eng.* **2021**, *42*, 103038. [CrossRef]
21. Chen, J.; Ren, Q.F.; Ding, Y.; Xiong, C.Y.; Guo, W.M. Synthesis of bifunctional composites Ag/BiOCl/diatomite: Degradation of tetracycline and evaluation of antimicrobial activity. *J. Environ. Chem. Eng.* **2021**, *9*, 106476. [CrossRef]
22. Shen, T.M.; Xu, H.X.; Miao, Y.; Ma, L.L.; Chen, N.C.; Xie, Q.L. Study on the adsorption process of Cd(II) by Mn-diatomite modified adsorbent. *Mater. Lett.* **2021**, *300*, 130087. [CrossRef]
23. Xiong, C.Y.; Ren, Q.F.; Chen, S.H.; Liu, X.Y.; Jin, Z.; Ding, Y. A multifunctional Ag_3PO_4/Fe_3O_4/Diatomite composites: Photocatalysis, adsorption and sterilization. *Mater. Today Commun.* **2021**, *28*, 102695. [CrossRef]
24. Sun, Y.X.; Zhou, J.B.; Liu, D.; Liu, X.J.; Leng, C.H. Highly efficient removal of tetracycline hydrochloride under neutral conditions by visible photo-Fenton process using novel $MnFe_2O_4$/diatomite composite. *J. Water Process. Eng.* **2021**, *43*, 102307. [CrossRef]
25. Dang, T.D.; Banerjee, A.N.; Tran, Q.T.; Roy, S. Fast degradation of dyes in water using manganese-oxide-coated diatomite for environmental remediation. *J. Phys. Chem. Solids* **2016**, *98*, 50–58. [CrossRef]
26. Wang, M.F.; Xu, H.L.; Huang, C.N.; Cui, Z.T.; Li, M.L.; Song, B.; Shao, G.; Wang, H.L.; Lu, H.X.; Zhang, R. Preparation of g-C_3N_4/diatomite composite with improved visible light photocatalytic activity. *Inorg. Chem. Commun.* **2021**, *129*, 108645. [CrossRef]
27. Ulloa-Ovares, D.; Rodríguez-Rodríguez, C.E.; Masís-Mora, M.; Durán, J.E. Simultaneous degradation of pharmaceuticals in fixed and fluidized bed reactors using iron-modified diatomite as heterogeneous Fenton catalyst. *Process. Saf. Environ. Prot.* **2021**, *152*, 97–107. [CrossRef]
28. Noreen, S.; Tahira, M.; Ghamkhar, M.; Hafiz, I.; Bhatti, H.N.; Nadeem, R.; Murtaza, M.A.; Yaseen, M.; Sheikh, A.A.; Naseem, Z.; et al. Treatment of textile wastewater containing acid dye using novel polymeric graphene oxide nanocomposites (GO/PAN, GO/PPy, GO/PSty). *J. Mater. Res. Technol.* **2021**, *14*, 25–35. [CrossRef]
29. Brodie, B.C., XIII. On the atomic weight of graphite. *Philos. Trans. R. Soc. Lond.* **1859**, *149*, 249–259. [CrossRef]
30. Chung, C.; Kim, Y.K.; Shin, D.; Ryoo, S.R.; Hong, B.H.; Min, D.H. Biomedical applications of graphene and graphene oxide. *Acc. Chem. Res.* **2013**, *46*, 2211–2224. [CrossRef]
31. Zhu, Y.; Murali, S.; Cai, W.; Li, X.; Suk, J.W.; Potts, J.R.; Ruoff, R.S. Graphene and graphene oxide: Synthesis, properties and applications. *Adv. Mater.* **2010**, *22*, 3906–3924. [CrossRef]
32. Zhang, Q.; Lu, W.; Wu, M.Y.; Qi, G.S.; Yuan, Y.; Li, J.; Su, H.L.; Zhang, H. Preparation and properties of cellulosenanofiber (CNF)/polyvinyl alcohol (PVA)/graphene oxide (GO): Application of CO_2 absorption capacity and molecular dynamics simulation. *J. Environ. Manag.* **2022**, *302*, 114044. [CrossRef]
33. Majumdar, S.; Ghosh, M.; Mukherjee, S.; Satpati, B.; Biswajit, D. DNA mediated graphene oxide (GO)-nanosheets dispersed supramolecular GO-DNA hydrogel: An efficient soft-milieu for simplistic synthesis of Ag-NPs@ GO-DNA and Gram +ve/−ve bacteria-based Ag-NPs@ GO-DNA-bacteria nano-bio composites. *J. Mol. Liq.* **2021**, *342*, 117482. [CrossRef]
34. Raza, M.M.H.; Khan, S.; Aalam, S.M.; Sadiq, M.; Sarvar, M.; Zulfequar, M.; Husain, S.; Ali, J. Study the electron field emission properties of plasma-based reduction of graphene oxide (GO): An ex-situ plasma approach. *Carbon Trends* **2021**, *5*, 100127. [CrossRef]

35. Grajek, H.; Jonik, J.; Witkiewicz, Z.; Wawer, T.; Purchaa, M. Applications of graphene and its derivatives in chemical analysis. *Crit. Rev. Anal. Chem.* **2020**, *50*, 445–471. [CrossRef]
36. Hassan, N.S.; Jalil, A.A.; Azami, M.S.; Rahman, A.F.A.; Firmansyah, M.L.; Nabgan, W. Photodegradation of bisphenol A from aqueous solution over reduced graphene oxide supported on tetragonal silica-zirconia nanocatalysts: Optimization using RSM. *Process. Saf. Environ.* **2021**, *156*, 496–507. [CrossRef]
37. Liu, M.S.; Wang, J.; Wang, X.; Zhu, W.X.; Yao, X.L.; Su, L.H.; Sun, J.; Yue, T.L.; Wang, J.L. Highly efficient and cost-effective removal of patulin from apple juice by surface engineering of diatomite with sulfur-functionalized graphene oxide. *Food Chem.* **2019**, *300*, 125111. [CrossRef]
38. Zhang, T.; Qian, C.Y.; Guo, P.R.; Gan, S.C.; Dong, L.Y.; Bai, G.; Guo, Q.Y. A novel reduced graphene oxide-attapulgite (RGO-ATP) supported Fe_2O_3 catalyst for heterogeneous Fenton-like oxidation of ciprofloxacin: Degradation mechanism and pathway. *Catalysts* **2020**, *10*, 189. [CrossRef]
39. Zhang, T.; Dong, L.Y.; Du, J.H.; Qian, C.Y.; Wang, Y. CuO and CeO_2 assisted Fe_2O_3/attapulgite catalyst for heterogeneous Fenton-like oxidation of methylene blue. *RSC Adv.* **2020**, *10*, 23431–23439. [CrossRef]
40. Chen, J.; Yao, B.; Li, C.; Shi, G.Q. An improved Hummers method for eco-friendly synthesis of graphene oxide. *Carbon* **2013**, *64*, 225–229. [CrossRef]
41. Xu, Z.H.; Jing, C.Y.; Li, F.S.; Meng, X.G. Mechanisms of photocatalytical degradation of monomethylarsonic and dimethylarsinic acids using nanocrystalline titanium dioxide. *Environ. Sci. Technol.* **2008**, *42*, 2349–2354. [CrossRef] [PubMed]
42. Wang, L.H.; Xu, H.D.; Jiang, N.; Wang, Z.M.; Jiang, J.; Zhang, T. Trace cupric species triggered decomposition of peroxymonosulfate and degradation of organic pollutants: Cu (III) being the primary and selective intermediate oxidant. *Environ. Sci. Technol.* **2020**, *54*, 4686–4694. [CrossRef] [PubMed]
43. Li, C.; Goetz, V.; Chiron, S. Peroxydisulfate activation process on copper oxide: Cu (III) as the predominant selective intermediate oxidant for phenol and waterborne antibiotics removal. *J. Environ. Chem. Eng.* **2021**, *9*, 105145. [CrossRef]
44. Chua, C.K.; Sofer, Z.; Pumera, M. Graphite oxides: Effects of permanganate and chlorate oxidants on the oxygen composition. *Chem. A Eur. J.* **2012**, *18*, 13453–13459. [CrossRef]
45. Li, M.; Shi, J.B. Mechanical and thermal performance assessment of paraffin/expanded vermiculite-diatomite composite phase change materials integrated mortar: Experimental and numerical approach. *Sol. Energy.* **2021**, *227*, 343–353. [CrossRef]
46. Cheng, G.; Liu, Y.L.; Wang, Z.G.; Zhang, J.L.; Sun, D.H.; Ni, J.Z. The GO/rGO–Fe_3O_4 composites with good water-dispersibility and fast magnetic response for effective immobilization and enrichment of biomolecules. *J. Mater. Chem.* **2012**, *22*, 21998–22004. [CrossRef]
47. Zhang, J.; Zhang, X.; Zhou, T.; Huang, Y.; Zhao, L.; Ren, Z.; Liu, X.; Chen, L.; Shao, Y.; Ma, X.; et al. Photocatalytic removal of hexavalent chromium by Fe doped g-C_3N_4 loaded with dispersed diatomite modified by electron beam bombardment. *J. Clean. Prod.* **2021**, *315*, 128219. [CrossRef]
48. Singhal, S.; Kaur, J.; Namgyal, T.; Sharma, R. Cu-doped ZnO nanoparticles: Synthesis, structural and electrical properties. *J. Physica. B.* **2012**, *407*, 1223–1226. [CrossRef]
49. Wang, J.X.; Zhang, G.Q.; Qiao, S.; Zhou, J.T. Magnetic Fe^0/iron oxide-coated diatomite as a highly efficient adsorbent for recovering phosphorus from water. *Chem. Eng. J.* **2021**, *412*, 128696. [CrossRef]
50. Tajmiri, S.; Hosseini, M.R.; Azimi, E. Combined photocatalytic-adsorptive removal of water contaminants using a biologically prepared CdS-diatomite nanocomposite. *Mater. Chem. Phys.* **2021**, *258*, 123913. [CrossRef]
51. Chio, J.; Oh, H.; Han, S.W.; Ahn, S.; Noh, J.; Park, J.B. Preparation and characterization of graphene oxide supported Cu, Cu_2O, and CuO nanocomposites and their high photocatalytic activity for organic dye molecule. *Curr. Appl. Phys.* **2017**, *17*, 137–145. [CrossRef]
52. Tang, G.Q.; Jiang, Z.G.; Li, X.F.; Zhang, H.B.; Dasari, A.; Yu, Z.Z. Three dimensional graphene aerogels and their electrically conductive composites. *Carbon* **2014**, *77*, 592–599. [CrossRef]
53. Kim, H.E.; Nguyen, T.T.M.; Lee, H.; Lee, C. Enhanced inactivation of Escherichia coli and MS_2 coliphage by cupric ion in the presence of hydroxylamine: Dual microbicidal effects. *Environ. Sci. Technol.* **2015**, *49*, 14416–14423. [CrossRef]
54. Pham, A.N.; Xing, G.; Miller, C.J.; Waite, T.D. Fenton-like copper redox chemistry revisited: Hydrogen peroxide and superoxide mediation of copper-catalyzed oxidant production. *J. Catal.* **2013**, *301*, 54–64. [CrossRef]
55. Ulanski, P.; Sonntag, C.V. Stability Constants and Decay of Aqua-Copper (III)–A Study by Pulse Radiolysis with Conductometric and Optical Detection. *Eur. J. Inorg. Chem.* **2000**, *2000*, 1211–1217. [CrossRef]
56. Deng, Y.; Handoko, A.D.; Du, Y.; Xi, S.; Yeo, B.S. In situ Raman spectroscopy of copper and copper oxide surfaces during electrochemical oxygen evolution reaction: Identification of CuIII oxides as catalytically active species. *ACS. Catal.* **2016**, *6*, 2473–2481. [CrossRef]
57. Lee, H.S.; Lee, H.J.; Sedlak, D.L.; Lee, C.H. PH-Dependent reactivity of oxidants formed by iron and copper-catalyzed decomposition of hydrogen peroxide. *Chemosphere* **2013**, *92*, 652–658. [CrossRef]
58. Chen, J.B.; Zhou, X.F.; Sun, P.Z.; Zhang, Y.L.; Huang, C.H. Complexation enhances Cu (II)-activated peroxydisulfate: A novel activation mechanism and Cu (III) contribution. *Environ. Sci. Technol.* **2019**, *53*, 11774–11782. [CrossRef]
59. Xing, G.W.; Pham, A.N.; Miller, C.J.; Waite, T.D. PH-dependence of production of oxidants (Cu(III) and/or HO·) by copper-catalyzed decomposition of hydrogen peroxide under conditions typical of natural saline waters. *Geochim. Cosmochim. Acta.* **2018**, *232*, 30–47. [CrossRef]

60. Gupta, B.; Gupta, A.K. Photocatalytic performance of 3D engineered chitosan hydrogels embedded with sulfur-doped C_3N_4/ZnO nanoparticles for Ciprofloxacin removal: Degradation and mechanistic pathways. *Int. J. Biol. Macromol.* **2022**, *198*, 87–100. [CrossRef]
61. Wang, A.Q.; Wang, H.; Deng, H.; Wang, S.; Shi, W.; Yi, Z.X.; Qiu, R.L.; Yan, K. Controllable synthesis of mesoporous manganese oxide microsphere efficient for photo-Fenton-like removal of fluoroquinolone antibiotics. *Appl. Catal. B* **2019**, *248*, 298–308. [CrossRef]

nanomaterials

Article

Development of Efficient and Recyclable ZnO–CuO/g–C_3N_4 Nanocomposite for Enhanced Adsorption of Arsenic from Wastewater

Qudrat Ullah Khan [1,2,†], Nabila Begum [3], Zia Ur Rehman [4], Afaq Ullah Khan [5,†], Kamran Tahir [6], El Sayed M. Tag El Din [7], Asma A. Alothman [8], Mohamed A. Habila [8], Dahai Liu [3,*], Patrizia Bocchetta [9] and Muhammad Sufyan Javed [10,*]

1 Greater Bay Area Institute of Precision Medicine (Guangzhou), Fudan University, Nansha District, Guangzhou 511458, China
2 Zhongshan-Fudan Joint Innovation Center, Zhongshan 528437, China
3 School of Medicine, Foshan University, Foshan 528000, China
4 Department of Chemistry, The University of Haripur, Haripur 22620, Pakistan
5 School of Chemistry and Chemical Engineering, Jiangsu University, 301 Xuefu Road, Zhenjiang 212013, China
6 Institute of Chemical Sciences, Gomal University Dera Ismail Khan, Dera Ismail Khan 29220, Khyber Pakhtunkhwa, Pakistan
7 Electrical Engineering Department, Faculty of Engineering & Technology, Future University in Egypt, New Cairo 11835, Egypt
8 Department of Chemistry, College of Science, King Saud University, Riyadh 11451, Saudi Arabia
9 Dipartimento di Ingegneria dell'Innovazione, Università del Salento, via Monteroni, 73100 Lecce, Italy
10 School of Physical Science and Technology, Lanzhou University, Lanzhou 730000, China
* Correspondence: seansean2014@126.com (D.L.); safisabri@gmail.com (M.S.J.)
† These authors contributed equally to this work.

Citation: Khan, Q.U.; Begum, N.; Rehman, Z.U.; Khan, A.U.; Tahir, K.; Tag El Din, E.S.M.; Alothman, A.A.; Habila, M.A.; Liu, D.; Bocchetta, P.; et al. Development of Efficient and Recyclable ZnO–CuO/g–C_3N_4 Nanocomposite for Enhanced Adsorption of Arsenic from Wastewater. *Nanomaterials* **2022**, *12*, 3984. https://doi.org/10.3390/nano12223984

Academic Editors: Christos A. Aggelopoulos, Danil N. Dybtsev and Sergio Brutti

Received: 20 August 2022
Accepted: 10 November 2022
Published: 12 November 2022

Abstract: Arsenic (III) is a toxic contaminant in water bodies, especially in drinking water reservoirs, and it is a great challenge to remove it from wastewater. For the successful extraction of arsenic (III), a nanocomposite material (ZnO–CuO/g–C_3N_4) has been synthesized by using the solution method. The large surface area and plenty of hydroxyl groups on the nanocomposite surface offer an ideal platform for the adsorption of arsenic (III) from water. Specifically, the reduction process involves a transformation from arsenic (III) to arsenic (V), which is favorable for the attachment to the –OH group. The modified surface and purity of the nanocomposite were characterized by SEM, EDX, XRD, FT–IR, HRTEM, and BET models. Furthermore, the impact of various aspects (temperatures, pH of the medium, the concentration of adsorbing materials) on adsorption capacity has been studied. The prepared sample displays the maximum adsorption capacity of arsenic (III) to be 98% at pH ~ 3 of the medium. Notably, the adsorption mechanism of arsenic species on the surface of ZnO–CuO/g–C_3N_4 nanocomposite at different pH values was explained by surface complexation and structural variations. Moreover, the recycling experiment and reusability of the adsorbent indicate that a synthesized nanocomposite has much better adsorption efficiency than other adsorbents. It is concluded that the ZnO–CuO/g–C_3N_4 nanocomposite can be a potential candidate for the enhanced removal of arsenic from water reservoirs.

Keywords: arsenic removal; ZnO–CuO/g–C_3N_4 nanocomposite; solution combustion; kinetic studies; adsorption

1. Introduction

Arsenic (III) is the most toxic and portable contaminant in nature, creating numerous environmental pollution problems worldwide as it can be effectively solubilized in groundwater [1,2]. Arsenic in water remains both in an organic as well as an inorganic state, whereas it mostly remains as arsenate ($H_2AsO_4^-$) and oxyanions: arsenite ($H_2AsO_3^-$) [3–7].

Several studies suggest that arsenic pollution can contribute to the development of different diseases, such as keratosis, melanosis, edema, and cancers of the skin, lungs, and bladder. Additionally, it has been implicated in contributing to the enlargement of the liver, kidneys, and spleen [8–11]. Consequently, the World Health Organization (WHO) has suggested that safe potable drinking water should have concentrations of arsenic no greater than 10 $\mu g\ L^{-1}$ [12,13]. However, concentrations of arsenic in polluted water have reached levels of 100–300 $\mu g\ L^{-1}$, which is 10–30 times higher than the maximum level recommended by WHO. This has the potential to cause severe damage to people's health [8,14], which is 10–30 times higher than the safety limitation and will seriously damage people's health. As a result, removing arsenic from contaminated water is a critical factor.

Up to now, different methods have been tested for the filtration of arsenic, such as reverse osmosis, adsorption, ultrafiltration, chemical precipitation, ion exchange, etc. [15–18]. Among said protocols, adsorption has many desired options for As filtration because of its effortlessness and huge efficacy [19]. However, sometimes, this protocol can be costly due to the preservation of adsorbent media. Nanotechnology solved these problems to a greater extent by providing cheap nano adsorbents with large surface areas and more specificity. Extensive efforts have been made for the adsorption of As on metal oxide nanoparticles, i.e., MgO [20], Fe_3O_4 [21], α-Fe_2O_3 [22], CeO_2 [23], aluminium oxides [24], etc. Moreover, it has been acknowledged that the good adsorption ability of adsorbents is due to desirable active sites and large surface areas. For instance, the main limitations associated with these oxides are their agglomeration during the reaction and lesser stability, affecting their efficacy to a greater extent [25]. Cao et al. prepared a CuO nanomaterial that was applied for As adsorption but, unfortunately, they obtained a low level of As adsorption, i.e., 5.7 mg/g [26]. Therefore, it is the need of the present and the future to introduce such nanomaterials which have high efficiency and are more economical for the disintegration of arsenic. At present, metal oxides (CuO, TiO_2, NiO, ZnO, etc.) and their combination with effective support (MCM–41, g–C_3N_4, etc.) have attracted great attention from researchers in water remediation purposes because of their easy preparations, huge surface areas, and well-organized porous structures [27]. These nanomaterials have special characteristics such as high reactivity, large surface areas, regeneration capabilities, and high selectivities, which are very important for environmental remediation [28]. Various nanoparticle-based materials, for example, zinc oxide, manganese oxide, ferric oxides, and titanium oxide, are listed as effective nano adsorbents that perform well as compared to other adsorbents used for the arsenic elimination from drinking water [29–33]. From this perspective, high reactivity, vast surface area, and a greater number of active sites could help remove contaminants from wastewater effectively. However, their toxicity despite affordable prices, including environmental concerns, remain major problems [34,35]. Additionally, excellent adsorption efficiency and reusing the nano adsorbent are other basic requirements for efficient and selective nanomaterials that can make them an efficient candidate for removing arsenic from wastewater. The literature survey suggests a better affinity of a different metal oxide-containing adsorbent towards the elimination of arsenic due to its better selectivity and high adsorption capability for both inorganic states of arsenic As(V) and As(III) from aqueous solutions [36,37]. On the other hand, the drawback related to these nano adsorbents is accumulation owing to its low energy barrier, which greatly decreases mobility, availability, and transfers to the polluted site for in situ adsorptions. Therefore, this shortcoming can be overcome by the impregnation method, surface coating, or doping on the surface of these nano adsorbents, which could be a superior choice for the elimination of arsenic and its derivatives from polluted water. Recently, carbon nitride (g–N_3C_4) and its composite with metal oxides have shown better performance because of their chemical, photophysical, and catalytic properties, simple synthesis, and high stability under harsh conditions for a variety of applications, especially with photo-/electro-catalysis, sensors, and bioimaging, etc. [38–41]. Therefore, we planned to design a metal-coated nanocomposite with g–C_3N_4 for As removal from waste water resources.

We have prepared ZnO–CuO/g–C_3N_4 nanocomposites in the present contribution. A new precipitation protocol was used for the synthesis of the said nanomaterial. The synthesized nanosheets were applied as an adsorbent for As adsorption. ZnO and CuO were individually synthesized by the same protocol and used for As adsorption. Results showed that ZnO–CuO/g–C_3N_4 nanocomposites have several times higher As removal efficiency than their individual counterparts. The impacts of various parameters (pH, temperature, material concentration) on the efficiency of ZnO–CuO/g–C_3N_4 nanocomposites were also studied. Kinetics studies and adsorption isotherm models were investigated to best explain the adsorption. Finally, the detailed mechanism of arsenic adsorption was also studied.

2. Experimental Section

2.1. Synthesis of ZnO–CuO Heterostructure

Initially, the ZnO–CuO heterostructure was synthesized through a solution combustion method. Initially, 20 mL of $Cu(NO_3)2.3H_2O$ at a 0.1 M concentration was mixed in a beaker with 100 mL of 0.1 M $Zn(NO_3)2.6H_2O$ and stirred for up to 30 min at an ambient temperature. After that, 10 mL of 2 M sucrose solution was added as fuel into the beaker and heated at 250 °C. Finally, the hot mixture was burnt after dehydration with a flame to obtain a delicate CuO–ZnO composite powder.

2.2. Synthesis of g–C_3N_4 and ZnO–CuO/g–C_3N_4 Composite

Graphitic resembling C_3N_4 was manufactured via thermal polymerization from melamine. Typically, to obtain a yellowish powder of g–C_3N_4, a particular amount of melamine was placed in an alumina crucible and annealed at 550 °C for roughly 2 h in an air atmosphere. In a beaker, 1 g of g–C_3N_4 and 0.2 g ZnO–CuO was mixed in 70 mL water and stirred for 3 h. After 3 h, the sample was filtered and dried at 80 °C for 12 h to obtain the new material ZnO–CuO/g–C_3N_4.

2.3. Batch Adsorption Experiments

We used sodium arsenite as a primary arsenic source to test the synthesized composite adsorption characteristics in the trials. The influence of interfering ions (nitrate, phosphate, carbonate, chloride, and sulfate), pH, temperature, and time interval on the adsorption amount was determined through a batch experiment. At the same time, the kinetic reactions were additionally investigated. Initially, the pH of the reaction mixture was balanced from around 2–10, utilizing HCl (0.1 M) and NaOH (0.1 M) solutions, whereas 50 μg/L, 100 μg/L, and 150, 200 μg/L concentrations of the sodium arsenite solutions were used. Following the state of the consolidation of meddling ions with 50 μg/L, 100 μg/L, and 150 μg/L, the impact of prepared contents on the penetration of sodium arsenite with a combination of 100 μg/L was determined. Likewise, to investigate the absorption dynamism, 50 μg/L, 100 μg/L, and 150 μg/L of sodium arsenite-concentrated chemicals were checked by fixing different sampling time durations at temperatures ranging from 20 to 80 °C. The absorption equilibrium record of sodium arsenite at a starting concentration of about 20 μg/L to 200 μg/L was examined through ZnO–CuO/g–C_3N_4 at various temperatures (20 to 80 °C). The adsorbent was cleaned thoroughly with 50 mL of 0.1 M NaOH solutions, washed many times with distilled water, and dried up in an oven at 60 °C for three hours to recover the synthesized adsorbent. Finally, the removal rate R(%) and adsorption quantity (qe (μg/L) of arsenic through the adsorbent was determined as follows:

$$q_e = (C_0 - C_e)V/m \quad (1)$$

$$\%R = (1 - C_e/C_0) \times 100 \quad (2)$$

Here, C_e (μg/L) and C_0 (μg/L) are the equilibrium concentration and the liquid phase initial of arsenic, m is the weight of the adsorbent utilized, and V is the volume of the liquid solution. The data obtained were processed for standard deviation and ANOVA in Excel to analyze the arsenic adsorption.

3. Results and Discussions

3.1. XRD and FT-IR Analysis

Figure 1a shows the X-ray diffraction (XRD) pattern of obtained samples. From these patterns, it can be seen that the peaks situated at 2θ = 31.82°, 34.43°, 36.22°, 47.61°, 56.61°, 63.01°, and 67.98° resemble the (100), (002), (101), (102), (110), (103), and (112) planes which can be readily indexed to hexagonal wurtzite structure of ZnO (JCPDS 36-1451), respectively [42]. On the other hand, the peaks at 13.3° (100) and 27.4° (002) are ascribed to the crystal structure of g–C_3N_4. Consequently, the XRD spectrum of the ZnO–CuO/g–C_3N_4 results incorporated all the common peaks of g–C_3N_4, ZnO, and CuO. Additionally, CuO shows some noticeable peaks at 35.7°, 38.6°, and 67.5°, corresponding to the (002), (200), and (311), in accordance with (JCPDS card no. 89-5899). Furthermore, the peak strength intensity of the typical g–C_3N_4 was continuously increased with an increase in the quantity of g–C_3N_4, indicating a decrease in the intensity of ZnO and CuO peaks, respectively. For the ZnO–CuO/g–C_3N_4 sample, the XRD pattern exhibited no basic pick for g–C_3N_4, which can be credited to the low substance of the g–C_3N_4 in the composite. This result provides more explicit evidence that no extra peaks were seen in all the patterns, which indicated the high purity of the synthesized materials.

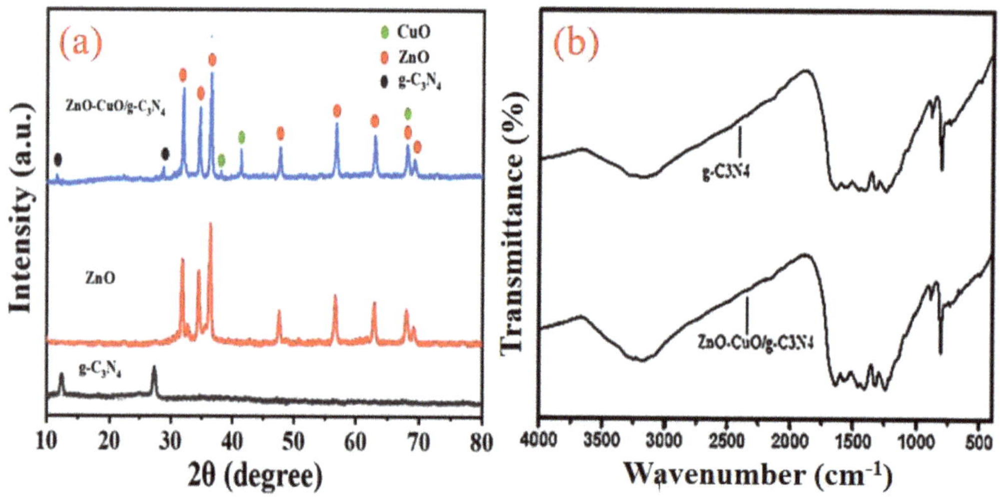

Figure 1. (**a**) XRD patterns of g–C_3N_4, ZnO and ZnO–CuO/g–C_3N_4 samples, (**b**) FTIR spectra of the g–C_3N_4 and prepared composite ZnO–CuO/g–C_3N_4.

For further analysis of a composite's surface functional groups and functional elements, we performed FT–IR spectroscopy. Figure 1b shows the FT–IR spectra of g–C_3N_4 and synthesized ZnO–CuO/g–C_3N_4 composite. The upper-level peak at around 810 cm^{-1} in the g–C_3N_4 test is associated with the twisting vibration attributes of heptazine moiety. At the same time, the peak observed at 3200 cm^{-1} reveals the O–H stretching vibration, suggesting the presence of moisture particles in CuO and ZnO materials. The broad group about 1200–1700 cm^{-1} in the g–C_3N_4 test is concerned with the ordinary stretching mode of C–N heterocycles, while another wideband about 3200 cm^{-1} in a similar case is associated with the extending vibration mode of the amine group [43]. In a composite ZnO–CuO/g–C_3N_4, the broad double peaks at 1631 and 1563 cm^{-1} are associated with C–N stretching vibration modes, whereas the peaks at 1251, 1323 and 1420 cm^{-1} are assigned to the aromatic C–N stretching. The broad peak around 3000 to 3500 cm^{-1} in the sample is associated with the adsorbed H_2O and N–H vibrations of the amine groups [44].

3.2. HR-TEM Analysis

The structure of the as-prepared samples was observed by HRTEM, as depicted in Figure 2. The HRTEM examination of ZnO–CuO shows that ZnO and CuO have sphere morphology, control size, and are highly dispersed in nature, which consistently aligns with the XRD results, as shown in Figure 2a. As per Figure 2b, the HRTEM picture of ZnO–CuO/g–C_3N_4 indicates its covered platelet-like morphology and plain paper-fold diluent sheet, which is identical to the design of the nanosheets of graphene [45]. The ZnO and CuO particles are dispersed uniformly with a small size on the face of g–C_3N_4 nanosheets. It confirms the successful synthesis of ZnO–CuO/g–C_3N_4 heterostructure composite [46].

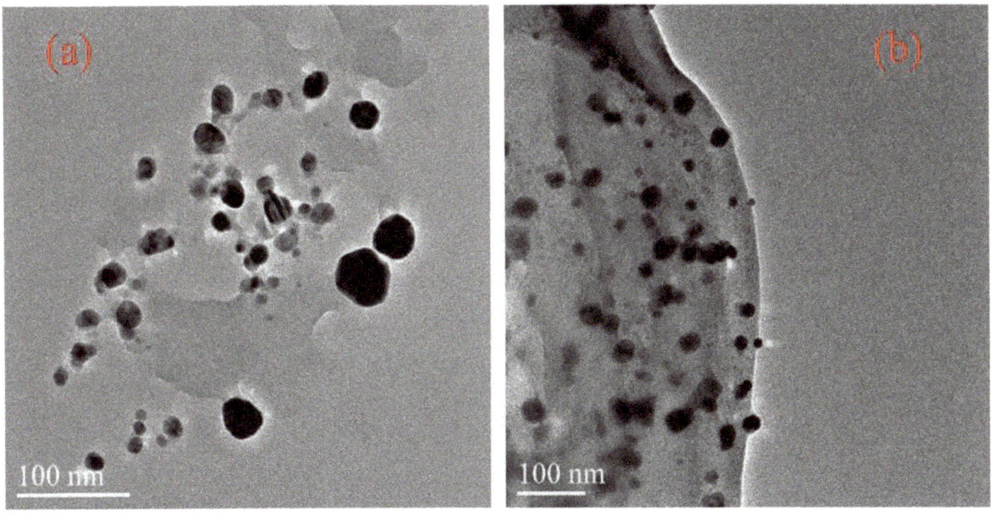

Figure 2. In HRTEM figure, (**a**) represents CuO–ZnO whereas (**b**) represents ZnO–CuO/g–C_3N_4 composites.

3.3. SEM and EDX Analysis

To test the fluctuation in material morphology in response to CuO and ZnO charging, EDX and SEM studies were applied to perceive the external structure and chemical configuration of ZnO–CuO/g–C_3N_4. Figure 3a represents the structure of the g–C_3N_4 as having sheet-like morphology with a large size and a higher degree of aggregation, whereas in the case of the ZnO–CuO/g–C_3N_4 composite, the tiny nanoparticles of ZnO and CuO are dispersed on the surface of the thin sheets to overcome the aggregation and enhance the surface area of the composite [47], as shown in Figure 3b. To validate the purity of the synthesized composite further, the EDX study was conducted. Figure 3c depicts that the existence of Zn, Cu, C, O, and N in the nanocomposite ensures the purity of the synthesized composite.

3.4. Nitrogen Adsorption-Desorption Study

Figure 4 depicts the N_2 adsorption-desorption isotherms and the Barrett-Joyner-Halenda pore-size dispersion bends of ZnO–CuO/g–C_3N_4 nanosheets. The Brunauer-Emmett-Teller (BET) explicit surface territory for CuO nanosheets ZnO–CuO/g–C_3N_4 is 268 $m^2\ g^{-1}$. The pore-size dispersions are at a maximum for the nanosheets and are about 4 nm each. The highly specific area of ZnO–CuO/g–C_3N_4 nanosheets might be because of the porous structure of nanosheets, which is good for electrochemical applications. ZnO–CuO/g–C_3N_4 nanosheets yield a large, exposed surface area designed for the adsorption of particles and charge transfer reactions. Figure 4B illustrates the pore size distribution of the synthesized ZnO–CuO/g–C_3N_4 nanomaterial. It is clear from the figure that most of the pores have a diameter between 8–12 nm, which corresponds to the mesoporous structure of the material [48].

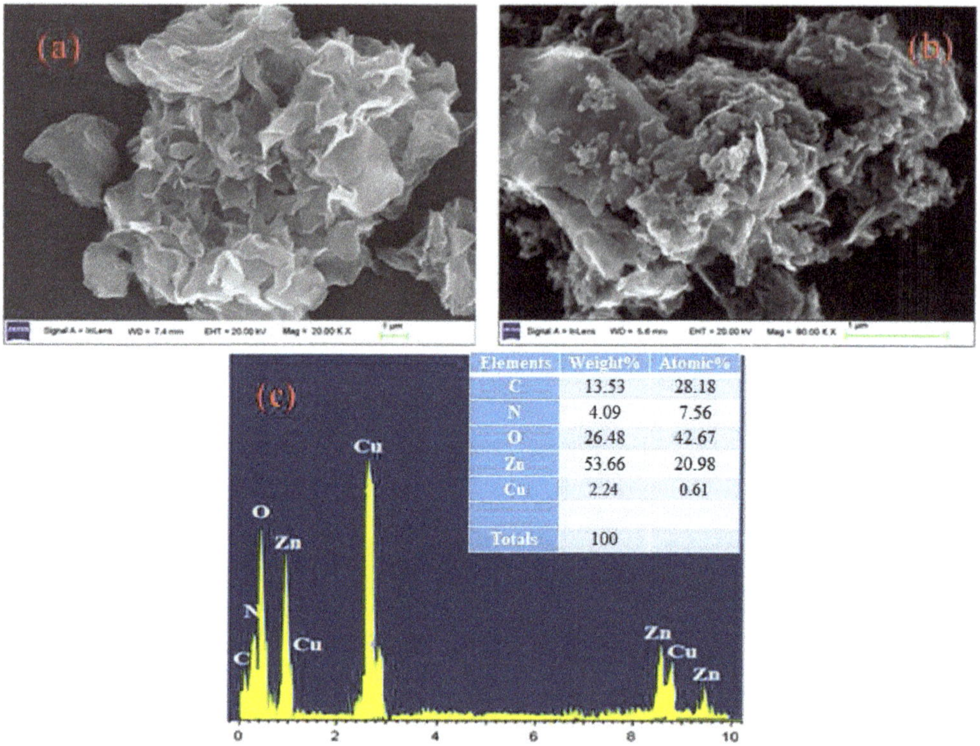

Figure 3. SEM images of (**a**) of g–C_3N_4, (**b**) ZnO–CuO/g–C_3N_4, and (**c**) EDS analysis of ZnO–CuO/g–C_3N_4 composite.

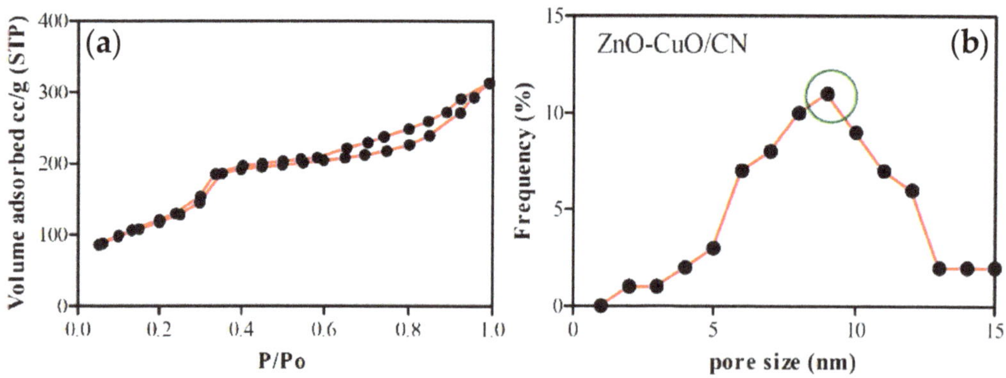

Figure 4. (**a**) N_2 adsorption-desorption isotherm curves of the synthesized ZnO–CuO/g–C_3N_4 and (**b**) their pore size distribution.

XPS Examination

XPS was applied to study the type of bonding and % age weight of each element present in the sample. Figure 5a,b represent the XPS analysis of ZnO and CuO, respectively. The result illustrated (Figure 5a) that two well-examined peaks are present at 1021.6 eV and

1034.3 eV, corresponding to Zn $2p_{3/2}$ and $2p_{1/2}$, respectively. The two very intense peaks originating at 933.7 eV and 953.8 eV are suggested for Cu $2p_{3/2}$ and $2p_{1/2}$, respectively (Figure 5b). Similarly, the XPS also confirmed the % weight of Zn and Cu, which are 26.7% and 31.22%, respectively.

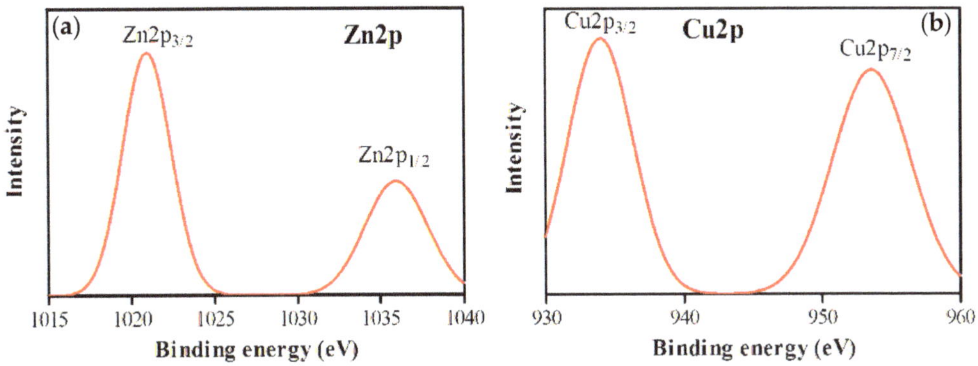

Figure 5. XPS analysis of synthesized nanocomposites, (**a**) Zn-2p and (**b**) Cu-2p.

3.5. Batch Adsorption Experiment

The water pH may impact the expulsion of arsenic by adsorbent material. To enhance the evacuation effectiveness of synthesized compounds in an actual application, the adsorption of three different samples of sodium arsenite (150, 100, 50 µg/L) was investigated under a range of pH, as it appears in Figure 6a. The shifting pattern of the curves revealed that acidic environments are beneficial for the removal of arsenic by ZnO–CuO/g–C_3N_4 material. In contrast, adsorption decreases at high pH levels, which is aligned with the prior studies [49]. It is additionally crucial that the expected pH value of the sodium arsenite solution is around 8; when the underlying pH is adapted to about 2.8, the adsorption capacity reports a total increasing tendency with a bit of increment adequacy. This is because, under corrosive conditions, an enormous quantity of H^+ ions can make adsorptive surfaces protonated and positively charged. Thus, it is easier to enhance the elimination of arsenic radicals by electrostatic forces. Although, when the pH ranged from basic up to 10, the adsorption capacity declined significantly. After being prewetted with water, the −OH ions may acquire the adsorption site with arsenite. The hydroxides of metal ions are created after adoption, which would stop the response from carrying on. In this way, examining pH supports choosing the ideal adsorbent in a reasonable application.

Taking into consideration the complicated chemical characteristics of normal water, the impacts of a few ordinary interfering ions (PO_4^{3-}, NO_3^-, SO_4^{2-}, CO_3^{2-} and Cl^-) with three initial dilutions (50, 100, 150 µg/L) on As(III) elimination by ZnO–CuO/g–C_3N_4 were tested, as can be observed in Figure 6b. Undoubtedly, the snooping of the five common ions decreased the adsorption limit, and the higher the number of active ions, the more significant the influence on adsorption capacity. Chloride ions have a minimal impact on the adsorption amount due to the additional framing of Cl^- ions by spheric complexes with a heterostructure composite. By correlation, the nearness of sulfate and carbonate affects the adsorption amount, particularly in the description that sulfate ions and carbonate ions convey progressively extra negative charges and occupy additional adsorption sites, which eventually decrease the removal proficiency of arsenic. Nitrate conveys a smaller amount of negative charge; therefore, its impacts on arsenic adsorption are firmly trailed by that of chloride ions. However, with high concentrations of conjunction anion quantities, seriously good adsorption develops stronger. Moreover, it is evident from the graph in Figure 6b that phosphate has the potential to interfere, and its existence seriously decreases the adsorption amount. In this examination, the concentration of contending anions was set much higher than those experienced in natural water. Consequently, even though confronting anions concentrations used in these experiments are unusually higher, they still have the adsorption ability for arsenic exclusion. Recyclability is one of the significant records to assess the application of adsorbents. Subsequently, the following study showed a sodium arsenite solution with a starting concentration of 50 µg/L as the trial object, and the adsorbent was 10 mg of ZnO–CuO/g–C_3N_4 that adsorbed in 24 h.

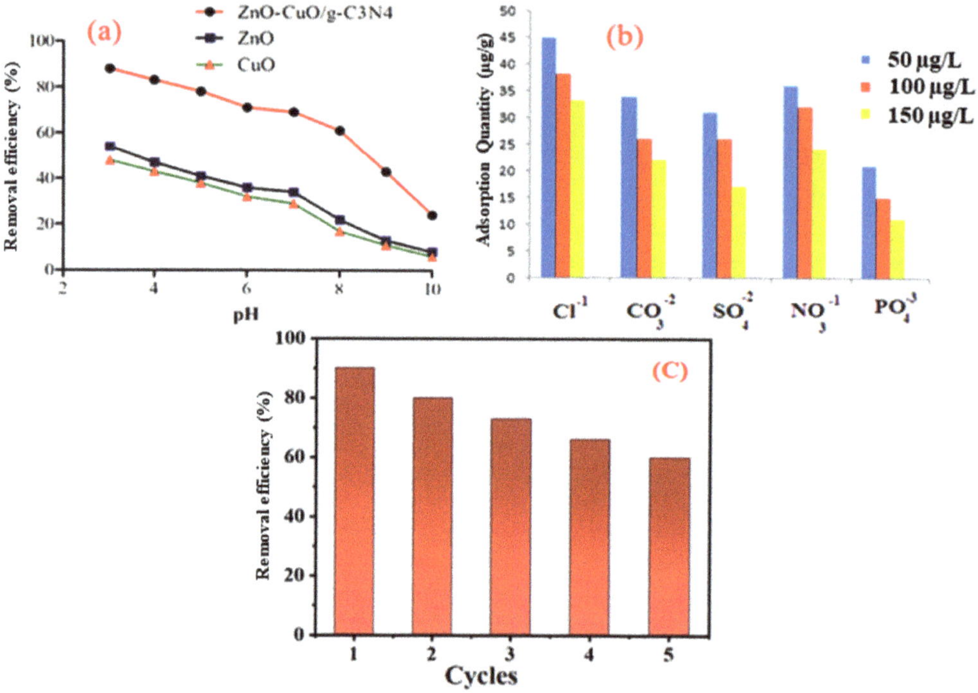

Figure 6. (**a**) Effect of pH (**b**) interfering ions on the removal of As(III) by ZnO–CuO/g–C_3N_4 composite, (**c**) removal rate (%) of As(III) by ZnO–CuO/g–C_3N_4 nanomaterial composites in five recycling steps.

On the other hand, a 0.1 M NaOH solution was used for the desorption of adsorbed materials, and the recycling experiments were repeated five times for the arsenic removal rate, as shown in Figure 6c, examining the reusability and stability of the prepared materials. Latterly, it tends to be the case that the exclusion percentage of the first cycle is as elevated as 90%. Still, the proficiency of the second exclusion cycle is reduced to about 80% because of several adsorption sites being filled and not desorbed. Furthermore, the elimination rates of the previous three cycles were not significantly different, showing that physio-adsorption may be dominant for the time being. The fifth cycle's As(IV) elimination rate can still reach above 60%, demonstrating that the synthesized materials, ZnO–CuO/g–C_3N_4, have excellent stability and recyclability and are projected to be functional in water purification.

3.6. *Effect of Physical Parameters on the Adsorption of Arsenic*

The effect of various physical parameters such as time, the concentration of nanomaterial, and temperature on the adsorption of arsenic was also examined at multiple time intervals in the presence of ZnO, CuO, and ZnO–CuO/g–C_3N_4 nanocomposites (Figure 7a). It was noticed that the adsorption capability of said nanomaterial improved with a rise in time by fixing the temperature at 50, pH at 4, and concentration of nanomaterial at 7 mg. 98% adsorption of As, which was achieved after 70 min of stirring.

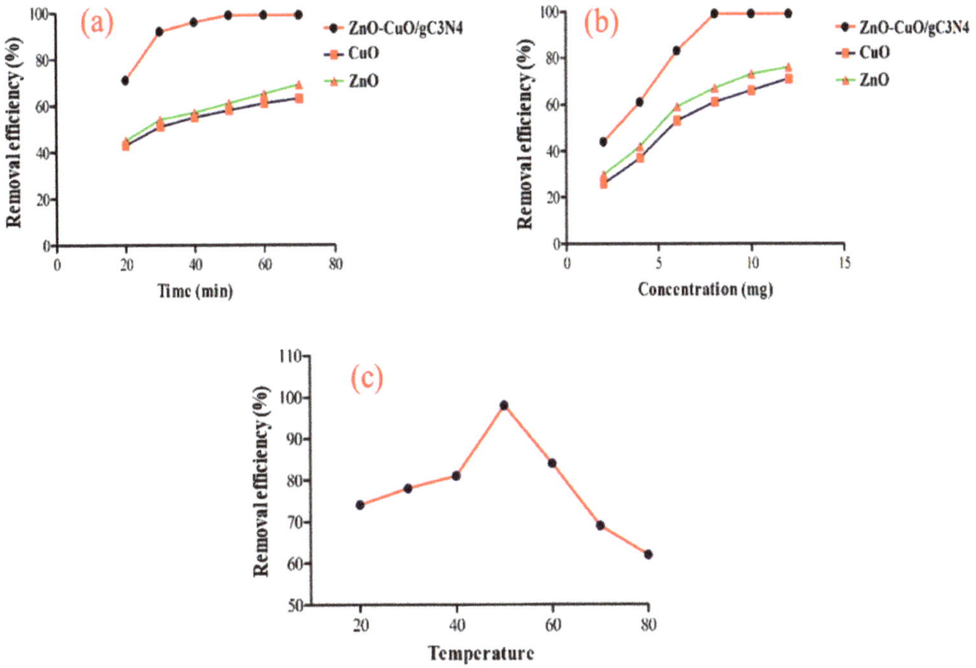

Figure 7. (**a**) Effect of time, (**b**) concentration of nanomaterial, (**c**) temperature on the adsorption of arsenic.

The arsenic adsorption capability of the as-synthesized ZnO, CuO, and ZnO–CuO/g–C_3N_4 nanocomposites were assessed at various nanomaterial amounts, i.e., 2 to 12 mg, as depicted in (Figure 7b). The result showed that As adsorption is directly related to the amount of nanomaterial used and a percent decrease (50–98%) of arsenic was observed from 2–7 mg of the nanomaterial used. Further, an increase in the amount of ZnO–CuO/g–C_3N_4 nanocomposites at specific optimized conditions does not affect the adsorption of arsenic because almost all arsenic is adsorbed at lower concentrations of the said nanomaterial. Less than 70% As was adsorbed by individual ZnO and CuO even at a concentration of 12 mg. Temperature also significantly affects As adsorption. In Figure 7c, the result showed that with an increase in temperature from 20 to 50 °C, the adsorption capability of the synthesized nanomaterials increases. However, beyond 50 °C, the adsorption capability of nanomaterials decreases sharply, which may be due to chemisorption fruitfully occurring at 50 °C. As adsorption also occurs at a lower temperature, physisorption also occurs side by side with chemisorption. All these results are well proven by Freundlich and Langmuir isotherms.

3.7. Adsorption Isotherms

To study the relationship between adsorption capacity and equilibrium concentration at different temperatures (20 to 80 °C), adsorption experimentation of sodium arsenite solution with starting concentration of 20–150 μg/L was performed for one day with an equimolar mass of adsorbent. On this premise, the popular Freundlich and Langmuir adsorption isotherm was utilized to fit the tested experimental data and to know the type of adsorption and higher arsenic adsorption ability; the results are shown in the form of the adsorption isotherm in Figure 8. Langmuir's calculations are dependent on the following perceptions: (1) The adsorbate accumulates on the upper surface of the adsorbent in the form of a monolayer; (2) limited adsorbent adsorption capacity; (3) the adsorbed particles do not interact with one another because they have the same reactive sites. The Freundlich model depicts multilayer adsorption and displays the variability of the adsorbent surface. The associated equations of the two models can be determined from Equations (3) and (4) [50,51].

$$Q_e = Q_m K_L C_e / 1 + K_L C_e \quad (3)$$

$$Q_e = KFC1/n\ e \quad (4)$$

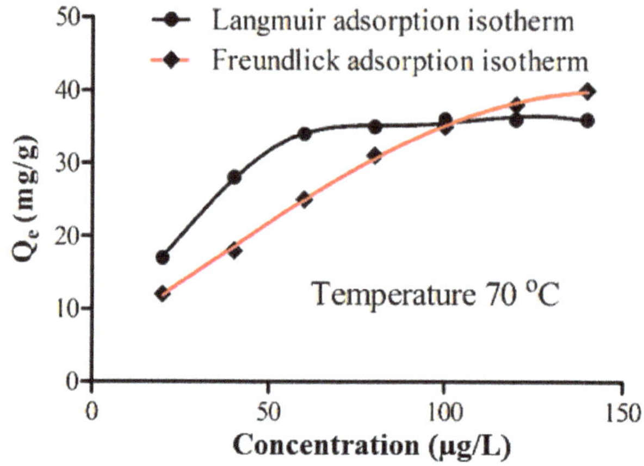

Figure 8. Freundlich and Langmuir isotherms of As(III) adsorption on ZnO–CuO/g–C_3N_4.

Here, C_e (μg/L) is the equilibrium concentration of adsorbate, Q_m (mg/g) and Q_e (mg/g) is the quantity of adsorbent absorbed at an equilibrium stage of C_e and the theoretical highest adsorption capacity, respectively; KF (mg/g) and K_L (μg/L) (μg/L) 1/n is the Freundlich constant and the Langmuir constant, correspondingly; and n is the heterogeneity factor. It tends to be naturally seen from Figure 8 that every one of the three curves displays a rising trend; however, there is no eternal ascending pattern. When the curve rises to a definite level, it could be flat due to the adsorption saturation. Additionally, the temperature also has a high impact on the adsorption ability. At a similar starting concentration, a temperature change can cause a difference in equilibrium concentration, affecting adsorption capabilities, and the law states that low temperatures are more favorable for adsorption. The boundaries of the Langmuir and Freundlich models are determined by association with Equation (1). Contrasting the R^2 of the two models, the R^2 fit by Langmuir (0.989) was higher than the R^2 fit by Freundlich (0.973), implying that the Langmuir model adequately exhibited the adsorption isotherms. The findings of this study revealed that arsenite was attached on the outside with confined and uniform sites, with a monolayer adsorption site. It can be naturally observed that acidic or neutral water is helpful for the adsorption of arsenic, which might be correlated with the diverse compositions of the prepared materials. Moreover, a comparison of different adsorbents with our designed ZnO–CuO/g–C_3N_4 adsorbent for arsenic removal is summarized in Table 1. The consequences recommended that ZnO–CuO/g–C_3N_4 was not just economically cheaper and ecologically safe in cases of raw ingredients, but also had a higher proficiency for arsenic removal.

Table 1. Comparison of the maximum arsenic adsorption capacity among the prepared ZnO–CuO/g–C_3N_4 nanocomposite and other adsorbents.

Adsorbents	**Adsorption Capacity** As ($mg \cdot g^{-1}$)	**References**
Reduced graphene oxide-supported mesoporous Fe_2O_3/TiO_2 nanoparticles	99.5	[52]
$CuFe_2O_4$ foam	85.4	[53]
Zirconium nanoscale carbon	110	[54]
(Layered double hydroxides/graphene oxide) nanocomposites	183.11	[55]
$MnFe_2O_4$	90	[56]
$CoFe_2O_4$	74	[56]
GO–$MnFe_2O_4$	207	[57]
Fe_3O_4	116.56	[58]
Ce–Fe Oxide–carbon nanotubes	30.96	[33]
Cellulose-based composites	83	[59]
Copper oxide (II) nanoparticles	88.3	[44]
ZnO–CuO/g–N_3C_4	97.56	Current work

3.8. Adsorption Kinetics

In adsorption kinetics, the relationship between the adsorption limit and time is investigated for various starting concentrations and temperatures of sodium arsenite solution. Moreover, Figure 9a displays the outcomes of adsorption amounts fluctuating at a certain temperature (70 °C) following the combination of a sodium arsenite solution with an underlying dilution of 150 μg/L consumed by a specific quantity of adsorptive material. Furthermore, it tends to be naturally observed that the adsorption volume expanded quickly in the initial 60 min and gradually expanded after the fixed 60 min. Still, the curve bends inclined toward the plane after 120 min. Since there were countless adsorption positions bringing adsorption from the start, arsenite particles responded to cupric oxide until the surface was fully occupied with plenty of hydroxyl functional clusters for adhesion. The place could be filled progressively with a gradual increase in adsorption quantity until all the sites were entirely occupied, leading to a saturated state. Likewise, various temperatures reported diverse adsorption capacities under similar conditions, and the standard is 70 °C, which is reliable with the above-stated results of the adsorption isotherm. Figure 9b shows the determined results for the change of adsorption quantity versus time intervals under multiple starting dilutions of 50 μg/L, 100 μg/L, and 150 μg/L at a temperature of 70 °C. At the start, the capacity of adsorption abruptly increases with time. It later slowly improves to the equilibrium state, and adsorption capability is larger for the higher concentration than for the lower concentration. It can be concluded that the capacity of adsorption increases at changing the degree of temperature and starting concentration with time. Adsorption kinetics may give valuable evidence for the entire adsorption method. For a deeper understanding of the impact of varying temperatures and to start focusing on the adsorption velocity, pseudo-first-order kinetics (Equation (5)) and 2nd-order kinetics (Equation (6)) were chosen to examine the experimental results extensively. Moreover, the kinetic equation of pseudo-first-order and pseudo-second-order is represented as follows:

$$\ln(Qe - Qt) = \ln Qe - K_1 t \quad (5)$$

$$t\ Qt = t\ Qe + 1\ K_2 Q_2\ e \quad (6)$$

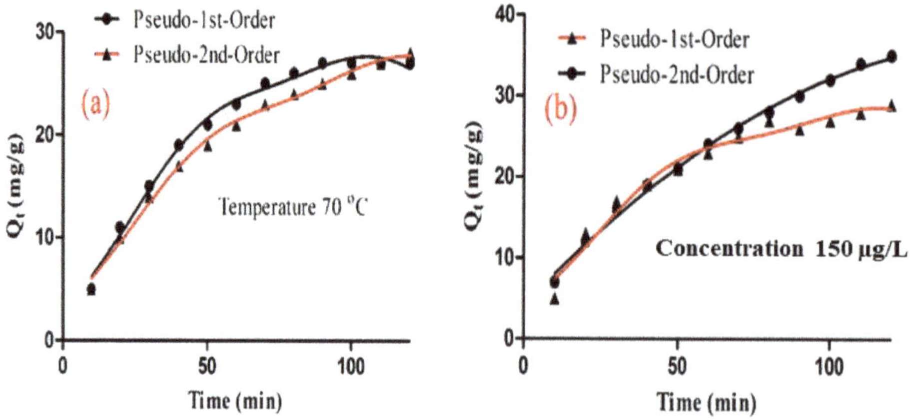

Figure 9. Kinetic study of As(III) removal by ZnO–CuO/g–C_3N_4 at (**a**) Temperature and (**b**) Concentration versus time intervals.

The quantities of As(III) adsorbed by adsorbents at equilibrium and at time t, respectively, are Q_e and Q_t (mg/g), and the rate constants for the pseudo-first-order and pseudo-second-order models, respectively, are k_1 (1/min) and k_2 (g/(mg min).

The value of R^2 shows that adsorption capacity was reported in a good arrangement in the presence of a pseudo-second-order dynamic model, where the connection temperature was already adjusted to 70 °C, and the starting focus of sodium arsenite was fixed at 50 µg/L and 100 µg/L. The outcome demonstrates that copper oxide is engaged with the reaction. The R^2 achieved via pseudo-first-order and pseudo-second-order dynamics are prominently dissimilar, demonstrating that typical adsorption is also found, primarily because of the copper oxide. Furthermore, a visible decline is seen in the constant rate of the pseudo-first-order kinetics model (K_1) and the pseudo-second-order kinetics model (K_2) with an increase in the temperature of the reaction. This demonstrates that either chemical or physical adsorption prevails, and the required time to bring the equilibrium in adsorption could increase in response to a rise in reaction temperature. In contrast, adsorption efficiency is improved at a lower temperature. On account of altering the starting concentrations, the constant rate of the pseudo-first-order kinetic model (K_1) is reported as 0.044 correspondingly, indicating that the concentration fluctuation has a slight impact on the typical adsorption process. The study of adsorption kinetic gives essential and valuable information to examine the process of adsorption efficiency and physicochemical responses.

3.9. Statistical Analysis

All the adsorption data were analyzed statistically to check the accessibility of the applied adsorption model and the efficiency of the synthesized adsorbent (ZnO–CuO/g–C_3N_4). A two-tailed *t*-test at a 5% significance level was used to confirm the optimum pH, and a Paired *t*-test was used to check the experiment's success.

3.9.1. Hypothesis Confirming Optimum pH of Adsorption of Arsenic (III)

Two hypotheses, null and alternate, were assumed to confirm the optimum pH by a two-tailed *t*-test at a 5% level. The maximum adsorption at various pH is given in Table 2.

Table 2. Maximum removal of arsenic (III) at various pH.

n	pH	% Removal (X_i)	$(X_i - X_{avg})^2$
1	3	98	915.06
2	4	92	588.06
3	5	84	264.06
4	6	73	27.56
5	7	65	7.56
6	8	61	45.56
7	9	43	612.56
8	10	26	1743.06
		$X_{avg} = 67.75$	$\sum(X_i - X_{avg})^2 = 4203.48$

Null Hypothesis = Optimum pH is equal to 3
Alternate Hypothesis = Optimum pH is not equal to 3
$T_{observed}$ was calculated using Equation (7):

$$t_{observed} = \frac{X_{avg} - \mu}{\sigma_s / \sqrt{n}} \quad (7)$$

In Equation (8), μ is the optimum pH, whereas σ_s is the standard deviation, and it was calculated by using Equation (8).

$$\sigma_s = \sqrt{[\sum\left(X_i - X_{avg}\right)^2 / (n-1)]} \quad (8)$$

$T_{observed}$ was found to be 0.92 and was compared with $t_{tabulated}$ (2.306). As $t_{observed}$ is less then $t_{tabulated,}$ the null hypothesis is accepted, and optimum pH = 3 for maximum adsorption is confirmed.

3.9.2. Hypothesis to Confirm the Success of the Experiment Using ZnO–CuO/g–C_3N_4 as Adsorbent

The success of the experiment was confirmed by proving that the concentration of arsenic (III) changes during the adsorption process. A Paired t-test was applied to the data given in Table 3 to test.

Table 3. Change in concentration of arsenic(III) during adsorption.

n	X_i (Arsenic Initial Concentration in ppm)	X_f (Arsenic Final Concentration in ppm)	$D_i = X_i - X_f$	$(D_i - D_{avg})^2$
1	20	3.5	16.5	1054.95
2	40	4.7	35.3	187.14
3	60	6.8	53.2	17.80
4	80	14.7	65.3	266.34
5	100	25.4	74.6	656.38
			$D_{avg} = 48.98$	$\sum(D_i - D_{avg})^2 = 2182.6$

Hypothesis. The two hypotheses were:
Null Hypothesis: No change in adsorbate concentration.
Alternate Hypothesis: During adsorption, adsorbate concentration changes.

$$t_{calculated} = \frac{D_{avg}}{\sigma_{diff}} \sqrt{n} \quad (9)$$

$$\sigma_{diff} = \frac{\sqrt{\left[\sum(D_i - D_{avg})^2\right]n}}{n-1} \tag{10}$$

σ_{diff} was calculated by Equation (5) 26.11 was found. In addition, the t value was calculated by Equation (4) which comes out to be 4.19, and it was compared with t tabulated (2.77). As $t_{calculated}$ is greater than the tabulation, the null hypothesis is rejected and the alternate hypothesis, arsenic (III) concentration changes during adsorption, is accepted.

3.10. Mechanism of As(III) Removal

The method of arsenic adsorption by material ZnO–CuO/g–C_3N_4 depends on the combined impact of redox reaction and physical interaction, and Scheme 1 shows the graphic design for the removal mechanism of arsenic. The preparation of a single CuO can result in severe accumulation, mainly decreasing the efficiency of the substance, but g–C_3N_4 based as the substrate not only resolved the environmental challenges affected by the single CuO and ZnO but also resolved the faults of accumulation, thus significantly enhancing the reactivity of materials and the adsorption proficiency. Moreover, the surface of g–C_3N_4 has plenty of –OH groups, which are beneficial for connecting copper ions. In the case of aqueous media, the progression of the hydrothermal process may result in copper ions formation, which can mostly create a CuO layer on the upper surface of g–C_3N_4 as per the "Seed growth method". Therefore, the g–C_3N_4 is an essential support for the CuO and ZnO to attain the highest proficiency. During the reactions, when the adsorbent comes in cross-contact with the arsenite ion, CuO oxidizes the arsenic trivalent (As (III)) into the arsenic pentavalent (As (V)). In the FT–IR spectra, the peak depth for –OH and CuO decreases, and enlightening hydroxyl is used during the adsorption. Accordingly, the partly oxidized arsenate ion might come in contact with the surface and respond with a –OH substitution, or it might be that As (III) directly attaches to the –OH. Convincingly, the determination of the mechanism indicates that the prepared ZnO–CuO/g–C_3N_4 can extract As from both physical and chemical reactions, and the combined impact of g–C_3N_4, ZnO, and CuO highly enhance the proficiency of arsenic removal. Another proposed mechanism is producing reactive oxygen species (ROS) in visible light on the surface of CuO and ZnO. The ROS produced may be hydroxyl radicals or superoxides. These ROS are easily obtained in the presence of ZnO, CuO, and g–C3N4 because all of these have shallow band gaps, and the ground electrons are easily subjected to an excited state even in ordinary visible light. After producing these oxygen species, they reacted with arsenic and adsorbed on the synthesized nanomaterial surface.

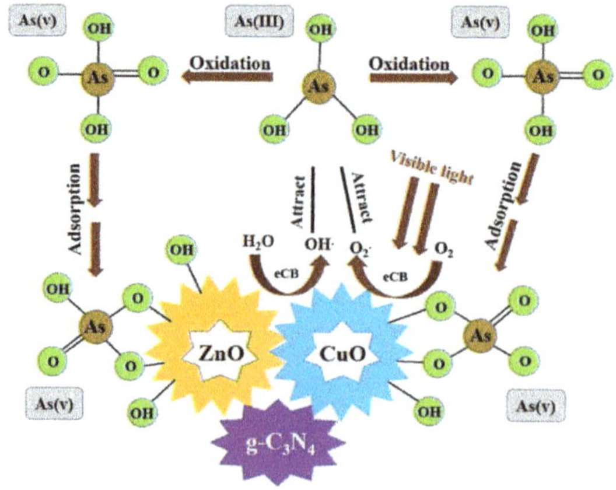

Scheme 1. Shows the schematic diagram for the removal mechanism of arsenic.

4. Conclusions

In brief, ZnO–CuO/g–C_3N_4 nanomaterial was synthesized by grafting ZnO–CuO onto g–C_3N_4. The shape, surface area, morphology, and stability of the synthesized nanostructures were confirmed by different analytical techniques. The result showed that the synthesized nanostructure could effectively remove arsenic from the water solution. Various factors affecting the adsorption efficiency of the said nanomaterials were also explored. The adsorption of arsenic increases with a decrease in pH, nanomaterials' concentration, and temperature (50 °C) was noted. Beyond this temperature, the adsorption efficacy of the as-synthesized nanomaterials decreased. Additionally, ZnO–CuO/g–C_3N_4 showed enhanced recycling activity in five consecutive cycles. The adsorption isotherm results showed that experimental data matches the Langmuir model well, suggesting that the adsorption approach is specific to a suitable site. The adsorption kinetic data exhibited that the maximum adsorption capacity of the as-synthesized composite by Langmuir and Freundlich was observed as 0.989 and 0.973 mg g^{-1}, respectively. Among the pseudo-second-order kinetic model and the pseudo-first-order kinetic model, there was no discernible difference in R^2. Consequently, our designed ZnO–CuO/g–C_3N_4 composites are a promising material for actual application in environmental remediation.

Author Contributions: Conceptualization, Q.U.K., N.B. and Z.U.R.; methodology, A.U.K.; software, K.T.; validation, E.S.M.T.E.D., A.A.A. and M.A.H.; formal analysis, D.L., P.B. and M.S.J.; investigation, E.S.M.T.E.D., A.A.A. and M.A.H.; resources, Q.U.K., N.B. and Z.U.R.; data curation, Q.U.K., N.B. and Z.U.R.; writing—original draft preparation, Q.U.K., N.B. and Z.U.R., E.S.M.T.E.D., A.A.A. and M.A.H.; writing—review and editing, E.S.M.T.E.D., A.A.A. and M.A.H.; visualization, Q.U.K., N.B. and Z.U.R.; supervision, D.L. and M.S.J.; project administration, P.B. and D.L.; funding acquisition, E.S.M.T.E.D., A.A.A. and M.A.H., M.S.J. and D.L. All authors have read and agreed to the published version of the manuscript.

Funding: This work was supported by research grants from the Natural Science Foundation of Shanghai (Grant 17ZR1402600), the Shanghai Science and Technology Innovation Action Plan, Belt & Road Young Scientist Program (Grant No. 19160745500), the Natural Science Foundation of Guangdong Province of China (Grant No. 2022A1515011368), the National Natural Science Foundation of China (Grant No. 81870307, 82270413) and The Key Projects of Department of Education of Guangdong Province of China (Grant No. 2022ZDZX2057, 2022ZXKC474). This work was funded by the Researchers Supporting Project Number (RSP-2021/243), King Saud University, Riyadh, Saudi Arabia.

Conflicts of Interest: The authors declare no conflict of interest.

References

1. Singh, R.; Singh, S.; Parihar, P.; Singh, V.P.; Prasad, S.M. Arsenic contamination, consequences and remediation techniques: A review. *Ecotoxicol. Environ. Saf.* **2015**, *112*, 247–270. [CrossRef] [PubMed]
2. Rathi, B.S.; Kumar, P.S. A review on sources, identification and treatment strategies for the removal of toxic Arsenic from water system. *J. Hazard. Mater.* **2021**, *418*, 126299. [CrossRef]
3. Jyothi, M.S.N.; Gayathri, S.; Gandhi, T.P.; Maliyekkal, S.M. Dissolved Arsenic in Groundwater Bodies: A Short Review of Remediation Technologies. In *Pollution Control Technologies*; Springer: Berlin/Heidelberg, Germany, 2021.
4. Madhukar, M.; Murthy, B.M.S.; Udayashankara, T.H. Sources of Arsenic in Groundwater and its Health Significance-A Review. *Nat. Environ. Pollut. Technol.* **2016**, *15*, 971–979.
5. Sarı, A.; Tuzen, M. Biosorption of As(III) and As(V) from aqueous solution by macrofungus (Inonotus hispidus) biomass: Equilibrium and kinetic studies. *J. Hazard. Mater.* **2009**, *164*, 1372–1378. [CrossRef] [PubMed]
6. Shrivastava, A. Dietary Arsenic Exposure: Sources and Risks. In *Arsenic Toxicity: Challenges and Solutions*; Kumar, N., Ed.; Springer Singapore: Singapore, 2021; pp. 95–125.
7. Fauser, P.; Sanderson, H.; Hedegaard, R.V.; Sloth, J.J.; Larsen, M.M.; Krongaard, T.; Bossi, R.; Larsen, J.B. Occurrence and sorption properties of arsenicals in marine sediments. *Environ. Monit. Assess.* **2013**, *185*, 4679–4691. [CrossRef] [PubMed]
8. Ahmad, S.; Wang, S.; Wu, W.; Yang, K.; Zhang, Y.; Tumukunde, E.; Wang, S.; Wang, Y. Functional Analysis of Peptidyl-prolyl cis-trans Isomerase from Aspergillus flavus. *Int. J. Mol. Sci.* **2019**, *20*, 2206. [CrossRef]
9. Giri, A.K.; Patel, R.K.; Mahapatra, S.S. Artificial neural network (ANN) approach for modelling of arsenic (III) biosorption from aqueous solution by living cells of Bacillus cereus biomass. *Chem. Eng. J.* **2011**, *178*, 15–25. [CrossRef]

10. Wang, S.; Mulligan, C.N. Speciation and surface structure of inorganic arsenic in solid phases: A review. *Environ. Int.* **2008**, *34*, 867–879. [CrossRef]
11. Wu, W.Z.; Ahmad, S.; Wang, S.; Zhang, Y.F.; Yang, H.; Wang, S.H.; Wang, Y. Expression and antibody preparation of Small Ubiquitin-like Modifier (SUMO) from Aspergillus flavus. In *IOP Conference Series: Earth and Environmental Science*; IOP Publishing: Bristol, UK, 2019; Volume 346, p. 012002.
12. Gomez-Caminero, A.; Howe, P.D.; Hughes, M.; Kenyon, E.; Lewis, D.R.; Moore, M.; Aitio, A.; Becking, G.C.; Ng, J.; International Programme on Chemical; et al. *Arsenic and Arsenic Compounds*; WHO: Geneva, Switzerland, 2001.
13. Monoj Kumar, M.; Garg, R. A comprehensive review on removal of arsenic using activated carbon prepared from easily available waste materials. *Environ. Sci. Pollut. Res.* **2017**, *24*, 13295–13306.
14. Zaw, M.; Emett, M.T. Arsenic removal from water using advanced oxidation processes. *Toxicol. Lett.* **2002**, *133*, 113–118. [CrossRef]
15. Nicomel, N.R.; Leus, K.; Folens, K.; Van Der Voort, P.; Du Laing, G. Technologies for Arsenic Removal from Water: Current Status and Future Perspectives. *Int J. Env. Res. Public Health* **2015**, *13*, 62. [CrossRef] [PubMed]
16. Mohan, D.; Pittman, U., Jr. Arsenic removal from water/wastewater using adsorbents—A critical review. *J. Hazard. Mater.* **2007**, *142*, 1–53. [CrossRef] [PubMed]
17. Kim, J.; Benjamin, M.M. Modeling a novel ion exchange process for arsenic and nitrate removal. *Water Res.* **2004**, *38*, 2053–2062. [CrossRef] [PubMed]
18. Farrell, J.; Wang, J.; O'Day, P.; Conklin, M. Electrochemical and spectroscopic study of arsenate removal from water using zero-valent iron media. *Environ. Sci. Technol.* **2001**, *35*, 2026–2032. [CrossRef] [PubMed]
19. Yamani, J.S.; Lounsbury, A.W.; Zimmerman, J.B. Towards a selective adsorbent for arsenate and selenite in the presence of phosphate: Assessment of adsorption efficiency, mechanism, and binary separation factors of the chitosan-copper complex. *Water Res.* **2016**, *88*, 889–896. [CrossRef]
20. Yu, X.-Y.; Luo, T.; Jia, Y.; Zhang, Y.-X.; Liu, J.-H.; Huang, X.-J. Porous hierarchically micro-/nanostructured MgO: Morphology control and their excellent performance in As (III) and As (V) removal. *J. Phys. Chem. C* **2011**, *115*, 22242–22250. [CrossRef]
21. Mou, F.; Guan, J.; Ma, H.; Xu, L.; Shi, W. Magnetic iron oxide chestnutlike hierarchical nanostructures: Preparation and their excellent arsenic removal capabilities. *ACS Appl. Mater. Interfaces* **2012**, *4*, 3987–3993. [CrossRef]
22. Cao, C.-Y.; Qu, J.; Yan, W.-S.; Zhu, J.-F.; Wu, Z.-Y.; Song, W.-G. Low-cost synthesis of flowerlike α-Fe_2O_3 nanostructures for heavy metal ion removal: Adsorption property and mechanism. *Langmuir* **2012**, *28*, 4573–4579. [CrossRef]
23. Guo, S.; Sun, W.; Yang, W.; Xu, Z.; Li, Q.; Shang, J.K. Synthesis of Mn_3O_4/CeO_2 hybrid nanotubes and their spontaneous formation of a paper-like, free-standing membrane for the removal of arsenite from water. *ACS Appl. Mater. Interfaces* **2015**, *7*, 26291–26300. [CrossRef]
24. Jeong, Y.; Fan, M.; Singh, S.; Chuang, C.-L.; Saha, B.; Van Leeuwen, J.H. Evaluation of iron oxide and aluminum oxide as potential arsenic (V) adsorbents. *Chem. Eng. Processing-Process. Intensif.* **2007**, *46*, 1030–1039. [CrossRef]
25. Li, W.; Chen, D.; Xia, F.; Tan, J.Z.; Huang, P.-P.; Song, W.-G.; Nursam, N.M.; Caruso, R.A. Extremely high arsenic removal capacity for mesoporous aluminium magnesium oxide composites. *Environ. Sci. Nano* **2016**, *3*, 94–106. [CrossRef]
26. Cao, A.-M.; Monnell, J.D.; Matranga, C.; Wu, J.-M.; Cao, L.-L.; Gao, D. Hierarchical nanostructured copper oxide and its application in arsenic removal. *J. Phys. Chem. C* **2007**, *111*, 18624–18628. [CrossRef]
27. Saddam, A.; Wang, H.; Liu, Y.; Peng, J.; Jiang, H. Variable universe fuzzy logic-based hybrid LFC control with real-time implementation. *IEEE Access* **2019**, *7*, 25535–25546.
28. Hristovski, K.; Baumgardner, A.; Westerhoff, P. Selecting metal oxide nanomaterials for arsenic removal in fixed bed columns: From nanopowders to aggregated nanoparticle media. *J. Hazard. Mater.* **2007**, *147*, 265–274. [CrossRef] [PubMed]
29. Awual, M.R.; Yaita, T.; Suzuki, S.; Shiwaku, H. Ultimate selenium (IV) monitoring and removal from water using a new class of organic ligand based composite adsorbent. *J. Hazard. Mater.* **2015**, *291*, 111–119. [CrossRef]
30. Awual, M.R.; El-Safty, S.A.; Jyo, A. Removal of trace arsenic (V) and phosphate from water by a highly selective ligand exchange adsorbent. *J. Environ. Sci.* **2011**, *23*, 1947–1954. [CrossRef]
31. Awual, M.R.; Asiri, A.M.; Rahman, M.M.; Alharthi, N.H. Assessment of enhanced nitrite removal and monitoring using ligand modified stable conjugate materials. *Chem. Eng. J.* **2019**, *363*, 64–72. [CrossRef]
32. Gupta, K.; Ghosh, U.C. Arsenic removal using hydrous nanostructure iron (III)–titanium (IV) binary mixed oxide from aqueous solution. *J. Hazard. Mater.* **2009**, *161*, 884–892. [CrossRef]
33. Chen, B.; Zhu, Z.; Ma, J.; Qiu, Y.; Chen, J. Surfactant assisted Ce–Fe mixed oxide decorated multiwalled carbon nanotubes and their arsenic adsorption performance. *J. Mater. Chem. A* **2013**, *1*, 11355–11367. [CrossRef]
34. Alka, S.; Shahir, S.; Ibrahim, N.; Ndejiko, M.J.; Vo, D.V.N.; Manan, F.A.J.J.o.C.P. Arsenic removal technologies and future trends: A mini review. *J. Clean. Prod.* **2021**, *278*, 123805. [CrossRef]
35. Lata, S.; Samadder, S. Removal of arsenic from water using nano adsorbents and challenges: A review. *J. Environ. Manag.* **2016**, *166*, 387–406. [CrossRef] [PubMed]
36. Huo, L.; Zeng, X.; Su, S.; Bai, L.; Wang, Y. Enhanced removal of As (V) from aqueous solution using modified hydrous ferric oxide nanoparticles. *Sci. Rep.* **2017**, *7*, 40765. [CrossRef] [PubMed]
37. Awual, M.R.; Jyo, A. Rapid column-mode removal of arsenate from water by crosslinked poly (allylamine) resin. *Water Res.* **2009**, *43*, 1229–1236. [CrossRef] [PubMed]

38. Barrio, J.; Mateo, D.; Albero, J.; García, H.; Shalom, M. A Heterogeneous Carbon Nitride–Nickel Photocatalyst for Efficient Low-Temperature CO_2 Methanation. *Adv. Energy Mater.* **2019**, *9*, 1902738. [CrossRef]
39. Barrio, J.; Volokh, M.; Shalom, M. Polymeric carbon nitrides and related metal-free materials for energy and environmental applications. *J. Mater. Chem. A* **2020**, *8*, 11075–11116. [CrossRef]
40. Fu, J.; Yu, J.; Jiang, C.; Cheng, B. g-C_3N_4-Based Heterostructured Photocatalysts. *Adv. Energy Mater.* **2018**, *8*, 1701503. [CrossRef]
41. Zhang, G.; Li, G.; Wang, X. Surface Modification of Carbon Nitride Polymers by Core–Shell Nickel/Nickel Oxide Cocatalysts for Hydrogen Evolution Photocatalysis. *ChemCatChem* **2015**, *7*, 2864–2870. [CrossRef]
42. Chen, X.; Wu, Z.; Liu, D.; Gao, Z. Preparation of ZnO Photocatalyst for the Efficient and Rapid Photocatalytic Degradation of Azo Dyes. *Nanoscale Res. Lett.* **2017**, *12*, 143. [CrossRef]
43. Bojdys, M.J.; Müller, J.O.; Antonietti, M.; Thomas, A.J.C.A.E.J. Ionothermal synthesis of crystalline, condensed, graphitic carbon nitride. *Chem. A Eur. J.* **2008**, *14*, 8177–8182. [CrossRef]
44. Zhu, B.; Xia, P.; Ho, W.; Yu, J. Isoelectric point and adsorption activity of porous g-C_3N_4. *Appl. Surf. Sci.* **2015**, *344*, 188–195. [CrossRef]
45. Javed, M.S.; Lei, H.; Wang, Z.; Liu, B.-T.; Cai, X.; Mai, W.J.N.E. 2D V_2O_5 nanosheets as a binder-free high-energy cathode for ultrafast aqueous and flexible Zn-ion batteries. *Nano Energy* **2020**, *70*, 104573. [CrossRef]
46. Javed, M.S.; Lei, H.; Shah, H.U.; Asim, S.; Raza, R.; Mai, W. Achieving high rate and high energy density in an all-solid-state flexible asymmetric pseudocapacitor through the synergistic design of binder-free 3D $ZnCo_2O_4$ nano polyhedra and 2D layered $Ti_3C_2T_x$-MXenes. *J. Mater. Chem. A* **2019**, *7*, 24543–24556. [CrossRef]
47. Mateen, A.; Javed, M.S.; Khan, S.; Saleem, A.; Majeed, M.K.; Khan, A.J.; Tahir, M.F.; Ahmad, M.A.; Assiri, M.A.; Peng, K.-Q. Metal-organic framework-derived walnut-like hierarchical Co-O-nanosheets as an advanced binder-free electrode material for flexible supercapacitor. *J. Energy Storage* **2022**, *49*, 104150. [CrossRef]
48. Javed, M.S.; Shaheen, N.; Hussain, S.; Li, J.; Shah, S.S.A.; Abbas, Y.; Ahmad, M.A.; Raza, R.; Mai, W.J.J.o.m.c.A. An ultra-high energy density flexible asymmetric supercapacitor based on hierarchical fabric decorated with 2D bimetallic oxide nanosheets and MOF-derived porous carbon polyhedra. *J. Mater. Chem. A* **2019**, *7*, 946–957. [CrossRef]
49. Ge, X.; Ma, Y.; Song, X.; Wang, G.; Zhang, H.; Zhang, Y.; Zhao, H. β-FeOOH Nanorods/Carbon Foam-Based Hierarchically Porous Monolith for Highly Effective Arsenic Removal. *ACS Appl. Mater. Interfaces* **2017**, *9*, 13480–13490. [CrossRef] [PubMed]
50. Liu, Y.; Peng, Y.; Zhang, T.; Qiu, F.; Yuan, D. Superhydrophobic, ultralight and flexible biomass carbon aerogels derived from sisal fibers for highly efficient oil–water separation. *Cellulose* **2018**, *25*, 3067–3078. [CrossRef]
51. Wang, Y.; Xu, J.; Xu, X.; Yang, D.; Zheng, X.; Pan, J.; Zhang, T.; Qiu, F.; Li, C. Mesoporous hollow silicon spheres modified with manganese ion sieve: Preparation and its application for adsorption of lithium and rubidium ions. *Appl. Organomet. Chem.* **2018**, *32*, e4182. [CrossRef]
52. Babu, C.M.; Vinodh, R.; Sundaravel, B.; Abidov, A.; Peng, M.M.; Cha, W.S.; Jang, H.-T. Characterization of reduced graphene oxide supported mesoporous Fe_2O_3/TiO_2 nanoparticles and adsorption of As (III) and As (V) from potable water. *J. Taiwan Inst. Chem. Eng.* **2016**, *62*, 199–208. [CrossRef]
53. Wu, L.-K.; Wu, H.; Liu, Z.-Z.; Cao, H.-Z.; Hou, G.-Y.; Tang, Y.-P.; Zheng, G.-Q. Highly porous copper ferrite foam: A promising adsorbent for efficient removal of As(III) and As(V) from water. *J. Hazard. Mater.* **2018**, *347*, 15–24. [CrossRef]
54. Mahanta, N.; Chen, J.P. A novel route to the engineering of zirconium immobilized nano-scale carbon for arsenate removal from water. *J. Mater. Chem. A* **2013**, *1*, 8636–8644. [CrossRef]
55. Wen, T.; Wu, X.; Tan, X.; Wang, X.; Xu, A. One-Pot Synthesis of Water-Swellable Mg-Al Layered Double Hydroxides and Graphene Oxide Nanocomposites for Efficient Removal of As(V) from Aqueous Solutions. *Acs Appl. Mater. Interfaces* **2013**, *5*, 3304–3311. [CrossRef] [PubMed]
56. Zhang, S.; Niu, H.; Cai, Y.; Zhao, X.; Shi, Y. Arsenite and arsenate adsorption on coprecipitated bimetal oxide magnetic nanomaterials: $MnFe_2O_4$ and $CoFe_2O_4$. *Chem. Eng. J.* **2010**, *158*, 599–607. [CrossRef]
57. Kumar, S.; Nair, R.R.; Pillai, P.B.; Gupta, S.N.; Iyengar, M.; Sood, A. Graphene oxide–$MnFe_2O_4$ magnetic nanohybrids for efficient removal of lead and arsenic from water. *ACS Appl. Mater. Interfaces* **2014**, *6*, 17426–17436. [CrossRef] [PubMed]
58. Feng, L.; Cao, M.; Ma, X.; Zhu, Y.; Hu, C. Superparamagnetic high-surface-area Fe_3O_4 nanoparticles as adsorbents for arsenic removal. *J. Hazard. Mater.* **2012**, *217*, 439–446. [CrossRef]
59. Mukherjee, S.; Kumar, A.A.; Sudhakar, C.; Kumar, R.; Ahuja, T.; Mondal, B.; Srikrishnarka, P.; Philip, L.; Pradeep, T. Sustainable and Affordable Composites Built Using Microstructures Performing Better than Nanostructures for Arsenic Removal. *Acs Sustain. Chem. Eng.* **2019**, *7*, 3222–3233. [CrossRef]

nanomaterials

Article

Preparation and Characterization of an Electrospun Whey Protein/Polycaprolactone Nanofiber Membrane for Chromium Removal from Water

Laura Cristina Ramírez-Rodríguez [1], María Ximena Quintanilla-Carvajal [2], Didilia Ileana Mendoza-Castillo [3,4], Adrián Bonilla-Petriciolet [4] and Carlos Jiménez-Junca [5,*]

[1] Maestría en Diseño y Gestión de Procesos Facultad de Ingeniería, Campus Universitario Puente del Común, Universidad de la Sabana, Km. 7 Autopista Norte, Chia 25001, Colombia
[2] Agroindustrial Processes Research Group, Campus Universitario Puente del Común, Universidad de La Sabana, Km. 7 Autopista Norte, Chia 25001, Colombia
[3] CONACYT, Ciudad de México 03940, Mexico
[4] Departamento de Ingeniería Química, Instituto Tecnológico de Aguascalientes, Aguascalientes 20256, Mexico
[5] Bioprospecting Research Group, Campus Universitario Puente del Común, Universidad de La Sabana, Km. 7 Autopista Norte, Chia 25001, Colombia
* Correspondence: carlos.jimenez@unisabana.edu.co; Tel.: +57-861-5555-25010

Abstract: Chromium pollution represents a worldwide concern due to its high toxicity and bioaccumulation in organisms and ecosystems. An interesting material to remove metal ions from water is a whey-protein-based material elaborated by electrospinning, which is an emerging method to produce adsorbent membranes with diverse applications. The aim of this study was to prepare an adsorbent membrane of whey protein isolate (WPI) and polycaprolactone (PCL) by electrospinning to remove chromium ions from water. The adsorbent membrane was synthesized by a central composed design denaturing WPI using 2-Mercaptoethanol and mixing it with PCL to produce electrospun nanofibers. The adsorbent membrane was characterized by denaturation, Scanning Electron Microscope, Fourier-Transform Infrared Spectroscopy, Contact Angle, Thermogravimetric Analysis, and X-ray Photoelectron Spectrometry. The adsorption properties of this membrane were assessed in the removal of chromium. The removal performance of the membrane was enhanced by an increase in temperature showing an endothermic adsorption process. The adsorption process of chromium ions onto the nanofiber membrane followed the Sips adsorption isotherm, while the adsorption kinetics followed a pseudo-second kinetics where the maximum adsorption capacity was 31.0 mg/g at 30 °C and pH 2. This work provides a novel method to fabricate a hybrid membrane with amyloid-type fibrils of WPI and PCL, which is a promising adsorbent to remove heavy metal ions from water.

Keywords: adsorption; chromium; electrospinning; hybrid membrane; nanofiber; polycaprolactone; whey

Citation: Ramírez-Rodríguez, L.C.; Quintanilla-Carvajal, M.X.; Mendoza-Castillo, D.I.; Bonilla-Petriciolet, A.; Jiménez-Junca, C. Preparation and Characterization of an Electrospun Whey Protein/Polycaprolactone Nanofiber Membrane for Chromium Removal from Water. *Nanomaterials* **2022**, *12*, 2744. https://doi.org/10.3390/nano12162744

Academic Editor: Christos A. Aggelopoulos

Received: 21 July 2022
Accepted: 7 August 2022
Published: 10 August 2022

1. Introduction

Water contamination from toxic heavy metals is one of the most serious worldwide environmental problems due to the rapid growth of industrial development and non-restrictive legislation [1]. Heavy metals are not biodegradable, exhibit high toxicity, bioaccumulated and bio-magnified in the food chain thus threatening the health of humans and ecological systems. Chromium (Cr) has been recognized as a potential risk to human health and the environment due to its high toxicity, carcinogenicity, mutagenicity and it is commonly found in wastewater of many industries [2]. According to the US-EPA and WHO, the maximum concentration for total chromium in drinking water is 100 μg/L and 50 μg/L [1], respectively. Similarly, the US-EPA and the European Union stated that the amount of chromium discharged to surface water should be below 2 mg/L for total chromium and 50 μg/L for Cr(VI) [3].

The removal of heavy metal ions from water should be performed efficiently, easily, without waste generation (e.g., sludge) and at low cost, and this topic represents an important environmental issue globally [4]. To reach the regulatory limits for water quality, several methods have been used to remove heavy metal ions from water including chemical precipitation, coagulation, ion exchange, electrodialysis, and among others [5,6]. Nowadays, a common method used for wastewater treatment is membrane filtration due to its easy implementation in large-scale applications where nanofiltration and reverse osmosis are used to remove heavy metals. However, these methods are high pressure-driven that demand high energy consumption due to the narrow pore size distribution of the membrane, which also tends to foul [7]. Alternatively, adsorption can be used to remove heavy metals due to its high efficiency, simplicity, versatility, the availability of different adsorbents besides their regeneration and recyclability of them [8].

To overcome the high energy consumption and fouling of the membrane filtration process in the removal of heavy metal ions, the adsorptive membrane technology has been a better option due to its convenient operation using membranes with higher pore size, which combines the adsorption potential of the material and the filtration performance of the membrane [9]. In recent years, nanotechnology has been used in adsorbents production to obtain high-efficacy adsorbents for pollutant removal [10]. For example, membrane-based adsorption technologies composed of electrospun nanofibers have been receiving widespread attention due to the advantages of high efficiency, relatively simple production, versatility, large specific-surface area, high porosity, and good structural stability, which make it a promising technology for water treatment [11].

Electrospinning is an emerging technique that has shown promising results because it is the simplest and most cost-effective method to produce a wide range of nanofibers in some cases smaller than 100 nm in diameter and is scalable for mass production [4]. To carry out the electrospinning process, several parameters must be considered such us polymer concentration, solution viscosity, electrical conductivity, and surface tension. Likewise, the operational parameters include electrical field strength, fluid flow rate and tip-to-collector distance [12]. When all conditions are met, electrostatic forces between the needle and the collector attract the solution, the Taylor cone is formed and finally the fibers are obtained (Figure 1).

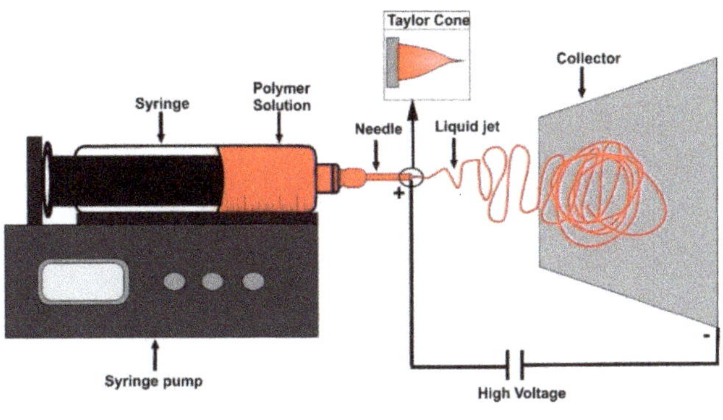

Figure 1. Illustration of an electrospinning apparatus.

Nanofibers produced by electrospinning for heavy metal removal have been gaining relevance due to their physical structure, good adsorption performance and favorable life cycle analysis [13]. Regarding chromium removal recently, Herath et al. elaborated polyacrylonitrile/ionic covalent organic framework hybrid nanofibers of 440 nm diameter by electrospinning to remove Cr(VI) from water, and the fibers presented an adsorption capacity of 173 mg/g [14]. Likewise, Sharafoddinzadeh et al. reported a maximum adsorption capacity of 225 mg/g to remove chromium from water using a polyacrylonitrile nanofibers of 165 nm of diameter made by electrospinning [15]. Ansari et al. prepared an electrospun zein/nylon-6 nanofiber membrane to remove Cr(VI) from water with an adsorption capacity of 4.73 mg/g [16].

Nanofibers elaborated by electrospinning are commonly made using polymers. Polycaprolactone (PCL) is widely used to fabricate nanofibers by electrospinning, because it bears high mechanical strength, easy processability, hydrophobicity and lack of toxicity [17]. PCL is known for its applications in tissue engineering owing to its biodegradability and biocompatibility, and has thus become a good choice for environmental applications due to its degradation time [18]. However, the use of PCL alone in adsorption materials is limited owing to its neutral charge and lack of functional groups [19]. Consequently, copolymerization and blending with other materials have been recently employed in order to improve its chemical, physical and mechanical properties [20].

Blends of polymers with other materials have many advantages owing to the combination of properties and the strong economic incentives arising from their use [19]. Consequently, a blend with amyloids was proposed to improve the chemical limitations of PCL, which are proteins in a fibrillar state that can be formed under certain conditions when denatured [21]. Hence, amyloid fibrils have been widely used recently for environmental remediation especially in the recovery of heavy metal ions owing to the high affinity of amyloids with these pollutants [22]. Recently, Ramírez-Rodríguez et al. [23] developed an adsorptive amyloid membrane to remove heavy metals from wastewater using whey amyloid fibrils and activated carbon, which is an efficient and economic technology to remove a wide range of heavy metals. Similarly, numerous studies support the promising potential of amyloids to remove heavy metals from water [24–26]. Nevertheless, due to the mechanical and physical properties of amyloids, it is difficult to handle alone. For this reason, most studies use a support to hold amyloids into it [27]. In this way, PCL is an excellent material to make a hybrid adsorption membrane of PCL and amyloids of whey to enhance the advantages of both materials.

Whey is a by-product of the cheese industry composed mainly of β-Lactoglobulin (60%) and α-lactalbumin (20%), which is a promising raw material to produce amyloids due to its high content of protein [12]. Therefore, Ahmed et al. [19] studied PCL/WPC nanofibers for pharmaceutical applications, where it presented good morphology, wettability, high porosity and degradation ability. This is evidence that the synergy of the high metal binding capacity of proteins with a high surface area material may lead to promising electrospun membranes for heavy metal ion removal. The aim of this study was to produce WPI/PCL membranes by electrospinning to remove Cr ions from water. This approach is an opportunity to give an added value to whey, which is a residue of the cheese industry. The estimations indicate that nine kilograms of whey are produced per one kilogram of cheese, and nowadays it is readily available in the dairy industry [28]. Furthermore, the use of electrospinning is a highly innovative method to produce amyloid-based materials allowing a targeted design of membranes for toxic ion removal.

2. Materials and Methods

2.1. Materials

Poly(epsilon-caprolactone) (PCL, Mn = 70–90 kDa) was obtained from Sigma Aldrich (St. Louis, MI, USA), Whey Protein Isolate (WPI) was purchased from Davisco Foods International Inc. (Eden Prairie, MN, USA) and it was 97% protein of which approximately 69% was β-Lactoglobulin and 22% was α-lactalbumin, activated carbon was obtained from Sigma Aldrich (USA) with a particle size of 150 μm. The 2-Mercaptoethanol (βME, 98%) was purchased from Bio-Rad Laboratories (Mississauga, ON, Canada). Tetrahydrofuran (THF) and Dimethyl Formamide (DMF) were obtained from AppliChem (AppliChem, Darmstadt, Germany), which were analytic grade and used as received.

2.2. Preparation of the Adsorptive Membrane

Polymeric dispersions of WPI and PCL were prepared according to a Central Composited Design (CCD) as reported in Table 1, which evaluated different WPI and PCL proportions, βME concentration and the collector distance of the electrospinning. Solutions of PCL and WPI with βME were dissolved in THF and DMF (7:3 v/v) separately and stirred overnight at room temperature to ensure complete dissolution [19]. The solutions of PCL and WPI with βME were mixed and completed with THF/DMF (7:3 v/v) to reach the desired final concentration and stirred for 5 h. Finally, the activated carbon was added at a concentration of 0.05% to the solution with the best results to analyze if it produced a change on the adsorption capacity of fibrils.

Table 1. Central Composite Design (CCD) to prepare the WPI/PCL hybrid adsorbent membrane for chromium removal.

Variable	Unit	Factors	Levels		
			−1	0	1
WPI	%	A	50	70	90
βME	%	B	0.1	0.65	1.2
Collector Distance	cm	C	7	10	13

α = 1 (star or axial point for orthogonal small CCD in the case of three independent variables) with 5 center points.

A horizontal electrospinning Fluidnatek® LE-10 (Bioinicia, Valencia, Spain) apparatus was used to produce the hybrid adsorbent membrane. The solution of WPI and PCL was pumped by a syringe pump operated at a flow rate of 4 mL/h, a voltage supply of 15 kV, collector distance from 7 to 13 cm [29]. The viscosity of the solutions was determined by Anton Paar MCR 502 rheometer (Anton Paar, Graz, Austria). The surface tension (ST) of the solutions was measured in the Sigma 700 tensiometer (Attension, Espoo, Finland) equipped with Wilhelmy plate at room temperature (19 °C) with 20 mL of sample, which was placed in a standard glass vessel to perform the measurement in triplicate.

The Design-Expert 10.0 software (Stat-Ease, Minneapolis, MN, USA) was used for the experimental design, response surface modeling, statistical regression analysis, process optimization and to evaluate the effect of the solution preparation on the viscosity, surface tension, fiber diameter and adsorption properties using a concentration of 100 mg/L of Cr(VI) at 30 °C, pH 3 for 24 h. The results were fitted via the response surface regression procedure using the following second-order polynomial Equation (1) and the models were evaluated following the criterion of desirability [30].

$$Y = \beta_0 + \sum_{i=1}^{n} \beta_i x_i + \sum_{i=1}^{n-1} \sum_{j=i+1}^{n} \beta_{ij} x_i x_j + \sum_{i=1}^{n} \beta_{ii} {x_i}^2 \quad (1)$$

where Y is the predicted response, β_0 is the regression coefficients, β_i is the linear coefficient, β_{ii} is the quadratic coefficients, β_{ij} is the interaction coefficients and X_i is the coded levels of independent variables, respectively.

2.3. *Characterization of the Adsorptive Membrane*

The characterization of the hybrid adsorbent membrane was carried out following the method proposed by Cong et al. [31]. The morphology of the hybrid membrane was analyzed by the fibril diameter using ImageJ and Tescan LYRA 3 Scanning Electron Microscopy (SEM) (TESCAN, Brno, Czech Republic) at an acceleration voltage of 4 kV [25], where all samples were gold-plated prior to observation. The surface chemistry of fibrils was determined by Fourier Transform Infrared Spectroscopy FTIR (Thermo Scientific®, Waltham, MA, USA) in the region of 400–1600 cm^{-1}. The denaturation of WPI as amyloid was determined by Congo Red assay where aqueous solutions of 15% of WPI with different concentrations (0.1–1.2% v/v) of βME were analyzed following the method proposed by Yakupova et al. [32]. The effect of denatured WPI on PCL over thermal and mechanical performance was studied with thermogravimetric assay (TGA) on a TGA/DSC Mettler Toledo STAR 1 System (Mettler Toledo, Columbus, OH, USA) at a rate of 10 °C/min under nitrogen atmosphere in the range of 30–600 °C according to ASTM D6370. The contact angle was determined using distilled water at 20 °C with the sessile drop method in a Drop Shape

Analysis System DSA (GH11, Krüss, Hamburg, Germany) according to ASTM-D7334-08 (2013). The average contact angle corresponded to three measurements for each hybrid membrane. X-ray Photoelectron Spectrometry (XPS) characterization was obtained on a Centeno-XPS/ISS/UPS X-ray Photoelectron Spectrometer (SPECS, Berlin, Germany) where the spectra were recorded using monochromatic Al Kα radiation (hv = 1486.6 eV), and the data analysis was performed with the CasaXPS program (Casa Software Ltd., Delray Beach, FL, USA).

2.4. Chromium Adsorption Capacity of the Adsorptive Membrane

To determine the adsorption capacity of the WPI/PCL hybrid membrane, the method of Sahebjamee et al. [33] was used where 20 g/L of the hybrid membrane was immersed in a heavy metal solution. Prior to the experiment, stock aqueous solutions of 1000 mg/L sodium chromate were adjusted to the desired pH using 0.1 M of HNO_3. The adsorption experiments were carried out at batch conditions using 120 rpm and 30 °C. Kinetic studies were performed with different Cr concentrations from 50 to 100 mg/L, the adsorption data were fitted to kinetic (see Table S1) and isothermal classical models (see Table S2). Adsorption isotherms were obtained at different conditions of pH (2–5) and temperature (20 and 30 °C) with Cr concentrations from 10 to 300 mg/L using an equilibrium time of 24 h. Finally, the resulting solution was centrifugated at 6000 rpm, filtrated and analyzed by Atomic Absorption Spectrophotometer ContrAA 700 (Analytik Jena, Jena, Germany). The adsorption capacity (qe, mg/g) was calculated with the following Equation (2). Blank experiments were performed to ensure that no adsorption occurred on the walls of the apparatus used.

$$qe = \frac{\left(C_i - C_f\right) * V}{m} \tag{2}$$

Thermodynamic parameters of Cr adsorption process were calculated using the Gibbs free energy (ΔG°, kJ/mol), see Equation (3). Standard enthalpy (ΔH°) and entropy (ΔS°) of hybrid membranes on Cr adsorption were calculated using the Van't Hoff approach with Equation (4). Equilibrium adsorption constants were calculated following a standard procedure reported in literature.

$$\Delta G = -RTlnK_c \tag{3}$$

$$lnK_c = \frac{-\Delta H^\circ}{RT} + \frac{\Delta S^\circ}{R} \tag{4}$$

2.5. Statistical Analysis

All statistical analyses were completed using SPSS software version 17.0. Analysis of variance (ANOVA) was conducted to determine differences between treatments. Significant differences were established with Fisher's least significant difference test with a significance level of 0.05. Experiments were performed in triplicate, and the results were reported as the mean and standard deviation of the measurements.

3. Results and Discussion

3.1. Preparation of the Adsorptive Membrane

To carry out the electrospinning process and to obtain the nanofibers with the desired properties, several parameters must be considered [34]. In this study, a CCD was established to investigate the effect of three influential factors: the proportion of WPI:PCL (%), the concentration of βME (%) and the collector distance (cm), which were used to obtain nanofibers with a thinner diameter (nm) and higher adsorption capacity (qe) for Cr(VI) removal (Table 2). Therefore, the solution properties such as viscosity and surface tension were considered since these properties are known to affect the electrostatic forces involved in the formation of the nanofibers [35,36].

Table 2. Central Composite Design (CCD), experimental and analytical response for the preparation of WPI/PCL hybrid membrane. The factors were A-WPI (%), B-βME (%) and C, Collector distance (cm).

Run	A	B	C	*qe* (mg/g)	Diameter (nm)	ST (mN/m)	Viscosity (Pa·s)	Beads
1	50	0.1	7	1.6	413	30.2	0.383	NO
2	70	1.2	10	1.44	52	30.8	0.048	YES
3	70	0.65	10	0.6	186	30.5	0.061	NO
4	70	0.65	10	1.02	188	30.6	0.057	NO
5	70	0.65	13	0.69	336	30.7	0.057	NO
6	50	0.65	10	0.81	350	30.5	0.323	NO
7	50	1.2	13	0.92	528	30.9	0.194	NO
8	70	0.1	10	0.45	299	30.3	0.059	YES
9	70	0.65	10	0.88	180	30.6	0.056	NO
10	70	0.65	10	1.42	170	30.5	0.050	NO
11	70	0.65	7	2.12	160	30.6	0.053	YES
12	90	0.1	13	1.56	68	30.7	0.007	YES
13	90	1.2	7	2.74	53	31.3	0.006	YES
14	90	0.65	10	1.83	48	31	0.005	YES
15	70	0.65	10	0.92	175	30.7	0.059	NO

The experimental design presented in Table 2 has 15 runs where the experimental results for the diameter, viscosity, surface tension, presence of beads and adsorption capacity are reported. It can be observed that the adsorption capacities, diameter, and viscosity varied strongly depending on the values of the influential factors. The experimental results of Table 2 were statistically analyzed by mean using ANOVA. Table 3 contains the fitting results of the second-order polynomial equation, the corresponding regression coefficients and the criterion of desirability. Statistical significance and accuracy of the models were confirmed via low *p*-value (<0.05), determination coefficient (R^2) close to 1, and a high value (>0.05) of lack of fit. According to these metrics, all models were statistically significant at a 95% probability level (*p*-value < 0.05), the lack of fit was not statistically significant (*p*-value < 0.05) indicating that the models were adequate. All models fitted well to the experimental data due to all R^2 values being over 0.94, which was according to the literature where it has been indicated that R^2 value must be higher than 0.7 to obtain a good fit to the experimental data [37]. The adjusted R^2 value for all models was above 0.8 thus confirming that the variations in the responses can be explained by the relationships obtained. Consequently, the models obtained in this study were useful to establish the best electrospinning preparation conditions of hybrid membranes of WPI and PCL for the Cr removal.

To obtain hybrid membranes of WPI-PCL with the best Cr removal properties, the diameter and adsorption capacity of the fibrils obtained by electrospinning were mainly evaluated. This study confirmed that the diameter and adsorption capacity were related due to the thinnest fiber diameter, which generated the highest value of adsorption capacity. This phenomenon was associated to a decrease in the diameter of the fibers leading to a membrane with a higher surface area [38,39]. Moreover, the diameter and *qe* were influenced significantly by the proportion of WPI:PCL (%), the concentration of βME (%) and the collector distance (cm). It was observed that the diameter decreased with the increase of the proportion of WPI (%) inversely proportional of PCL (%) and decrease in the concentration of βME (%). This can be explained via the composition effect on the viscosity and surface tension of the solution, where thinner fibers were obtained (experiments 12–14) using solutions with the lowest values of viscosity (0.005–0.007 Pa·s) and the intermediate values of surface tension (30.7–31.3 mN/m) were also obtained.

Table 3. Results of ANOVA for the variables analyzed in the optimization design to prepare WPI/PCL hybrid membrane.

	qe (mg/g)			Diameter (nm)			ST (mN/m)			Viscosity (Pa·s)		
	SS	df	*p*-Value	SS	df	*p*-Value	SS	df	*p*-Value	SS	df	*p*-Value
Model	5.22	9	0.0162	288,599	9	0.0001	1.03	9	0.0042	0.18	9	0.0001
A	0.52	1	0.0422	45,602	1	0.0001	0.10	1	0.0149	0.05	1	0.0001
B	0.49	1	0.0455	30,505	1	0.0001	0.15	1	0.0065	0.00	1	0.1856
C	1.02	1	0.0126	15,488	1	0.0001	0.01	1	0.3084	0.00	1	0.5547
AB	0.08	1	0.3239	4107	1	0.0002	0.00	1	0.6927	0.00	1	0.0001
AC	0.18	1	0.1644	29,403	1	0.0001	0.00	1	0.4217	0.00	1	0.0002
BC	0.00	1	0.7908	3888	1	0.0002	0.00	1	0.8538	0.00	1	0.0099
A^2	0.30	1	0.0939	1058	1	0.0048	0.05	1	0.0393	0.03	1	0.0001
B^2	0.00	1	0.8376	30	1	0.4542	0.00	1	0.4108	0.00	1	0.0578
C^2	0.47	1	0.0483	12,494	1	0.0001	0.01	1	0.2815	0.00	1	0.114
Lack of Fit	0.01	1	0.9408	3	1	0.8387	0.00	1	0.7419	52	1	0.173
Pure Error	0.35	4		225	4		0.03	4		75	4	
R^2		0.94			0.99			0.96			0.99	
R^2 Adj		0.82			0.99			0.9			0.99	

It was noted that a high proportion of WPI (90%) and low proportion of PCL (10%) were needed to obtain low viscosity values, and the concentration of βME (%) did not have a significant effect on the solution viscosity. Therefore, the solution viscosity was influenced mainly by the amount of PCL owing to its high molecular weight, which lead to a greater chain entanglement between PCL monomers and, consequently, an increment of the solution viscosity [40]. Furthermore, a range of viscosity values between 0.008 cP to 0.056 Pa·s has been reported in the literature as a starting point to obtain thin electrospinning fibrils without beads [41], where various authors have concluded that depending on the nature of the components of the spinning solution, the concentration and viscosity are essential to obtain structured fibers without defects such as beads [42]. In this study, the membrane with the highest adsorption capacity and thinner diameter was obtained at low viscosities, which agreed with other studies that used WPI and PEO, and reported the formation of fibers using low viscosity solutions with values ranging between 0.001 and 0.004 Pa·s.

Moreover, surface tension is another property involved in fibril formation that, in balance with the hydrostatic pressure and electrical forces, leads a controlled formation of the spinning jet producing fibers without defects [34]. Consequently, the results obtained by CCD demonstrated that the preparation of thin fibrils without beads requires a middle value (30 mN/m) of the surface tension, which was in agreement with the range of surface tension values of 19–70 mN/m reported by Ricaurte and Quintanilla Carvajal [41] to obtain thin nanofibers. To positively influence the surface tension according to CCD results, the proportion of WPI and PCL must be equilibrated and the concentrations of βME must be high (1.2%). Many authors stated that the surface tension of the solution can be influenced mainly by the solvent, which was true in this case as the surface tension of the DMF and THF were 35 and 28 mN/m and the surface tension of the solution of DMF/THF (3:7) was 30 mN/m [34,40,43]. Moreover, the entanglements of the polymer and the interaction with the WPI could modify the effect of solution surface tension, where WPI can behave as a surface-active agent [28] or it can decrease the surface tension [43] depending on the manner whey proteins interact with the polymer. Furthermore, the use of a reducing agent such as βME affects the surface tension of the solution as it induces the denaturation of whey proteins by reducing disulfide linkages changing some properties of the native proteins such as the surface tension [44].

In addition, the collector distance presented a significant effect on the diameter and the adsorption capacity for Cr removal on the hybrid membrane, where the collector distance must be 7 cm to obtain thin fibers with a high adsorption capacity. That is why, the high evaporation rate of the solvents allowed the formation of fibers at short distances, obtaining a high amount of thin fibers [45]. Moreover, it is important to find the distance between the tip and the collector to obtain fibers with a balance among the electrostatic forces and the solution properties, to prevent the formation of beads [42]. Drosou et al. [12] studied the effect of the collector distance to synthesize fibers of WPI and pullulan. These authors concluded that the fibers presented more bead defects with an increase in the collector distance. Thus, a decrease in the electrostatic field and jet splitting was obtained with an increment in the collector distance and using a fixed voltage.

The results obtained by the CCD were used to establish the best preparation conditions using the criterion of desirability (Table 4) with the aim of obtaining the fibers by electrospinning with the highest adsorption capacity and the lowest diameter.

Table 4. Best preparation conditions to produce WPI/PCL hybrid membrane based on the results of adsorption capacity, fiber diameter, viscosity, and surface tension. Validated values were expressed as mean ± SD of triplicate.

Optimal Preparation Conditions	
WPI (%)	88
2-mercaptoethanol (%)	1.2
Collector Distance (cm)	7.1
q (mg/g) predicted	2.73
q (mg/g) validated	2.59 ± 0.21
Error (%)	5.0
Diameter predicted (nm)	39.1
Diameter Validated (nm)	38.3 ± 2.3
Error (%)	2.2
Viscosity predicted (cP)	5.9
Viscosity validated (cP)	5.7 ± 0.4
Error (%)	4.3
Surface Tension predicted (mN/m)	31.2
Surface Tension validated (mN/m)	30.5 ± 0.1
Error (%)	2.4

The results showed that the adsorption capacity was favored at high amounts of WPI (~90%), βME (~1%) and a short distance (~7 cm). These results demonstrated that high amounts of WPI and βME displayed the best results. This is because whey proteins are rich in functional groups, which in this case, have been desaturated with βME. Comparing WPI with PCL alone, it lacks functional groups and, in consequence, it does not have the potential to remove heavy metals by itself. Moreover, a short distance promoted the elaboration of ultrathin nanofibers as discussed before.

Whey proteins have been studied owing to their high nutritional value and some authors have studied their interactions with metal ions because of their capacity to serve as carriers for metal complexes [46]. Whey protein is composed mainly of β-Lactoglobulin and α-Lactalbumin, where β-lactoglobulin is well known for its interaction with metal ions due to its free cysteine group (Cys-121) that acts as a binding site for metal ions of the d-block such as copper, silver and mercury [24]. Likewise, α-lactalbumin has a structural affinity for metal ions, which may replace the calcium site of the protein. Furthermore, the interaction between proteins and metal ions results in metalloproteins and metallocomplexes, where metalloproteins are formed by coordination bonds between metal ions and functional groups of amino acids, such as carboxyl [47]. Metallocomplex is an artificial system that lead to a binding of metal ions with proteins that interact through weak interactions such as hydrogen bonding, electrostatic, Wan der Waals forces, or donor–acceptor bonds. This binding arises via adsorption or intraparticle diffusion, where the metal ion of a metalloprotein is immersed in a protein structure forming a natural system such as hemoglobin [48].

In this study, the addition of βME was essential to promote the adsorption of Cr into the membrane, which without the addition of this reductant agent, the resulted membrane did not adsorb any metal ion. βME is an agent that is used to reduce the disulfide bonds of proteins leading to a tautomerization and breaking up of the protein's quaternary structure, where an excess of βME holds the thiol groups of proteins in their reduced state [49]. Nguyen et al. studied the effect of βME on the denaturation of whey proteins and concluded that thiol reagents such as βME initiate thiol–disulfide exchange reactions with the disulfide bonds on α-lactalbumin and β-Lactoglobulin that lead to irreversibly denatured proteins [50]. Moreover, the presence of coordination bonds is desired to promote metal–protein interactions, which allows the metal to be inserted into the structure of the protein. For example, the free thiol group of the cysteine of β-Lactoglobulin has to be deprotonated for involvement in the metal coordination as well as tyrosine which can be deprotonated to produce a phenolate oxygen donor atom that can act as a ligand for metals of the d-block [47]. βME is quite a strong hydrogen-bond acceptor, capable of coordinating with many metal ions such as As, Ni, Pb and Zn [51]. The results demonstrated that βME is a critical factor in the adsorption capacity of the membrane, which denaturized the whey proteins of the WPI making it suitable for heavy metal removal. Furthermore, it was demonstrated the βME did not have an effect on the PCL at the evaluated conditions because fibrils produced with PCL alone and PCL with the higher concentration of βME evaluated did not present Cr removal.

Effect of the Total Solids and Activated Carbon Addition on the Preparation of the Membrane

The membrane obtained by the CCD was achieved using 15% of total solids (TS), ~90% of WPI, ~1% βME and a collector distance of ~7 cm, which corresponded to the conditions that generated the best adsorption capacity and the thinner diameter (Table 4). The presence of beads was observed in the fibrils of the membrane (see Figure 2D). Note that it is well known that the properties of the spun solution affect the morphology of the fibers [41]. To overcome the presence of beads, this study considered that the concentration of the spun solution affected the stretching of the charged jet. Because there must be enough entangled polymer chains to form fibers that can reach the collector and, if this does not occur, the entangled polymer chains will break and the fragments can cause the formation of beads or beaded nanofibers [42]. For this reason, this study evaluated the effect of the content of the total solids ranging from 15% to 21% on the shape of the fibers (Table 5) to improve their shape and avoid bead formation.

The results demonstrated (see Table 5) that the fibers obtained using 18% of total solids on the solution presented the best results in terms of the adsorption capacity and there were no beads on the fibers. The adsorption capacity of these samples was improved by 3.6 times for Cr removal in comparison to the fibers prepared using 15%. Moreover, it was very difficult to obtain fibers using 21% of total solids due to the properties of the solution such as its high viscosity which did not allow the electrospinning process to occur. The enhancement of the adsorption capacity of the membrane prepared with a solution with

18% of solids was greater than the membrane made with 15% of solids due to the absence of beads on the fibers, which led a material almost without defects and with a homogeneous surface area promoting a better affinity between the surface and heavy metal. In contrast, the membrane obtained with a solution with 18% of total solids demonstrated fibers with a thicker diameter, without beads and an equilibrated surface area [52].

Table 5. Effect of the total solids on the adsorption capacity, fibril diameter and bead formation of WPI/PCL hybrid membrane.

Total Solids %	qe (mg/g)	Diameter (nm)	Viscosity (Pa·s)	Surface Tension (mN/m)	Beads Formation
15%	2.4	36.2	0.008	30.3	YES
18%	8.61	62.2	0.034	30.5	NO
21%	-	-	0.089	30.7	-

All values are presented as means (n = 3).

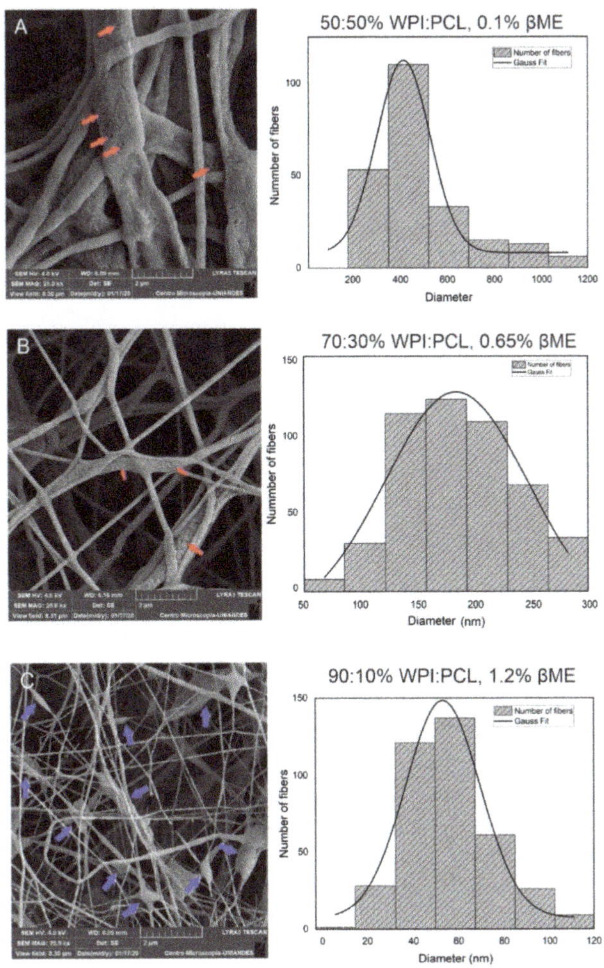

Figure 2. *Cont.*

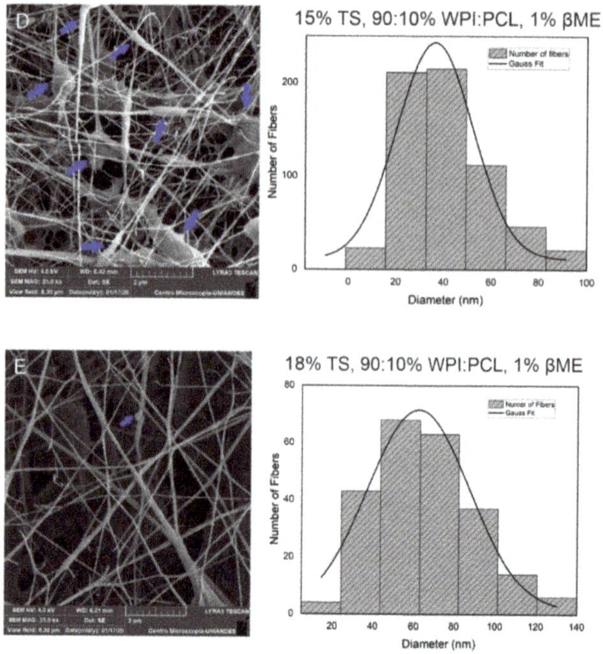

Figure 2. Morphological analysis of hybrid membrane of WPF and PCL by SEM. These results demonstrated the diameter change with the composition of % WPI:PCL and % βME at 7 cm of collector distance. (**A**) 50:50%, 0.1%; (**B**) 70:30%, 0.65%; (**C**) 90:10%, 1.2%; (**D**) membrane at the best conditions 15% of total solids, 90:10%, 1% (**E**) 18% of total solids, 90:10%, 1%. Graphical arrows in red represent the micropores in the fibers and blue arrows represent beads.

3.2. *Characterization of Membrane*

3.2.1. SEM Characterization

To understand the morphology of the fibers obtained by the CCD, SEM characterization was used. The fiber diameter was analyzed varying the composition of PCL and WPI in the fiber solution at a collector distance of 7 cm. The results confirmed a fiber morphology in all samples and showed that thick fibers with a diameter of 413 nm were obtained using a solution with a high concentration of PCL at a proportion of 50:50% WPI:PCL (see Figure 2A). Decreasing the amount of PCL in the solution to 70:30% (Figure 2B) and 90:10% WPI:PCL (Figure 2C), the diameter of the fibers decreased obtaining diameter values of 160 and 53 nm, which represented an increase on the surface area [53]. However, the presence of beads increased by increasing the proportion of WPI in the spun solution where the same phenomenon was observed by Drosou et al. [12]. These authors elaborated electrospinning fibrils using a composite pullulan and WPI and noted that the fibrils presented beads using WPI above 70%.

As discussed, the viscosity, conductivity and surface tension of the spun solution are critical parameters that are affected by the polymer concentration, which also have an effect on the fiber diameter and the presence of beads in the fiber [41]. In this case, the polymer proportion on the fiber had a significant effect on the diameter of the fibers due to the amount and length of the polymer chain represented by the polymer weight that determines the amount of the polymer entanglement [54]. Thus, when the chain entanglements of polymer are high, the entanglements between the polymer and WPI within the solution maintain the jet formation during electrospinning process and bead-free fibers are obtained [43].

On the other hand, the diameter and presence of beads in the fibers were influenced by the concentration of βME in the spun solution. Note that the viscosity of the solution is modulated with the addition of βME owing to βME-induced unfolding of the globular proteins of WPI by the polypeptide chain entanglement and chain–chain interaction with the PCL. This enhanced the elasticity of the solution jet during the electrospinning process [28]. The same result was observed by Kabay et al. [55], who modified the viscosity of a BSA protein solution with the addition of βME and it was possible to spin the protein solution without the addition of a polymer, which would not have been possible with the protein alone.

The presence of some micropores was observed on the surface of the nanofibers obtained as shown in the Figure 2A,B. This result means that the membrane, besides having ultra-thin fibers that increase the surface area of the material, contains some micropores that also increase the surface area favoring the adsorption process. This is because a fiber with a smaller diameter decreases the particle size, which means that the surface area per volume increases and more fibers can occupy the same volume [38,39]. Analyzing the increase of a total solids concentration in the spun solution, Figure 2D showed that the solution with 18% of solids did not present beads into the fibers in comparison to the fibers obtained with the spun solution of 15% (see Figure 2C), which showed the presence of beads in the fibers. Thus, the formation of beads on the solution of 15% of solids was due to the low concentration of WPI and PCL and, in consequence, the solution viscosity was low. As stated, it is well known that viscosity is a critical factor in the formation of well-defined fibers because this property is related to polymer chain entanglements in the solution produced by the change of the polymer and solvent composition [39] which, in consequence, affect the viscoelastic response, charge relaxation times, and solvent evaporation rate [56]. Thus, continuous and homogenous fibers were obtained with an equilibrium of the composition, viscosity, conductivity, surface tension as well as process parameters, such as voltage, feed flow rate, and tip–collector distance [41].

3.2.2. FT-IR

The FT-IR analysis was used to identify the functional groups of the adsorptive membrane compared with their raw materials such as WPI and PCL to perform a qualitative characterization of the membrane composition. The FT-IR spectra (see Figure 3) of all samples showed a band at 3328 cm^{-1}, which corresponded to stretching vibrations of -OH due to the presence of water [57]. The spectra of WPI and membrane showed characteristic bands of the peptide bonds of whey proteins at 1643 cm^{-1} related to a primary amide group (C=O, C-N) [43], and the band at 1546 cm^{-1} representing the secondary amide group (N-H, C-N) [58]. This demonstrated that the membrane contained the main functional groups of the WPI indicating that the fibers were mainly composed of WPI, which was the component with an affinity to heavy metals.

The spectra of PCL alone and the membrane presented the characteristic PCL band of the carbonyl stretching vibration at 1723 cm^{-1} [59], the bands at 2956 and 2907 cm^{-1} represented the symmetric and asymmetric stretching vibration of a methylene group (CH_2) [17], the absorption bands at 1298 (C-O bands tension), 1246 (C-O-C symmetrical tension) and 1192 cm^{-1} (OC-O stretching) were associated with the C-O stretching vibrations on the crystalline phase of PCL [60]. The analysis of the FT-IR spectra showed that the membrane of PCL and WPI presented the characteristic bands of the functional groups of both components, thus demonstrating the successful blending of the components within the spun fibers. A similar result was obtained by Ahmed et al. [19] for the case of fibrils of whey protein and PCL for the antibiotic release. These authors confirmed a successful doping of whey protein in the fibrils, which proved that the electrospinning process only mixed both components and did not change the chemical structure of PCL and whey protein.

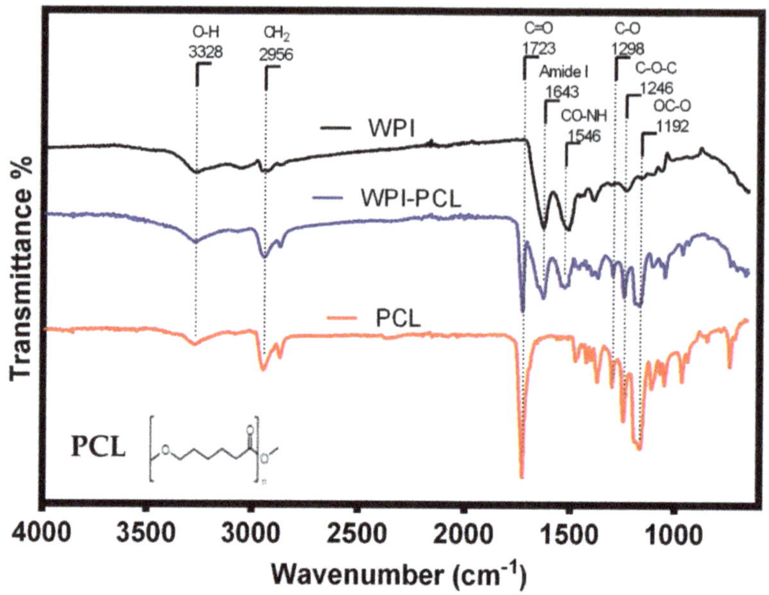

Figure 3. FTIR spectra of the hybrid membrane of WPI:PCL (90:10, blue line), WPI (black line) and PCL (red line).

Furthermore, the results suggested the βME did not influence the chemical structure of the PCL and there was no appreciable difference between the spectra of WPI alone and the WPI in the electrospun fibers. Moreover, the used organic solvents did not have an effect on whey protein, similar to the study reported by Ahmed et al. [19].

3.2.3. Congo Red Assay

Congo Red (CR) assay was used to qualitatively identify amyloids and to determine the effect of the addition of βME on WPI that is a denaturant agent of proteins. CR is a specific dye that binds with the crossed-β-pleated sheet structure common to a variety of amyloid fibrils, thus causing an increase in the absorption from 490 nm and when bonded to a native protein to 510 nm when CR is bonded to amyloids [61]. CR analysis (see Figure 4) showed the absorption spectra of this dye alone and in solution with a WPI (14%) and presented a maximum absorption peak at 490 nm. The solution of WPI (14%) treated with different concentrations of βME showed an increase in the absorption band to 510 nm demonstrating the presence of amyloid structures in the solution. Moreover, there was an increment of the absorbance at 510 nm when the concentration of βME increased in the solution of WPI, which was due to an increase of the β-sheet content relative to the native state of the proteins [62].

These results confirmed the presence of amyloid fibrils in the solutions of WPI treated with βME, where this effective reductant agent of disulfide bonds opened the tertiary structure of globular proteins changing whey proteins from their native states to an amyloid-like structure [55]. The formation of new extended structures with strong intermolecular and disulfide covalent bonds owing to the unfolding of whey proteins by the reduction of the disulfide bonds generated to a membrane of WPI amyloids with higher affinity for heavy metals compared to a membrane of WPI without the addition of βME [24]. These results were in agreement with the findings of Kabay et al. [63]. These authors prepared a protein membrane of BSA by electrospinning using βME in the electrospun solution to reduce its disulfide bonds breaking its tertiary structure enhancing its supportive properties, thus making the BSA solution spinnable as a natural polymer without the use of a co-polymer.

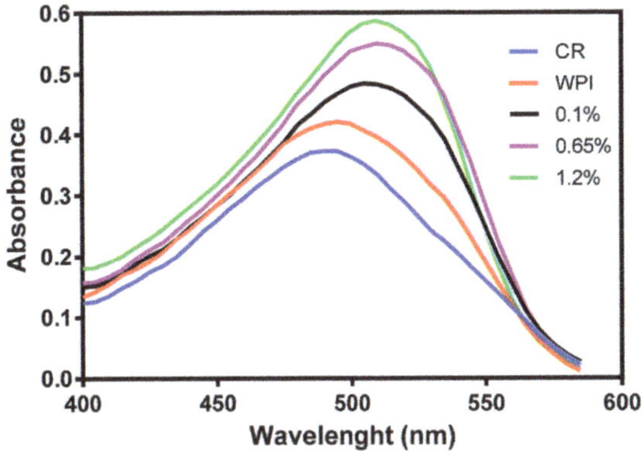

Figure 4. Congo red absorption spectra. Blue line represents red Congo dye as reference, red line represents an aqueous solution of WPI at 15% without βME, black, purple and green lines represent WPI in aqueous solution with different compositions of βME.

3.2.4. Contact Angle

The effect of hydrophilicity or hydrophobicity on the membrane surfaces was analyzed using contact angle measurements via changing the proportion of WPI and PCL on the spinning solution. The contact angle of the adsorptive membranes is shown in Figures 5 and 6, where the contact angle of the membrane of PCL alone presented a contact angle greater than 90° demonstrating its hydrophobic surface and low wettability in accordance with Lin and Razali [64]. On the other hand, the membranes with 50:50, 70:30, and 90:10 had a contact angle of 75°, 51°, and 42°, respectively. They presented a contact angle below 90° indicating that the membranes had a hydrophilic surface that increased with an increment of WPI on the membrane due to the polarity on the membrane surface produced by the presence of hydroxyl and amide groups on the surface [60].

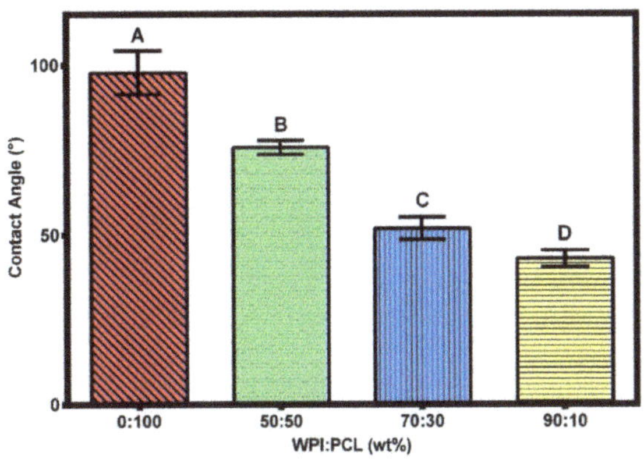

Figure 5. Contact angle of the hybrid membrane of PCL-WPI prepared with different compositions of WPI:PCL (wt%). Error bars are standard deviations (n = 3). Different letters indicate significant differences between different treatments at *p*-level = 0.05 based on the least significant difference (LSD) test.

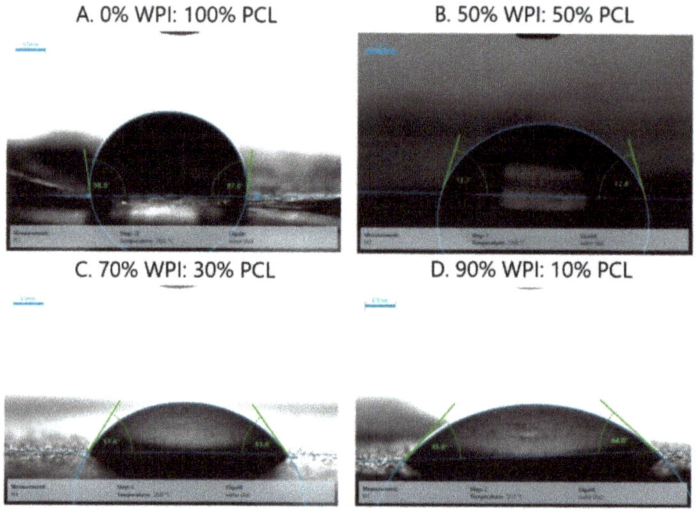

Figure 6. Pictures of the contact angle of the hybrid membrane of PCL-WPI prepared with different compositions of WPI:PCL (wt%).

To produce a suitable adsorptive membrane for membrane filtration, it is important to have an equilibrium between the hydrophilicity and hydrophobicity on the surface of the membrane. Therefore, hydrophilic surfaces are desired because the contact of the membrane with the aqueous solution is facilitated at a smaller contact angle and, in consequence, higher adsorption capacities to remove heavy metals are obtained [65]. Moreover, the presence of hydrophilic groups accelerates the permeation of water molecules through the membrane [66]. However, a super hydrophilic membrane with a contact angle lower than 20° is not desired because it renders the membrane unstable due to its solubility in water [67]. A hydrophobic membrane such as a membrane of PCL alone is not desired because it is well known that the adsorption capacity is not favorable due to the low contact between the water molecules and the membrane; hence, it challenges the affinity of heavy metals with the membrane [65]. These results demonstrated that the membrane obtained at the best results showed a hydrophilic character thus being suitable for heavy metal removal.

3.2.5. Thermogravimetric Analysis

Thermogravimetric analysis (TGA) was carried out to determine the degradation temperature of the membrane compared to its main components. This analysis provided information about the weight loss and thermal profile of the adsorptive membrane, PCL and WPI, see Figure 7. The thermal degradation of PCL occurred in one step between 327–450 °C where two types of reactions took place: the random chain scission and unzipping from hydroxyl leading to the formation of ε-caprolactone [20,68]. The DTG curve of PCL showed that the degradation temperature of PCL was at 400 °C with a weight loss of 70%, and almost complete degradation occurred at 460 °C with a weight loss of 96%, which is in agreement with Seyedsalehi et al. [69].

TGA analysis showed that the mass loss of the WPI and the membrane occurred in two and three stages, respectively. The first weight stage below 200 °C was due to the evaporation of moisture in WPI and the membrane sample, and the water loss weight percentages were 4 and 6%, respectively. The second decomposition peak corresponded to the main thermal degradation zone with about 30–40% of the weight loss between 290–310 °C and 300–315 °C for WPI and the membrane, respectively, which could be attributed to the breakage of peptide bonds and the decomposition of molecular amine units of WPI thus generating an exothermic change in the DTA [70]. The membrane presented a third decomposition peak at 361 °C due to the decomposition of the PCL of the membrane into its monomers of ε-caprolactone. Finally, the membrane presented degradation temperatures similar to the WPI because it was composed of 88% WPI, but there was a low increment in the degradation temperatures due to the composition of PCL into the membrane. The third degradation stage was confirmed by the mass loss percentage of the PCL degradation. In this way, it was important to know the degradation behavior of the membrane due to temperature was an important variable for carrying out the adsorption process, and could be a relevant factor when choosing a material to produce a filtration membrane in large-scale applications

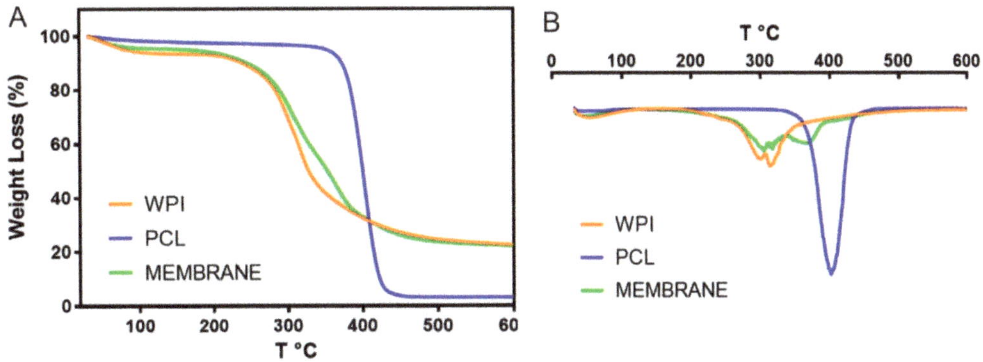

Figure 7. Thermogravimetric (**A**) and Derivate thermogravimetric (**B**) of the pure PCL, pure WPI, and the adsorptive membrane.

3.3. *Chromium Adsorption*

3.3.1. Kinetics

The adsorption kinetics were studied to evaluate the adsorption rate between the surface of the hybrid membrane of PCL-WPI and Cr ions in the solution at concentrations of 50 and 100 mg/L. Figure 8 shows that the hybrid membrane adsorbed a large amount of Cr in about 60 min, which could be due to the presence of many adsorption sites on the adsorbent surface [17]. However, the adsorption of Cr decreased over time by the occupation of the active sites of the membrane and the adsorption equilibrium was achieved after 3.5 h.

These experimental data were fitted using the kinetics classic models of pseudo-first, pseudo-second-order and Intra-particle diffusion. Note that the first-order kinetic model is commonly used to model the first adsorption step at the initial period and the second-order kinetic model is suitable to describe the whole adsorption process involving two stages for the pollutant removal where the first one was fast and reached the equilibrium quickly, the second was slower, and continued for a long contact time [71]. The intraparticle diffusion model involves three steps, which consist in a film diffusion, the intraparticle diffusion and the attachment of the adsorbate that occurs rapidly and it is not a rate limiting step. The pseudo-second-order model was the best to fit the experimental data with $R^2 > 0.97$ with modeling errors lower than 9%, see Table 6. Thus, it can be concluded that the rate of the adsorption process relied on two adsorption processes where Cr(VI) ions were adsorbed on the surface of the membrane and also possibly reduced to Cr(III) [72]. Considering that pseudo-second order represented the adsorption process better on the hybrid membrane, this suggests that chemisorption may be the rate controlling mechanism [73]. Moreover, the low R^2 value of the intraparticle diffusion model was evidence to conclude that the diffusion occurred relatively quickly from the solution, and Cr adsorption on the hybrid membrane of WPI-PCL did not follow this model. Therefore, it could be expected that the adsorption may be an external surface process in the absence of internal diffusion [74].

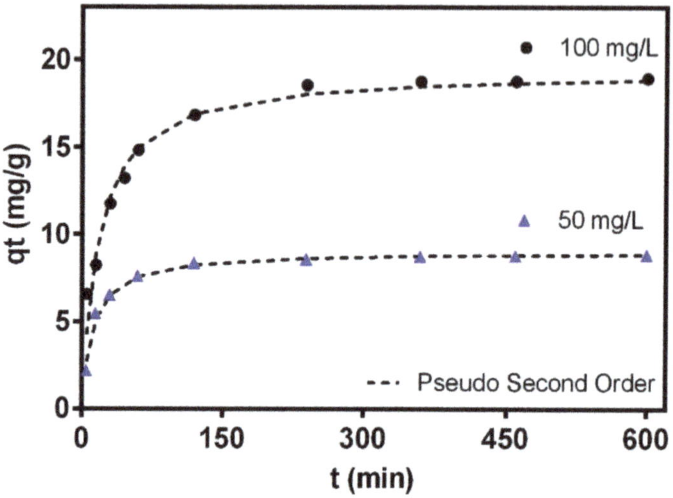

Figure 8. Kinetics for the chromium adsorption on the hybrid membrane at initial concentrations of 50 ppm and 100 ppm, 30 °C and pH 2.

Table 6. Adsorption kinetic rates for chromium removal from aqueous solution using the hybrid membrane of WPI-PCL. Experimental conditions: pH 2 and 30 °C.

Model	Parameters	50 mg/L	100 mg/L
Pseudo First Order	q_{te} (mg/g)	8.57	18.34
	K1 (mg/min)	0.05	0.03
	R^2	0.97	0.94
Pseudo Second Order	q_{te} (mg/g)	9.03	19.40
	K2 (mg/min)	0.01	0.002
	R^2	0.99	0.97
Intraparticle diffusion	Kd (mg/(g $min^{1/2}$)	0.10	0.33
	C (mg/g)	5.38	10.30
	R^2	0.49	0.63

The chromium removal rates ranged from 0.002 and 0.01 mg/min thus indicating the pollutant diffusion rate from the liquid phase to the outer surface of the hybrid membrane. The increment of the adsorption capacity concerning the initial concentration of Cr confirmed that the driving force of the concentration can overcome the mass transfer resistance of Cr ions between the aqueous and solid phases [75].

3.3.2. Isotherms

The chromium adsorption isotherms using the hybrid membrane of WPI-PCL at different temperature and pH are reported in Figure 9.

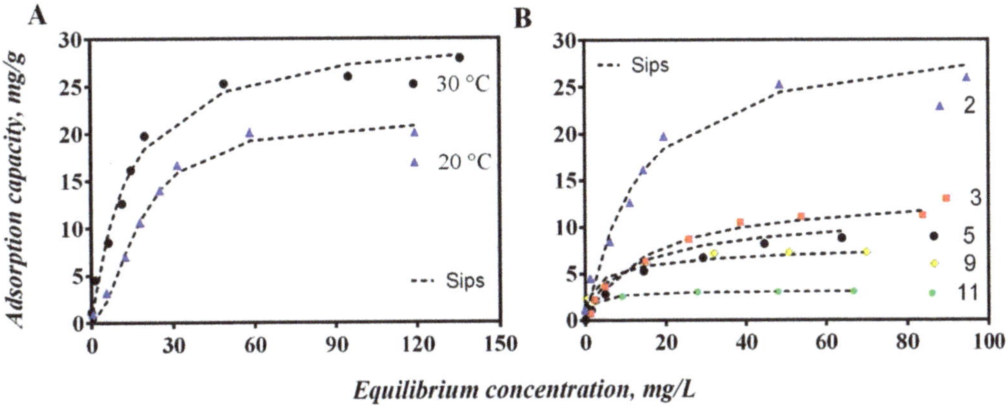

Figure 9. Isotherms for chromium adsorption from aqueous solutions using an adsorptive membrane of WPI-PCL. (**A**) Effect of temperature at pH 2 and (**B**) effect of pH at 30 °C.

The adsorption capacity increased significantly with the decrease of solution pH from 3 to 2 and reached a maximum value of 31 mg/g at pH 2, see Figure 9B. This could be explained by the fact that the protonation of the membrane of WPI-PCL was enhanced at highly acidic conditions. Thus, when the pH of the solution decreased below to the PZC of the membrane that was pH 3, the surface was positively charged by the protonation of amino groups of the WPI proteins, which resulted in positively charged surfaces favoring the adsorption of anionic species such as $HCrO_4^-$ [76]. The same behavior was observed by Jin et al. [77]. These authors prepared a keratin/PET nanofiber membrane with a PZC at pH 3.77, thus indicating that the surface of K-PET-5 was positively charged at pH 3 and attracted anions such as $HCrO_4^-$. It is important to remark that PCL alone did not present affinity by heavy metal ions due to its hydrophobic character and even when it has a carbonyl group that could be protonable at low pH, it did not present affinity by Cr or another heavy metal [78].

Adsorption equilibrium experiments were performed at different temperatures to analyze the Cr removal, see Figure 9A. The results demonstrated that the removal performance of the hybrid membrane of WPI increased with the increase of temperature obtaining the maximum adsorption capacity at 30 °C and thus confirming the temperature-dependent nature of this adsorption process [79]. This behavior demonstrated that the adsorption process was endothermic where the estimated enthalpy was 45.45 kJ/mol, see Table 7. These results indicated that the adsorption process of this membrane to remove Cr could be attributed to a physico-chemical adsorption instead of a purely physical or chemical adsorption process [80]. As stated, the functional groups of the whey proteins in the membrane interacted with Cr ions through electrostatic interactions and coordinated bonds [47]. The Gibbs free energy for the Cr adsorption on the hybrid membrane of WPI-PCL was negative (see Table 7), which indicated that the adsorption process was spontaneous [81]. The entropy was positive indicating the increase of randomness mainly on Cr adsorption at the solid/solution interface during the separation process where ion replacement reactions occurred [80].

Table 7. Thermodynamic parameters for the chromium adsorption on the hybrid membrane of PCL and WPI.

T (°C)	ΔG°, kJ/mol	ΔH°, kJ/mol	ΔS°, kJ/mol K
20	−2.17	45.45	0.16
30	−3.80		

The experimental adsorption isotherms were correlated with classical isotherm equations such as Langmuir, Freundlich and Sips. Results of this modeling analysis are reported in Table 8 and Table S3. Sips equation was the best model for fitting the Cr isotherms of the WPI-PCL hybrid membrane with $R^2 > 0.96$ and modeling errors lower than 7%. Sips isotherm is a model that includes the features of Langmuir and Freundlich models. It reduces to the Freundlich isotherm at low concentrations and predicts a monolayer adsorption capacity characteristic of the Langmuir isotherm at higher concentrations [82].

Table 8. Results of data correlation of chromium adsorption isotherms of WPI-PCL adsorptive membrane using the Sips model at pH 5.

Temperature (°C)	pH	Sips			
		q_s (mg/g)	Ks (L^{ns}/mg^{ns})	ns	R^2
20	2	21.5	0.056	1.8	0.99
30	2	31.0	0.074	1.0	0.98
	3	13.2	0.074	1.0	0.99
	5	12.3	0.077	0.7	0.98
	9	9.0	0.220	0.5	0.96
	11	3.1	0.580	1.0	0.99

The heterogeneity factor (n_s) of Sips model indicates the heterogeneity of the surface sites that carry out the adsorption process. It indicates a homogeneous surface if its value is equal to 1, the Sips model simplifies to the Langmuir model, while a significant difference from 1 indicates a heterogenous surface [83]. Results of this parameter may suggest that Cr ions were weakly bonded to the surface of the adsorbent thus confirming the electrostatic interactions between the adsorbent and adsorbate and also the presence of different binding sites on the adsorbent surface [84]. The K_L parameter indicates the level of interaction between the adsorbate and adsorbent, where a value between 0 and 1 describes favorable adsorption of Cr onto the hybrid membrane [85].

3.3.3. Mechanism

To gain further insight into the possible Cr adsorption mechanism, FTIR and XPS analyses of the membrane after the metal adsorption were performed. FTIR spectrum (see Figure 10) showed the changes in the functional groups of the membrane before and after the adsorption of Cr at pH 5, 3 and 2. FTIR spectra of the membranes after adsorption at pH 5 did not show significant differences compared with the membrane before the adsorption due to the low quantity of Cr loaded on the sample surface. However, the membrane after the Cr adsorption at pH 2 and 3 showed that the bands related to the amides at 1668 and 1561 cm^{-1} had a lower intensity and even disappeared in comparison with the raw membrane, which indicated that the amide groups were involved in this adsorption process. Considering that the membrane was under acidic conditions, the possible adsorption mechanism could be due to the amide bonds of this membrane being hydrolyzed into carboxyl and amino groups, which also explained the increase in the intensity of the carbonyl group at 1722 cm^{-1} [77]. Moreover, at pH 2 and 3 the amino group of the proteins could be protonated into NH_3 favoring the electrostatic interaction with $HCrO_4^-$, as stated by Jin et al. [77]. The same was confirmed by XPS in Figure 11E with the appearance of NH_3 peak after the adsorption of Cr.

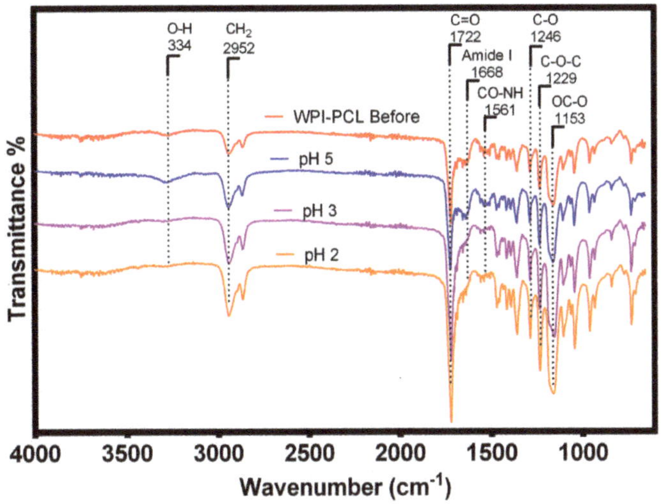

Figure 10. FTIR spectra of the WPI-PCL hybrid membrane before and after the chromium adsorption at pH 5, 3 and 2.

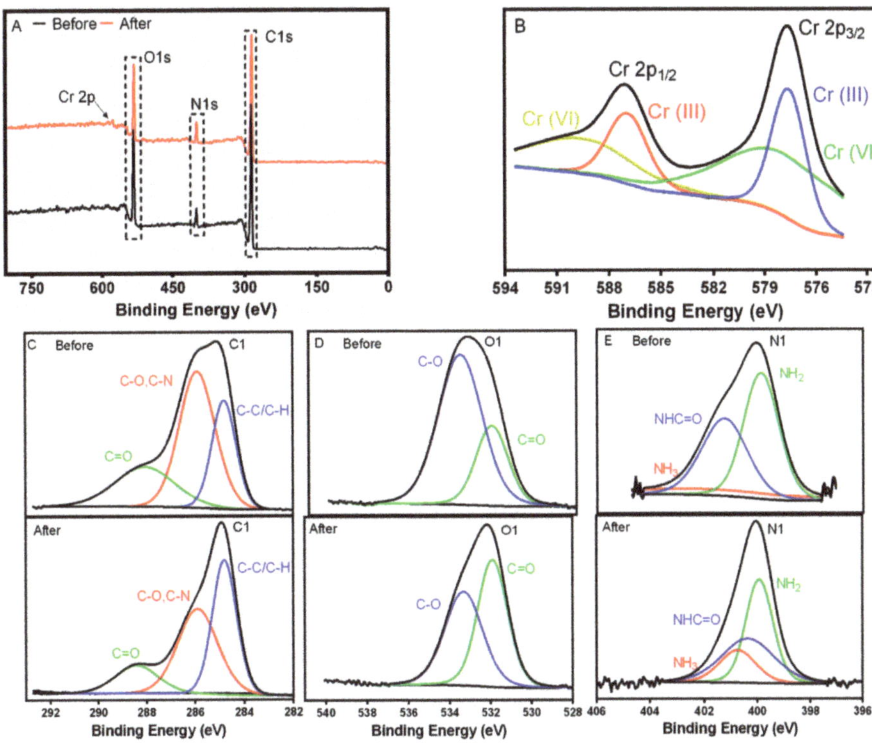

Figure 11. XPS analysis of the WPI-PCL hybrid membrane before and after chromium adsorption. (**A**) Survey spectra, (**B**) Cr high-resolution XPS spectra after adsorption, (**C**) C 1s, (**D**) O 1s and (**E**) N 1s before and after adsorption.

The XPS results (see Figure 11A) of the membrane display peaks of O 1s, C 1s and N 1s. The membrane obtained after adsorption showed a new peak at 577 eV of Cr 2p, which was subdivided in two different peaks belonging to Cr(VI) and Cr(III) species [86]. The high-resolution Cr 2p XPS (see Figure 11B) spectra showed the typical two broad peaks at 587.4 and 577.9 eV, which were assigned to Cr 2p1/2 and Cr 2p3/2, respectively. The peak of Cr 2p1/2 could be deconvoluted into two peaks with binding energies of 589.6 and 586.9 eV for Cr(III) and Cr(VI), respectively [87]. Likewise, the Cr 2p3/2 XPS spectrum was deconvoluted into two peaks of Cr(III) and Cr(VI) species at 579.3 and 577.6 eV, which indicated the coexistence of Cr(VI) and Cr(III) species. This evidence confirmed that the membrane reduced partially Cr(VI) into Cr(III) [88].

The high-resolution C 1s, O 1s and N1s XPS spectra were also analyzed (see Figure 11C–E. The C 1s spectrum of the membrane was deconvoluted into C=O at 288.7 eV, C-O/C-N at 286.1 and C-C/C-H at 284.8 eV. The relative content of C-C/C-H decreased after the Cr adsorption and the relative content of C-O increased in the membrane (see Table S4). These results may indicate that carbon in the membrane could be oxidized by Cr(VI) and Cr may be reduced to Cr(III) as explained by Yuan et al. [87]. Moreover, O1s spectrum was deconvoluted into C-O at 533.6 eV and C=O at 532.1 eV, and a decrease in the binding energy of O 1s after Cr adsorption was observed and, as a result, it increased the electron density. This finding may indicate a complexation between O and Cr, which was also confirmed by the increase in C=O group and the decrease of C-O group (see Table S4). As stated by Rodzik et al. [47], the proteins interact with metal ions and metalloproteins are formed by coordination bonds between metal ions and functional groups of amino acids, such as carboxyl [47]. Finally, a decrease in the concentration of nitrogen was observed (see Table S5) and the concentration of NH_3 increased in the membrane after the adsorption confirming the electrostatic interaction with $HCrO_4^-$, in agreement with FTIR results.

3.3.4. Comparative Study

Table 9 shows a comparison of the adsorption properties of the hybrid membrane obtained in this study with other hybrid membranes obtained by electrospinning reported in the literature. The adsorption capacity of WPI-PCL hybrid membrane was highly competitive compared to the membrane of zein/nylon-6 nanofiber membrane [16], PVDF/silica thiol nanofiber membrane [89], PCL/Clay electrospun fibers [20], cerium oxide nanoparticles embedded in polyacrylonitrile [90] and citric-acid-incorporated cellulose mats [91], which presented adsorption capacities of 4.73, 15.1, 24.57, 28.09 and 20 mg/g. Some of these studies have affirmed that the low Cr removal was due to the high amount of low-affinity polymers in the fibers such as PVA [20].

Table 9. Comparison of adsorption capacities of different membrane-based adsorbents for the chromium removal.

Adsorbent	pH	Ci (mg/L)	*qe* (mg/g)	Equilibrium Time	Reference
Zein/nylon-6 nanofiber membrane	2	5–25	4.73	60 min	[16]
Polyacrylonitrile/ionic covalent organic framework nanofibers	3	1–1000	173	4 h	[14]
PCL/Clay and PVA/Clay Electrospun Fibers	5	200	24.57	-	[20]
Cerium oxide Nanoparticles embedded in polyacrylonitrile nanofiber	6	20	28.09	3 h	[90]
Cellulose acetate (CA), acetone/DMAc, Citric Acid	2	20–60	20	2 h	[91]
ZIF-8@ZIF-8/polyacrylonitrile nanofibers	2	1–500	39.68	90 min	[92]
PAN-GO-Fe_3O_4 hybrid nanofibers	3	50	124.3	70 min	[93]
keratin/PET nanofiber	3	20–120	75	4 h	[77]
PVDF/silica thiol nanofiber	2	31	15.1	60 min	[89]
Hybrid membrane of WPI-PCL	2	10 to 300	31.0	3.5 h	This study

In addition, there are competitive membranes made by electrospinning reported in the literature such as ZIF-8@ZIF-8/polyacrylonitrile nanofibers [92], keratin/PET nanofiber membrane [77], and polyacrylonitrile/ionic covalent organic framework nanofibers [14] that presented adsorption capacities of 39.68, 75 mg/g and 173 mg/g. In addition, the higher performance of keratin/PET nanofiber membrane could be due to the use of a polymer such as PET that alone showed an adsorption capacity of 27.27 mg/g, but it is well known that PET biodegrades at a slower rate than PCL [94]. It is important to note that the equilibrium time depends on agitation speed, particle size and temperature so that these parameters can be modified to reduce the equilibrium time of the membrane regarding the application.

4. Conclusions

In this work, a WPI-PCL hybrid membrane was prepared by electrospinning and applied to adsorb Cr from an aqueous solution. The effect of the collector distance and composition of βME, WPI, and PCL on the adsorption capacity of this hybrid membrane was studied, thus demonstrating that the addition of βME allowed the functionalization of PCL with amyloid-type WPI proteins with high affinity to heavy metal ions. SEM characterization of the hybrid membrane showed a fiber morphology with ultrathin fibers with an average fiber diameter of 62.2 nm, without the presence of beads. The presence of amyloid-type proteins and the functional groups of WPI and PCL in the electrospun fibers were confirmed by FTIR spectra and Red Congo assay, respectively. The hydrophilic nature of the membrane was confirmed by the contact angle measurement, proving that the hydrophilic surface of the membrane enhanced the affinity of Cr ions by the hydroxyl and amide groups of WPI. TGA analysis allowed us to conclude that the hybrid membrane presented degradation temperatures similar to WPI because its WPI composition was 88%, but also presented a low increment on the degradation temperatures due to the PCL composition. These characteristics could be relevant for further applications.

The adsorption capacities of the hybrid membrane increased with an increase in temperature showing an endothermic adsorption process. The best performance of this membrane was obtained at strong acidic conditions (pH 2) demonstrating that its PZC had a strong influence on heavy metal removal by electrostatic interactions. Adsorption experiments indicated that the hybrid membrane exhibited a reasonably high adsorption performance to remove Cr ions from aqueous solutions with a good equilibrium time compared with other membranes made by electrospinning and reported in the literature. FTIR and XPS analyses confirmed that Cr(VI) was reduced to Cr(III) by the membrane under acidic conditions and the Cr removal could be due to a combination of electrostatic interactions, redox reactions, and the formation of metallocomplexes. This work provides a novel method to fabricate a hybrid membrane with an amyloid-type fibril made of WPI and PCL, with notable characteristics to be used as potential material to remove heavy metals from water. This material holds promise for application in membrane technology at industrial scale.

Supplementary Materials: The following data are available online at https://www.mdpi.com/article/10.3390/nano12162744/s1, Table S1: Classical kinetic models, Table S2: Classical isotherm models, Table S3: Results of data correlation of adsorption isotherms to classical models for chromium removal from aqueous solutions using and WPI-PCL adsorptive membrane, Table S4: Contributions of individual chemical moieties in the high-resolution spectra of the WPI-PCL adsorptive membrane before and after chromium adsorption, Table S5: Results of the atomic concentration of the general XPS before and after chromium adsorption by the membrane.

Author Contributions: L.C.R.-R.: Conceptualization, Investigation, Performed the experiments, Writing—Original Draft under the supervision of C.J.-J.; M.X.Q.-C.: Conceptualization, Writing—Review and Editing; D.I.M.-C.: Formal analysis, Writing—Review and Editing; A.B.-P.: Formal analysis, Writing—Review and Editing; C.J.-J.: Project administration, Conceptualization, Methodology, Funding acquisition, Writing—Review and Editing. All authors have read and agreed to the published version of the manuscript.

Funding: This research was funded by Universidad de La Sabana under grant number GL 0843-ING 228-2019.

Institutional Review Board Statement: Not applicable.

Informed Consent Statement: Not applicable.

Data Availability Statement: Not applicable.

Acknowledgments: Authors gratefully acknowledge Martha Isabel Cobo Angel for the use of the ChemBET Pulsar TPR/TPD apparatus. The authors also acknowledge Universidad de La Sabana for the financial support. Laura Cristina Ramirez-Rodriguez would personally like to acknowledge Universidad de La Sabana for the Teaching Assistant Scholarship for her master's studies.

Conflicts of Interest: The authors declare no conflict of interest.

References

1. Zhang, S.; Shi, Q.; Korfiatis, G.; Christodoulatos, C.; Wang, H.; Meng, X. Chromate Removal by Electrospun PVA/PEI Nanofibers: Adsorption, Reduction, and Effects of Co-Existing Ions. *Chem. Eng. J.* **2020**, *387*, 124179. [CrossRef]
2. Li, Y.; Xia, M.; An, F.; Ma, N.; Jiang, X.; Zhu, S.; Wang, D.; Ma, J. Superior Removal of Hg (II) Ions from Wastewater Using Hierarchically Porous, Functionalized Carbon. *J. Hazard. Mater.* **2019**, *371*, 33–41. [CrossRef] [PubMed]
3. Singh, A.; Malaviya, P. Chromium Phytoaccumulation and Its Impact on Growth and Photosynthetic Pigments of Spirodela Polyrrhiza (L.) Schleid. on Exposure to Tannery Effluent. *Environ. Sustain.* **2019**, *2*, 157–166. [CrossRef]
4. Zhu, F.; Zheng, Y.M.; Zhang, B.G.; Dai, Y.R. A Critical Review on the Electrospun Nanofibrous Membranes for the Adsorption of Heavy Metals in Water Treatment. *J. Hazard. Mater.* **2021**, *401*, 123608. [CrossRef]
5. Carolin, C.F.; Kumar, P.S.; Saravanan, A.; Joshiba, G.J.; Naushad, M. Efficient Techniques for the Removal of Toxic Heavy Metals from Aquatic Environment: A Review. *J. Environ. Chem. Eng.* **2017**, *5*, 2782–2799. [CrossRef]
6. Rodriguez-Narvaez, O.M.; Peralta-Hernandez, J.M.; Goonetilleke, A.; Bandala, E.R. Treatment Technologies for Emerging Contaminants in Water: A Review. *Chem. Eng. J.* **2017**, *323*, 361–380. [CrossRef]
7. Hezarjaribi, M.; Bakeri, G.; Sillanpää, M.; Chaichi, M.J.; Akbari, S.; Rahimpour, A. Novel Adsorptive PVC Nanofibrous/Thiol-Functionalized TNT Composite UF Membranes for Effective Dynamic Removal of Heavy Metal Ions. *J. Environ. Manag.* **2021**, *284*, 111996. [CrossRef] [PubMed]
8. Hamad, A.A.; Hassouna, M.S.; Shalaby, T.I.; Elkady, M.F.; Elkawi, M.A.A.; Hamad, H.A. Electrospun Cellulose Acetate Nanofiber Incorporated with Hydroxyapatite for Removal of Heavy Metals. *Int. J. Biol. Macromol.* **2020**, *151*, 1299–1313. [CrossRef]
9. Maleš, L.; Fakin, D.; Bračič, M.; Gorgieva, S. Efficiency of Differently Processed Membranes Based on Cellulose as Cationic Dye Adsorbents. *Nanomaterials* **2020**, *10*, 642. [CrossRef]
10. El-Aswar, E.I.; Ramadan, H.; Elkik, H.; Taha, A.G. A Comprehensive Review on Preparation, Functionalization and Recent Applications of Nanofiber Membranes in Wastewater Treatment. *J. Environ. Manag.* **2022**, *301*, 113908. [CrossRef]
11. Thamer, B.M.; Aldalbahi, A.; Moydeen, A.M.; Rahaman, M.; El-Newehy, M.H. Modified Electrospun Polymeric Nanofibers and Their Nanocomposites as Nanoadsorbents for Toxic Dye Removal from Contaminated Waters: A Review. *Polymers* **2021**, *13*, 20. [CrossRef] [PubMed]
12. Drosou, C.; Krokida, M.; Biliaderis, C.G. Composite Pullulan-Whey Protein Nanofibers Made by Electrospinning: Impact of Process Parameters on Fiber Morphology and Physical Properties. *Food Hydrocoll.* **2018**, *77*, 726–735. [CrossRef]
13. Zakaria, A.F.; Kamaruzaman, S.; Rahman, N.A. Electrospun Polyacrylonitrile/Lignin/Poly(Ethylene Glycol)-Based Porous Activated Carbon Nanofiber for Removal of Nickel(II) Ion from Aqueous Solution. *Polymers* **2021**, *13*, 3590. [CrossRef] [PubMed]
14. Herath, A.; Salehi, M.; Jansone-Popova, S.; Salehi, M. Production of Polyacrylonitrile/Ionic Covalent Organic Framework Hybrid Nanofibers for Effective Removal of Chromium(VI) from Water. *J. Hazard. Mater.* **2022**, *427*, 128167. [CrossRef]
15. Sharafoddinzadeh, D.; Salehi, M.; Jansone-Popova, S.; Herath, A.; Bhattacharjee, L. Advance Modification of Polyacrylonitrile Nanofibers for Enhanced Removal of Hexavalent Chromium from Water. *J. Appl. Polym. Sci.* **2022**, *139*, 52169. [CrossRef]
16. Ansari, S.; Ahmed, N.; Mahar, R.B.; Khatri, Z.; Khatri, M. Fabrication and Characterization of Electrospun Zein/Nylon-6 (ZN6) Nanofiber Membrane for Hexavalent Chromium Removal. *Environ. Sci. Pollut. Res. Int.* **2022**, *29*, 653–662. [CrossRef]
17. Irandoost, M.; Pezeshki-Modaress, M.; Javanbakht, V. Removal of Lead from Aqueous Solution with Nanofibrous Nanocomposite of Polycaprolactone Adsorbent Modified by Nanoclay and Nanozeolite. *J. Water Process Eng.* **2019**, *32*, 100981. [CrossRef]
18. Tran-Ly, A.N.; Ribera, J.; Schwarze, F.W.M.R.; Brunelli, M.; Fortunato, G. Fungal Melanin-Based Electrospun Membranes for Heavy Metal Detoxification of Water. *Sustain. Mater. Technol.* **2020**, *23*, e00146. [CrossRef]
19. Ahmed, S.M.; Ahmed, H.; Tian, C.; Tu, Q.; Guo, Y.; Wang, J. Whey Protein Concentrate Doped Electrospun Poly(Epsilon-Caprolactone) Fibers for Antibiotic Release Improvement. *Colloids Surf. B Biointerfaces* **2016**, *143*, 371–381. [CrossRef]
20. Roque-Ruiz, J.H.; Cabrera-Ontiveros, E.A.; Torres-Pérez, J.; Reyes-López, S.Y. Preparation of PCL/Clay and PVA/Clay Electrospun Fibers for Cadmium (Cd^{+2}), Chromium (Cr^{+3}), Copper (Cu^{+2}) and Lead (Pb^{+2}) Removal from Water. *Water Air Soil Pollut.* **2016**, *227*, 286. [CrossRef]

21. Knowles, T.P.J.; Mezzenga, R. Amyloid Fibrils as Building Blocks for Natural and Artificial Functional Materials. *Adv. Mater.* **2016**, *28*, 6546–6561. [CrossRef] [PubMed]
22. Morris, K.; Serpell, L. From Natural to Designer Self-Assembling Biopolymers, the Structural Characterisation of Fibrous Proteins and Peptides Using Fibre Diffraction. *Chem. Soc. Rev.* **2010**, *39*, 3445–3453. [CrossRef] [PubMed]
23. Ramírez-Rodríguez, L.C.; Barrera, L.E.D.; Quintanilla-Carvajal, M.X.; Mendoza-Castillo, D.I.; Bonilla-Petriciolet, A.; Jiménez-Junca, C. Preparation of a Hybrid Membrane from Whey Protein Fibrils and Activated Carbon to Remove Mercury and Chromium from Water. *Membranes* **2020**, *10*, 386. [CrossRef] [PubMed]
24. Bolisetty, S.; Mezzenga, R. Amyloid-Carbon Hybrid Membranes for Universal Water Purification. *Nat. Nanotechnol.* **2016**, *11*, 365–371. [CrossRef]
25. Yu, X.; Sun, S.; Zhou, L.; Miao, Z.; Zhang, X.; Su, Z.; Wei, G. Removing Metal Ions from Water with Graphene–Bovine Serum Albumin Hybrid Membrane. *Nanomaterials* **2019**, *9*, 276. [CrossRef]
26. Zhang, Q.; Zhang, S.; Zhao, Z.; Liu, M.; Yin, X.; Zhou, Y.; Wu, Y.; Peng, Q. Highly Effective Lead (II) Removal by Sustainable Alkaline Activated β-Lactoglobulin Nanofibrils from Whey Protein. *J. Clean. Prod.* **2020**, *255*, 120297. [CrossRef]
27. Li, C.; Adamcik, J.; Mezzenga, R. Biodegradable Nanocomposites of Amyloid Fibrils and Graphene with Shape-Memory and Enzyme-Sensing Properties. *Nat. Nanotechnol.* **2012**, *7*, 421–427. [CrossRef]
28. Vega-Lugo, A.C.; Lim, L.T. Effects of Poly(Ethylene Oxide) and PH on the Electrospinning of Whey Protein Isolate. *J. Polym. Sci. Part B Polym. Phys.* **2012**, *50*, 1188–1197. [CrossRef]
29. Oktar, F.N.; Su, S.; Ozbek, B.; Yücel, S.; Kazan, D.; Gunduz, O. Production and Characterization of Whey Protein Concentrate (WPC) Based Nano-Fibers. *Mater. Sci. Forum* **2018**, *923*, 47–51. [CrossRef]
30. Ricaurte, L.; Tello-Camacho, E.; Quintanilla-Carvajal, M.X. Hydrolysed Gelatin-Derived, Solvent-Free, Electrospun Nanofibres for Edible Applications: Physical, Chemical and Thermal Behaviour. *Food Biophys.* **2020**, *15*, 133–142. [CrossRef]
31. Ge, J.C.; Wu, G.; Yoon, S.K.; Kim, M.S.; Choi, N.J. Study on the Preparation and Lipophilic Properties of Polyvinyl Alcohol (Pva) Nanofiber Membranes via Green Electrospinning. *Nanomaterials* **2021**, *11*, 2514. [CrossRef] [PubMed]
32. Yakupova, E.I.; Bobyleva, L.G.; Vikhlyantsev, I.M.; Bobylev, A.G. Congo Red and Amyloids: History and Relationship. *Biosci. Rep.* **2019**, *39*, 1495–1506. [CrossRef] [PubMed]
33. Sahebjamee, N.; Soltanieh, M.; Mousavi, S.M.; Heydarinasab, A. Removal of Cu^{2+}, Cd^{2+} and Ni^{2+} Ions from Aqueous Solution Using a Novel Chitosan/Polyvinyl Alcohol Adsorptive Membrane. *Carbohydr. Polym.* **2019**, *210*, 264–273. [CrossRef] [PubMed]
34. Bera, B. Literature Review on Electrospinning Process (A Fascinating Fiber Fabrication Technique). *Imp. J. Interdiscip. Res.* **2016**, *2*, 972–984.
35. Nieuwland, M.; Geerdink, P.; Brier, P.; Van Den Eijnden, P.; Henket, J.T.M.M.; Langelaan, M.L.P.; Stroeks, N.; Van Deventer, H.C.; Martin, A.H. Food-Grade Electrospinning of Proteins. *Innov. Food Sci. Emerg. Technol.* **2013**, *20*, 269–275. [CrossRef]
36. Sullivan, S.T.; Tang, C.; Kennedy, A.; Talwar, S.; Khan, S.A. Electrospinning and Heat Treatment of Whey Protein Nanofibers. *Food Hydrocoll.* **2014**, *35*, 36–50. [CrossRef]
37. Olivato, J.B.; Nobrega, M.M.; Müller, C.M.O.; Shirai, M.A.; Yamashita, F.; Grossmann, M.V.E. Mixture Design Applied for the Study of the Tartaric Acid Effect on Starch/Polyester Films. *Carbohydr. Polym.* **2013**, *92*, 1705–1710. [CrossRef]
38. Fetz, A.E.; Fantaziu, C.A.; Smith, R.A.; Radic, M.Z.; Bowlin, G.L. Surface Area to Volume Ratio of Electrospun Polydioxanone Templates Regulates the Adsorption of Soluble Proteins from Human Serum. *Bioengineering* **2019**, *6*, 78. [CrossRef]
39. Nasouri, K.; Shoushtari, A.M. Effects of Diameter and Surface Area of Electrospun Nanocomposite Fibers on Electromagnetic Interference Shielding. *Polym. Sci. Ser. A* **2017**, *59*, 718–725. [CrossRef]
40. Colmenares-Roldán, G.J.; Quintero Martínez, Y.; Agudelo Gómez, L.M.; Rodríguez Vinasco, L.F.; Hoyos Palacio, L.M. Influence of the Molecular Weight of Polymer, Solvents and Operational Condition in the Electrospinning of Polycaprolactone. *Rev. Fac. Ing. Univ. Antioq.* **2017**, *84*, 35–45. [CrossRef]
41. Ricaurte, L.; Quintanilla Carvajal, M.X. Use of Electrospinning Technique to Produce Nanofibres for Food Industries: A Perspective from Regulations to Characterisations. *Trends Food Sci. Technol.* **2019**, *85*, 92–106. [CrossRef]
42. Haider, A.; Haider, S.; Kang, I.K. A Comprehensive Review Summarizing the Effect of Electrospinning Parameters and Potential Applications of Nanofibers in Biomedical and Biotechnology. *Arab. J. Chem.* **2018**, *11*, 1165–1188. [CrossRef]
43. Luo, C.J.; Stride, E.; Edirisinghe, M. Mapping the Influence of Solubility and Dielectric Constant on Electrospinning Polycaprolactone Solutions. *Macromolecules* **2012**, *45*, 4669–4680. [CrossRef]
44. Colín-Orozco, J.; Zapata-Torres, M.; Rodríguez-Gattorno, G.; Pedroza-Islas, R. Properties of Poly (Ethylene Oxide)/Whey Protein Isolate Nanofibers Prepared by Electrospinning. *Food Biophys.* **2014**, *10*, 134–144. [CrossRef]
45. Aceituno-Medina, M.; Lopez-Rubio, A.; Mendoza, S.; Lagaron, J.M. Development of Novel Ultrathin Structures Based in Amaranth (*Amaranthus Hypochondriacus*) Protein Isolate through Electrospinning. *Food Hydrocoll.* **2013**, *31*, 289–298. [CrossRef]
46. Gao, A.; Xie, K.; Song, X.; Zhang, K.; Hou, A. Removal of the Heavy Metal Ions from Aqueous Solution Using Modified Natural Biomaterial Membrane Based on Silk Fibroin. *Ecol. Eng.* **2017**, *99*, 343–348. [CrossRef]
47. Rodzik, A.; Pomastowski, P.; Sagandykova, G.N.; Buszewski, B. Interactions of Whey Proteins with Metal Ions. *Int. J. Mol. Sci.* **2020**, *21*, 2156. [CrossRef]
48. Pryshchepa, O.; Sagandykova, G.N.; Pomastowski, P.; Railean-Plugaru, V.; Król, A.; Rogowska, A.; Rodzik, A.; Sprynskyy, M.; Buszewski, B. A New Approach for Spontaneous Silver Ions Immobilization onto Casein. *Int. J. Mol. Sci.* **2019**, *20*, 3864. [CrossRef]

49. Mahmood Albakaa, A.R. Determination of 2-Mercaptoethanol by Potentiometric Titration with Mercury (II) Chloride. *Chem. Sci. J.* **2016**, *7*. [CrossRef]
50. Nguyen, N.H.A.; Streicher, C.; Anema, S.G. The Effect of Thiol Reagents on the Denaturation of the Whey Protein in Milk and Whey Protein Concentrate Solutions. *Int. Dairy J.* **2018**, *85*, 285–293. [CrossRef]
51. Muñoz, M.I.; Aller, A.J.; Littlejohn, D. The Bonding of Heavy Metals on Nitric Acid-Etched Coal Fly Ashes Functionalized with 2-Mercaptoethanol or Thioglycolic Acid. *Mater. Chem. Phys.* **2014**, *143*, 1469–1480. [CrossRef]
52. Mascarenhas, B.C.; Tavares, F.A.; Paris, E.C. Functionalized Faujasite Zeolite Immobilized on Poly(Lactic Acid) Composite Fibers to Remove Dyes from Aqueous Media. *J. Appl. Polym. Sci.* **2019**, *137*, 48561. [CrossRef]
53. El-aziz, A.M.A.; El-Maghraby, A.; Taha, N.A. Comparison between Polyvinyl Alcohol (PVA) Nanofiber and Polyvinyl Alcohol (PVA) Nanofiber/Hydroxyapatite (HA) for Removal of Zn^{2+} ions from Wastewater. *Arab. J. Chem.* **2017**, *10*, 1052–1060. [CrossRef]
54. Turan, D.; Gibis, M.; Gunes, G.; Baier, S.K.; Weiss, J. The Impact of the Molecular Weight of Dextran on Formation of Whey Protein Isolate (WPI)–Dextran Conjugates in Fibers Produced by Needleless Electrospinning after Annealing. *Food Funct.* **2018**, *9*, 2193–2200. [CrossRef] [PubMed]
55. Kabay, G.; Meydan, A.E.; Can, G.K.; Demirci, C.; Mutlu, M. Controlled Release of a Hydrophilic Drug from Electrospun Amyloid-like Protein Blend Nanofibers. *Mater. Sci. Eng. C* **2017**, *81*, 271–279. [CrossRef]
56. Garg, K.; Bowlin, G.L. Electrospinning Jets and Nanofibrous Structures. *Biomicrofluidics* **2011**, *5*, 013403. [CrossRef]
57. Gbassi, G.; Yolou, F.; Sarr, S.; Atheba, P.; Amin, C.; Ake, M. Whey Proteins Analysis in Aqueous Medium and in Artificial Gastric and Intestinal Fluids. *Int. J. Biol. Chem. Sci.* **2012**, *6*, 1828–1837. [CrossRef]
58. O'Loughlin, I.B.; Kelly, P.M.; Murray, B.A.; Fitzgerald, R.J.; Brodkorb, A. Concentrated Whey Protein Ingredients: A Fourier Transformed Infrared Spectroscopy Investigation of Thermally Induced Denaturation. *Int. J. Dairy Technol.* **2015**, *68*, 349–356. [CrossRef]
59. Zhang, Y.; Ullah, I.; Zhang, W.; Ou, H.; Domingos, M.; Gloria, A.; Zhou, J.; Li, W.; Zhang, X. Preparation of Electrospun Nanofibrous Polycaprolactone Scaffolds Using Nontoxic Ethylene Carbonate and Glacial Acetic Acid Solvent System. *J. Appl. Polym. Sci.* **2020**, *137*, 48387. [CrossRef]
60. Li, X.; Wang, C.; Yang, S.; Liu, P.; Zhang, B. Electrospun PCL/Mupirocin and Chitosan/Lidocaine Hydrochloride Multifunctional Double Layer Nanofibrous Scaffolds for Wound Dressing Applications. *Int. J. Nanomed.* **2018**, *13*, 5287–5299. [CrossRef]
61. Maity, S.; Pal, S.; Sardar, S.; Sepay, N.; Parvej, H.; Begum, S.; Dalui, R.; Das, N.; Pradhan, A.; Halder, U.C. Inhibition of Amyloid Fibril Formation of β-Lactoglobulin by Natural and Synthetic Curcuminoids. *New J. Chem.* **2018**, *42*, 19260–19271. [CrossRef]
62. Bolder, S.; Sagis, L.; Venema, P.; Van Der Linden, E. Effect of Stirring and Seeding on Whey Protein Fibril Formation. *J. Agric. Food Chem.* **2007**, *55*, 5661–5669. [CrossRef] [PubMed]
63. Kabay, G.; Kaleli, G.; Sultanova, Z.; Ölmez, T.T.; Şeker, U.Ö.Ş.; Mutlu, M. Biocatalytic Protein Membranes Fabricated by Electrospinning. *React. Funct. Polym.* **2016**, *103*, 26–32. [CrossRef]
64. Lin, W.C.; Razali, N.A.M. Temporary Wettability Tuning of PCL/PDMS Micro Pattern Using the Plasma Treatments. *Materials* **2019**, *12*, 644. [CrossRef]
65. Efome, J.E.; Rana, D.; Matsuura, T.; Lan, C.Q. Metal-Organic Frameworks Supported on Nanofibers to Remove Heavy Metals. *J. Mater. Chem. A* **2018**, *6*, 4550–4555. [CrossRef]
66. Liu, X.; Feng, P.; Zhang, L.; Chen, Y. Mussel-Inspired Method to Decorate Commercial Nanofiltration Membrane for Heavy Metal Ions Removal. *Polym. Adv. Technol.* **2020**, *31*, 665–674. [CrossRef]
67. Ullah, S.; Hashmi, M.; Hussain, N.; Ullah, A.; Sarwar, M.N.; Saito, Y.; Kim, S.H.; Kim, I.S. Stabilized Nanofibers of Polyvinyl Alcohol (PVA) Crosslinked by Unique Method for Efficient Removal of Heavy Metal Ions. *J. Water Process Eng.* **2020**, *33*, 101111. [CrossRef]
68. Tsotetsi, T.A.; Mochane, M.J.; Motaung, T.E.; Gumede, T.P.; Linganiso, Z.L. Synergistic Effect of EG and Cloisite 15A on the Thermomechanical Properties and Thermal Conductivity of EVA/PCL Blend. *Mater. Res.* **2017**, *20*, 109–118. [CrossRef]
69. Seyedsalehi, A.; Daneshmandi, L.; Barajaa, M.; Riordan, J.; Laurencin, C.T. Fabrication and Characterization of Mechanically Competent 3D Printed Polycaprolactone-Reduced Graphene Oxide Scaffolds. *Sci. Rep.* **2020**, *10*, 22210. [CrossRef]
70. Erdem, B.G.; Kaya, S. Production and Application of Freeze Dried Biocomposite Coating Powders from Sunflower Oil and Soy Protein or Whey Protein Isolates. *Food Chem.* **2021**, *339*, 127976. [CrossRef]
71. Guo, Z.; Kang, Y.; Liang, S.; Zhang, J. Detection of Hg(II) in Adsorption Experiment by a Lateral Flow Biosensor Based on Streptavidin-Biotinylated DNA Probes Modified Gold Nanoparticles and Smartphone Reader. *Environ. Pollut.* **2020**, *266*, 115389. [CrossRef] [PubMed]
72. Acharya, S.; Sahoo, S.; Sonal, S.; Lee, J.H.; Mishra, B.K.; Nayak, G.C. Adsorbed Cr(VI) Based Activated Carbon/Polyaniline Nanocomposite: A Superior Electrode Material for Asymmetric Supercapacitor Device. *Compos. Part B Eng.* **2020**, *193*, 107913. [CrossRef]
73. Bagbi, Y.; Sarswat, A.; Mohan, D.; Pandey, A.; Solanki, P.R. Lead (Pb^{2+}) Adsorption by Monodispersed Magnetite Nanoparticles: Surface Analysis and Effects of Solution Chemistry. *J. Environ. Chem. Eng.* **2016**, *4*, 4237–4247. [CrossRef]
74. Ghanbarian, M.; Nabizadeh, R.; Nasseri, S.; Shemirani, F.; Mahvi, A.H.; Beyki, M.H.; Mesdaghinia, A. Potential of Amino-Riched Nano-Structured $MnFe_2O_4$@cellulose for Biosorption of Toxic Cr(VI): Modeling, Kinetic, Equilibrium and Comparing Studies. *Int. J. Biol. Macromol.* **2017**, *104*, 465–480. [CrossRef]

75. Aktas, D.; Dizge, N.; Yatmaz, H.C.; Caliskan, Y.; Ozay, Y.; Caputcu, A. The Adsorption and Fenton Behavior of Iron Rich Terra Rosa Soil for Removal of Aqueous Anthraquinone Dye Solutions: Kinetic and Thermodynamic Studies. *Water Sci. Technol.* **2017**, *76*, 3114–3125. [CrossRef]
76. Fellenz, N.; Perez-Alonso, F.J.; Martin, P.P.; García-Fierro, J.L.; Bengoa, J.F.; Marchetti, S.G.; Rojas, S. Chromium (VI) Removal from Water by Means of Adsorption-Reduction at the Surface of Amino-Functionalized MCM-41 Sorbents. *Microporous Mesoporous Mater.* **2017**, *239*, 138–146. [CrossRef]
77. Jin, X.; Wang, H.; Jin, X.; Wang, H.; Chen, L.; Wang, W.; Lin, T.; Zhu, Z. Preparation of Keratin/PET Nanofiber Membrane and Its High Adsorption Performance of Cr(VI). *Sci. Total Environ.* **2020**, *710*, 135546. [CrossRef]
78. Hota, G.; Kumar, B.R.; Ng, W.J.; Ramakrishna, S. Fabrication and Characterization of a Boehmite Nanoparticle Impregnated Electrospun Fiber Membrane for Removal of Metal Ions. *J. Mater. Sci.* **2008**, *43*, 212–217. [CrossRef]
79. Zou, H.; Lv, P.F.; Wang, X.; Wu, D.; Yu, D.G. Electrospun Poly(2-Aminothiazole)/Cellulose Acetate Fiber Membrane for Removing Hg(II) from Water. *J. Appl. Polym. Sci.* **2017**, *134*, 2–9. [CrossRef]
80. Liu, Y.; Liu, Y.J. Biosorption Isotherms, Kinetics and Thermodynamics. *Sep. Purif. Technol.* **2008**, *61*, 229–242. [CrossRef]
81. Gowraraju, N.D.; Jagadeesan, S.; Ayyasamy, K.; Olasunkanmi, L.O.; Ebenso, E.E.; Subramanian, C. Adsorption Characteristics of Iota-Carrageenan and Inulin Biopolymers as Potential Corrosion IGGnhibitors at Mild Steel/Sulphuric Acid Interface. *J. Mol. Liq.* **2017**, *232*, 9–19. [CrossRef]
82. Majd, M.M.; Kordzadeh-Kermani, V.; Ghalandari, V.; Askari, A.; Sillanpää, M. Adsorption Isotherm Models: A Comprehensive and Systematic Review (2010−2020). *Sci. Total Environ.* **2021**, *812*, 151334. [CrossRef] [PubMed]
83. Guimarães, V.; Rodríguez-Castellón, E.; Algarra, M.; Rocha, F.; Bobos, I. Influence of PH, Layer Charge Location and Crystal Thickness Distribution on U(VI) Sorption onto Heterogeneous Dioctahedral Smectite. *J. Hazard. Mater.* **2016**, *317*, 246–258. [CrossRef] [PubMed]
84. Zúñiga-Muro, N.M.; Bonilla-Petriciolet, A.; Mendoza-Castillo, D.I.; Reynel-Ávila, H.E.; Duran-Valle, C.J.; Ghalla, H.; Sellaoui, L. Recovery of Grape Waste for the Preparation of Adsorbents for Water Treatment: Mercury Removal. *J. Environ. Chem. Eng.* **2020**, *8*, 103738. [CrossRef]
85. Abdus-Salam, N.; Buhari, M. Adsorption of Alizarin and Fluorescein Dyes onto Palm Seeds Activated Carbon: Kinetic and Thermodynamic Studies. *J. Chem. Soc. Pak.* **2016**, *38*, 604–614.
86. Xiang, L.; Lin, J.; Yang, Q.; Lin, S.; Chen, S.; Yan, B. Facile Preparation of Hierarchical Porous Polydopamine Microspheres for Rapid Removal of Chromate from the Wastewater. *J. Leather Sci. Eng.* **2020**, *2*, 22. [CrossRef]
87. Yuan, Z.; Cheng, X.; Zhong, L.; Wu, R.; Zheng, Y. Preparation, Characterization and Performance of an Electrospun Carbon Nanofiber Mat Applied in Hexavalent Chromium Removal from Aqueous Solution. *J. Environ. Sci.* **2019**, *77*, 75–84. [CrossRef]
88. Mortazavian, S.; Saber, A.; Hong, J.; Bae, J.H.; Chun, D.; Wong, N.; Gerrity, D.; Batista, J.; Kim, K.J.; Moon, J. Synthesis, Characterization, and Kinetic Study of Activated Carbon Modified by Polysulfide Rubber Coating for Aqueous Hexavalent Chromium Removal. *J. Ind. Eng. Chem.* **2019**, *69*, 196–210. [CrossRef]
89. Celine, S.; Varun, S.; Chandran, A.M.; Mural, P.K.S. Electrospun PVDF/Silica Thiol Nanofiber for Chromium Exclusion. *Mater. Today Proc.* **2021**, *47*, 1461–1465. [CrossRef]
90. Vazquez-Velez, E.; Lopez-Zarate, L.; Martinez-Valencia, H. Electrospinning of Polyacrylonitrile Nanofibers Embedded with Zerovalent Iron and Cerium Oxide Nanoparticles, as Cr(VI) Adsorbents for Water Treatment. *J. Appl. Polym. Sci.* **2019**, *137*, 48663. [CrossRef]
91. Zhang, D.; Xu, W.; Cai, J.; Cheng, S.Y.; Ding, W.P. Citric Acid-Incorporated Cellulose Nanofibrous Mats as Food Materials-Based Biosorbent for Removal of Hexavalent Chromium from Aqueous Solutions. *Int. J. Biol. Macromol.* **2020**, *149*, 459–466. [CrossRef] [PubMed]
92. Yang, X.; Zhou, Y.; Sun, Z.; Yang, C.; Tang, D. Effective Strategy to Fabricate ZIF-8@ZIF-8/Polyacrylonitrile Nanofibers with High Loading Efficiency and Improved Removing of Cr(VI). *Colloids Surf. A Physicochem. Eng. Asp.* **2020**, *603*, 125292. [CrossRef]
93. Sahoo, S.K.; Panigrahi, G.K.; Sahoo, J.K.; Pradhan, A.K.; Purohit, A.K.; Dhal, J.P. Electrospun Magnetic Polyacrylonitrile-GO Hybrid Nanofibers for Removing Cr(VI) from Water. *J. Mol. Liq.* **2021**, *326*, 115364. [CrossRef]
94. Fan, F.; da Silva, M.A.C.; Moraes, C.R.; Dunham, A.D.; HogenEsch, H.; Turner, J.W.; Lannutti, J.J. Self-Reinforcing Nanoscalar Polycaprolactone-Polyethylene Terephthalate Electrospun Fiber Blends. *Polymer* **2020**, *202*, 122573. [CrossRef]

nanomaterials

Article

Nanostructured Metal Oxide Sensors for Antibiotic Monitoring in Mineral and River Water

Cátia Magro [1,2,*], Tiago Moura [2], Joana Dionísio [1], Paulo A. Ribeiro [3], Maria Raposo [3] and Susana Sério [3,*]

1 School for International Training, World Learning Inc., Brattleboro, VT 05302, USA; joana.dionisio@sit.edu
2 Department of Physics, NOVA School of Science and Technology, NOVA University Lisbon, 2829-516 Caparica, Portugal; ta.moura@campus.fct.unl.pt
3 Laboratory of Instrumentation, Biomedical Engineering and Radiation Physics (LIBPhys-UNL), Department of Physics, NOVA School of Science and Technology, NOVA University Lisbon, 2829-516 Caparica, Portugal; pfr@fct.unl.pt (P.A.R.); mfr@fct.unl.pt (M.R.)
* Correspondence: catia.magro@sit.edu (C.M.); susana.serio@fct.unl.pt (S.S.)

Abstract: Antibiotics represent a class of pharmaceuticals used to treat bacterial infections. However, the ever-growing use of antibiotics in agriculture and human and veterinary medicine has led to great concern regarding the outbreak of microbe strains resistant to antimicrobial drugs. Azithromycin, clarithromycin, and erythromycin are macrolides, a group of molecules with a broad spectrum of antibiotic properties, included in the second EU watchlist of emerging pollutants which emphasizes the importance of understanding their occurrence, fate, and monitoring in aquatic environments. Thus, the aim of this study was to develop sensors based on nanostructured thin films deposited on ceramic substrates with gold interdigitated electrodes, to detect azithromycin, clarithromycin, and erythromycin in water matrices (mineral and river water). Impedance spectroscopy was employed as the transducing method for the devices' electrical signal, producing multivariate datasets which were subsequently analyzed by principal component analysis (PCA). The PCA plots for mineral water demonstrated that ZnO- and TiO_2-based sensors produced by DC magnetron sputtering either with 50% or 100% O_2 in the sputtering chamber, were able to detect the three macrolides in concentrations between 10^{-15} M and 10^{-5} M. In river water, the PCA discrimination presented patterns and trends, between non-doped and doped, and sorting the different concentrations of azithromycin, clarithromycin, and erythromycin. Considering both matrices, by applying the e-tongue concept, sensitivity values of 4.8 ± 0.3, 4.6 ± 0.3, and 4.5 ± 0.3 per decade to azithromycin, clarithromycin, and erythromycin concentration, respectively, were achieved. In all cases, a resolution of 1×10^{-16} M was found near the 10^{-15} M concentration, the lowest antibiotic concentration measured.

Keywords: environmental monitoring; antibiotics; impedance spectroscopy; nanostructured sensors; metal oxides; electronic tongue

Citation: Magro, C.; Moura, T.; Dionísio, J.; Ribeiro, P.A.; Raposo, M.; Sério, S. Nanostructured Metal Oxide Sensors for Antibiotic Monitoring in Mineral and River Water. *Nanomaterials* **2022**, *12*, 1858. https://doi.org/10.3390/nano12111858

Academic Editor: Camelia Bala

Received: 5 May 2022
Accepted: 27 May 2022
Published: 29 May 2022

1. Introduction

The Water Research Group predicted that by 2030 more than 160% of the total available water volume in the world will be needed to satisfy global water requirements [1,2]. Water scarcity is already a reality in several countries, leading to severe economic, social, and environmental consequences, which will only be aggravated in the near future [3–5]. Climate change, population growth, changing consumption patterns, and the expansion of irrigated agriculture will further heighten this shortage and intensify the need to find alternatives to save freshwater [4–6]. Although the efficacy of wastewater treatment plants (WWTPs) has been improved in past decades, the conventional procedures applied are still not able to remove or monitor emerging pollutants (EPs); therefore, these are the main entrance vectors of these compounds in surface waters [7]. Emerging pollutants "are chemicals or microorganisms that are not commonly monitored but have the potential to enter the environment and cause adverse ecological/human health effects" [8]. Their physicochemical properties determine their persistence and bioaccumulation in the environment [7].

Water resources polluted with antibiotics increase the risk of developing antibiotic resistant bacteria and genes [9]. Azithromycin (AZI), clarithromycin (CLAR), and erythromycin (ERY) are substances with antibacterial effects for a variety of pathogenic bacteria, which have previously been detected in the final effluent of WWTPs [10–12]. These three macrolides were included in the latest version of the surface water watchlist, a list which proposes the need for the monitoring of 15 substances in water by EU Member States [10]. Currently, the most frequently used approach to monitor such compounds is offline monitorization, which involves sample collection followed by chemical analysis, making it difficult for real-time quantification [13]. A promising direction for environmental aqueous matrix monitoring is the application of multisensory systems [14]. Electronic tongues (e-tongues) are systems that use an array of non-specific sensors linked to data processing methods, to interpret complex responses using advanced chemometric tools in order to relate its analytical meaning [13,14]. Additionally, working electrodes in the e-tongue array can be covered with thin coatings, which highly contribute to the sensitivity and stability of the sensor's devices. Selecting the type of film or sensorial layer are critical decisions to perform reliable qualitative and quantitative analysis [15]. Nanostructured TiO_2 is biocompatible and non-toxic and presents a large specific surface area, exceptional electron transfer properties, and strong adsorption ability for organic and biological molecules, making it an attractive material for the development of sensor devices [16,17]. Zinc oxide is an inorganic semiconductor with strong chemical, mechanical, and thermal stability, as well as other optical and electrical properties which make it interesting for electronic, optoelectronic, and laser technology applications [18,19]. Facure et al. [20] developed an electronic tongue based on graphene hybrid nanocomposites for the detection of organophosphate (OP) pesticides in mineral and tap water matrices. The resistance data, analyzed by principal component analysis, showed the ability of these sensors to detect concentrations of OP of 0.1 nM. To test a similar thesis, in 2021, Magro et al. [21] used an array of sputtered thin films based on multi-walled carbon nanotubes and titanium dioxide, to identify triclosan concentrations ranging from 10^{-15} to 10^{-5} M in water and milk-based solutions. More recently, aligned TiO_2 nanorod arrays decorated with closely interconnected Au/Ag nanoparticles for near-infrared surface-enhanced Raman scattering (SERS) active sensors, was used for the detection of ciprofloxacin antibiotic and chloramphenicol in environmental water samples, with detection limits of 10^{-9} M and 10^{-8} M, respectively [22].

Following the potential of those materials and the need for monitoring EP contaminants, particularly antibiotics, the aim of this study was to develop nanostructured sensors capable of discriminating concentrations in the range of 10^{-15} M to 10^{-5} M of AZI, CLAR, and ERY in mineral and river water matrices. Hence, the e-tongue concept was applied to an array of sensors composed of ceramic supports with gold interdigitated electrodes coated with thin films of titanium dioxide (TiO_2) and zinc oxide (ZnO), deposited by DC magnetron sputtering.

2. Materials and Methods

All chemicals employed were of analytical grade or chemical grade (Sigma–Aldrich, St. Louis, MO, USA). The argon (Ar), oxygen (O_2), and nitrogen (N_2) gas, with $\geq$99.9% purity, were all acquired from Air Liquide (Algés, Portugal). Standards of AZI, CLAR, and ERY were purchased from Sigma-Aldrich (St. Louis, MO, USA). Ultra-pure water (resistivity of 20.4 MΩ cm at 24 °C) was obtained using a Direct-Q 3 UV system from Millipore (Bedford, MA, USA).

For this study, two experimental aqueous matrices were used: commercial Portuguese mineral water (MW) and river water (RW) collected from Tagus River at Porto Brandão, Caparica, Portugal. Table 1 presents the pH and electrical conductivity values measured for the two matrices, which were obtained using pH Prolab 1000 apparatus (Schott Instruments GmbH, Merseyside, UK).

Table 1. pH and electrical conductivity values measured for mineral and river water.

	Mineral Water	**River Water**
pH	5.875 ± 0.001	7.536 ± 0.001
Electrical Conductivity (mS/cm)	11.62 ± 0.01	52.43 ± 0.01

The target compound solutions with concentrations ranging from 10^{-15} M to 10^{-5} M were prepared by performing sequential dilutions of a mother solution with a concentration of 10^{-4} M. All the solutions were prepared with an experimental matrix/MeOH (9:1). Moreover, a combination of experimental matrix/MeOH (9:1), containing no dissolved macrolides, was used as the blank standard (0 M). The samples were kept in a refrigerator unit between being prepared and their use in impedance spectroscopy measurements. The river water samples collected from Tagus River were stored in a freezing compartment.

To produce the nanostructured sensors, thin films of metal oxides were deposited on the sensorial area of ceramic substrates devices acquired from Metrohm DropSens (Oviedo, Asturias, Spain). Those were composed of two interdigitated electrodes (IDEs) with two connection tracks, all made of gold, on a ceramic substrate. The device's dimensions were 22.8 mm (length) × 7.6 mm (width) × 1 mm (thickness), and each internal "finger" had a 200 μm width, which also corresponded to the spacing between fingers. Prior to the deposition of the thin films, all sensors were cleaned with ethanol and ultrapure water. Thereafter, the substrates were dried with compressed nitrogen gas (99% purity, Air Liquide, Algés, Portugal). Subsequently, impedance spectroscopy assays were performed on the uncoated, unused sensors exposed to surrounding air, to assure similarity between the devices.

TiO_2 and ZnO thin films were produced by reactive DC magnetron sputtering, using titanium and zinc targets (Goodfellow, Cambridge, UK, 99.99%), respectively, as well as argon (Air Liquid, Paris, France, 99.99%) and oxygen (Gás Piedense gases, Setúbal, Portugal, 99.99%). To achieve a base pressure of 10^{-4}–10^{-5} Pa (before introducing the sputtering gas) a turbomolecular pump (Pfeiffer TMH 1001, Pfeiffer Vacuum GmbH, Asslar, Germany) was utilized. Before the sputter-deposition of the films, a movable shutter was placed between the target and the supports. The target was pre-sputtered in an Ar atmosphere for 1 min to remove the target surface oxidation. The target-to-support distance was maintained at 100 mm. For each sensor device, the sputtering was performed both in a 100% O_2 and 50:50 O_2/Ar atmospheres and the other deposition parameters were kept constant, differing for each oxide, as summarized in Table 2.

Table 2. Sputtering thin film production characteristics.

	ZnO_50	**ZnO_100**	**TiO_2_50**	**TiO_2_100**
O_2 (%)	50%	100%	50%	100%
Ar (%)	50%	0%	50%	0%
Working Pressure (Pa)	0.8	0.8	0.8	0.8
Power (W)	300	290	1010	1020
Voltage (V)	462	275	302	376
Current (A)	0.65	1.04	3.34	2.70
Time (min)	5	5	15	15

The surface morphology was studied with a scanning electron microscope (ThermoFisher Scientific (Waltham, MA, USA) model Phenom ProX G6), operating at 15 keV. A palladium–gold thin film (~20 nm thickness) was coated on the film's surface before SEM analysis to prevent charge build-up.

The electrical response of the sensors was assessed by measuring the impedance spectra through a Solartron 1260 Impedance/Gain-Phase Analyzer coupled to a 1296A Dielectric Interface (Solartron Analytical, AMETEK scientific instruments, Berwyn, PA, USA) by sweeping the frequency of the applied signal in the range of 1 Hz to 1 MHz.

The amplitude of the stimulus was set to 25 mV due to the short spacing between the interdigitated electrodes. The parameters for the assays were processed through SMaRT Impedance Measurement Software (v. 3.3.1, AMETEK scientific instruments, Berwyn, PA, USA). Every measurement began with a 30 s delay, which allowed the system to stabilize, and was repeated twice. The impedance spectroscopy assays were performed in an ascending order of macrolide concentration. Each assay was performed in duplicate and at room temperature (~23 °C). The electrical response was replicated with a similar sensor, i.e., a sensor with a similar thin film (deposited at the same time) used as a sensitive layer, to validate the behavior of the thin films and study the reproducibility of the results measured by the nanostructured sensor devices.

The principal component analysis (PCA) was carried out considering the normalized impedance spectroscopy data (Z-score normalization: $z = \frac{x-\mu}{\sigma}$, with μ and σ as the mean value and standard deviation of the samples, respectively). Thus, an array of the nanostructured sensors constituted by all produced thin films was assessed as an e-tongue for AZI, CLAR, and ERY macrolides in both matrices.

3. Results and Discussion

3.1. Surface Morphology Characterization

The surface morphology of the zinc oxide and titanium dioxide thin films was analyzed by scanning electron microscopy; the obtained SEM images are depicted in Figure 1. From the SEM analysis, it can be observed that the surface of ZnO thin films is rougher than the surface of TiO_2 films, exhibiting a granular surface microstructure, more evident for the film deposited with 50% O_2. By increasing the oxygen percentage to 100% (i.e., only reactive gas in the chamber) the surface of the ZnO films became smoother, with fewer irregularities (Figure 1a,b; and Figure S7 in the Supplementary Materials with 30,000× magnification).

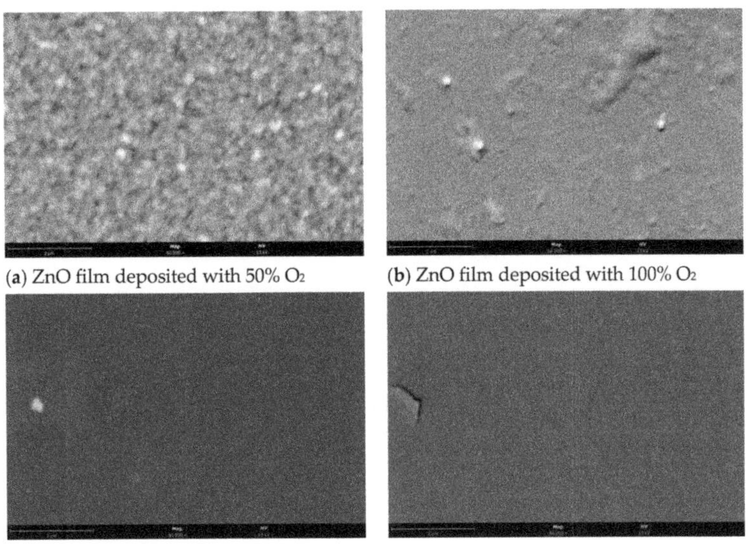

(**a**) ZnO film deposited with 50% O_2 (**b**) ZnO film deposited with 100% O_2

(**c**) TiO_2 film deposited with 50% O_2 (**d**) TiO_2 film deposited with 100% O_2

Figure 1. *Cont.*

(e) SEM image of the uncoated ceramic substrate.

Figure 1. SEM images with 60,000× magnification for (**a**) ZnO 50% (60,000× magnitude); (**b**) ZnO 100%; (**c**) TiO_2 50%; (**d**) TiO_2 100% (scale bar size: 2 μm) and 30,000× magnification for (**e**) uncoated ceramic substrate (scale bar size: 5 μm).

Regarding TiO_2 (Figure 1c,d; and in Figure S7 with 30,000× magnification) both produced films exhibit similar characteristics displaying relatively uniform and smooth surface [23]. In Figure 1e, an SEM image is presented of the uncoated ceramic substrate. Although it displays quite a rough surface, it does not seem to influence the observed morphology of the thin films.

3.2. *Impedimetric Measurements: Sputtered Thin Films*

When immersed in aqueous samples, the electrical behavior of thin films deposited on the IDE is influenced by its characteristics as a sensitive layer, the properties of the double layer formed on the interface of the thin film solution, and the attributes of the bulk electrolyte [24]. Moreover, the electrical behavior is analogous to an equivalent circuit with components that represent the thin film, double layer, and electrolyte; thus, it may be interpreted by the electrical properties of that circuit [25]. Prior to the analysis of the results, the reproducibility was evaluated. Thus, it may be observed by the Tables S1–S4 in the Supplementary Materials that higher values of standard deviation could be explained by the slight heterogeneity which may be experienced upon the sputtering of the thin films onto the substrate. Although the pairs of films were deposited at the same time, the substrates' positions, regarding the magnetron cathode, may differ slightly; therefore, this may lead to very small discrepancies in the thickness of the films that justify the observed standard deviation. Nonetheless, the results were consistent and reproducible, presenting the same trends on the curves for impedance data. Accordingly, the impedance spectra measured by the different thin film sensors immersed in MW and RW samples of the three macrolides are presented in the subsequent sections.

3.2.1. ZnO Thin Films Deposited with 50% and 100% O_2

In Figure 2, the response of ZnO films produced with 50% and 100% of O_2 is presented, with respect to the different antibiotics in MW.

In the low-conductivity matrix, the sensor coated with the ZnO thin film deposited with 50% O_2 showed a clear pattern of increasing impedance with the increases in AZI, CLAR, and ERY concentrations in the entire frequency range studied (Figure 2). Nonetheless, for ERY, the range of sensitivity is more evident in higher-frequency values. As shown in Figure 2, the electrical measurements obtained with the ZnO sensors produced with 100% O_2 exhibited a similar behavior for all antibiotics, showing a more pronounced sensitivity between 1 and 10 Hz in CLAR samples. By fixing the impedance in relation to compound concentrations, the devices that presented larger sensitivity for CLARY and ERY were ZnO sensors produced with 100% O_2 and 50% O_2, respectively (Figures S3 and S5). Those sensors exhibited identical trends of monotonic functions with inverse tendency.

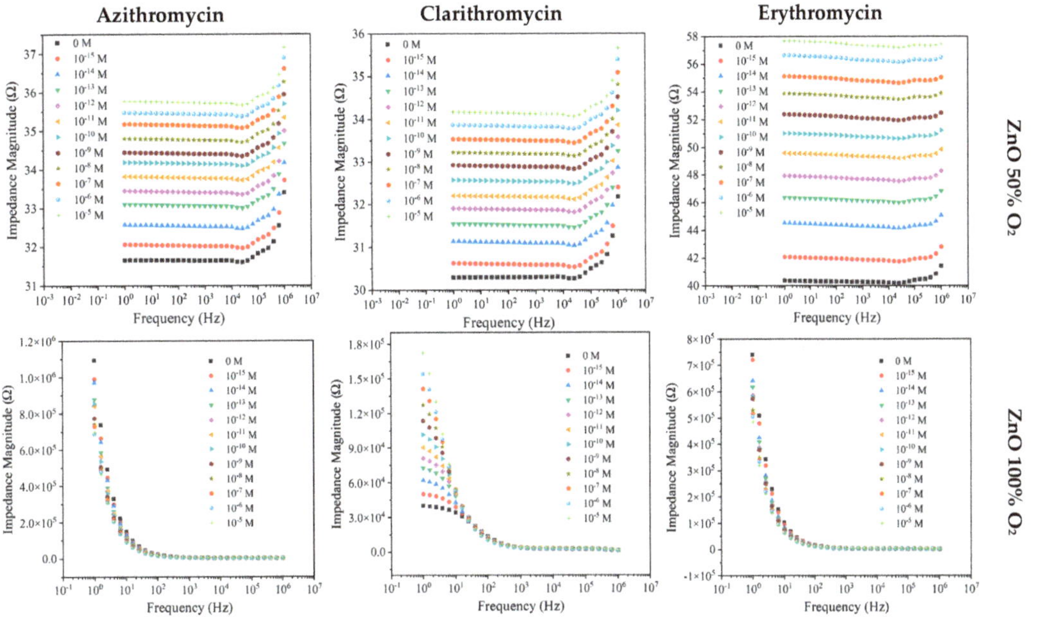

Figure 2. Impedance spectra of the ZnO sensor devices deposited with 50% and 100% O_2 measured at different frequencies, when immersed in MW at different AZI, CLA, and ERY concentrations.

Measuring the target compounds in river water means the change in the conductivity of the media and its pH; therefore, this may influence the impedance behavior (Figure 3).

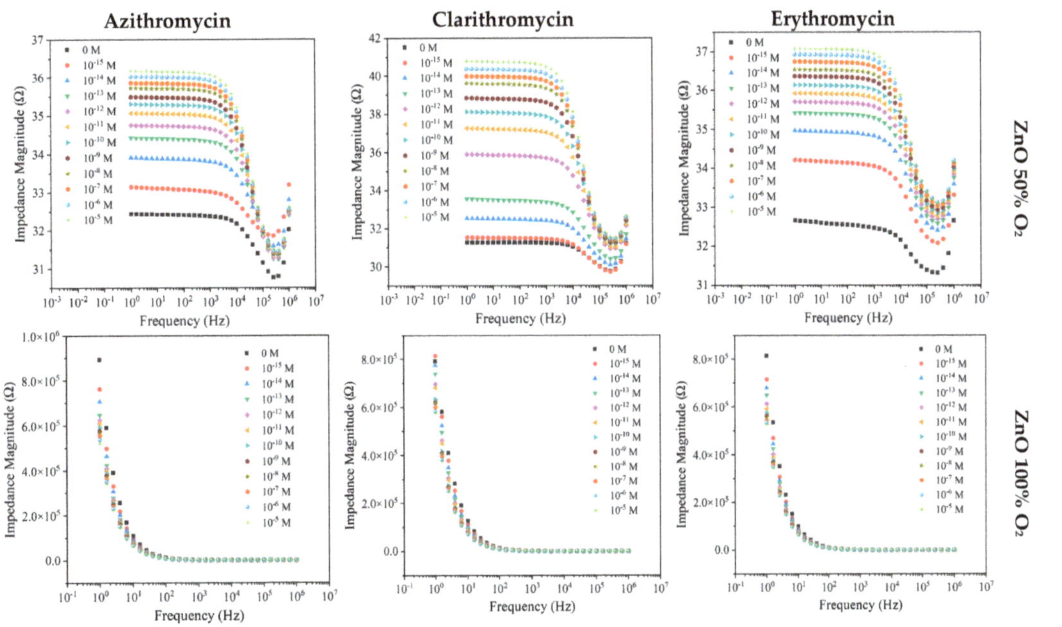

Figure 3. Impedance spectra of the ZnO sensors devices deposited with 50% and 100% O_2 measured at different frequencies, when immersed in RW at different AZI, CLAR, and ERY concentrations.

Changing the matrix did not show variation in the trend of the impedance, which increased with the increase in concentration. Nevertheless, for all antibiotics, a more pronounced sensitivity in the 1–10 Hz frequency range (Figure 3) is observed, due to the impedimetric curve change at 10,000 Hz. Additionally, the fluctuation of the impedance values achieved for ZnO sensors deposited with 50% O_2 narrowed from 10^{-5} M to 10^{-12} M, as compared with the amplitude from 0 M to 10^{-12} M For ZnO (50% O_2), by fixing the impedance at 6.31×10^4 and 3.98×10^2 Hz in relation to compound concentrations, resulted in an increasing normalized impedance trend for both AZI and CLAR (Figures S2 and S4). Regarding ERY, the ZnO sensor produced with 50% of O_2 presented a normalized impedance increase between 10^{-15} and 10^{-8} M; however, the device reached saturation point, which can be seen by the approximately constant impedance tendency (Figure S6). In the case of the ZnO (100% O_2) sensor, when the frequency was fixed at 1.58 Hz for ERY and 1 Hz for both AZI and CLAR (Figures S2, S4 and S6), all the compounds under study displayed a decreasing pattern, proving that this sensor exhibits higher sensitivity in the studied concentration range.

3.2.2. TiO_2 Thin Films Produced with 50% O_2 and 100% O_2

The electrical responses of TiO_2 films produced with 50% and 100% of O_2 to the different antibiotics in MW matrices are depicted in Figure 4.

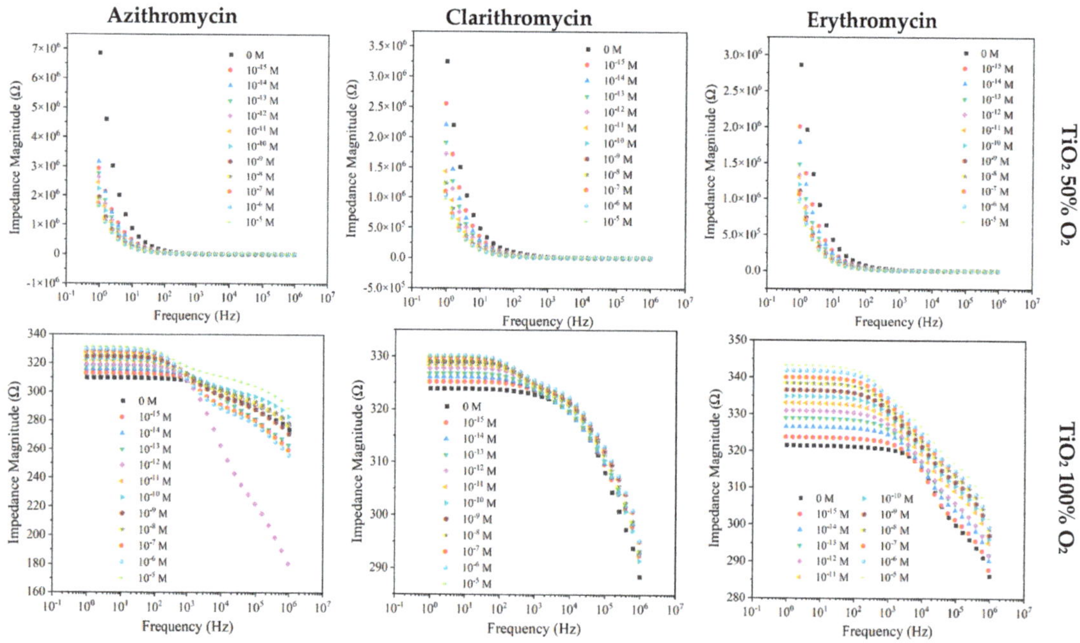

Figure 4. Impedance spectra of the TiO_2-based sensors device produced with 50% and 100% O_2 measured at different frequencies, when immersed in MW at different AZI, CLAR, and ERY concentrations.

When the considered matrix was MW, similar footprints regarding AZI, CLAR, and ERY were observed for the TiO_2 (50% O_2) sensor, with the measured impedance magnitude presenting significant sensitivity at lower frequencies (1–100 Hz) with a decreasing trend (Figure 4). The TiO_2 sensor produced with 100% O_2 exhibited increased impedance with the increases in AZI, CLAR, and ERY concentrations, displaying substantial sensitivity in the frequency range of 1 Hz to 1 kHz, more pronounced in the case of ERY samples (Figure 4). The impedance measured for TiO_2 thin film sensors was normalized in relation to the blank concentration (Figure S1). The results achieved with AZI samples show how the sensors with TiO_2 acted in contrast to one another, with TiO_2 (100% O_2) thin films showing an increasing impedance trend and TiO_2 (50% O_2) presenting a decreasing impedance trend. Both types of sensors presented significant sensitivity to the distinct AZI concentrations. A decreasing trend was also observed in TiO_2 (50% O_2) for CLAR until the concentration of 10^{-6} M, where it seemed to have reached a sensor saturation (Figure S3). Likewise, this sensor saturation was observed for the ERY compound above 10^{-7} M (Figure S5). For CLAR and ERY, the TiO_2 (100% O_2) sensor exhibited a dispersion in results, evidencing a lower sensitivity to ERY.

Impedance spectra of RW samples with AZI, CLAR, or ERY show similar patterns to those observed in the MW samples for a TiO_2 sensor produced with 50% O_2, displaying a more pronounced sensitivity in the 1 Hz to 10 Hz frequency range (Figure 5).

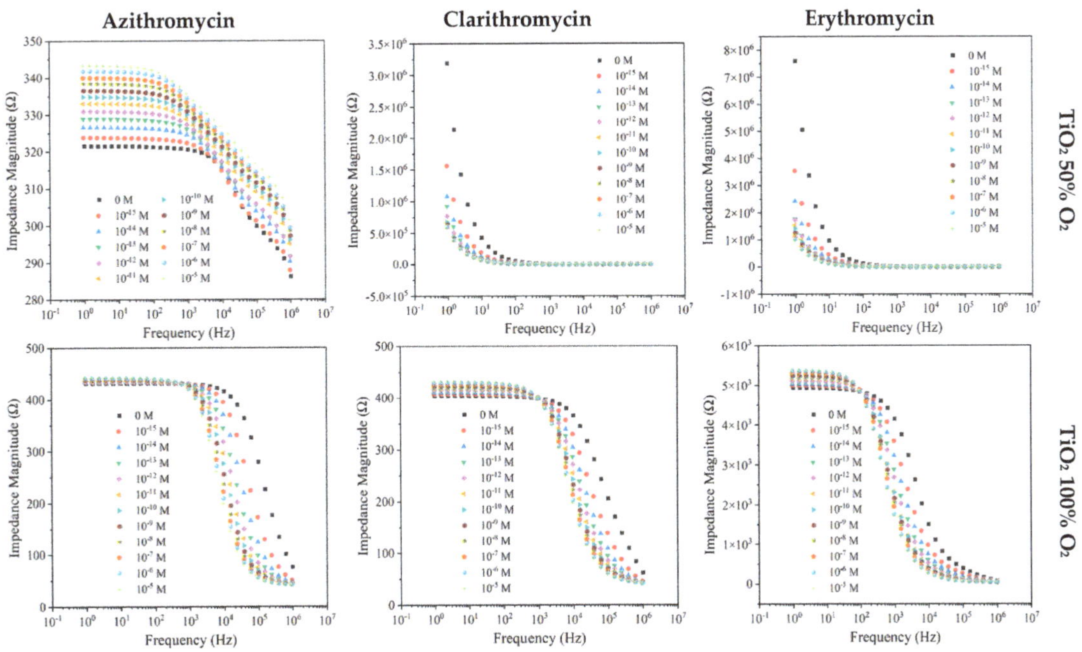

Figure 5. Impedance spectra at different frequencies of the TiO_2-based sensor devices produced with 50% and 100% O_2, when immersed in RW at different AZI, CLA, and ERY concentrations.

Concerning the TiO_2 sensor produced with 100% O_2, the device showed sensitivity to two distinct frequency ranges, for AZI, CLAR, and ERY. The impedance values increased between 1 Hz and 100 Hz and 1 Hz and 1 kHz for AZI and CLAR, respectively. Above 1 MHz for AZI and 1 kHz for CLAR, a decreasing impedance trend was observed. As previously mentioned, by normalizing the impedance values to a fixed frequency of 1 Hz for AZI and ERY and 10^6 Hz for CLAR (Figures S2, S4 and S6), it could be observed that the results achieved with the TiO_2 device produced with 50% O_2 only showed sensitivity for AZI and ERY samples between 10^{-15} and 10^{-8} M. On the other hand, for CLAR samples, the TiO_2 (100% O_2) sensor displayed a more pronounced sensitivity in comparison with the TiO_2 (50% O_2) sensor. Finally, the function that fit the data measured by the TiO_2 (100% O_2) follows an impedance increasing trend.

3.3. *Electronic Tongue Concept*

Principal Component Analysis: Sensor Sensitivity and Resolution

Following the analysis of the electrical response and capabilities of the individual nanostructured sensors, the electronic tongue concept was assessed through PCA, and applied to the impedance, capacitance, resistance, reactance, and loss tangent dataset of an array of sensors composed by all the thin film sensors produced in the present study. The e-tongue sensor was used to distinguish macrolide concentrations in both experimental matrices. The resulting PCA plots are presented in Figure 6.

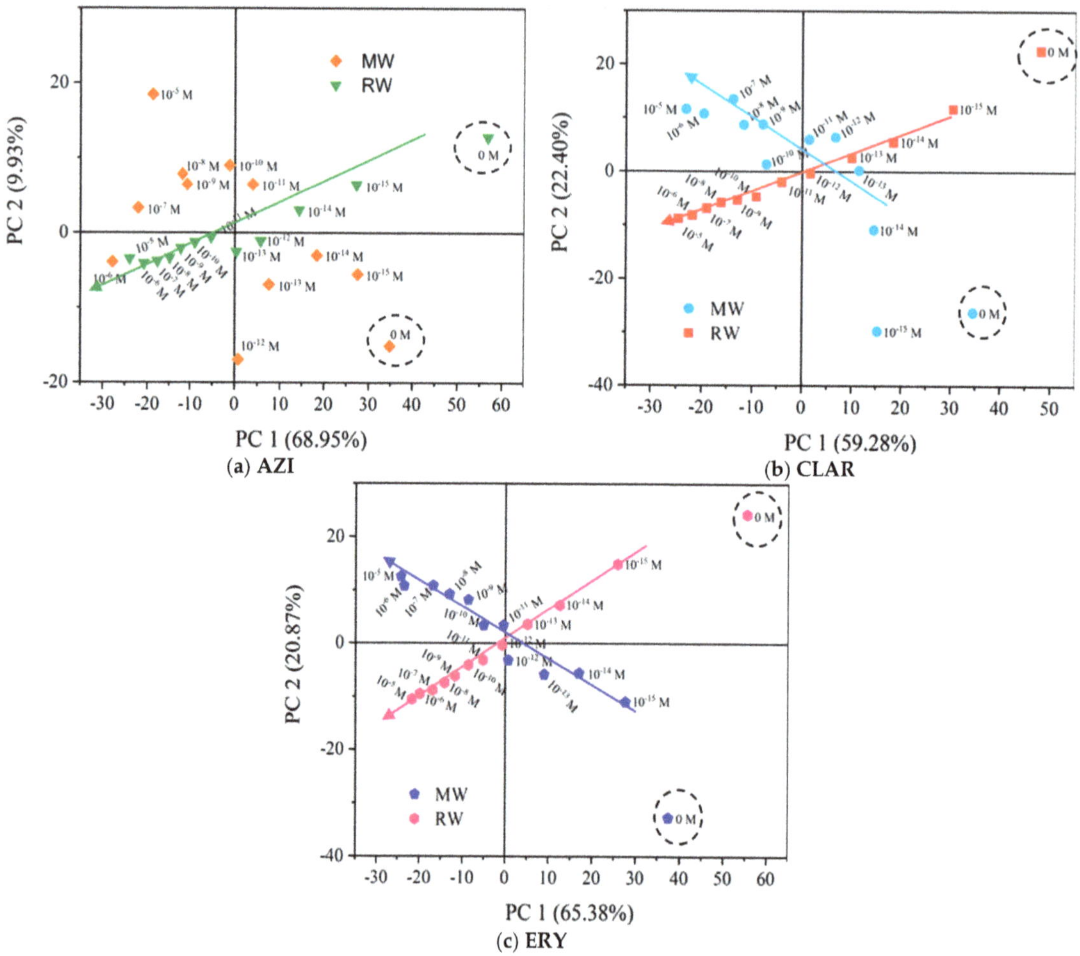

Figure 6. PCA score plot for the electronic tongue concept for AZI (**a**), CLAR (**b**), and ERY (**c**) in ranges of concentration from 0 M to 10^{-15} M to 10^{-5} M, when immersed in MW and RW.

Overall, the first two principal components, PC 1 and PC 2, accounted for above 79% of the total variance in the three electronic tongues. From Figure 6a, a clear decreasing trend in PC 1 with respect to AZI concentration in RW was observed. In contrast, although the PCA plot discriminated the different AZI concentrations in MW, no trends or patterns were observed. Regarding the e-tongue for CLAR, the PCA plot of the e-tongue sensor (Figure 6b) very clearly distinguishes different concentrations in RW across the main axis, PC 1. Within MW, a pattern of PC 1 is also observed. In both matrices, 0 M is in the equivalent quadrant of the lowest concentration, which stresses the ability to distinguish different concentrations of CLAR. Figure 6c shows a clear trend along the main axis, similar to, although more pronounced when compared with the AZI and CLAR PCA plots. Lastly, the three PCA plots showed that the e-tongue device could discriminate between non-doped samples and samples spiked with AZI, CLAR, and ERY, regardless of the matrix.

Considering the trends and patterns found in the PCA plots (Figure 6) of the data measured by the various thin film sensors, the principal component (PC1) that presented the most notable and significant tendency with respect to the antibiotics' concentration, in each case, was plotted as a function of concentration. Thus, because it was observed as a linear trend of the principal component factor score, it was possible to use the PC1 data to determine the sensors' sensitivity for the three macrolides and their resolution.

The sensitivity corresponded to the slope of the linear function that adequately fitted the plotted data. The resolution was found near the smallest concentration of the implicit linear range (C_S), through the following equation, considering the minimum value that could be measured and the error associated with the sensitivity:

$$\Delta logC = logC - logCs$$

where $\Delta logC = \frac{error}{sensitivity}$; therefore, the resolution is equal to $C - Cs$.

The factor scores of PC 1 obtained for both matrices as a function of AZI, CLAR, and ERY concentration and its operational parameters (linear range information, sensitivity, and resolution) are presented in Figure 7. As can be observed in Figure 7, the sensor array chosen in this paper showed similar sensitivity and equal resolution for all the macrolides under study.

Moreover, a PCA plot was constructed to understand the capability of the proposed e-tongue to distinguish between the three antibiotics in the two matrices (Figure 8).

The observations depicted in Figure 8 are in line with previous results [15,24], which have highlighted the essential role of ionic media and pH–pKa relationships in compound adsorption or non-absorption onto thin films. In the present study, the target compounds were in non-ionized form in mineral water; in river water they could be found in both forms (ionized and non-ionized). Thus, there are two outlines in both Figure 8a,b. In Figure 8(a1), although the ranges of concentrations are typically in the same quadrant, namely in the second and third areas, no trend is observed. In contrast, in Figure 8(b1), there is a clear trend and pattern between the detection of the target compounds. Nevertheless, with those two e-tongues it was possible to attain a device with sensitivity of 5.0 ± 0.2 and -4.8 ± 0.3 per decade in MW and RW spiked with AZI, CLAR, and ERY. For both e-tongues, resolutions of 7.0×10^{-17} M and 1×10^{-16} M were achieved. The proposed e-tongue presents a lower detection limit when compared with other promising sensors developed for the monitoring of antibiotics and/or organic compounds: (1) In 2013, Zhang et al. developed multi-walled carbon nanotubes (MWCNTs) for azithromycin detection with a detection limit in a water sample of 1.1×10^{-9}–6×10^{-9} M [26]; Khanna et al. (2018) tested a MnO_2 electrochemical sensor to detect BPA and achieved a detection limit of 0.66 μM [27]; in 2020, Ayankojo et al. derived a molecularly imprinted polymer with a portable electrochemical transducer, a screen-printed electrode, and could detect erythromycin in tap water at concentrations lower than 0.1 nM [28]. Thus, reflecting upon the data observed in Figure 8, the superior ability of the e-tongue to present patterns and trends in higher pH and conductivity matrices is clear, as was the case in RW. Finally, it is evident that the analogous characteristics and morphology may explain the data observed in a matrix where there are 50% of ionized molecules. This confirms that these types of sensors, working as an e-tongue, may be considered as a measuring technique; therefore, they could be further used to develop an optimized antibiotic device working as a monitoring tool on aqueous matrices with different complexities.

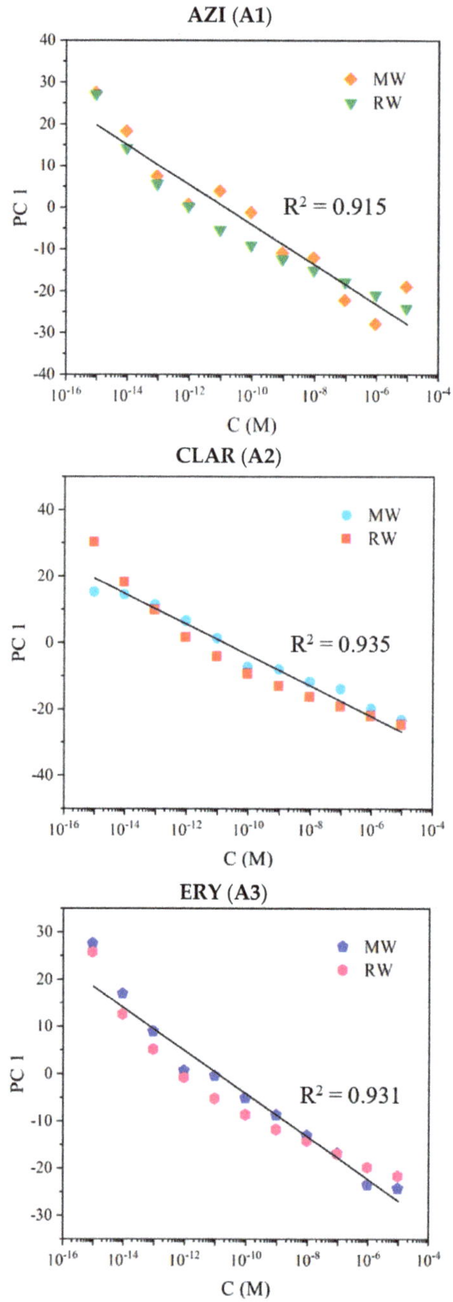

(A)

Intercept	-51.9
Slope	-4.8
Error	0.3
C_H	1×10^{-5}
C_S	1×10^{-15}
C	8.6×10^{-16}
Resolution	1×10^{-16}

Intercept	-50.0
Slope	-4.6
Error	0.3
C_H	1×10^{-5}
C_S	1×10^{-15}
C	8.8×10^{-16}
Resolution	1×10^{-16}

Intercept	-49.7
Slope	-4.5
Error	0.3
C_H	1×10^{-5}
C_S	1×10^{-15}
C	8.7×10^{-16}
Resolution	1×10^{-16}

(B)

Figure 7. (**A**) PC1 factor scores as a function of AZI (**A1**), CLAR (**A2**), and ERY (**A3**) concentrations (0 M to 10^{-15} M to 10^{-5} M), distinguished with the electronic tongue build-up with TiO_2- and ZnO-based sensors, when immersed in MW and RW; (**B**) linear range, sensitivity, and resolution for each macrolide.

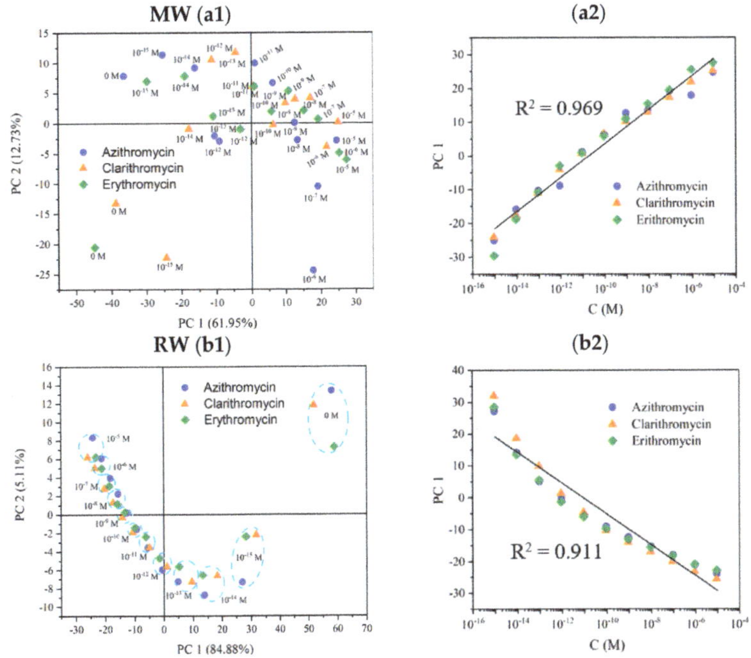

(a3)

Intercept	54.0
Slope	5.0
Error	0.2
C_H	1×10^{-15}
C_S	1×10^{-15}
C	9.3×10^{-16} M
Resolution	7×10^{-17}

(b3)

Intercept	−53.5
Slope	−4.8
Error	0.3
C_H	1×10^{-15}
C_S	1×10^{-15}
C	8.8×10^{-16}
Resolution	1×10^{-16}

Figure 8. Principal component analysis (PCA) score plot for the electronic tongue concept for AZI, CLA, and ERY in the range of concentrations from 0 M to 10^{-15} M to 10^{-5} M, when immersed in MW (**a1**) and RW (**b1**); PC1 factor scores as a function of AZI, CLAR, and ERY concentrations (0 M to 10^{-15} M to 10^{-5} M) when immersed in MW (**a2**) and RW (**b2**); linear range, sensitivity, and resolution for each macrolide in MW (**a3**) and RW (**b3**).

4. Conclusions

The development of an array of nanostructured TiO_2 and ZnO sensors for the detection of AZI, CLAR, and ERY in concentrations between 10^{-15} M and 10^{-5} M in MW and RW matrices has been proposed. Impedance spectra measured by sensors coated with ZnO and TiO_2 thin films deposited with 50% and 100% O_2 by DC magnetron sputtering showed different footprints according to the experimental matrix. However, the electrical behaviors of the different thin film sensors were identical for the three macrolides. Overall, the nanostructured ZnO and TiO_2 sensors produced impedance spectra with distinct trends and patterns regarding macrolide concentrations.

Normalized impedance, measured at fixed frequencies by the different devices as a function of concentration, was used to determine monotone functions of macrolide concentrations that allowed operational measurements of the analytes. The ZnO thin film showed a better ability to work as a detection sensor for the target compounds in comparison with TiO_2. This may be due to its higher roughness and surface area that allowed the adsorption of the compound and further detection with impedance systems. Additionally, ZnO seemed to present more robustness to pH fluctuations and matrix characteristics (good results in both matrices), as well as greater sensitivity to a higher range of concentrations and frequencies.

Additionally, the electronic tongue concept was applied to the full range of sensors under study. The e-tongue sensor's PCA plots indicated that the proposed device could discriminate between the two experimental matrices. The e-tongue was further capable of detecting AZI, CLAR, and ERY in both matrices, and quantify the three macrolides in RW in the complete concentration range. Thus, by applying the e-tongue concept, sensitivities of 4.8 ± 0.3, 4.6 ± 0.3, and 4.5 ± 0.3 per decade to AZI, CLAR, and ERY concentration were achieved, respectively. In all cases, a resolution of 1×10^{-16} M was found near the 10^{-15} M concentration, the smallest concentration that could be quantified. Lastly, it was revealed that the proposed e-tongue, with a sensitivity of -4.8 ± 0.3 per decade and a resolution of 1×10^{-16} M, could recognize the different ranges of concentrations with a clear trend, but not a specific molecule in river water.

Supplementary Materials: The following supporting information can be downloaded at: https://www.mdpi.com/article/10.3390/nano12111858/s1, Figure S1. Normalized impedance as a function of azithromycin concentration in MW at fixed frequencies: 1, 1.58, 2.51, and 2.51×10^3 Hz measured by the TiO_2 (100% O_2) sensor, TiO_2 (50% O_2) sensor, ZnO (100% O_2) sensor, and ZnO (50% O_2) sensor, respectively; Figure S2. Normalized impedance as a function of azithromycin concentration in RW at fixed frequencies: 1, 1.58, 3.98×10^3, and 6.31×10^4 Hz measured by the TiO_2 (50% O_2) sensor, ZnO (100% O_2) sensor, TiO_2 (100% O_2) sensor, and ZnO (50% O_2) sensor, respectively; Figure S3. Normalized impedance as a function of clarithromycin concentration in MW at fixed frequencies: 1, 1, 2.51, and 2.51 Hz measured by the TiO_2 (50% O_2) sensor, TiO_2 (100% O_2) sensor, ZnO (50% O_2) sensor, and ZnO (100% O_2) sensor, respectively; Figure S4. Normalized impedance as a function of clarithromycin concentration in RW at fixed frequencies: 1, 3.98 × 102, 2.51×10^3 and 10^6 Hz measured by the ZnO (100% O_2) sensor, ZnO (50% O_2) sensor, TiO_2 (100% O_2) sensor, and TiO_2 (50% O_2) sensor, respectively; Figure S5. Normalized impedance as a function of erythromycin concentration in MW at fixed frequencies: 1, 1.58×10^2 and 2.51×10^3 Hz measured by the TiO_2 (50% O_2) sensor, ZnO (50% O_2) sensor, ZnO (100% O_2), and TiO_2 (100% O_2) sensor, respectively; Figure S6. Normalized impedance as a function of erythromycin concentration in RW at fixed frequencies: 1, 10, 2.51×10^4, and 10^6 Hz measured by the TiO_2 (50% O_2) sensor, TiO_2 (100% O_2) sensor, ZnO (50%O_2) sensor, and ZnO (100% O_2) sensor, respectively. Figure S7. SEM images with 30,000 times magnification for (a) ZnO 50%; (b) ZnO 100%; (c) TiO_2 50%; (d) TiO_2 100%. Table S1. Normalized Impedance data (Ω) reproducibility of the TiO_2 based sensors devices produced with 100% O_2, when immersed in MW and RW at different AZI, CLA and ERY concentrations. Table S2. Normalized Impedance data (Ω) reproducibility of the TiO_2 based sensors devices produced with 50% O_2, when immersed in MW and RW at different AZI, CLA and ERY concentrations. Table S3. Normalized Impedance data (Ω) reproducibility of the ZnO based sensors devices produced with 100% O_2, when immersed in MW and RW at different AZI, CLA and ERY concentrations. Table S4. Normalized Impedance data (Ω) reproducibility of the ZnO based sensors devices produced with 50% O_2, when immersed in MW and RW at different AZI, CLA and ERY concentrations.

Author Contributions: Conceptualization, C.M. and S.S.; methodology, C.M. and S.S.; software, C.M. and T.M.; validation, C.M., P.A.R., M.R. and S.S.; formal analysis, C.M., P.A.R., M.R. and S.S.; investigation, C.M. and T.M.; resources, P.A.R., M.R. and S.S.; data curation, C.M. and T.M.; writing—original draft preparation, C.M. and J.D.; writing—review and editing, P.A.R., M.R. and S.S.; supervision, C.M., M.R. and S.S.; funding acquisition, C.M., M.R. and S.S. All authors have read and agreed to the published version of the manuscript.

Funding: This research was supported in part by Fundação para a Ciência e a Tecnologia (FCT), Portugal, through the research center Grants No. UIDB/FIS/04559/2020 and No. UIDP/FIS/04559/2020 (LIBPhys), from FCT/MCTES, Portugal.

Institutional Review Board Statement: Not applicable.

Informed Consent Statement: Not applicable.

Data Availability Statement: Not applicable.

Acknowledgments: This research was supported in part by Fundação para a Ciência e a Tecnologia (FCT), Portugal, through the research center Grants No. UIDB/FIS/04559/2020 and No. UIDP/FIS/04559/2020 (LIBPhys), from FCT/MCTES, Portugal.

Conflicts of Interest: The authors declare no conflict of interest.

References

1. Water Resources Group. Charting Our Water Future—Economic Frameworks to Inform Decision Making. 2009. Available online: https://www.mckinsey.com/~{}/media/mckinsey/dotcom/client_service/sustainability/pdfs/charting%20our%20water%20future/charting_our_water_future_full_report_.ashx (accessed on 4 May 2022).
2. Vo, P.T.; Hao Ngo, H.; Guo, W.; Zhou, J.L.; Nguyen, D.; Listowski, A.; Wang, X.C.; City, M.; Nam, V. A Mini-Review on The Impacts of Climate Change on Wastewater Reclamation and Reuse. *Sci. Total Environ.* **2014**, *494*, 9–17.
3. Hristov, J.; Barreiro-Hurle, J.; Salputra, G.; Blanco, M.; Witzke, P. Reuse of treated water in European agriculture: Potential to address water scarcity under climate change. *Agric. Water Manag.* **2021**, *251*, 106872. [CrossRef] [PubMed]
4. Liu, J.; Yang, H.; Gosling, S.N.; Kummu, M.; Flörke, M.; Pfister, S.; Hanasaki, N.; Wada, Y.; Zhang, X.; Zheng, C.; et al. Water scarcity assessments in the past, present, and future. *Earth's Future* **2017**, *5*, 545–559. [CrossRef] [PubMed]
5. Mekonnen, M.M.; Hoekstra, A.Y. Sustainability: Four billion people facing severe water scarcity. *Sci. Adv.* **2016**, *2*, e1500323. [CrossRef]
6. Lavrnić, S.; Zapater-Pereyra, M.; Mancini, M.L. Water Scarcity and Wastewater Reuse Standards in Southern Europe: Focus on Agriculture. *Water Air Soil Pollut.* **2017**, *228*, 251. [CrossRef]
7. Kulkarni, P.; Olson, N.D.; Raspanti, G.A.; Goldstein, R.E.R.; Gibbs, S.G.; Sapkota, A.; Sapkota, A.R. Antibiotic concentrations decrease during wastewater treatment but persist at low levels in reclaimed water. *Int. J. Environ. Res. Public Health* **2017**, *14*, 668. [CrossRef]
8. Geissen, V.; Mol, H.; Klumpp, E.; Umlauf, G.; Nadal, M.; van der Ploeg, M.; van de Zee, S.E.A.T.M.; Ritsema, C.J. Emerging pollutants in the environment: A challenge for water resource management. *Int. Soil Water Conserv. Res.* **2015**, *3*, 57–65. [CrossRef]
9. Sanganyado, E.; Gwenzi, W. Antibiotic resistance in drinking water systems: Occurrence, removal, and human health risks. *Sci. Total Environ.* **2019**, *669*, 785–797. [CrossRef]
10. Rodriguez-Mozaz, S.; Vaz-Moreira, I.; Della Giustina, S.V.; Llorca, M.; Barceló, D.; Schubert, S.; Berendonk, T.U.; Michael-Kordatou, I.; Fatta-Kassinos, D.; Martinez, J.L.; et al. Antibiotic residues in final effluents of European wastewater treatment plants and their impact on the aquatic environment. *Environ. Int.* **2020**, *140*, 105733. [CrossRef]
11. Thai, P.K.; Ky, L.X.; Binh, V.N.; Nhung, P.H.; Nhan, P.T.; Hieu, N.Q.; Dang, N.T.T.; Tam, N.K.B.; Anh, N.T.K. Occurrence of antibiotic residues and antibiotic-resistant bacteria in effluents of pharmaceutical manufacturers and other sources around Hanoi, Vietnam. *Sci. Total Environ.* **2018**, *645*, 393–400. [CrossRef]
12. Wei, L.; Qin, K.; Zhao, N.; Noguera, D.R.; Qiu, W.; Zhao, Q.; Kong, X.; Zhang, W.; Kabutey, F.T. Transformation of erythromycin during secondary effluent soil aquifer recharging: Removal contribution and degradation path. *J. Environ. Sci.* **2017**, *51*, 173–180. [CrossRef] [PubMed]
13. Mimendia, A.; Gutiérrez, J.M.; Leija, L.; Hernández, P.R.; Favari, L.; Muñoz, R.; del Valle, M. A review of the use of the potentiometric electronic tongue in the monitoring of environmental systems. *Environ. Model. Softw.* **2010**, *25*, 1023–1030. [CrossRef]
14. Nuñez, L.; Cetó, X.; Pividori, M.I.; Zanoni, M.V.B.; Valle, M. Development and application of an electronic tongue for detection and monitoring of nitrate, nitrite and ammonium levels in waters. *Microchem. J.* **2013**, *110*, 273–279. [CrossRef]
15. Magro, C.; Mateus, E.P.; Paz-Garcia, J.M.; Sério, S.; Raposo, M.; Ribeiro, A.B. Electronic Tongue Coupled to an Electrochemical Flow Reactor for Emerging Organic Contaminants Real Time Monitoring. *Sensors* **2019**, *19*, 5349. [CrossRef]
16. Singh, N.; Reza, K.K.; Ali, M.A.; Agrawal, V.V.; Biradar, A.M. Self assembled DC sputtered nanostructured rutile TiO_2 platform for bisphenol A detection. *Biosens. Bioelectron.* **2015**, *68*, 633–641. [CrossRef]
17. Ali, M.A.; Srivastava, S.; Solanki, P.R.; Varun Agrawal, V.; John, R.; Malhotra, B.D. Nanostructured anatase-titanium dioxide based platform for application to microfluidics cholesterol biosensor. *Appl. Phys. Lett.* **2012**, *101*, 084105.
18. Kolodziejczak-Radzimska, A.; Jesionowski, T. Zinc oxide-from synthesis to application: A review. *Materials* **2014**, *7*, 2833–2881. [CrossRef]
19. Khadher, A.; Farooqui, M.; Mohsin, M.; Rabbani, G. Metal Oxide Thin Films: A Mini Review. *J. Adv. Sci. Res.* **2016**, *7*, 1–8.
20. Facure, M.H.M.; Mercante, L.A.; Mattoso, L.H.C.; Correa, D.S. Detection of trace levels of organophosphate pesticides using an electronic tongue based on graphene hybrid nanocomposites. *Talanta* **2017**, *167*, 59–66. [CrossRef]
21. Magro, C.; Sardinha, M.; Ribeiro, P.A.; Raposo, M.; Sério, S. Magnetron sputtering thin films as tool to detect triclosan in infant formula powder: Electronic tongue approach. *Coatings* **2021**, *11*, 336. [CrossRef]
22. Jiao, A.; Cui, Q.; Li, S.; Li, H.; Xu, L.; Tian, Y.; Ma, H.; Zhang, M.; Liu, X.; Chen, M. Aligned TiO_2 nanorod arrays decorated with closely interconnected Au/Ag nanoparticles: Near-infrared SERS active sensor for monitoring of antibiotic molecules in water. *Sens. Actuators B Chem.* **2022**, *350*, 130848. [CrossRef]
23. Sério, S.; Melo Jorge, M.E.; Maneira, M.J.P.; Nunes, Y. Influence of O_2 partial pressure on the growth of nanostructured anatase phase TiO_2 thin films prepared by DC reactive magnetron sputtering. *Mater. Chem. Phys.* **2011**, *126*, 73–81. [CrossRef]
24. Magro, C.; Zagalo, P.; Pereira-da-Silva, J.; Mateus, E.P.; Ribeiro, A.B.; Ribeiro, P.; Raposo, M. Polyelectrolyte Based Sensors as Key to Achieve Quantitative Electronic Tongues: Detection of Triclosan on Aqueous Environmental Matrices. *Nanomaterials* **2020**, *10*, 640. [CrossRef] [PubMed]
25. Faris, B.K.; Hassan, A.A.; Aziz, S.B.; Brza, M.A.; Abdullah, A.M.; Abdalrahman, A.A.; Abu Ali, O.A.; Saleh, D.I. Impedance, Electrical Equivalent Circuit (EEC) Modeling, Structural (FTIR and XRD), Dielectric, and Electric Modulus Study of MC-Based Ion-Conducting Solid Polymer Electrolytes. *Materials* **2021**, *15*, 170. [CrossRef] [PubMed]

26. Zhang, K.; Lu, L.; Wen, Y.; Xu, J.; Duan, X.; Zhang, L.; Hu, D.; Nie, T. Facile synthesis of the necklace-like graphene oxide-multi-walled carbon nanotube nanohybrid and its application in electrochemical sensing of Azithromycin. *Anal. Chim. Acta* **2013**, *787*, 50–56. [CrossRef] [PubMed]
27. Khanna, M.; Roy, S.; Kumar, R.; Wadhwa, S.; Mathur, A.; Dubey, A.K. MnO_2 Based Bisphenol-A Electrochemical Sensor Using Micro-Fluidic Platform. *IEEE Sens. J.* **2018**, *18*, 2206–2210. [CrossRef]
28. Ayankojo, A.G.; Reut, J.; Ciocan, V.; Öpik, A.; Syritski, V. Molecularly imprinted polymer-based sensor for electrochemical detection of erythromycin. *Talanta* **2020**, *209*, 120502. [CrossRef] [PubMed]

nanomaterials

Article

Synthesis and Characterization of Ag/ZnO Nanoparticles for Bacteria Disinfection in Water

Julia de O. Primo [1,2], Dienifer F. Horsth [1,2], Jamille de S. Correa [1], Arkaprava Das [2], Carla Bittencourt [2,*], Polona Umek [3], Ana Guilherme Buzanich [4], Martin Radtke [4], Kirill V. Yusenko [4], Cristina Zanette [5] and Fauze J. Anaissi [1]

1 Departamento de Química, Universidade Estadual Do Centro-Oeste, Guarapuava 85040-080, Brazil; juliadeoliveira@unicentro.edu.br (J.d.O.P.); dhorsth@unicentro.edu.br (D.F.H.); jscorrea@unicentro.edu.br (J.d.S.C.); anaissi@unicentro.edu.br (F.J.A.)
2 Chimie des Interactions Plasma-Surface (ChIPS), Research Institute for Materials Science and Engineering, University of Mons, 7000 Mons, Belgium; arkaprava.das@umons.ac.be
3 Solid State Physics Department, Jožef Stefan Institute, 1000 Ljubljana, Slovenia; polona.umek@ijs.si
4 Federal Institute for Materials Research and Testing (BAM), 12205 Berlin, Germany; ana.buzanich@bam.de (A.G.B.); martin.radtke@helmholtz-berlin.de (M.R.); kirill.yusenko@bam.de (K.V.Y.)
5 Departamento de Engenharia de Alimentos, Universidade Estadual do Centro-Oeste, Guarapuava 85040-080, Brazil; czanette@unicentro.br
* Correspondence: carla.bittencourt@umons.ac.be

Abstract: In this study, two green synthesis routes were used for the synthesis of Ag/ZnO nanoparticles, using cassava starch as a simple and low-cost effective fuel and Aloe vera as a reducing and stabilizing agent. The Ag/ZnO nanoparticles were characterized and used for bacterial disinfection of lake water contaminated with *Escherichia coli* (*E. coli*). Characterization indicated the formation of a face-centered cubic structure of metallic silver nanoparticles with no insertion of Ag into the ZnO hexagonal wurtzite structure. Physicochemical and bacteriological analyses described in "Standard Methods for the Examination of Water and Wastewater" were used to evaluate the efficiency of the treatment. In comparison to pure ZnO, the synthesized Ag/ZnO nanoparticles showed high efficiencies against *Escherichia coli* (*E. coli*) and general coliforms present in the lake water. These pathogens were absent after treatment using Ag/ZnO nanoparticles. The results indicate that Ag/ZnO nanoparticles synthesized via green chemistry are a promising candidate for the treatment of wastewaters contaminated by bacteria, due to their facile preparation, low-cost synthesis, and disinfection efficiency.

Keywords: zinc oxide; silver nanoparticles; water disinfection

Citation: Primo, J.d.O.; Horsth, D.F.; Correa, J.d.S.; Das, A.; Bittencourt, C.; Umek, P.; Buzanich, A.G.; Radtke, M.; Yusenko, K.V.; Zanette, C.; et al. Synthesis and Characterization of Ag/ZnO Nanoparticles for Bacteria Disinfection in Water. *Nanomaterials* **2022**, *12*, 1764. https://doi.org/10.3390/nano12101764

Academic Editor: Christos A. Aggelopoulos

Received: 5 April 2022
Accepted: 17 May 2022
Published: 22 May 2022

Publisher's Note: MDPI stays neutral with regard to jurisdictional claims in published maps and institutional affiliations.

1. Introduction

Water is a key resource for the existence of life on Earth; therefore, water access is considered one of the most critical humanitarian goals. Nevertheless, according to World Health Organization (WHO) [1] in 2020, only 50% of the population had access to safely managed drinking water, while the remaining 50% still lacked access to drinking water sources and safely managed services. The same source reports that, on a global scale, two billion people use drinking water sources contaminated with feces, and the microbial contamination of drinking water may transmit diseases such as diarrhea, cholera, dysentery, polio, and typhoid, which is estimated to cause 485,000 diarrheal deaths each year.

Drinking water legislation requires microbial, physical, and chemical water parameters to be met to ensure adequate public health conditions [2]. Knowing that, the bacterial contamination of water is a concern, as it causes diseases that can be life-threatening upon ingestion or exposure. Various traditional water disinfectant methods, along with chemical (chlorination, ozonation) and physical (ultraviolet radiation) methods, are effective, and among these, chlorine disinfection has been used to prevent infections for more than a century [3,4]. Although the disinfection methods currently used in drinking water treatment can effectively control microbial pathogens, effective disinfection results in the formation of harmful disinfection byproducts (DBPs). For example, the water industry uses free chlorine, chloramines, and ozone for disinfection; these products can react with various components presents in water to form DBPs [3,5,6].

Given the limitations of conventional disinfection methods, advances in nanotechnology have the potential to generate novel treatment capabilities that may allow economic utilization of unconventional water sources to expand the water supply [6,7]. Depending on their chemical composition, nanomaterials are promising alternatives offering leapfrogging opportunities to develop next-generation water supply systems due to their large specific surface areas and high reactivity, which make them excellent adsorbents, catalysts, sensors, and as recent studies have shown, antimicrobial agents. For example, silver nanoparticles are an antibacterial nanomaterial with strong oxidant properties that are relatively inert in water [5,7].

The effective antibacterial activity of silver (Ag) nanoparticles (NPs) for a wide spectrum of bacteria, viruses, protozoa, algae, and yeasts opens up opportunities for their use in water disinfection [8]. Ag NPs have been widely used against Gram-negative bacteria due to their higher activity against them. However, Ag NPs in contact with the environment during water treatment may pose a health risk, as the level of toxicity to the end user is not well known. For water disinfection, Ag NPs can be combined with particles of inorganic oxides to prevent the silver leaching into the water due the formation of water-soluble species. Among the inorganic particles that have been tested against several pathogenic and non-pathogenic bacteria, zinc oxide (ZnO) has been reported to have high antimicrobial activity [3,9,10]. ZnO particles are also chemically stable, have large surface areas, and have strong antibacterial properties, even in very small quantities [4,11]; and most importantly ZnO particles have been reported to be non-toxic to human cells [12]. Compared to organic materials, inorganic materials such as ZnO have greater durability, selectivity, and heat resistance [6].

Both Ag and ZnO particles present satisfactory antibacterial results; however, when used as individual components, they tend to agglomerate, which reduces their effectiveness [3]. The use of noble metals such as silver (Ag) to modify the ZnO surface can increase the chemical stability and antibacterial properties [13]. In recent years, methods for obtaining nano-Ag/ZnO have been reported for different applications, such as photocatalysis [14,15], controlled drug release [16], and antimicrobial activity [17].

In this study, ZnO particles decorated with silver nanoparticles were synthesized via two low-toxicity synthesis routes: route (I) is a one-step synthesis that uses starch as a complexing gelling and combustion power, and route (II) has two steps synthesis. In the first step pure ZnO nanoparticles are synthesized, and in the second step Ag decoration on their surface is performed using Aloe vera leaf extract as the reducing agent for Ag^+. The samples were investigated using X-ray Diffraction (XRD), Raman Spectroscopy, X-ray photoelectron spectroscopy XPS, X-ray Absorption Near Edge Structure spectroscopy (XANES), and Extended X-Ray Absorption Fine Structure spectroscopy(EXAFS). XPS was used to evaluate the near surface chemistry of the nanoparticles because the X-ray absorption (XAS) measurements were performed in transmission mode; therefore, they may not indicate the presence of very thin oxide layers on the particle's surface that can be detected by XPS. The morphology of the samples was determined by field-emission scanning electron microscopy (SEM) and transmission electron microscopy (TEM). The atomic distances from the Zn and Ag atoms were calculated. The antimicrobial activity of the synthesized Ag/ZnO nanoparticles was tested using contaminated water from a lake with Gram-negative *Escherichia coli*, to evaluate the efficacy of these nanomaterials when applied for disinfection during water treatment.

2. Materials and Methods

2.1. Materials

All reagents used in the synthesis of the samples were of analytical grade and were used as received. The inorganic salts used were zinc nitrate hexahydrate ($Zn(NO_3)_2{\cdot}6H_2O$), 98%, NEON, Suzano, Brazil) and silver nitrate ($AgNO_3$, 99.8%, Quimex, Cotia, Brazil. All solutions were prepared with deionized water. Natural Cassava starch in the form of colloidal suspension was used as fuel [18]. Aloe vera leaves were harvested in the São José region of Parana-Brazil (Parana, Brazil).

2.2. Synthesis of ZnO Using Cassava Starch: Pure and Ag/ZnO (Route I)

Pure and Ag/ZnO were obtained by the gelatinization method. First a white matrix was prepared from 200 g of starch in 1000 g of water; second, the $Zn(NO_3)_2{\cdot}6H_2O$ in the proportion of 98% to 2% (w/w) was added. This suspension (zinc/starch) was kept under stirring at room temperature (RT) until complete dissolution of the emulsion. After 60 min of mechanical stirring (600 rpm), the suspension was calcined in a muffle furnace at 750 °C for 1 h to obtain the pure ZnO [18].

The Ag/ZnO nanoparticles were prepared by adding silver (I) ions (10% w/w) to the zinc solution containing starch and zinc nitrate hexahydrate. Then, 9.28 g of $Zn(NO_3)_2{\cdot}6H_2O$ and 0.32 g of $AgNO_3$ were added to the starch matrix during the formation of the emulsion. After 60 min under of mechanical stirring (600 rpm), the suspension was calcined in a muffle furnace at 750 °C for 1 h (Figure 1).

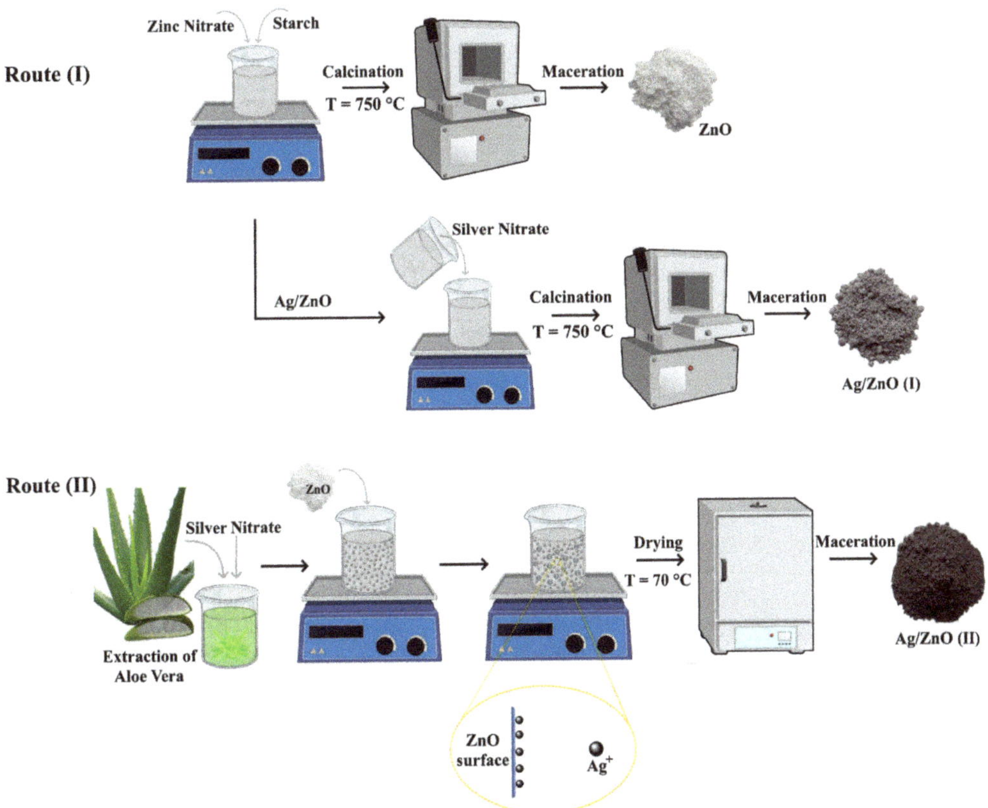

Figure 1. Schematic diagram of the synthesis routes used.

2.3. Synthesis of Ag/Zinc Oxide (Route II)

To obtain the extract of Aloe gel, 5 g of Aloe vera leaves were thoroughly washed, then the leaves were finely cut and boiled in 100 mL of deionized water. Next, the leaves and mucilaginous gel were crushed using a pistil and a ceramic mortar to obtain the complete extract. Finally, the solution was washed and filtered, and the resulting Aloe vera gel broth extract was stored under refrigeration (2 °C) for further experiments.

For the synthesis of silver nanoparticles, 0.5 g of $AgNO_3$ was added to 50 mL of Aloe vera extract, related to (10% *w*/*w*) the ZnO. The solution was kept under constant magnetic stirring for 12 h. Afterwards, 3 g of ZnO obtained in route (I) was added to the solution and kept under constant magnetic stirring for 4 h. The obtained product was centrifuged, washed with water and ethanol, and dried at 70 °C in an oven for 12 h (Figure 1). The obtained material was labeled Ag/ZnO (II).

2.4. Characterization Techniques

The crystalline phases were identified by powder X-ray diffraction (XRD) performed on a Bruker model D2 Phaser with Cu Kα radiation (λ = 1.5418 Å), with scans in 2θ from 20° to 70° and step rate of 0.2 °/sec. The crystallite size (***D***) was estimated by the Scherrer equation [19]:

$$D = \frac{k\lambda}{\beta\cos(\theta)} \quad (1)$$

k is the form factor, λ is the wavelength of incident X-rays, β is the full width at half-maxima (FWHM) of the maximum intensity peak, and θ is the Bragg angle of the maximum intensity peak.

Raman spectra were recorded using a Micro-Raman system (Senterra Bruker Optik GmbH, Massachusetts, USA), λ = 532 nm, laser power 5 mW, time 10 s, resolution 4 cm^{-1}. The morphology of the samples was examined using a field-emission scanning electron microscope (JSM − 7600F, JEOL, Tokyo, Japan) and a transmission electron microscope (JEM − 1011, JEOL, Tokyo, Japan). For SEM analysis, specimens were prepared by dispersing the powder samples in a small amount of deionized water, and a drop of the dispersion was deposited on a polished surface of an Al sample holder. Prior to the SEM investigation, an ca. 4 nm thick carbon layer was deposited on the specimens to reduce the charging effect. Energy dispersive X-ray (EDX) mapping and line analyses were preformed using SEM (Verios 4G HP, ThermoFischer, Waltham, MA, USA).

Specimens for TEM investigation were prepared by dispersing the prepared samples of NPs ultrasonically in methanol, and a drop of the dispersed solution was deposited onto a lacy carbon film supported by a copper grid.

The oxidation state and composition of the chemical elements at the nanoparticle surface were evaluated by X-ray photoelectron spectroscopy (XPS) (Versaprobe PHI 5000, from Physical Electronics, Chanhassen, MN, USA), equipped with a monochromatic Al Kα X-ray source. The XPS spectra were collected at a take-off angle of 45° with respect to the electron energy analyzer, and the spot size was 200 µm. Pass energy (PE) of 20 eV was used for the high-energy resolution spectra (Zn 2p, O 1s, Ag 3d, and C 1s). The spectra were analyzed using the CASA-XPS software, and the binding energy of the XPS spectra was calibrated using the C 1s peak at 284.6 eV [20]. Multipack version 9.8 software (ULVAC-PHI, 2017, Chigasaki, Japan) was used to evaluate the relative composition of the elements.

In order to investigate the local structural properties of Ag NPs in ZnO, the extended X-ray absorption fine structure (EXAFS) spectra were recorded with standard transmission setup at the Zn K edge (9.6586 keV) and Ag K edge (25.5140 keV) at BAMline at the Helmholtz–Zentrum Berlin (BESSY II, Berlin, Germany) [21]. The X-ray beam was monochromatized with a Si(111) double crystal monochromator ($\Delta E/E = 2 \times 10^{-4}$). The energy was scanned in 10 eV steps until 20 eV before the edge, followed by 0.5 eV steps around the edge until 100 eV, and then 1.5 eV steps until 200 eV. From then on, acquisition in 0.04 A^{-1} equidistant k-steps was carried out until k = 16 A^{-1}. Samples in powder form were mixed with hexagonal BN powder to get enough transmission and pressed between two Kapton foils to get 1 mm thick pellets. All measurements were performed with standard metal foil, i.e., Ag and Zn foil, for calibration, and repeated twice to check the reproducibility of the Ag K and Zn K-edge spectra. Initial data processing with background removal and Fourier transformation of EXAFS oscillation was performed with Athena software of the IFEFFIT package. Simulation of EXAFS spectra was accomplished with FEFF and fitted in Artemis software of the same package [22].

2.5. Water Disinfection

For the study, samples of contaminated water were collected during February 2022 from a lake located on the UNICENTRO, Campus Cedeteg, in the city of Guarapuava, Parana, Brazil. The samples were prepared in 100 mL sterile plastic flasks containing sodium thiosulfate pellets that are used for neutralization of chlorine present in the lake water.

2.5.1. Preliminary Tests

The amount of the synthesized nanoparticles required for the disinfection process was optimized by testing different concentrations of Ag/ZnO particles. The disinfection process consisted of direct contact between the contaminated water (100 mL) and the prepared ZnO and Ag/ZnO samples (0.25, 0.5 and 1 g) for 15 min under constant agitation (300 rpm) at RT. Subsequently, filtration was carried out on filter paper, with a few seconds waiting time for the sample to settle in the filter. The treated water was collected in a sealed and sterilized bottle for microbiological and physical-chemical analysis at the Water Analyses Laboratory UNICENTRO.

2.5.2. Water Disinfection Ability of Nanoparticles

Leveraging the optimization results, the water disinfection test was performed according to Section 2.5.1. Physicochemical and microbiological analyses were performed on the contaminated water and after treatment with the particles. The physicochemical and bacteriological analyses were carried out according to the methodologies described in "Standard Methods for the Examination of Water and Wastewater" [23]. The evaluated parameters were total coliforms (TC), pH, turbidity, and *Escherichia coli*. The methods used for analysis of these parameters are shown in Table 1.

Table 1. Methods used for analysis of the parameters evaluated in the present study.

Parameter Analyzed	Method
Total coliforms	SMEWW 9223 B–Enzymatic Substrate Coliform Test
Escherichia coli	SMEWW 9223 B
pH	SMEWW4500H+B—Eletrometric Methods
Turbidity	SMEWW 2130 B. Nephelometric Method

3. Results and Discussion

3.1. XRD

The crystal structures of ZnO and Ag/ZnO samples were determined using X-ray diffraction (XRD). The diffraction patterns of the three samples (Figure 2) have peaks characteristic of the hexagonal wurtzite structure of ZnO (ICDD card number 01-075-9742). According to some prior reports, the incorporation of Ag ions into the ZnO lattice can be substituted for Zn^{2+} or as an interstitial atom [24,25]. In the case of silver replacing Zn^{2+}, a corresponding peak shift would be expected in the XRD; however, for our samples, shifts were not observed, indicating that there was no substitutional doping. For Ag/ZnO (II), substitutional doping was not expected because the ZnO particles were already formed when Ag^+ were added.

Additional diffraction peaks at $2\theta = 38.1°$ (111), 44.2° (200), and 64.3° (311) are shown in the diffractogram of Ag/ZnO (I), corresponding to (111), (200), and (220) planes of the face-centered cubic (fcc) structure of Ag (ICDD card number 04-0783) [13,26,27]. For the sample obtained by route II, only a low intensity peak centered at 38.1° can be observed

(Figure 2b). The absence or low intensity of other peaks characteristic of Ag indicates a low concentration and/or small Ag NPs [24,28,29]. This peak alone may be associated with the Ag fcc structure or Ag_2O. Further, EXAFS analysis was performed to determine the local structure.

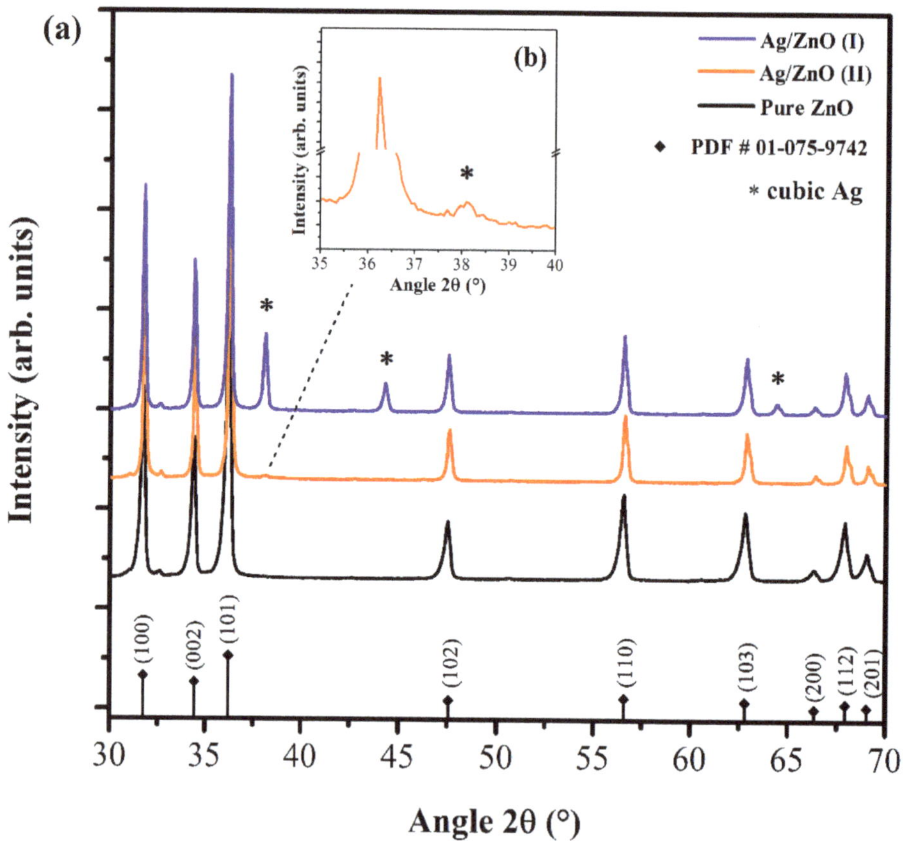

Figure 2. (**a**) XRD patterns of ZnO and Ag/ZnO prepared by route I and II; (**b**) the amplified XRD pattern in the range of 2θ = 35 – 40 for the sample Ag/ZnO (II).

The crystallite sizes of the samples were calculated from the X-ray line broadening using Scherrer's equation [19]. The average calculated values were 23.4 nm for pure ZnO, 32.5 nm for Ag/ZnO (I), and 33.9 nm for Ag/ZnO (II). This increase in crystallite size can be attributed to the Ag nanoparticles anchored to the surface of ZnO [14].

3.2. Raman Spectroscopy

In order to investigate the ZnO defects and chemical bonding at the Ag/ZnO interface, Raman spectra of all samples were measured (Figure 3). The Raman spectrum of the pure ZnO sample (Figure 3a) consisted of four predominant bands located at about 99, 380, 437, and 573 cm^{-1}, which correspond to the fundamental phonon modes of ZnO wurtzite structure [18], in accordance with the XRD results. The Ag/ZnO spectra (Figure 3b,c) show characteristic peaks centered at 99 and 435 cm^{-1}, which can be assigned the fundamental phonon modes E_2^{low} and E_2^{high} of hexagonal wurtzite ZnO, respectively [30,31]. The A_1(LO) polar branches appeared at about 555 cm^{-1} for the samples decorated with Ag. Due to the deposition of Ag on ZnO, this mode was broadened and shifted toward lower energy. Such a shift and broadening of A_1(LO) phonon mode can be attributed to the scattering contributions with the phonons away from the center of Brillouin zone [32]. The A_1(LO) phonon mode is commonly assigned to the oxygen vacancies (V_o) [32–34], zinc interstitial (Zn_i) defects in ZnO [31,34], or defect complexes of combined V_o and Zn_i in the host lattice [34]. Additionally, a broad Raman peak positioned at about 483 cm^{-1} only appeared in the spectra of the samples containing Ag, which was assigned as the interfacial surface phonon mode in the literature [35,36]. This phonon mode was also observed for ZnO loaded with cobalt [36], and thus could not be a local vibrational mode associated with Ag cations.

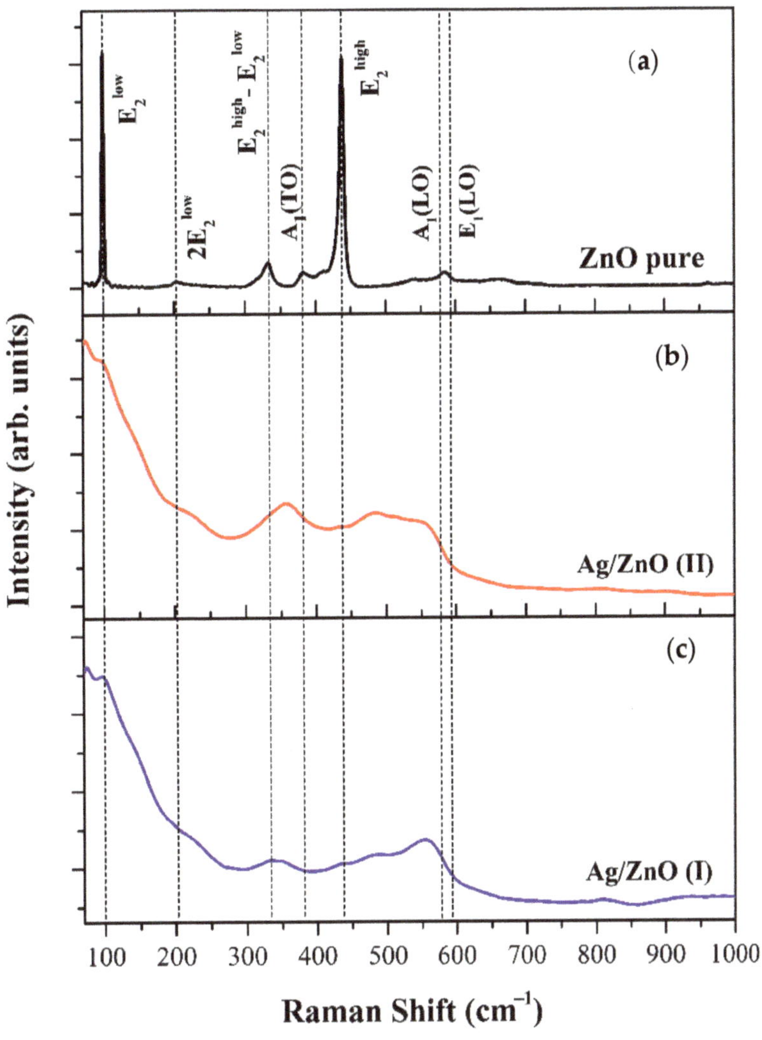

Figure 3. Raman spectra of synthesized (**a**) pure ZnO, (**b**) Ag/ZnO (I), and (**c**) Ag/ZnO (II) using starch and Aloe Vera Barbadensis miller extract.

3.3. SEM

An investigation with SEM was used to assess the morphology and size of ZnO particles (Figure 4). The particles in all three samples had no particular shape, and their size was in the range of 100 to 350 nm. A comparison of the SEM images shown in Figure 4a,c with that in Figure 4b also revealed that the presence of Ag^+ in the reaction mixture of Ag/ZnO (I) had no effect on the morphology and size of the ZnO particles.

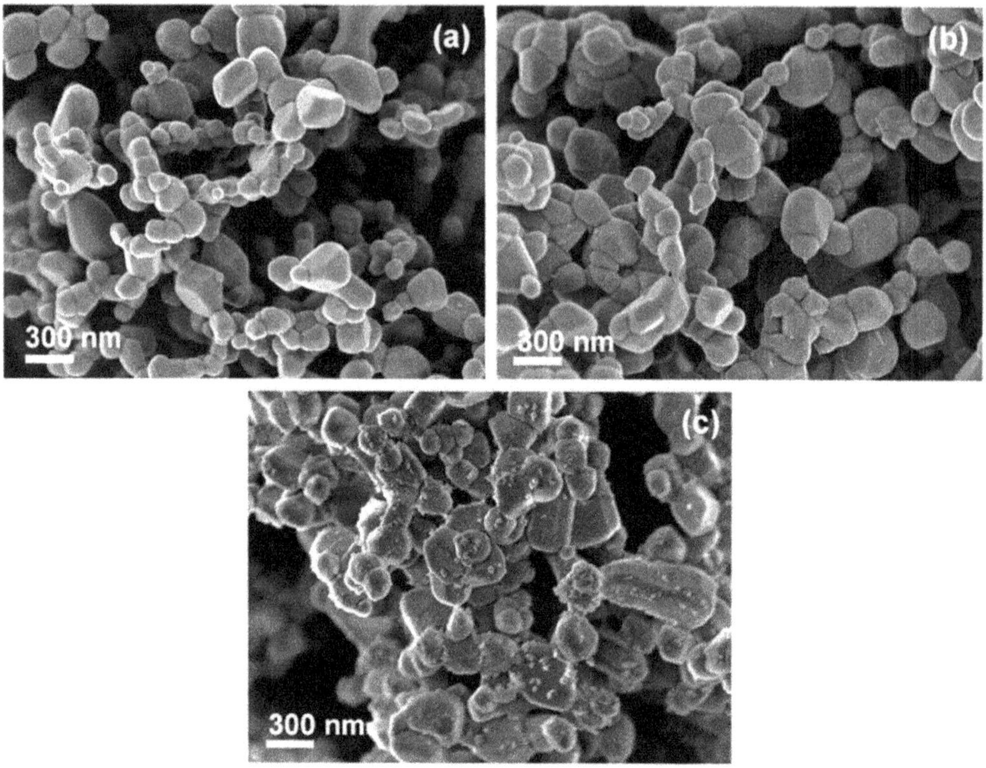

Figure 4. SEM images of pure ZnO (**a**), Ag/ZnO (I) (**b**), and Ag/ZnO (II) (**c**).

To distinguish ZnO and Ag particles and to determine the Ag NP density distribution, the Ag-loaded ZnO samples were visualized with a back scattered electron (BSE) mode of SEM (BSE-SEM). First, the particles in bright contrast BSE-SEM images were identified as Ag NPs (Figures S1 and S2). A comparison of BSE-SEM images of both Ag/ZnO samples (Figure S3, see Supplementary Materials) revealed that: (i) the distribution of Ag NPs was denser in Ag/ZnO (II) and (ii) the Ag NPs in Ag/ZnO (I) were bigger (up to 300 nm) (Figure S1).

3.4. TEM

TEM was used to further examine the ZnO and Ag particles and determine the size of Ag NPs in Ag/ZnO (II). No amorphous layer around ZnO particles was observed (Figure 5 and Figure S4), indicating that during calcination, organic species were removed successfully. ZnO particles were crystalline (see inset of Figure 5a). The average measured d-spacing of 0.26 nm well corresponds with the (002) planes of the hexagonal wurtzite structure of ZnO (ICDD card number 01-075-9742) [37], and is in accordance with the results of Raman and XRD analyses. In Figure 5c, besides the presence of larger ZnO particles, smaller particles with diameters from 10 to 40 nm can be observed (arrows in Figure 5c are pointing to these particles). The measured inter-planner lattice spacing was 0.25 nm (Figure S4) and agrees well with (111) planes of fcc Ag.

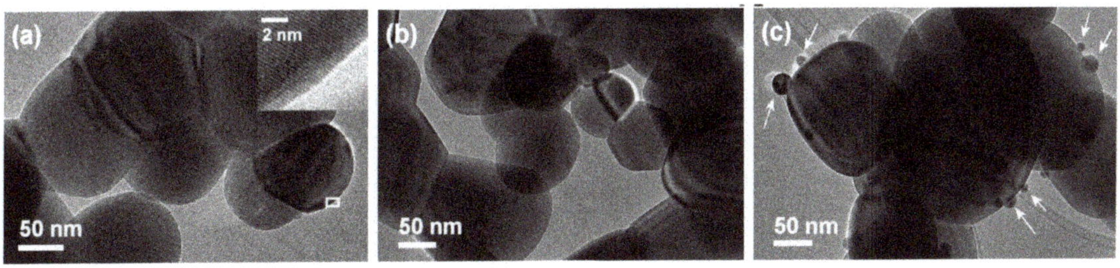

Figure 5. TEM images of (**a**) ZnO pure, (**b**) Ag/ZnO (I), and (**c**) Ag/ZnO (II). The white rectangle in the (**a**) defines the area of the inset. Arrows in the image (**c**) point toward Ag nanoparticles.

3.5. XPS

The chemical bonding states of Ag decorated ZnO samples were evaluated by analyzing the XPS spectra (Figure 6). The core-level XPS spectra of Zn (2p), Ag (3d), and O (1s) are shown in Figure 6a–c. In Figure 6a, the doublet peak with components at ~1021.0 and ~1044.0 eV for both Ag/ZnO samples were assigned to Zn $2p_{3/2}$ and Zn $2p_{1/2}$, respectively. These values agree with the binding energies of Zn 2p3/2 and Zn 2p1/2 in stoichiometric ZnO, which is attributed to Zn^{2+} state oxidation [38,39].

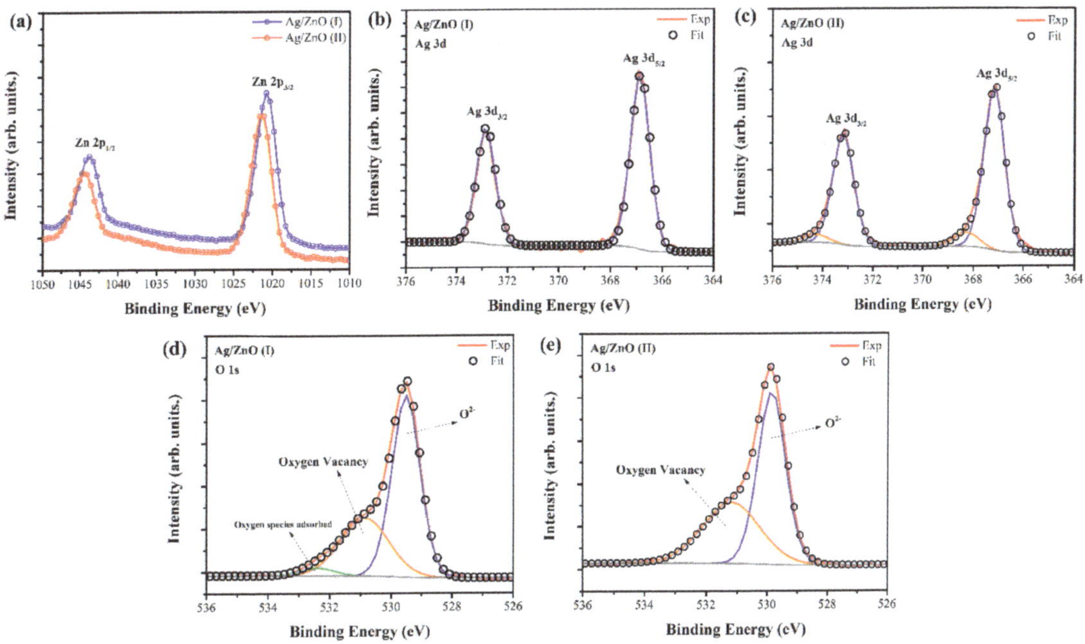

Figure 6. XPS spectra of Ag/ZnO particles obtained by two different synthesis routes. High-resolution regional XPS spectra of (**a**) Zn (2p), (**b**,**c**) Ag (3d), and (**d**,**e**) O (1s).

Figure 6b shows the spectrum recorded in the region of the Ag 3d core level. For the sample Ag/ZnO (I), the components of the Ag 3d core levels are centered at 366.8 eV (Ag $3d_{5/2}$) and 372.8 eV (Ag $3d_{3/2}$). The spin-orbit splitting of 6 eV indicates that Ag atoms were mainly in the metallic state Ag^0 [39]. The Ag 3d core-level binding energy in the Ag/ZnO (I) sample was shifted to the lower binding energy compared to the corresponding value of the pure metallic Ag (Ag $3d_{5/2}$ is about 368.2 eV). This was associated with the interaction between Ag NPs on the surfaces of ZnO nanocrystals, which led to the adjustment of Fermi level [39,40]. These results are in good agreement with the presence of the fcc metallic Ag crystal structure obtained by XRD. As can be seen, the Ag $3d_{5/2}$ spectrum of the Ag/ZnO (II) sample was fitted by two components centered at 367.1 and 368.2 eV, corresponding to Ag_2O and Ag^0 electronic states, respectively [37–39,41]

The O (1s) core-level spectra illustrated in Figure 6d were fitted with three components centered at 529.5, 530.8, and 532.6 eV for sample (I), and two components centered on 529.8 and 531.2 eV for sample (II) (Figure 6f). The peak with the lower binding energy (~529.0 eV) was attributed to O^{2-} ions participating in the Zn–O bonding in the wurtzite structure of the hexagonal ZnO [42,43]. The component centered at ~531.0 eV is associated with photoelectrons emitted from O^{2-} ions in oxygen-deficient regions in the matrix of ZnO [42]. As shown in Figure 6c, the sample obtained in synthesis route (I) had one more component at 532.5 eV, which was attributed to oxygen species chemisorbed on the surface of the Ag/ZnO [44]. The adsorbed oxygen species in Ag/ZnO (I) may affect the physical and chemical properties of the material.

3.6. EXAFS

The local environments of the Ag and Zn atoms in the samples were probed by EXAFS spectra of Ag K edge and Zn K edge. The EXAFS oscillations occur due to the interference between outgoing and backscattered photoelectron wave. The scattering signal $\chi(E)$ is defined as follows [45]:

$$\chi(E) = \frac{\mu(E) - \mu_0(E)}{\Delta\mu_0\,(E_0)} \tag{2}$$

E_0 is the tabulated energy of an absorption edge, $\Delta\mu_0\,(E_0)$ is step or edge jump between pre- and post-edge lines, $\mu_0\,(E_0)$ is the mass absorption coefficient of the isolated atom, and $\mu(E)$ is the mass absorption coefficient of the element in the sample. For EXAFS interpretation, it is convenient to have the data in k-space, which follows the mathematical relation.

$$k = \sqrt{\frac{2m(E - E_0)}{\hbar^2}} \tag{3}$$

m represents the electron mass in the Equation (3). In the Fourier transformed plot, the actual distance of the oscillation from the photo-absorbing atom is plotted. The background removal, normalization, and Fourier transformation of the k^2 weighted $\chi(k)$ spectra were performed with Athena software from the IFEFFIT package [22]. Simulation of the Ag K edge and Zn K edge was performed with FEFF within Artemis software. From the XRD pattern in Figure 1, we already had an idea about the existing phases. Therefore, considering the concerned crystallographic information file (CIF), the different scattering paths were obtained mainly for the first coordination shell around the central Ag and Zn atoms. With the cubic Ag and hexagonal wurtzite CIF file, the Ag–Ag, Zn–O, and Zn–Zn paths were considered for simulation of Ag K and Zn K edges, respectively. Debye–Waller factor (σ^2), nearest neighbor coordination number (CN), and bond distance (c) were considered as variable parameters during simulations. We calculated the amplitude reduction factor (s_0^2) from the simulation of standard metallic foil's (Ag and Zn) EXAFS spectrum and used the same value for simulating the EXAFS profile for Ag-K and Zn-K edges of the doped oxides samples.

Figure 7a,c shows the normalized XANES spectra for the Ag K edge and Zn K edge. Figure 7b,d portrays the $\chi(k).k^2$ EXAFS spectra for Ag K edge and Zn K edge. In Figure 7a, no changes are visible in the intensity or scattering features of the Ag K edge for both samples as compared to the Ag foil, which indicates the same local environment around the photo-absorbing Ag atom, and proves the presence of a metallic Ag phase. There was no change observed in the Zn K edge in XANES region for both samples (Figure 7c). The fitting of the Ag K edge in Figure 8a with the cubic Ag cif file indicates that there is no variation in Ag in coordination number between Ag/ZnO(I) and Ag/ZnO(II) samples. The same is the case for the Zn K edge in Figure 8c, showing that coordination number does not vary and no charge transfer took place due to any kind of cationic substitution between Ag and Zn. The Zn K edge fitting was performed with the hexagonal wurtzite cif file. The ionic radii of Ag^{2+} (0.94 Å) and Zn^{2+} (0.60 Å) are significantly different [46]. The change in Gibbs' free energy with cationic substitution might be higher as compared to the bond formation energy of Ag–Ag bonding. Therefore, it is easier for Ag to make a bond with another nearby Ag atom rather than cationic substitution, resulting in a metallic Ag phase in both of the samples. The amplitude reduction factor (s_0^2) was calculated with a standard metallic foil of Ag K edge and Zn K edge. For the Ag K edge and Zn K edge, the s_0^2 values were found to be 0.7 and 1, respectively, which were kept fixed for all fits. Figure 8a–d portrays the EXAFS fitting for magnitude and the real part of the $\chi(R)$ spectra for Ag K and Zn K edges. From this simulation, we calculated the CN, R, and σ^2 values, i.e., for Ag–Ag, Zn–O, and Zn–Zn single scattering paths. The k-range for Fourier transformation was 3 to 11.5 A^{-1} during the simulation of the Zn K edge. The same for the Ag K edge was 3 to 9.5 A^{-1}. The R ranges were 1.85–3.15 Å and 1–3.65 Å for Ag and Zn fits, respectively.

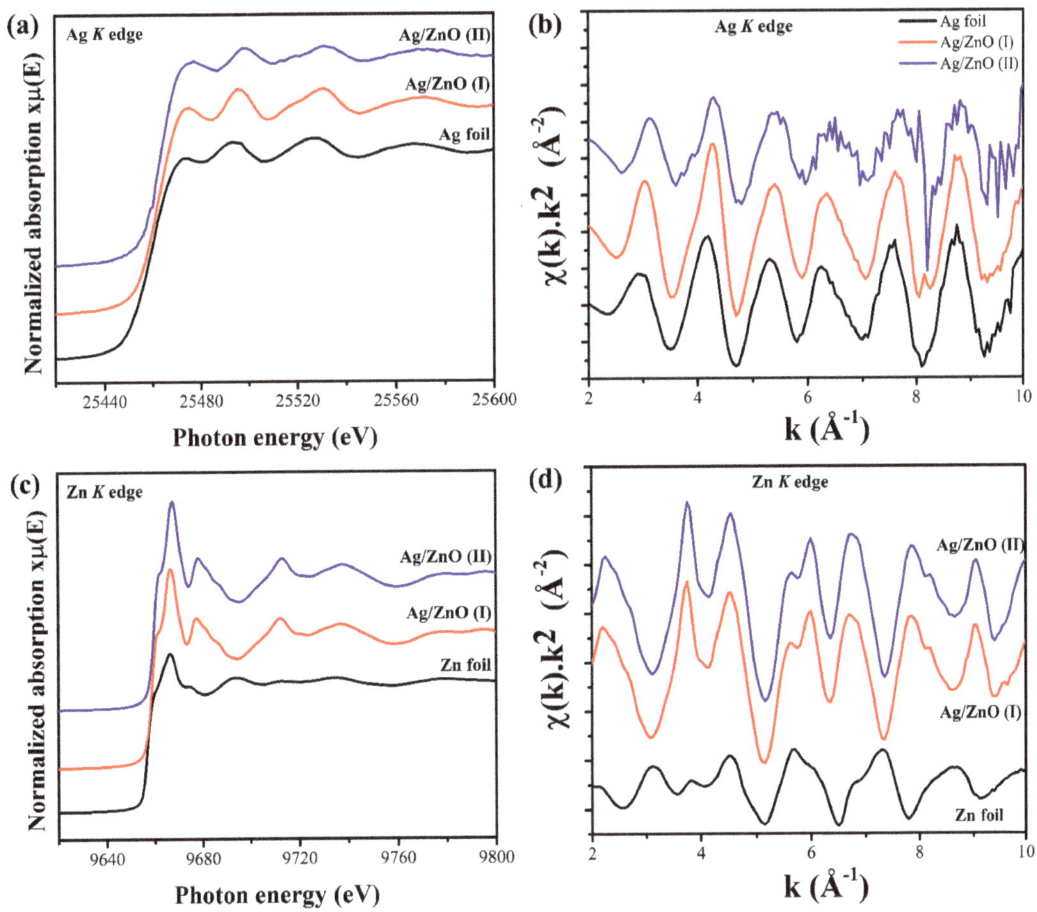

Figure 7. Ag loaded ZnO NPs. (**a**,**c**) Normalized XANES data for Ag K edge and Zn K edge. (**b**,**d**) k^2-weighted $\chi(k)$ spectra for Ag K edge and Zn K edge. XANES and $\chi(k)$ were both shifted vertically and stacked for observational clarity.

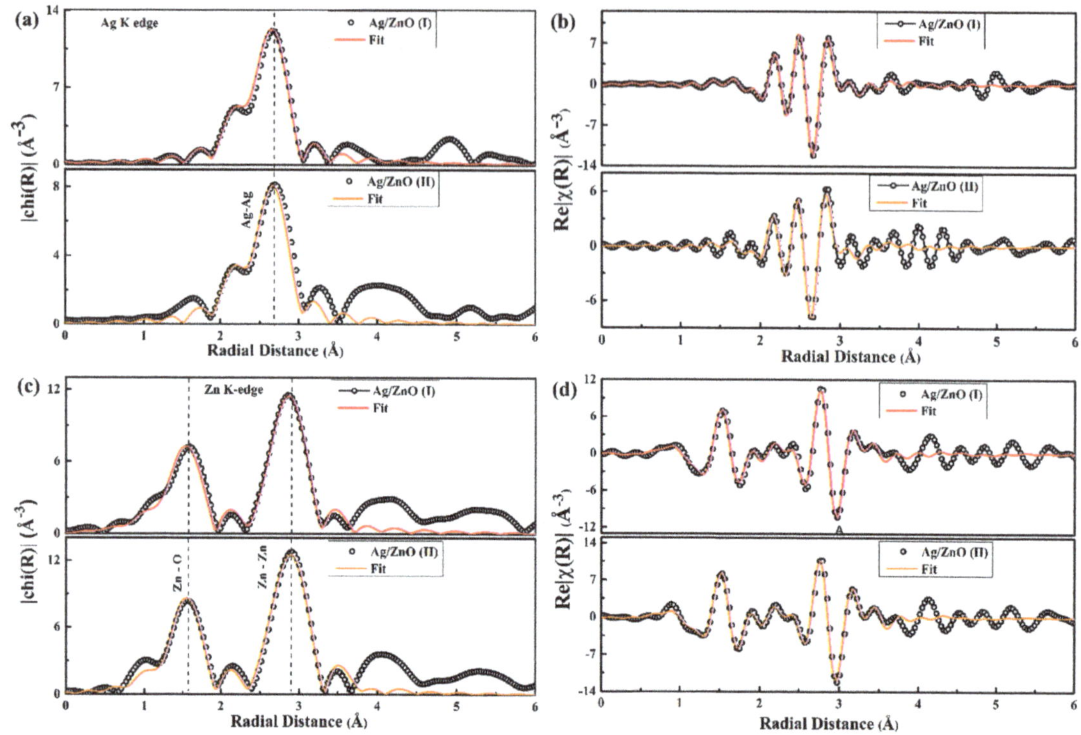

Figure 8. (**a**,**c**) Magnitude and (**b**,**d**) real component of χ(R) for Ag/ZnO (I) and Ag/ZnO (II) samples.

Using the fcc Ag cif file, the EXAFS fitting of experimental Ag K edges for Ag/ZnO (I) and Ag/ZnO(II) samples is shown in Figure 8a. For the Ag/ZnO (I) sample, the Ag atom was coordinated with the 11.9 nearest Ag atoms with a bond length of 2.87 (0) Å. The standard fcc metallic Ag has fcc structure and 12 as the first coordination number (CN). This implies high crystallinity and no ionic substitution for Ag unit cells in the Ag/ZnO (I) sample, as this had the same CN as the standard fcc Ag unit cell. However, for the Ag/ZnO (II) sample, the CN was reduced to 8.9. However, the bond lengths (2.86 (0) Å) were the same. The reduced CN indicates the presence of Ag vacancies in the first coordination shell. The presence of such vacancies might be the reason behind tiny increment in σ^2 values from 0.008 (0) A^2 to 0.009 (0) A^2. The presence of cationic interstitials or vacancies increases the σ^2 value [47]. Therefore, both samples, Ag/ZnO (I) and Ag/ZnO (II), indicate the presence of metallic Ag with both of the synthesis methods.

The real part of $\chi(R)$ R spectra for the Zn K edge in Figure 8c indicates two major peaks corresponding to the nearest Zn–O coordination and second nearest Zn–Zn coordination. The modelling for the Zn K edge was performed with the hexagonal wurtzite ZnO cif file. The R values for Zn–O and Zn–Zn do not indicate any significant changes in Ag/ZnO (I) or Ag/ZnO (II) samples with different synthesis routes. The CN values for Zn–O coordination were 2.9 (4) and 2.9 (3) for Ag/ZnO (I) and Ag/ZnO (II) samples, which are different from the ideal ZnO lattice values. In ideal wurtzite ZnO cif file Zn–O CN is 4. The reduced Zn–O CN suggests the presence of V_o in both of the samples. These fitting results regarding the presence of V_o are in complete agreement with our retrospective O 1s edge XPS analysis. The second nearest neighbor, i.e., Zn–Zn CN, was 11.9 for both samples, alongside no noteworthy change in the Zn–Zn R value. The σ^2 values also remained the same for Zn–Zn coordination in both samples. All results from fitting of the Ag K edge with fcc Ag structure and the Zn K edge with wurtzite ZnO structure are mentioned in Table 2.

Table 2. The obtained values of the first shell, coordination number (CN), bond length(R) and Debye–Waller σ^2 factor for nearest neighbor (Ag–Ag), (Zn–O), and (Zn–Zn) scattering from EXAFS fitting. The number in the parenthesis indicates the uncertainty in the last digit.

Sample	Shell	C_N (1st Shell)	R (Å)	$\sigma^2_{Ag-Ag}(\text{Å}^2)$	$\sigma^2_{Zn-O}(\text{Å}^2)$	$\sigma^2_{Zn-Zn}(\text{Å}^2)$
Ag/ZnO (I)	Ag–Ag	11.9 (0)	2.87 (0)	0.008 (0)	-	-
	Zn–O	2.9 (4)	1.96 (0)	-	0.004 (2)	-
	Zn–Zn	11.9 (0)	3.22 (0)	-	-	0.01 (0)
Ag/ZnO (II)	Ag–Ag	8.9 (0)	2.86 (0)	0.009 (0)	-	-
	Zn–O	2.9 (3)	1.96 (0)	-	0.003 (1)	-
	Zn–Zn	11.9 (0)	3.23 (3)	-	-	0.01 (0)

3.7. Water Disinfection

For optimization of the sample mass used in the disinfection of the water, preliminary tests were carried out using 0.25, 0.5, and 1.0 g of the particles. Considering both the microbiological and physicochemical parameters, the best performance was obtained using 0.5 g of sample because it met the requirements for absence of total coliforms and *Escherichia Coli* (Table 3) and kept the disinfected water sample within the pH and turbidity limits accepted by the standard Standard Methods For The Examination of Water and Wastewater [23]. For the larger (1 g) and smaller (0.25 g) masses, the pH and turbidity of the water after disinfection were not satisfactory.

Table 3. Results validated according to the Standard Methods For The Examination of Water and Wastewater—23th Edition—SMEWW.

Sample	Parameters			
	Total Coliforms (MPN/100 mL)	*Escherichia coli* (MPN/100 mL)	pH	Turbidity (NTU)
Contaminated Water	>2419.6	59.4	7.5	28.1
Pure ZnO	Presence	Presence	8.2	1.5
Ag/ZnO (I)	<1 (Absence)	<1 (Absence)	7.7	1.9
Ag/ZnO (II)	<1 (Absence)	<1 (Absence)	7.5	3.1

The results in Table 3 are in accordance with the Standard Methods For The Examination of Water and Wastewater—23th Edition—SMEWW [23]. According to this method, the microbiological parameter limit for total coliforms and *Escherichia coli* in the water is "<1 or absence MPN/100 mL," and as for the physicochemical parameters, the pH must be between 5.0 to 9.0 and the turbidity between null to 5.0. The effluent showed the count of > 2419.6 MPN/100 mL of total coliforms and 59.4 MPN/100 mL for *E. coli*. Figure 9 shows the effluent before and after treatment using ZnO and Ag/ZnO samples. After treatment with the pure ZnO sample, there was not total elimination of total coliforms or *E. coli* from the water. After disinfection with the pure ZnO, the pH increased to 8.2 and turbidity decreased to 95% of the original value (Table 3). After using the Ag/ZnO nanoparticles, no significant change in pH was observed, and the turbidity decreased to 93% or 90% after disinfection with Ag/ZnO (I) or Ag/ZnO (II), respectively (Table 3). It is known that various factors compromise the bacterial efficiency, such as turbidity, chloride, presence of organic matter, and pH [48]; however, the parameters assessed (pH and turbidity) did not affect the efficacy of silver-containing materials relative to total coliforms and *E. coli*, as the samples Ag/ZnO (I) and Ag/ZnO-Ag (II), physicochemical parameters, and microbiological parameters were within the limits permitted by the standard [23]. In this context, the absence of *E. coli* (<1 MPN/100 mL) indicated that the water after disinfection with Ag/ZnO nanoparticles did not present a sanitation risk [49].

Figure 9. Pictures of the wastewater before and after disinfection tests using: (**a**) pure ZnO, (**b**) Ag/ZnO (I), and (**c**) Ag/ZnO (II).

According to the results, ZnO exhibited higher antibacterial efficiency when decorated with Ag NPs, suggesting that Ag NPs promote the antibacterial efficiency of ZnO, and this can be explained by the widely known bacterial effect of Ag nanoparticles [50,51]. As shown in the TEM analyses, the Ag/ZnO (I) was composed of Ag NPs at the nanometer scale. Several works report on the sizes of Ag particles in their antimicrobial properties against *E. coli* [52,53]. These results indicated that Ag/ZnO exhibited strong antimicrobial activity against *E. coli* compared to pure ZnO.

The mechanisms of antibacterial activity of ZnO and Ag NPs have not yet been clearly elucidated. However, some hypotheses have been proposed for the antimicrobial property of ZnO, such as the generation of reactive oxygen species (ROS) on the surfaces of the particles, release of zinc ions, membrane damage, and alteration of protein and nucleic acid functions [51,54–56]. On the other hand, AgNPs' Ag^{+} ions may be attached to the negatively charged proteins and nucleic acids, leading to structural changes within the cell wall and membrane [57], Ag NPs penetrating inside the bacterial cell, resulting in DNA damage, and adhesion of nanoparticles to the surface altering the membrane properties [5,57]; and dissolution of AgNPs may release Ag^{+} ions [5,50,58], which may be disruptive to the cell walls. Considering the efficiency in disinfection performance, these materials obtained by green syntheses are potential alternative disinfectant agents to chlorine and its derivatives, which despite being widely used, has harmful impacts on human health [59].

Characterization of the Ag/ZnO Samples after Water Disinfection

XPS analyses of Ag/ZnO samples were carried out before and after their use in the disinfection treatment to investigate if the Ag and Zn (at%) concentrations varied, and we then evaluated the leaching of silver ions in water. Table 4 summarizes the results. Taking into consideration the error in the evaluation, we can observe that the relative amount of Ag did not vary significantly.

Table 4. Relative concentration of elements evaluated by XPS.

Sample	Elements	Atomic Concentration (%)	
		Before	After
Ag/ZnO (I)	C1s	5.6	8.2
	O1s	46.0	43.0
	Zn 2p3	47.4	47.8
	Ag 3d	1.0	1.0
Ag/ZnO (II)	C1s	11.9	10.3
	O1s	43.0	45.0
	Zn 2p3	43.3	43.0
	Ag 3d	1.8	1.7

4. Conclusions

Pure ZnO and Ag-loaded ZnO were successfully synthesized by an eco-friendly method using Cassava starch and Aloe vera leaves. Aloe vera was used as a reducing agent and the starch in metal complexing and fuel for the reaction. The XRD results confirm that the decoration with silver did not alter the structural properties of pure ZnO that remained pure hexagonal wurtzite. The presence of Ag NPs on the surface of ZnO particles was confirmed by TEM and XPS. XANES and EXAFS denote the presence of metallic Ag NPs with both synthesis methods.

The synthesized pure ZnO and Ag-loaded ZnO through different methods were applied to wastewater disinfection. The Ag/ZnO suspensions showed a bactericidal effect against *E. coli* and total coliforms, exhibiting potential to be also used as an effective prophylactic agent for the disinfection of wastewater.

Supplementary Materials: The following supporting information can be downloaded at: https://www.mdpi.com/article/10.3390/nano12101764/s1. Figure S1: BSE-SEM image of Ag/ZnO (I) (a), enlarged area of line analysis (b) and SEM-EDX line scan profiles of Zn (magenta), O (green) and Ag (orange) (c). Orange line in image a indicate the area of line analysis. Figure S2: BSE-SEM image of Ag/ZnO (II) with EDXS elemental maps of Zn, O and Ag. Scale bar on the elemental maps is 250 nm. Orange frame in BSE-SEM image indicates area of analysis. Figure S3: SEM images of Ag/ZnO (I) (a) and Ag/ZnO (II) (b) captured with a backscattered electron (BSE) detector. Figure S4: Characteristic TEM images of Ag-loaded ZnO sample prepared by ex situ doping (Ag/ZnO (II)). The white rectangular frame in the image a indicates area of (b). The round particle in (b) is crystalline. The average d-spacing is 0.25 nm and corresponds to (004) planes of hexagonal Ag (ICSD No. 01-870-0598).

Author Contributions: Conceptualization, J.d.O.P., F.J.A. and C.B.; methodology, J.d.O.P.; validation, J.d.O.P. and D.F.H.; formal analysis, J.d.O.P., C.B. and A.D.; investigation, J.d.O.P., K.V.Y., C.B., F.J.A., P.U., C.Z., A.D. and D.F.H.; resources, C.B. and F.J.A.; writing—original draft preparation, J.d.O.P., J.d.S.C. and C.B.; writing—review and editing, C.B., P.U., A.D., C.Z., A.G.B., M.R. and K.V.Y.; visualization, J.d.O.P., A.D. and C.B.; supervision, F.J.A. and C.B.; project administration, F.J.A. and C.B.; funding acquisition, F.J.A. and C.B. All authors have read and agreed to the published version of the manuscript.

Funding: J. de O.P. appreciates the Coordenação de Aperfeiçoamento de Pessoal de Nível Superior-Brasil (CAPES)—Finance Code 001 for a graduate scholarship and the FNRS for mobility grant (2021/V6/5/003–JG/MF–726). A.D. appreciates WBI Excellence post-doctoral scholarship (Belgium). C.B. is a Research Associate of the FRS-FNRS, Belgium. C.B., J.P., and A.D. thank the Belgian Fund for Scientific Research under the FRFC contract EQP 40002995 (PHOTOFUN). F.J.A. is thankful for a CNPq Productivity grant (308625/2019-6), and the grants CNPq–427127/2018-1 and Fundação Araucária–CBPA-001/2016.

Data Availability Statement: Not applicable.

Acknowledgments: The authors are grateful to the Water Analyses Laboratory—Central de Análises/UNICENTRO for the analyses. We thank the Helmholtz-Zentrum Berlin für Materialien und Energie for the allocation of synchrotron radiation beamtime.

Conflicts of Interest: The authors declare no conflict of interest.

References

1. Drinking-Water. Available online: www.who.int/news-room/fact-sheets/detail/drinking-water (accessed on 28 March 2022).
2. Lopes, M.P.; Matos, C.T.; Pereira, V.J.; Benoliel, M.J.; Valério, M.E.; Bucha, L.B.; Rodrigues, A.; Penetra, A.I.; Ferreira, E.; Cardoso, V.V.; et al. Production of Drinking Water Using a Multi-Barrier Approach Integrating Nanofiltration: A Pilot Scale Study. *Sep. Purif. Technol.* **2013**, *119*, 112–122. [CrossRef]
3. Motshekga, S.C.; Ray, S.S.; Onyango, M.S.; Momba, M.N.B. Preparation and Antibacterial Activity of Chitosan-Based Nanocomposites Containing Bentonite-Supported Silver and Zinc Oxide Nanoparticles for Water Disinfection. *Appl. Clay Sci.* **2015**, *114*, 330–339. [CrossRef]
4. Motshekga, S.C.; Sinha Ray, S.; Maity, A. Synthesis and Characterization of Alginate Beads Encapsulated Zinc Oxide Nanoparticles for Bacteria Disinfection in Water. *J. Colloid Interface Sci.* **2018**, *512*, 686–692. [CrossRef] [PubMed]

5. Li, Q.; Mahendra, S.; Lyon, D.Y.; Brunet, L.; Liga, M.V.; Li, D.; Alvarez, P.J.J. Antimicrobial Nanomaterials for Water Disinfection and Microbial Control: Potential Applications and Implications. *Water Res.* **2008**, *42*, 4591–4602. [CrossRef]
6. Dimapilis, E.A.S.; Hsu, C.-S.; Mendoza, R.M.O.; Lu, M.-C. Zinc Oxide Nanoparticles for Water Disinfection. *Sustain. Environ. Res.* **2018**, *28*, 47–56. [CrossRef]
7. Qu, X.; Alvarez, P.J.J.; Li, Q. Applications of Nanotechnology in Water and Wastewater Treatment. *Water Res.* **2013**, *47*, 3931–3946. [CrossRef]
8. Pathak, S.P.; Gopal, K. Evaluation of Bactericidal Efficacy of Silver Ions on Escherichia Coli for Drinking Water Disinfection. *Environ. Sci. Pollut. Res.* **2012**, *19*, 2285–2290. [CrossRef]
9. Bao, Q.; Zhang, D.; Qi, P. Synthesis and Characterization of Silver Nanoparticle and Graphene Oxide Nanosheet Composites as a Bactericidal Agent for Water Disinfection. *J. Colloid Interface Sci.* **2011**, *360*, 463–470. [CrossRef]
10. Lalley, J.; Dionysiou, D.D.; Varma, R.S.; Shankara, S.; Yang, D.J.; Nadagouda, M.N. Silver-Based Antibacterial Surfaces for Drinking Water Disinfection—An Overview. *Curr. Opin. Chem. Eng.* **2014**, *3*, 25–29. [CrossRef]
11. Liu, Y.; He, L.; Mustapha, A.; Li, H.; Hu, Z.Q.; Lin, M. Antibacterial Activities of Zinc Oxide Nanoparticles against Escherichia Coli O157:H7. *J. Appl. Microbiol.* **2009**, *107*, 1193–1201. [CrossRef]
12. Colon, G.; Ward, B.C.; Webster, T.J. Increased Osteoblast and DecreasedStaphylococcus Epidermidis Functions on Nanophase ZnO and TiO_2. *J. Biomed. Mater. Res. Part A* **2006**, *78A*, 595–604. [CrossRef]
13. Panchal, P.; Paul, D.R.; Sharma, A.; Choudhary, P.; Meena, P.; Nehra, S.P. Biogenic Mediated Ag/ZnO Nanocomposites for Photocatalytic and Antibacterial Activities towards Disinfection of Water. *J. Colloid Interface Sci.* **2020**, *563*, 370–380. [CrossRef] [PubMed]
14. Alharthi, F.A.; Alghamdi, A.A.; Al-Zaqri, N.; Alanazi, H.S.; Alsyahi, A.A.; Marghany, A.E.; Ahmad, N. Facile One-Pot Green Synthesis of Ag–ZnO Nanocomposites Using Potato Peeland Their Ag Concentration Dependent Photocatalytic Properties. *Sci. Rep.* **2020**, *10*, 20229. [CrossRef]
15. Seong, S.; Park, I.-S.; Jung, Y.C.; Lee, T.; Kim, S.Y.; Park, J.S.; Ko, J.-H.; Ahn, J. Synthesis of Ag-ZnO Core-Shell Nanoparticles with Enhanced Photocatalytic Activity through Atomic Layer Deposition. *Mater. Des.* **2019**, *177*, 107831. [CrossRef]
16. Pokrowiecki, R.; Wojnarowicz, J.; Zareba, T.; Koltsov, I.; Lojkowski, W.; Tyski, S.; Mielczarek, A.; Zawadzki, P. Nanoparticles and Human Saliva: A Step towards Drug Delivery Systems for Dental and Craniofacial Biomaterials. *Int. J. Nanomed.* **2019**, *14*, 9235–9257. [CrossRef] [PubMed]
17. Zare, M.; Namratha, K.; Alghamdi, S.; Mohammad, Y.H.E.; Hezam, A.; Zare, M.; Drmosh, Q.A.; Byrappa, K.; Chandrashekar, B.N.; Ramakrishna, S.; et al. Novel Green Biomimetic Approach for Synthesis of ZnO-Ag Nanocomposite; Antimicrobial Activity against Food-Borne Pathogen, Biocompatibility and Solar Photocatalysis. *Sci. Rep.* **2019**, *9*, 8303. [CrossRef]
18. De, O.; Primo, J.; Bittencourt, C.; Acosta, S.; Sierra-Castillo, A.; Colomer, J.-F.; Jaerger, S.; Teixeira, V.C.; Anaissi, F.J. Synthesis of Zinc Oxide Nanoparticles by Ecofriendly Routes: Adsorbent for Copper Removal from Wastewater. *Front. Chem.* **2020**, *8*, 1100. [CrossRef]
19. Mote, V.; Purushotham, Y.; Dole, B. Williamson-Hall Analysis in Estimation of Lattice Strain in Nanometer-Sized ZnO Particles. *J. Theor. Appl. Phys.* **2012**, *6*, 6. [CrossRef]
20. Das, J.; Pradhan, S.K.; Sahu, D.R.; Mishra, D.K.; Sarangi, S.N.; Nayak, B.B.; Verma, S.; Roul, B.K. Micro-Raman and XPS Studies of Pure ZnO Ceramics. *Phys. B Condens. Matter* **2010**, *405*, 2492–2497. [CrossRef]
21. Riesemeier, H.; Ecker, K.; Görner, W.; Müller, B.R.; Radtke, M.; Krumrey, M. Layout and First XRF Applications of the BAM Line at BESSY II. *X-Ray Spectrom.* **2005**, *34*, 160–163. [CrossRef]
22. Newville, M.; Ravel, B.; Haskel, D.; Rehr, J.J.; Stern, E.A.; Yacoby, Y. Analysis of Multiple-Scattering XAFS Data Using Theoretical Standards. *Phys. B Condens. Matter* **1995**, *208–209*, 154–156. [CrossRef]
23. Rice, E.W.; Baird, R.B.; Eaton, A.D. *Standard Methods for the Examination of Water and Wastewater Method 2540 D. Total Suspended Solids*, 23rd ed.; American Public Health Association (APHA), American Water Works Association (AWWA), Water Environment Federation (WEF): Washington, DC, USA, 1926.
24. Karunakaran, C.; Rajeswari, V.; Gomathisankar, P. Enhanced Photocatalytic and Antibacterial Activities of Sol–Gel Synthesized ZnO and Ag-ZnO. *Mater. Sci. Semicond. Process.* **2011**, *14*, 133–138. [CrossRef]
25. Georgekutty, R.; Seery, M.K.; Pillai, S.C. A Highly Efficient Ag-ZnO Photocatalyst: Synthesis, Properties, and Mechanism. *J. Phys. Chem. C* **2008**, *112*, 13563–13570. [CrossRef]
26. Guo, Y.; Yang, X.; Ruan, K.; Kong, J.; Dong, M.; Zhang, J.; Gu, J.; Guo, Z. Reduced Graphene Oxide Heterostructured Silver Nanoparticles Significantly Enhanced Thermal Conductivities in Hot-Pressed Electrospun Polyimide Nanocomposites. *ACS Appl. Mater. Interfaces* **2019**, *11*, 25465–25473. [CrossRef] [PubMed]
27. Wei, Y.; Wang, X.; Yi, G.; Zhou, L.; Cao, J.; Sun, G.; Chen, Z.; Bala, H.; Zhang, Z. Hydrothermal Synthesis of Ag Modified ZnO Nanorods and Their Enhanced Ethanol-Sensing Properties. *Mater. Sci. Semicond. Process.* **2018**, *75*, 327–333. [CrossRef]
28. Kokate, M.; Garadkar, K.; Gole, A. Zinc-Oxide-Silica-Silver Nanocomposite: Unique One-Pot Synthesis and Enhanced Catalytic and Anti-Bacterial Performance. *J. Colloid Interface Sci.* **2016**, *483*, 249–260. [CrossRef]
29. Asamoah, R.B.; Annan, E.; Mensah, B.; Nbelayim, P.; Apalangya, V.; Onwona-Agyeman, B.; Yaya, A. A Comparative Study of Antibacterial Activity of CuO/Ag and ZnO/Ag Nanocomposites. *Adv. Mater. Sci. Eng.* **2020**, *2020*, 1–18. [CrossRef]
30. Srithar, A.; Kannan, J.C.; Senthil, T.S. Kannan Preparation and Characterization of Ag Doped ZnO Nanoparticles and Its Antibacterial Applications. *J. Adv. Chem.* **2017**, *13*, 6273–6279.

31. Hamidian, K.; Sarani, M.; Sheikhi, E.; Khatami, M. Cytotoxicity Evaluation of Green Synthesized ZnO and Ag-Doped ZnO Nanoparticles on Brain Glioblastoma Cells. *J. Mol. Struct.* **2022**, *1251*, 131962. [CrossRef]
32. Gayathri, S.; Ghosh, O.S.N.; Sathishkumar, S.; Sudhakara, P.; Jayaramudu, J.; Ray, S.S.; Viswanath, A.K. Investigation of Physicochemical Properties of Ag Doped ZnOnanoparticles Prepared by Chemical Route. *Appl. Sci. Lett.* **2015**, *1*, 8–13.
33. Zeferino, R.S.; Flores, M.B.; Pal, U. Photoluminescence and Raman Scattering in Ag-Doped ZnO Nanoparticles. *J. Appl. Phys.* **2011**, *109*, 014308. [CrossRef]
34. Youn, C.J.; Jeong, T.S.; Han, M.S.; Kim, J.H. Optical Properties of Zn-Terminated ZnO Bulk. *J. Cryst. Growth* **2004**, *261*, 526–532. [CrossRef]
35. Zhou, H.; Chen, L.; Malik, V.; Knies, C.; Hofmann, D.M.; Bhatti, K.P.; Chaudhary, S.; Klar, P.J.; Heimbrodt, W.; Klingshirn, C.; et al. Raman Studies of ZnO:Co Thin Films. *Phys. Status Solidi* **2007**, *204*, 112–117. [CrossRef]
36. Jusserand, B.; Cardona, M. Raman Spectroscopy of Vibrations in Superlattices. In *Light Scattering in Solids V*; Springer: Berlin/Heidelberg, Germany, 1989; pp. 49–152.
37. Cheng, Z.; Song, H.; Zhang, X.; Cheng, X.; Xu, Y.; Zhao, H.; Gao, S.; Huo, L. Non-Enzymatic Nitrite Amperometric Sensor Fabricated with near-Spherical ZnO Nanomaterial. *Colloids Surf. B Biointerfaces* **2022**, *211*, 112313. [CrossRef] [PubMed]
38. Cheng, Y.; Wang, W.; Yao, L.; Wang, J.; Liang, Y.; Fu, J. Insights into Charge Transfer and Solar Light Photocatalytic Activity Induced by the Synergistic Effect of Defect State and Plasmon in Au Nanoparticle-Decorated Hierarchical 3D Porous ZnO Microspheres. *Appl. Surf. Sci.* **2019**, *494*, 959–968. [CrossRef]
39. Yıldırım, Ö.A.; Unalan, H.E.; Durucan, C. Highly Efficient Room Temperature Synthesis of Silver-Doped Zinc Oxide (ZnO:Ag) Nanoparticles: Structural, Optical, and Photocatalytic Properties. *J. Am. Ceram. Soc.* **2013**, *96*, 766–773. [CrossRef]
40. Zheng, Y.; Zheng, L.; Zhan, Y.; Lin, X.; Zheng, Q.; Wei, K. Ag/ZnO Heterostructure Nanocrystals: Synthesis, Characterization, and Photocatalysis. *Inorg. Chem.* **2007**, *46*, 6980–6986. [CrossRef]
41. Chauhan, A.; Verma, R.; Kumari, S.; Sharma, A.; Shandilya, P.; Li, X.; Batoo, K.M.; Imran, A.; Kulshrestha, S.; Kumar, R. Photocatalytic Dye Degradation and Antimicrobial Activities of Pure and Ag-Doped ZnO Using Cannabis Sativa Leaf Extract. *Sci. Rep.* **2020**, *10*, 7881. [CrossRef]
42. Chen, M.; Wang, X.; Yu, Y.H.; Pei, Z.L.; Bai, X.D.; Sun, C.; Huang, R.F.; Wen, L.S. X-Ray Photoelectron Spectroscopy and Auger Electron Spectroscopy Studies of Al-Doped ZnO Films. *Appl. Surf. Sci.* **2000**, *158*, 134–140. [CrossRef]
43. Al-Gaashani, R.; Radiman, S.; Daud, A.R.; Tabet, N.; Al-Douri, Y. XPS and Optical Studies of Different Morphologies of ZnO Nanostructures Prepared by Microwave Methods. *Ceram. Int.* **2013**, *39*, 2283–2292. [CrossRef]
44. Zhang, Q.; Xie, G.; Xu, M.; Su, Y.; Tai, H.; Du, H.; Jiang, Y. Visible Light-Assisted Room Temperature Gas Sensing with ZnO-Ag Heterostructure Nanoparticles. *Sens. Actuators B Chem.* **2018**, *259*, 269–281. [CrossRef]
45. Koningsberger, D.C.; Prins, R. *X-ray Absorption: Principles, Applications, Techniques of EXAFS, SEXAFS and XANES*; Wiley: New York, NY, USA, 1988.
46. Shannon, R.D. Revised Effective Ionic Radii and Systematic Studies of Interatomic Distances in Halides and Chalcogenides. *Acta Crystallogr. Sect. A* **1976**, *32*, 751–767. [CrossRef]
47. Das, A.; Rajput, P.; Kaur, A.; Balasubramanian, C.; Kanjilal, D.; Jha, S.N. e_g–t_{2g} Sub Band Splitting via Crystal Field and Band Anticrossing Interaction in $Ni_xCd_{1-x}O$ Thin Films. *Thin Solid Film.* **2021**, *736*, 138908. [CrossRef]
48. Fan, M.; Gong, L.; Huang, Y.; Wang, D.; Gong, Z. Facile Preparation of Silver Nanoparticle Decorated Chitosan Cryogels for Point-of-Use Water Disinfection. *Sci. Total Environ.* **2018**, *613*, 1317–1323. [CrossRef] [PubMed]
49. Kayser, G.; Moriarty, P.; Fonseca, C.; Bartram, J. Domestic Water Service Delivery Indicators and Frameworks for Monitoring, Evaluation, Policy and Planning: A Review. *Int. J. Environ. Res. Public Health* **2013**, *10*, 4812–4835. [CrossRef] [PubMed]
50. Bhardwaj, A.K.; Sundaram, S.; Yadav, K.K.; Srivastav, A.L. An Overview of Silver Nano-Particles as Promising Materials for Water Disinfection. *Environ. Technol. Innov.* **2021**, *23*, 101721. [CrossRef]
51. Sawai, J.; Kojima, H.; Igarashi, H.; Hashimoto, A.; Shoji, S.; Takehara, A.; Sawaki, T.; Kokugan, T.; Shimizu, M. Escherichia Coli Damage by Ceramic Powder Slurries. *J. Chem. Eng. Jpn.* **1997**, *30*, 1034–1039. [CrossRef]
52. Gogoi, S.K.; Gopinath, P.; Paul, A.; Ramesh, A.; Ghosh, S.S.; Chattopadhyay, A. Green Fluorescent Protein-Expressing Escherichia c Oli as a Model System for Investigating the Antimicrobial Activities of Silver Nanoparticles. *Langmuir* **2006**, *22*, 9322–9328. [CrossRef]
53. Morones, J.R.; Elechiguerra, J.L.; Camacho, A.; Holt, K.; Kouri, J.B.; Ramírez, J.T.; Yacaman, M.J. The Bactericidal Effect of Silver Nanoparticles. *Nanotechnology* **2005**, *16*, 2346–2353. [CrossRef]
54. Applerot, G.; Lipovsky, A.; Dror, R.; Perkas, N.; Nitzan, Y.; Lubart, R.; Gedanken, A. Enhanced Antibacterial Activity of Nanocrystalline ZnO due to Increased ROS-Mediated Cell Injury. *Adv. Funct. Mater.* **2009**, *19*, 842–852. [CrossRef]
55. Brayner, R.; Ferrari-Iliou, R.; Brivois, N.; Djediat, S.; Benedetti, M.F.; Fiévet, F. Toxicological Impact Studies Based on Escherichia Coli Bacteria in Ultrafine ZnO Nanoparticles Colloidal Medium. *Nano Lett.* **2006**, *6*, 866–870. [CrossRef] [PubMed]
56. Patrinoiu, G.; Hussien, M.D.; Calderón-Moreno, J.M.; Atkinson, I.; Musuc, A.M.; Ion, R.N.; Cimpean, A.; Chifiriuc, M.C.; Carp, O. Eco-Friendly Synthesized Spherical ZnO Materials: Effect of the Core-Shell to Solid Morphology Transition on Antimicrobial Activity. *Mater. Sci. Eng. C* **2019**, *97*, 438–450. [CrossRef] [PubMed]
57. Abduraimova, A.; Molkenova, A.; Duisembekova, A.; Mulikova, T.; Kanayeva, D.; Atabaev, T.S. Cetyltrimethylammonium Bromide (CTAB)-Loaded SiO_2–Ag Mesoporous Nanocomposite as an Efficient Antibacterial Agent. *Nanomaterials* **2021**, *11*, 477. [CrossRef]

58. Tian, Y.; Qi, J.; Zhang, W.; Cai, Q.; Jiang, X. Facile, One-Pot Synthesis, and Antibacterial Activity of Mesoporous Silica Nanoparticles Decorated with Well-Dispersed Silver Nanoparticles. *ACS Appl. Mater. Interfaces* **2014**, *6*, 12038–12045. [CrossRef] [PubMed]
59. Mazhar, M.A.; Khan, N.A.; Ahmed, S.; Khan, A.H.; Hussain, A.; Rahisuddin; Changani, F.; Yousefi, M.; Ahmadi, S.; Vambol, V. Chlorination Disinfection By-Products in Municipal Drinking Water-A Review. *J. Clean. Prod.* **2020**, *273*, 123159. [CrossRef]

nanomaterials

Article

Treatment of Wastewater, Phenols and Dyes Using Novel Magnetic Torus Microreactors and Laccase Immobilized on Magnetite Nanoparticles

Paula Andrea Peñaranda [1,2], Mabel Juliana Noguera [1], Sergio Leonardo Florez [1], Johana Husserl [2], Nancy Ornelas-Soto [3], Juan C. Cruz [4] and Johann F. Osma [1,*]

1 Department of Electrical and Electronic Engineering, Universidad de los Andes, Cra. 1E No. 19a-40, Bogota 111711, Colombia; pa.penaranda1711@uniandes.edu.co (P.A.P.); mj.noguera10@uniandes.edu.co (M.J.N.); sl.florez10@uniandes.edu.co (S.L.F.)
2 Department of Civil and Environmental Engineering, Universidad de los Andes, Cra. 1E No. 19a-40, Bogota 111711, Colombia; jhusserl@uniandes.edu.co
3 Laboratorio de Nanotecnología Ambiental, Escuela de Ingeniería y Ciencias, Tecnológico de Monterrey, Monterrey 64849, Mexico; ornel@tec.mx
4 Department of Biomedical Engineering, Universidad de los Andes, Cra. 1E No. 19a-40, Bogota 111711, Colombia; jc.cruz@uniandes.edu.co
* Correspondence: jf.osma43@uniandes.edu.co; Tel.: +57-1-339-4949

Abstract: In this work, the design, manufacture, and testing of three different magnetic microreactors based on torus geometries (i.e., one-loop, two-horizontal-loop, and two-vertical-loop) is explored to increase the enzyme-based transformation of dyes by laccase bio-nanocomposites, improve the particle suspension, and promote the interaction of reagents. The laccase enzyme was covalently immobilized on amino-terminated silanized magnetite nanoparticles (laccase-magnetite). The optimal configuration for the torus microreactor and the applied magnetic field was evaluated in silico with the aid of the CFD and particle tracing modules of Comsol Multiphysics®. Eriochrome Black T (EBt) dye was tested as a biotransformation model at three different concentrations, i.e., 5 mg/L, 10 mg/L, and 20 mg/L. Phenol oxidation/removal was evaluated on artificial wastewater and real wastewater. The optimal catalytic performance of the bionanocomposite was achieved in the range of pH 4 to 4.5. A parabolic movement on the particles along the microchannels was induced by the magnetic field, which led to breaking the stability of the laminar flow and improving the mixing processes. Based on the simulation and experiments conducted with the three geometries, the two-vertical-loop microreactor demonstrated a better performance mainly due to larger dead zones and a longer residence time. Additionally, the overall dye removal efficiencies for this microreactor and the laccase-magnetite bionanocomposite were 98.05%, 93.87%, and 92.74% for the three evaluated concentrations. The maximum phenol oxidation with the laccase-magnetite treatment at low concentration for the artificial wastewater was 79.89%, while its phenol removal efficiency for a large volume of real wastewater was 17.86%. Treatments with real wastewater were carried out with a larger volume, equivalent to 200 biotransformation (oxidation) operating cycles of those carried out with dyes or phenol. Taken together, our results indicate that the novel microreactors introduced here have the potential to process wastewaters rich in contaminant dyes in continuous mode with efficiencies that are attractive for a potential large-scale operation. In this regard, future work will focus on finding the requirements for scaling-up the processes and evaluating the involved environmental impact indexes, economic performance, and different device geometries and processing schemes.

Keywords: dye; phenol; removal; bionanocomposites; laccase; magnetic nanoparticles; microreactor; torus

Citation: Peñaranda, P.A.; Noguera, M.J.; Florez, S.L.; Husserl, J.; Ornelas-Soto, N.; Cruz, J.C.; Osma, J.F. Treatment of Wastewater, Phenols and Dyes Using Novel Magnetic Torus Microreactors and Laccase Immobilized on Magnetite Nanoparticles. *Nanomaterials* **2022**, *12*, 1688. https://doi.org/10.3390/nano12101688

Academic Editor: Christos A. Aggelopoulos

Received: 31 March 2022
Accepted: 6 May 2022
Published: 15 May 2022

1. Introduction

Water pollution is one of the main environmental problems that mankind is facing over the coming few decades [1,2]. Industrial and agricultural processes are thought to be major causing agents due to the continuous discharge of large volumes of wastewaters to these aquatic environments [3]. Some of the most common discharged pollutants include pathogens, excess nutrients, suspended solids and sediments, pesticides, plastics, fertilizers, acids, detergents, pharmaceuticals, phenols, minerals, dyes and pigments, and heavy metals [4]. Phenolic compounds are mainly found in wastewater of many industries such as coal conversion, resin, plastic, and petroleum refineries. These compounds are toxic pollutants in industrial waste imposing risk to human health, and some of them are suspected carcinogens. Phenols and cresols are highly corrosive and toxic and can cause damage to the respiratory system, scarring of the skin, damage to gastrointestinal tracts, kidney failure, hematological changes, and nervous system depression [2,5]. The annual industrial production of synthetic or azo dyes approaches 70 million tons [6]. These xenobiotic chemicals are not normally encountered in nature and show high solubility in water and consequently, stand as one of the major sources of water pollution [7]. Due to their production ease, azo dyes are ubiquitously found around the world in many industries including textiles, leather goods, paper, plastics, foodstuffs, cosmetics, and candles [8]. Despite the industry's efforts to couple wastewater treatment processes to their continuous production process, 90% of reactive textile dyes entering activated sludge sewage treatment plants will pass through unchanged [9]. As a result, between 30 to 150 thousand tons of dyes are discharged into water bodies, soil, and aquatic ecosystems annually [6].

The main methods used to treat water effluents containing azo compounds or phenolic compounds are coagulation/flocculation and precipitation, adsorption, flotation, membrane filtration, chlorine disinfection, bioflocculation, ion pair extraction, ultrasonic mineralization, electrolysis, ion exchange, advanced oxidation processes, sonication, photocatalysis, and ozonation [2,6,7]. Some of these processes, however, are difficult to operate, rely on costly feedstocks, require complex instrumentation and control schemes, exhibit limited versatility, and are negatively affected by other wastewater pollutants. Moreover, they can generate genotoxic or hazardous byproducts [10,11]. One attractive alternative to overcome some of these issues is the incorporation of enzyme-based biocatalysts as active components into treatment processes. This approach facilitates the degradation of the recalcitrant organic compounds, including the azo dyes and phenols, without introducing any extra toxic components, processing an ample range of pollutant concentrations, requiring shorter treatment times, and with low operation costs [12]. One of the most attractive families of enzymes for treatment of azo dyes are laccases [13]. These enzymes are oxidoreductases capable of oxidizing phenolic compounds into phenoxyl radicals, with the aid of four copper electrons in their structures [7,14].

Laccases have proven to catalyze the oxidation of a wide range of potential substrates, and they have high activity and specificity toward phenolic and non-phenolic aromatic compounds [15,16]. Nevertheless, free laccases are environmentally sensitive and difficult to recover from reaction media; have low long-term stability, costly isolation and purification processes, and limited large-scale applications due to their disposable use; and their catalytic activity shows a marked decrease in harsh environments [14,16]. Enzyme immobilization on different types of supports provides a technological route to compensate for many of these drawbacks, improve their stability, and allow their reuse in a cyclic reaction scheme [17,18].

Recently, laccases have been successfully immobilized on microscopic supports such as porous glasses [19,20], TiO_2 [21], membranes [22], and microspheres [23]. Main immobilization methods include adsorption, self-immobilization, covalent binding, mesh embedding, microencapsulated embedding, and two-step combination. Yet, compared with conventional immobilization carriers, nanostructured materials are regarded as promising supports due to their small size, high surface area, and large surface-to-volume ratio [16]. MagnetiteM (Fe_3O_4) nanoparticles (MNPs) stand out as promising supports due to their large surface areas, superparamagnetism, ease of separation and recovery by applying external magnetic fields, and well-defined surface properties and morphology [24,25]. MNPs also stand out for their low toxicity, mature synthesis technology, and the possibility to be recycled without major changes in properties [16,19,26–28]. Thus far, MNPs have been exploited for the immobilization of several enzymes, including lipases, lactase, and glucosidase for degradation of pollutants [29].

Recent developments have focused on enabling incorporation of free and immobilized laccases into continuous treatment processes, which might be beneficial for low-cost industrial applications [13,30,31]. In this regard, continuous microreactor devices have been recently developed by taking advantage of important advances in the microfluidics field. One of the most attractive features of these devices is that they can carry out chemical processes with low reagent consumption due to the small sample volumes handled. There are different device configurations and peripherals that have been developed to assemble systems capable of complying with different analysis and functions including sampling, sample processing and in-line real-time monitoring, and processing of the collected data. With the advent of easier and cheaper ways of manufacturing at the microscale, the field of microfluidics has had an exponential growth and therefore has reached the sufficient maturity for its incursion into industrialization routes [32]. One of such tools is CFD simulations of the fluid flow and transport of objects within the microsystems [33]. With the simulation results, manufacturing takes shorter times and favor only prototypes with the highest performance, which can be further optimized with much less investment [34].

Here, we explore the design and manufacture of microreactors with toroidal topologies to enable the enzyme-based oxidation of phenols and dyes. We hypothesize that such microreactors are suitable for maximizing biotransformation processes due to the absence of dead volume, the efficient mixture of reagents, and the possibility of continuous reaction within the toroidal loop [35]. A first attempt to find an optimal configuration for the torus microreactor was explored in silico by analyzing mixing patterns and fluid dynamics. Low-cost device prototyping was conducted in polymethyl methacrylate using a laser cutting system and commercially available fittings for the assembly and subsequent testing. The microreactor's potential for biotransformation of dyes was evaluated for laccases covalently immobilized magnetite (Fe_3O_4). The model reactions were the oxidation of the commercially available dye Eriochrome Black T (EBt) as well as phenol. To maintain the nanoparticles suspended during the treatment process and maximize contact between the components; one or two permanent magnets were coupled to the microreactor. Finally, the extent transformation of the azo molecules was examined in a real wastewater after measuring some water quality parameters and evaluating the removal of phenolic compounds from this water.

2. Materials and Methods

2.1. Materials

Polymethyl methacrylate (PMMA) sheets, also known as acrylic sheets, were purchased from Surtiacrylicos (Bogota, Colombia), Methyl methacrylate based-glue Veracril ® was obtained from New Stetic (Guarne, Colombia), Ethanol (96%) was acquired from Expert Clean Colombia (Bogota, Colombia), and 345 mT Neodymium cylindrical magnets (ϕ: 6 mm × h 7 mm) were purchased from Constructor Mil Imanes (Bogota, Colombia). 2,2-azino-bis(3-ethylbenzothiazoline-6) sulphonic acid (ABTS), glutaraldehyde (25%), sodium hydroxide (NaOH) (98%), tetramethylammonium hydroxide (TMAH) (25%), and 3-aminopropyl-triethoxysilane (APTES) (98%) were purchased from Sigma-Aldrich (Burlington, MA, USA). Iron (II) chloride tetrahydrate (98%) ($FeCl_2$*$4H_2O$), iron (III) chloride hexahydrate (97%) ($FeCl_3$*$6H_2O$), phenol crystallized (99,5%) (C_6H_6O) (phenol) and dye Eriochrome Black T (EBt) (C.I. 14645) were obtained from PanReac AppliChem (Castellar del Valles, Spain).

2.2. Laccase

Laccases from *Pycnoporus Sanguineus* CS43 (EC 1.10.3.2) were obtained from tomato medium as described elsewhere [36]. In brief, mycelia were separated from the tomato medium supernatant after 10 days of culture by membrane filtration (0.2 μm RC Whatman). Then, it was concentrated by ultrafiltration with a tangential-flow filter (Membrane cut-off of 10 kDa, Sartorius Sartojet, Göttingen, Germany). The ultra-filtered sample was purified with a DEAE-cellulose ion exchange column eluted with 20 to 300 mM phosphate buffer pH 6.0 at a flow rate of 2 mL/min. The obtained laccase cocktail was collected and concentrated using an Amicon ultrafiltration cell (Membrane cut-off of 10 kDa, Merk Millipore, Burlington, MA, USA).

2.3. Synthesis of Magnetite Nanoparticles

Magnetite nanoparticles (Magnetite) were synthesized via coprecipitation by mixing 20 mL of 1 M $FeCl_2$ and 20 mL 2 M $FeCl_3$ under agitation at 1500 revolutions per minute (rpm) and 90 °C. Subsequently, 40 mL of 8 M NaOH and 40 mL 2% (v/v) of TMAH were added to the mixture during 3.5 h at a flow rate of 12 mL/h. The obtained magnetite nanoparticles were purified by magnetic separation with the aid of a strong permanent magnet, then washed thoroughly with 2% (v/v) TMAH, and finally sonicated for 100 min using a VibraCell sonication system (Sonics, Newtown, CT, USA). The synthesized Magnetite exhibited an average hydrodynamic diameter of 88.59 nm with a polydispersity index of 0.182 as determined by Dynamic Light Scattering (DLS) analysis with the aid of a Zetasizer Nano ZS, (Malvern, UK) as previously reported by Lopez-Barbosa et al. (2020) [37]. Microscopic inspection of the Magnetite structure, morphology, and shape was achieved via transmission electron microscopy (TEM) in a Tecnai F30 instrument (FEI Company, Fremont, CA, USA)

2.4. Enzyme Immobilization

2.4.1. Enzyme Immobilization on Magnetite Nanoparticles

A nanoparticle suspension (5 mg/mL) was buffered using a NaOH 1M solution until reaching pH 11, then sonicated for 10 min. Next, 50 μL of 2% (v/v) TMAH was pipetted, and the resulting reaction mixture sonicated for 10 min. Silanization was carried out by adding 50 μL of 2% (v/v) APTES, and then the mixture was sonicated again for 20 min. Subsequently, 50 μL of 2% (v/v) glutaraldehyde was added to the mixture as the crosslinker and left to react for 30 min. Finally, 50 μL of 960 U/L laccase enzyme was added and left overnight to immobilize on the surface of the MNP via covalent bonding. The resulting bionanocomposites (laccase-magnetite) were recovered with the aid of a strong permanent magnet and washed thoroughly with Milli Q water. Laccase immobilization on magnetite was confirmed by Fourier transform infrared spectroscopy (FT-IR) aided by an A250/D FTIR-ATR (Bruker, Bremen, Germany). The resulting laccase-magnetite exhibited an immobilization ratio of 0.0096 U/g of laccase activity per MNP mass, and 13 μg/g of laccase mass per MNP mass.

2.4.2. Effect of pH and Temperature on Enzymatic Activity

Stability of free (free-laccase) and immobilized laccase (laccase-magnetite) bionanocomposites was examined in phosphate–citrate buffer solutions at pH values of 2.0, 3.0, 4.0, 4.5, 5.0, 6.0, 7.0, and 10.0, at 25 °C and at different temperatures ranging from 30 to 70 °C, at pH 4.0. Laccase activity, of free and immobilized laccase, was determined spectrophotometrically as described by Moilanen et al. [11] at 436 nm using a GENESYS 10S UV-Vis v4.004 2L5R078128 (Thermo Fisher Scientific, Waltham, MA, USA). One activity unit was defined as the amount of enzyme that oxidized 1 μmol of ABTS per min. The activities were expressed in U/L. All measurements were carried out in triplicate.

2.5. Microreactors Design and Manufacture

2.5.1. Microreactor Geometry Design and Simulation

Three different geometries for the torus microreactor were studied in silico via Comsol Multiphysics 5.3® (Stockholm, Sweden). Figure 1 shows such geometries, namely, one-loop, two-horizontal-loop, and two-vertical-loop microreactors. All microreactors were designed to work under continuous operation, with a wastewater continuous input flow through the inlet that travels through the microchannel until reaching the outlet. During its trajectory, wastewater meets with the laccase-magnetite bionanocomposites that are contained inside the microchannel by the effect of the magnetic field of a permanent magnet or magnets coupled to the systems, depending on the design. Wastewater is then treated by the laccase-magnetite and continue its trajectory to the outlet of the microreactor, while laccase-magnetite is retained inside the microreactor. To study the behavior of each microreactor design, the Computational Fluid Dynamics (CFD) and Magnetic Field no current (MF) modules of Comsol® were coupled to simultaneously simulate the hydrodynamics and impact of the imposed magnetic fields, respectively. To evaluate the impact of magnetic fields on the transport of the individual nanoparticles, the Particle Tracing (PT) module was implemented solely for the one-loop microreactor.

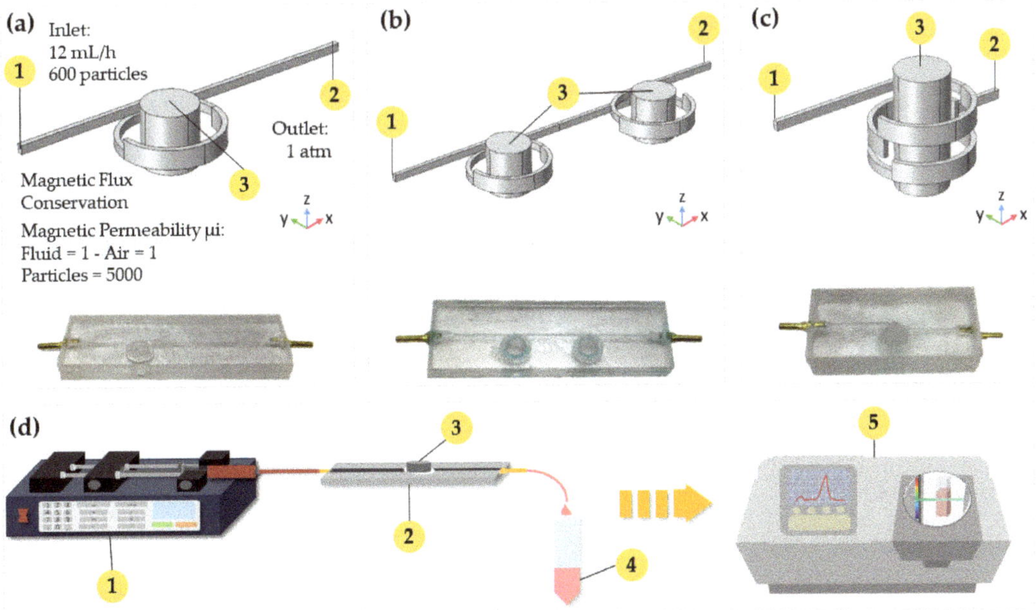

Figure 1. (**a**) Simulation geometry and actual picture of one-loop microreactor. (**b**) Simulation geometry and actual picture of the two-horizontal-microreactor. (**c**) Simulation geometry and actual picture of the two-vertical-loop microreactor. (**d**) Schematic of the experimental setup for the biotransformation process. The numbers 1, 2, 3, 4, and 5 in the figure correspond to the solution injection, microreactor, permanent magnet, treated sample, and absorbance analysis, respectively.

The laminar flow module (Equation (1)) was used here to describe the fluid flow due to the low Reynolds number (Re) calculated for the microreactors (Re = 4.4). The inflow to the microreactor was 12 mL/h, while the output pressure was set to 1 atmosphere. Density and viscosity of water were assumed as the properties of the flowing fluid. The non-slip boundary condition was imposed at the walls of the microchannels. The imposed boundary conditions are summarized in Figure 1. MF was used to simulate the magnetic field generated by a neodymium permanent magnet (Equations (2) and (3)). An additional domain representing the surrounding air was added to the computational domain. Air and water were assumed with magnetic permeability of 1. These two physics were solved simultaneously in a stationary study with the direct solver PARDISO:

$$0 = -\nabla P + \mu \nabla^2 V \tag{1}$$

$$B = \mu_0 \mu_r H \tag{2}$$

$$B = \mu_0 \mu_r H + B_R \tag{3}$$

In these equations, ∇P is the pressure gradient in the fluid, μ is the viscosity of the fluid, and V is the velocity of the fluid. For the equations of MF, B is a magnetic flux density, μ_0 is the magnetic permeability of the vacuum, μ_r is the magnetic permeability of the fluid, H is the magnetic field, and B_R is the remanent flux density of the neodymium magnet. The PT module was used to study the dynamics of the magnetic particles within the microreactor. The particle diameter was assumed as 1.2 µm with a density of 5180 kg/m^3. The particle size was decided according to our own previous experimentation that demonstrated the presence of agglomerations of NPs forming clusters (data not shown). Additionally, the magnetic permeability of the particle was set to 5000 H/m. The particles were delivered at the main inlet of the device. The PT was solved in a time-dependent study for 10 s using 0.1 s steps. In this case, a projected conjugate gradient iterative solver was chosen due to the high demand for computational resources. Additionally, a parametric analysis was performed to evaluate the impact of the remanent flux density of the neodymium magnet on the particles transport. Equations (4)–(7) describe the forces experienced by a particle. This study considered the interactions between the particles and the fluid (drag force) and the magnetic attraction between particles. Particle-particle interactions, lifting force, and Brownian motion were disregarded to avoid dealing with unnecessary complexity that provides little extra information at the expense of exceedingly large computational resources:

$$F_D = \frac{1}{\tau_P} m_P (U - V) \tag{4}$$

$$\tau_P = \frac{\rho_P d_P^2}{18\mu} \tag{5}$$

$$F_{mp} = 2\pi r_p^3 \mu_0 \mu_r k \nabla H^2 \tag{6}$$

$$k = \frac{(\mu_{rp} - \mu_r)}{(\mu_{rp} + 2\mu_r)} \tag{7}$$

In these equations F_D is the drag force, m_P is the mass of the particle, U is the velocity of the particle, V is the fluid velocity, ρ_P is the density of the particle, and d_p is the particle diameter. F_{mp} is the magnetophoretic force, r_P is the particle radius, and μ_{rp} is the magnetic permeability of the particle. As described before, the meshing was subjected to a convergence analysis to determine the minimum number of elements necessary to arrive to a meaningful solution. For this study, five random measurement points in the microchannel were selected along the computational domain, and the change in magnitude of speed was evaluated as the number of mesh elements was increased. As a convergence criterion, it was determined that the velocity magnitude change obtained for a meshing level and the next one was below 3%. An unstructured mesh with 60,000 tetrahedral elements was therefore generated. A stationary study was then run with the direct solver PARDISO that allows to parallelize processes solving large symmetric or structurally symmetric dispersed linear systems of equations in shared memory multiprocessors [38].

2.5.2. Manufacture of the Torus Microreactors

Device prototyping was conducted by engraving the microchannels (1 mm of depth) on PMMA slides of 2.5 mm thickness with an area of 75 mm × 25 mm. The proposed torus geometries were cut with the aid of a laser cutter system Speedy 100, 60 W (TROTEC, Germany). Proper assembly and sealing were achieved by gluing the slides with ethanol followed by heating at 105 °C for 8 min. The device was then maintained under constant pressure with the aid of a home-made press. Inlets and outlets were equipped with commercially available fittings to facilitate connection to pumping devices (e.g., syringe pumps) (See Figure 1).

2.6. *Dye Decolorization Studies*

2.6.1. Experimental Tests for Decolorization of Dyes

EBt dye was used as the model compound to study dye treatment using MNP or laccase-magnetite inside the microreactors at pH 5.5 and three different dye concentrations, namely, 5 mg/L, 10 mg/L, and 20 mg/L. The setup consisted of a microreactor previously loaded with magnetite or laccase-magnetite, neodymium permanent magnets placed in the corresponding slots, and 5 mL of Ebt solution, at previously described concentrations, pumped through the inlet at a constant rate of 12 mL/h for 25 min. Prior to the decolorization experiments, 5 mg of MNP or laccase-magnetite (representing 4.85×10^{-5} U of laccase activity) was introduced to each microreactor in the presence of the neodymium permanent magnets, of 349.23 mT, to retain the bionanocomposites at the reaction loops. Samples were analyzed spectrophotometrically by measuring the absorbance peak at 545 nm, which is the maximum absorbance for EBt, and the absorbance area of the visible spectrum in the range between 400 and 700 nm (See Figure 1). All measurements were carried out in triplicate.

$$\% \, Removal = \frac{Abs_{Dye\ Control} - Abs_{Sample}}{Abs_{Dye\ Control}} \quad (8)$$

The dye decolorization was calculated as removal percentage according to Equation (8), where the $Abs_{Dye\ Control}$ is the average absorbance (at the absorbance peak or the area under the absorbance curve, respectively) of the dye at each concentration prior to entering the microreactor, and the Abs_{Sample} is the absorbance at the outlet of each replica.

2.6.2. Treatment of Artificial Wastewater

Artificial wastewater (AW) (pH 4.42) containing three different concentrations of phenol, namely, 5 mg/L, 10 mg/L, and 20 mg/L was prepared. A total of 5 mg of laccase-magnetite (representing 4.85×10^{-5} U of laccase activity) was introduced to each microreactor in the presence of the neodymium permanent magnets to retain the bionanocomposites at the reaction loops. Each microreactor was then infused with 5 mL of each artificial wastewater solution at a constant rate of 12 mL/h for 25 min. Effluent samples were analyzed spectrophotometrically by measuring the absorbance peak at 270 nm, which is the maximum absorbance for phenol, and the absorbance area in the range between 190 and 1100 nm. All measurements were carried out in triplicate.

Phenol oxidation was calculated as a percentage of change in the phenol concentration, according to Equation (9). Where the $Abs_{AW\ Control}$ is the average absorbance (at the absorbance peak or area under the absorbance curve, respectively) of phenol at each concentration previous to entering the microreactor, and the Abs_{Sample} is the absorbance of each replica, while $Abs_{Total\ conversion}$ is the absorbance of the total oxidized phenol. A value of zero in the percentage corresponds to a situation of no noticeable change in the phenol contents while negative values represent a reduction in such a concentration mainly due to physical phenomena (e.g., adsorption). Additionally, positive values might be explained by the polymerization of phenol upon oxidation by the Laccase molecules.

$$\%\ Oxidized\ Phenol = \frac{Abs_{Sample} - Abs_{AW\ Control}}{Abs_{Total\ conversion} - Abs_{AW\ Control}} \quad (9)$$

2.6.3. Treatment of Real Wastewater

Real Wastewater (RW) samples were collected from the laboratories of the School of Medicine at los Andes University (Bogota, Colombia). RW was then filtered through filter paper Munktell grade 3 HW with a pore diameter of 110 nm (Ahlstrom, Finland) to remove solid particles and some colloids that could interfere with the process. RW was then classified as untreated-RW (U-RW) and characterized. In contrast, treated-RW (T-RW) corresponds to U-RW after treatment through the laccase-magnetite system inside the two-vertical-loop microreactor. A total of 5 mg of laccase-magnetite (representing 4.85×10^{-5} U of laccase activity) was introduced to the microreactor prior to treatment, as explained previously. One liter of U-RW was pumped into the microreactor at a constant rate of 12 mL/h. T-RW was then collected at the outlet of the microreactor.

Samples of U-RW and T-RW were analyzed to quantify phenol, biochemical oxygen demand (BOD), NH_3-nitrogen, and total Kjeldahl nitrogen concentrations. BOD was analyzed using the Standard Method 5210B; NH_3-nitrogen was analyzed by Standard Method 4500-NH_3-C; total nitrogen was evaluated by Standard Method 4500-N. Phenol was selected as a role contaminant, as it is susceptible to oxidation by the immobilized laccase molecules. Nitrogen was analyzed due the presence of amine groups in the chemical structure of immobilization linkers. Finally, the biochemical oxygen demand was quantified to estimate the amount of dissolved organic matter present in the samples, which can be partially oxidized by laccase enzymes.

The removal ratio for each of the parameters above was calculated according to Equation (10), where the X_{U-RW} is the value obtained for each parameter in the untreated sample (U-RW), and X_{T-RW} is the value of the same parameter after treatment (T-RW).

$$Removal\ Ratio\ [\%] = \frac{(X_{U-RW} - X_{T-RW})}{X_{U-RW}} \quad (10)$$

3. Results and Discussion

3.1. Characterization of Laccase Immbolized on Magnetite

FTIR was carried out to confirm the immobilization of laccase on magnetite nanoparticles (laccase-magnetite) (Figure 2). Bare magnetite exhibited two bands at 1620 and 3160 cm^{-1} related to the vibration of the hydroxyl groups on the surface of the magnetite ν (Fe-OH) [39,40]. Laccase-magnetite nanobiocomposites presented a peak at 3350 cm^{-1} attributed to the NH stretching vibration of laccase amines and a peak at 1080 cm^{-1} characteristic of the Si-O-Si bond resulting after silanization [39].

We evaluated the impact of pH and temperature on enzyme activity (Figure 3). Optimal activity for the free-laccase was observed under acidic conditions (maximum activity at pH 2). Similar optimal values between pH 2–4 were reported previously [41]. In contrast, for pH values of 6 and above, the activity decreases and becomes almost negligible. In the case of the bionanocomposites, maximum activity was observed at pH 4, followed by a considerable decrease for pH values of 6 and above. The change in the optimal pH for Laccase-Magnetite to higher values is likely caused by an uneven concentration of the H^+ and OH^- ions between the support matrix and the bulk solution [33]. Regarding the effects of temperature on the activity of the enzyme, both the immobilized laccase and the free laccase showed more than 85% of relative activity between 40 and 60 °C; in both cases (free-laccase and laccase-magnetite), there is a pronounced deactivation effect above 60 °C, being slightly higher for the immobilized laccase, losing around 15–30% of the relative activity.

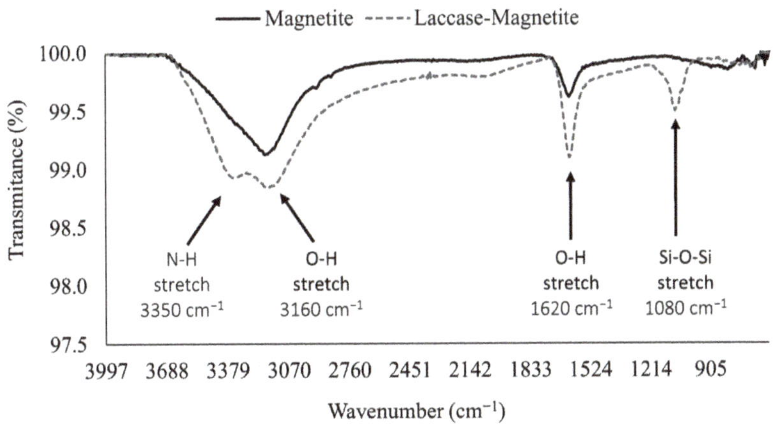

Figure 2. FTIR analysis of magnetite nanoparticles and laccase-magnetite nanobiocomposites, showing the identified peaks at 1080, 1620, 3160 and 3350 cm^{-1}.

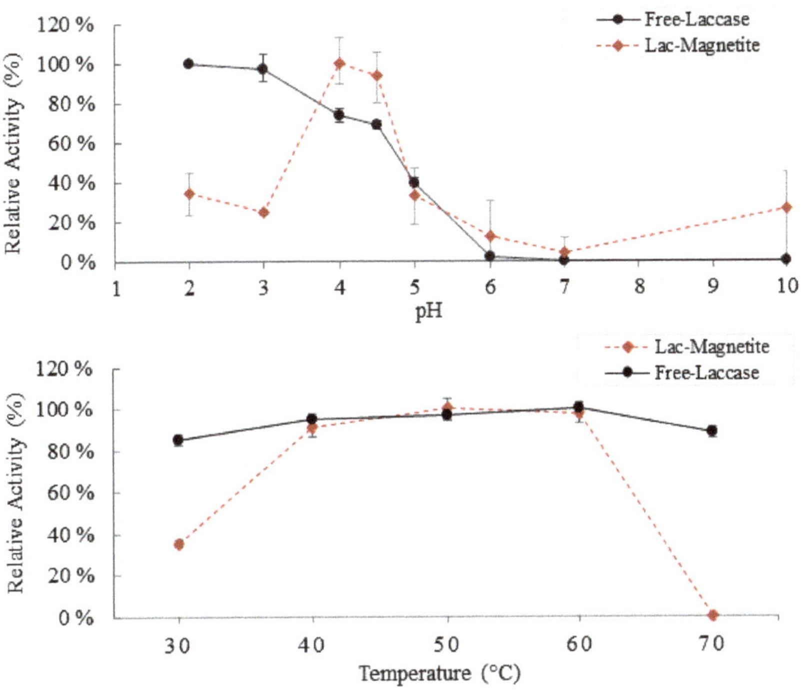

Figure 3. Effect of pH and temperature on enzyme activity: (•) free-laccase and (♦) laccase-magnetite.

3.2. Microreactor Geometry Design and Simulation

Figure 4 shows the convergence plot for a one-loop microreactor. All the simulations were carried out with a minimum of 60,000 mesh elements for the fluid domain to assure the system convergence.

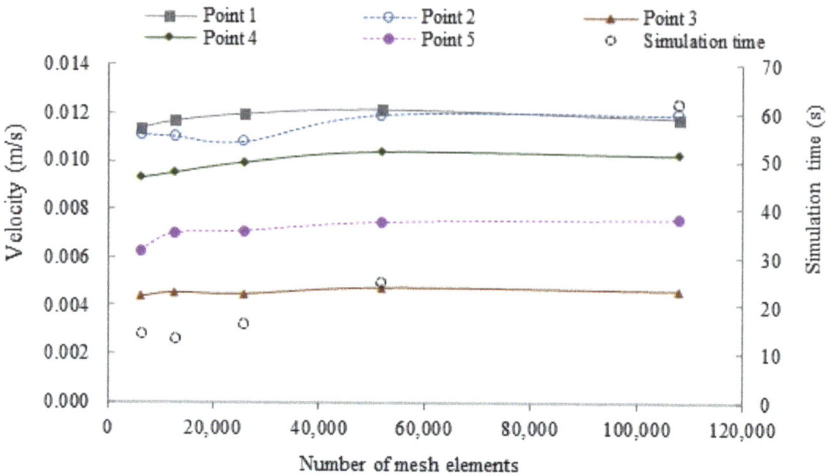

Figure 4. Mesh convergence analysis for the one-loop torus microreactor. Five points along the computational domain were selected for the analysis and plotted as a function of number of mesh elements. Additionally, the simulation time for the different levels of meshing is presented as dotted circles.

The velocity profiles and magnetic field fluxes for the three proposed devices are shown in Figure 5. Dead zones of low velocities (blue color) are observed in the three configurations; however, the two-vertical-loop microreactor shows a larger dead zone compared to the other 2 devices. This can be explained by the relatively important changes in height as the fluid passes through the loops. In addition, the fluid decelerates because of changes in the cross-sectional area of the microchannels. This velocity reduction is beneficial for biotransformation purposes as the contact time of the bionanocomposites with the fluid is likely to be significantly exacerbated.

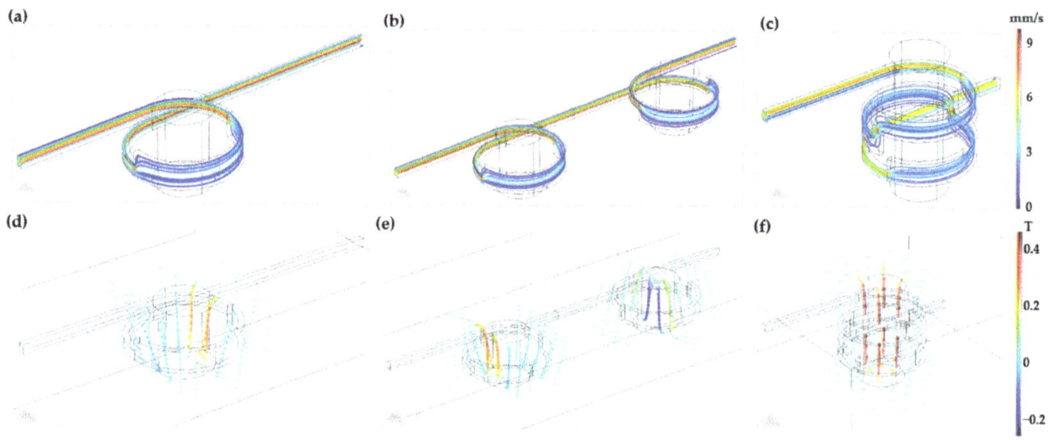

Figure 5. (**a**) Velocity profile of one-loop microreactor. (**b**) Velocity profile of two-horizontal-loop microreactor. (**c**) Velocity profile of two-vertical−loop microreactor. (**d**) Magnetic field flux of one-loop microreactor. (**e**) Magnetic field flux of two-horizontal-loop microreactor. (**f**) Magnetic field flux of two-vertical-loop microreactor.

The magnetic field flux results show a uniform distribution around the magnet for the case of the single loop (Figure 5d). For the two magnets arranged in a horizontal configuration (Figure 5e), there is no interaction between the field lines of the magnets. Finally, the configuration of two-vertical-loop (Figure 5f) showed the greatest field interaction, and consequently superior chances to retain the bionanocomposites within the loops.

The time evolution of the particle's transport within the one-loop microreactor was studied under varying intensities of the applied magnetic field (i.e., 0, 50, 100, 200, 300, 350, 500, and 1000 mT). The simulations were conducted for a total time of 30 s with data collection at 1, 5, 15 and 25 s (See Figure 6). For these simulations, each particle trajectory is random and driven by the exerted drag and magnetophoretic forces. As the particles are attracted to the magnet and due to the interplay of involved forces in the loop, they follow a parabolic trajectory until finally sticking to the wall of the microreactor channel. For some particles, the magnetic field will not be enough to retain them within the loop. These unique trajectories have been exploited by others for the separation and manipulation of magnetic particles within microchannels [42]. In our case, the movement is enough to perturb the laminar flow, thereby generating mixing patterns that are useful to promote the intimate interaction of the nanoparticles with other components present in the solution. This was also evidenced by a better suspension of the nanoparticles as the magnetic field was increased.

Total retention of the particles was achieved for the fields of 500 and 1000 mT, as the particles remain statically attached to the walls of the microreactor. To support this result, the particle's loss and retention ratio were analyzed by counting the particles leaving the system through the microreactor's outlet during the 30 s of simulation (See Table 1). The magnetic field applied experimentally by the magnet was of 349.23 mT, which is close to the one simulated at 350 mT. In this case, the particle's retention approached 96.83% while the loss ratio was 3.17%. The actual particle's loss ratio obtained experimentally was between 13–20%, which was closer to the results obtained in silico at 300 mT where the retention ratio was 82.67%. In this case, the loss ratio approached 17.33%. The reduction in the actual strength of the magnetic field can be explained by the marked difference in the medium surrounding the fluid computationally (i.e., air) and the actual medium (i.e., PMMA walls), which attenuates the applied magnetic field.

Table 1. Retention ratio analysis at the microreactor's outlet after 30 s of simulation.

Remanent Flux Density ((mT))	Particle's Loss Ratio (%)	Particle's Retention Ratio (%)
0	93.67	6.33
50	91.17	8.83
100	85.00	15.00
200	56.00	44.00
300	17.33	82.67
350	3.17	96.83
500	0.00	100.00
1000	0.00	100.00

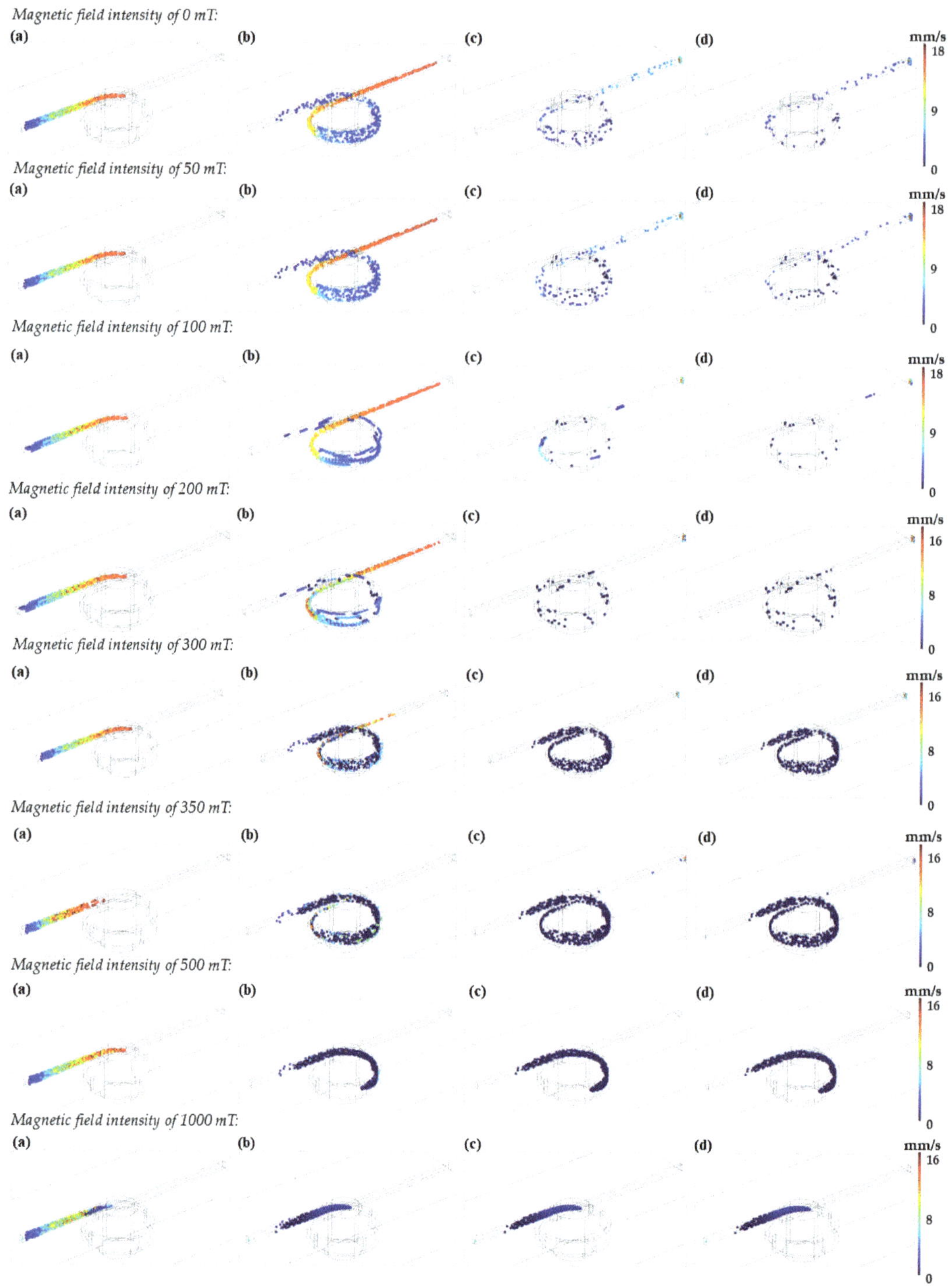

Figure 6. (**a**) Particle distribution in the one-loop microreactor after 1 s. (**b**) Particle distribution in the one-loop microreactor after 5 s. (**c**) Particle distribution in the one-loop microreactor after 15 s. (**d**) Particle distribution in the one-loop microreactor after 25 s.

3.3. Decolorization and Degradation Studies

3.3.1. Experimental Tests for Decolorization of Dyes and Degradation of Phenols

Potential for dye retention on the surface of the microreactor´s walls was estimated by pumping the dye solution into the system in the absence of magnetite. The average retention obtained was 24.3%, 52.1%, and 26.2% for the three dye concentrations evaluated (i.e., 5, 10, and 20 mg/L). Figure 7 shows the average removal of EBt for each device, condition, and concentration evaluated. Data are presented for the two types of analysis implemented in this work, i.e., absorbance peak and area under the absorbance spectrum. For almost all treatments, maximum removal/decolorization was achieved with the laccase-magnetite except for one-loop at 10 mg/L and two-horizontal loop at 20 mg/L, where bare magnetite nanoparticles were more effective in removing the dye. We attributed this superior performance to the adsorption of the dye molecules on the surface of the nanoparticles. The maximum removal of dye was 98.74% (4.9 mg/g dye/laccase-magnetite) for the two-vertical-loop microreactor. The average removals for laccase-magnetite systems based on absorbance (545 nm) (Figure 7a) were 91.08% (10.5 mg/g dye/laccase-magnetite), 74.38% (6.9 mg/g dye/laccase-magnetite) and 96.21% (11.1 mg/g dye/laccase-magnetite) for the one-loop, two-horizontal-loop, and two-vertical-loop microreactors, respectively. The dye removals for the same treatments but with quantification based on the area under the absorbance spectrum (Figure 7b) approached 85.62% (10.0 mg/g dye/laccase-magnetite), 62.44% (4.7 mg/g dye/laccase-magnetite), and 93.56% (10.7 mg/g dye/laccase-magnetite) for the three microreactors. Figure 8 shows a TEM micrograph of the magnetite, describing a size diameter ranging from 10 to 20 nm approximately, suggesting that the hydrodynamic diameter calculated via DLS is based on the agglomeration of nanoparticles in solution. In addition, Figure 8b,c show pictures of the operation of the one-loop microreactor during the Ebt removal process. Taken together, these results strongly suggest that the microreactor with the best performance in terms of the dye removal was the two-vertical-loop system. Additionally, in the case of the laccase-magnetite treatment, the overall (absorbance peak and area under the absorbance spectrum) removal efficiencies for this microreactor approached 98.05% (4.9 mg/g dye/laccase-magnetite), 93.87% (9.3 mg/g), and 92.74% (18.5 mg/g) for the low, medium, and high dye concentrations (5, 10, and 20 mg/L). The initial ratio between mass of laccase and dye for each experiment, corresponded to 2.6, 1.3 and 0.65 μg/mg, and the ratio between activity of laccase and dye mass was 0.00192, 0.00096 and 0.00048 U/mg, for the dye concentrations mentioned above.

Phenol removal was evaluated using the two-vertical-loop system (best performing microreactor from the three evaluated), at three initial phenol concentrations (5, 10 and 20 mg/L). Data (Figure 7c,d) are presented for the two types of analysis implemented in this work i.e., absorbance peak and area under the absorbance spectrum. The maximum phenol biotransformation was approximately 75% (3.75 mg/g phenol/laccase-magnetite) for the lowest phenol concentration, followed by approximately 50% (10 mg/g) at the highest concentration, and approximately 35% (3.5 mg/g) at the middle concentration. The initial ratio between mass of laccase and phenol for each experiment corresponded to 2.6, 1.3, and 0.65 μg/mg, and the ratio between activity of laccase and phenol mass was 0.00192, 0.00096, and 0.00048 U/mg, for the phenol concentrations mentioned above. Further research is needed to identify the potential intermediates that result from the transformation of the dye or phenol, their potential toxicity, and the potential need to include additional systems carrying other enzymes to complete the biodegradation process.

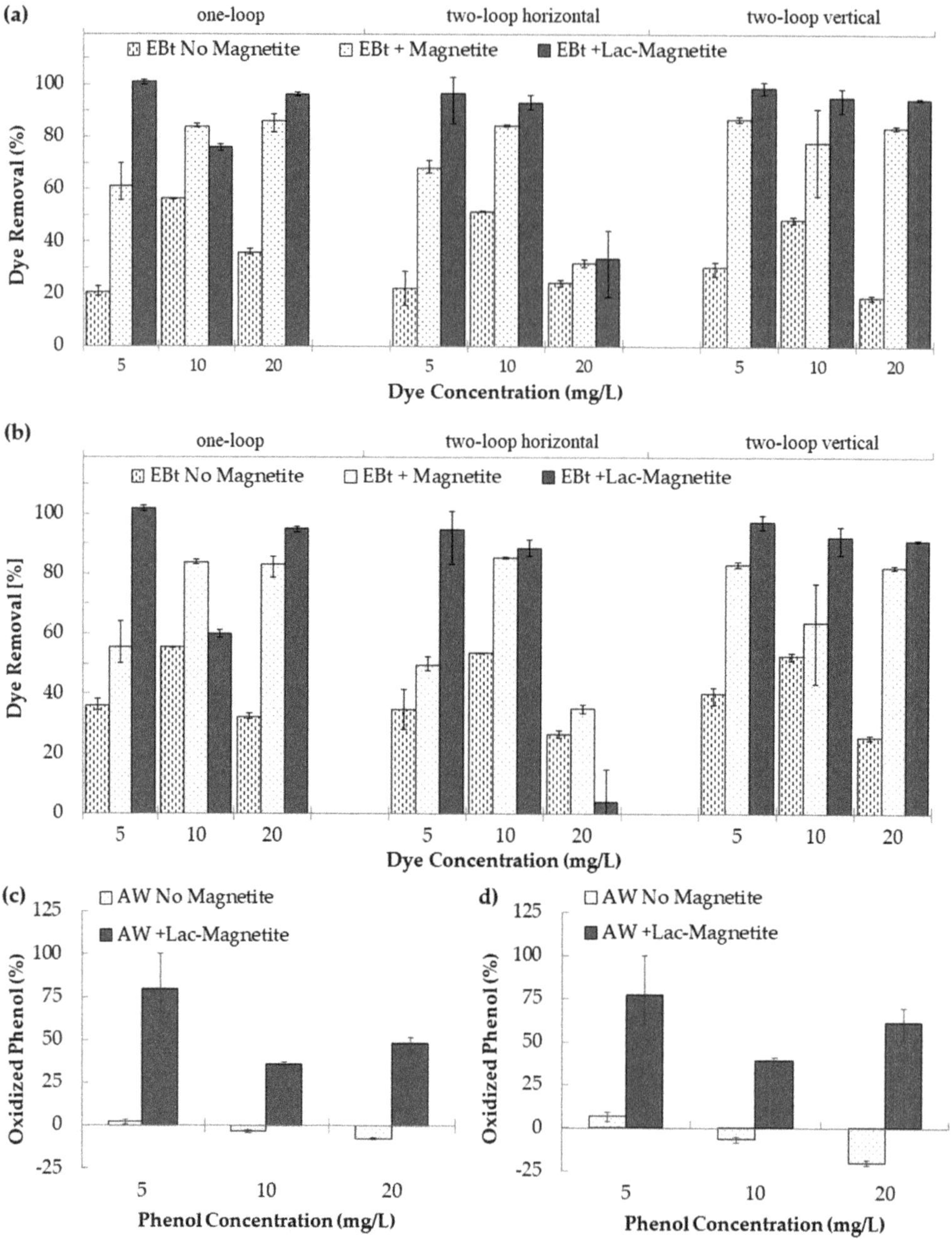

Figure 7. (**a**) Removal of Eriochrome Black T for the three microreactor configurations as estimated by absorbance peak. (**b**) Removal of Eriochrome Black T for the three microreactor configurations as estimated by absorbance area. (**c**) Percentage of oxidized phenol for the two-vertical-loop microreactor as estimated by absorbance peak. (**d**) Percentage of oxidized phenol for the two-vertical-loop microreactor as estimated by absorbance area.

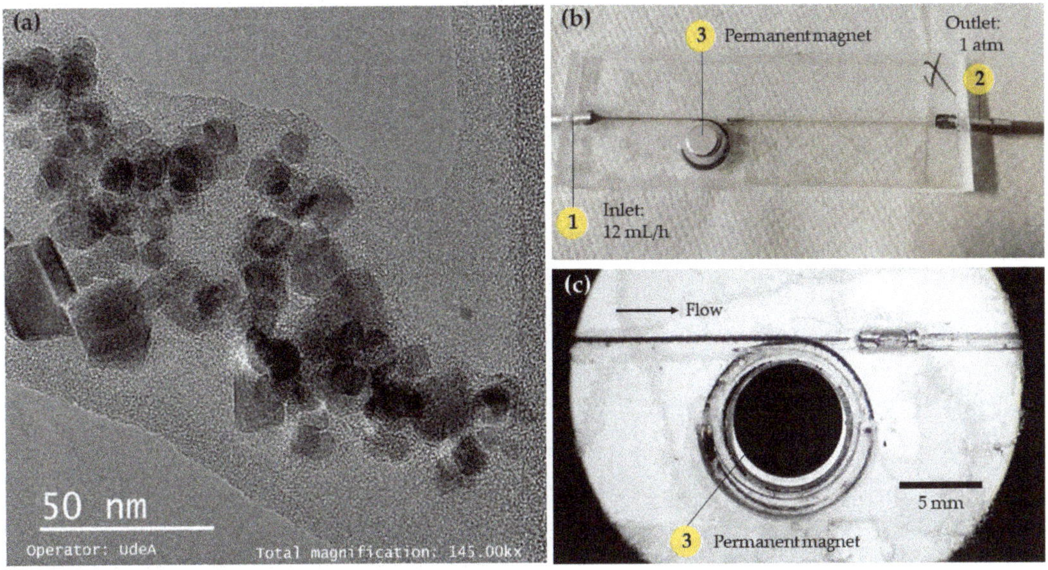

Figure 8. (**a**) Transmission electron microscopy micrograph of magnetite nanoparticles. (**b**) Picture of the one-loop microreactor during Eriochrome Black T dye treatment. (**c**) Picture of the one-loop microreactor during Eriochrome Black T dye treatment under optical microscope (10X magnification).

3.3.2. Treatment of Real Wastewater

Phenol containing real wastewater (1 L) was treated through the two-vertical-loop microreactor in the presence of laccase-magnetite. Table 2 shows the measured water quality parameters before (U-RW) and after (T-RW) treatment. The removal of phenol reached 17.86% (700 mg/g phenol/Laccase-Magnetite), achieving a final concentration of 16.10 mg/L, which is close to but still greater than the Drinking Water Equivalent Level maximum (DWEL) in the U.S. i.e., of 11 mg/L [43] as imposed by the United States Environmental Protection Agency (EPA) regulations. Relatively low oxidation of phenol was expected due to the presence of very high concentrations of biodegradable organic matter (BOD) in real wastewater and its simultaneous oxidation by the laccase-magnetite system. It also needs to be considered that the initial ratio between mass of laccase and phenol for this experiment was only 0.00325 μg/mg, and the ratio between activity of laccase and phenol mass was 0.00000024 U/mg, representing 800-fold less than the ratio used for the dye and artificial wastewater experiments at the lower concentrations, or 200-fold reduction for the highest dye or artificial wastewater phenol concentration.

Recent studies have evaluated systems for treating large volumes of phenol-contaminated wastewater. These reports indicate that the most successful phenol remediation strategies for real wastewater rely on batch biodegradation processes [24]. For instance, Garg et al. reported the use of low-purity peroxidases extracts for the removal of phenol from textile industry wastewater in a batch process with an efficiency of 94.95 $\pm$ 0.82%. Additionally, the same process led to 91.49 $\pm$ 1.54% removal efficiency in leather industry wastewater [44]. Hayati et al. reported a removal ratio of 85.9% (TOC) and 92.1% (COD), with GZnTi nanocomposites [45]. However, studies that evaluate the transformation of phenol in continuous systems are rather scarce. Therefore, the observed reductions of phenol in our continuous system are promising and encourage us to continue exploring further improvements to the microreactors in continuous flow for future contributions. Moreover, to our knowledge, no other microreactor system has been identified in the literature capable of continuous operation for phenol-contaminated water treatment that treats large volumes of real wastewater and that also offers the possibility for in-line and real-time monitoring.

Table 2. Characterization of the large volume samples of real wastewater treated with the developed bionanocomposites.

Parameter	Sample			Removal Ratio (%)	Sorption Capacity (mg/g)
	U-RW	T-RW	Units		
Phenols	19.60	16.10	mg Phenol/L	17.86	700
Biochemical oxygen demand (BOD)	385.00	345.00	mg O_2/L	10.39	8000
NH_3 nitrogen	0.60	0.50	mg N/L	16.67	20
Kjeldahl nitrogen	12.40	9.90	mg N/L	20.16	500

As there are amine groups present in the molecules used as linkers to complete the immobilization process, the nitrogen content becomes a parameter of interest. According to the results of two different techniques, i.e., NH_3-nitrogen and Kjeldahl nitrogen, removal reached 16.67% and 20.16%, respectively, indicating that there was organic nitrogen oxidation due to laccase, but also that the nitrogen present in the linkers of the immobilization method were not released or lost during the treatment. This indicates that the functionalization of the laccase-magnetite was successful. Additionally, as the emission limit value (ELV) for nitrogen in wastewater by the regulation presented by the United States Environmental Protection Agency (EPA), is 10 mg/L and 15 mg/L [46], for NH_3-nitrogen and Kjeldahl nitrogen, respectively; the treatment is likely helping in the nitrogen removal (final concentrations of 0.50 and 9.90 mg/L) as well. In the case of the biochemical oxygen demand (BOD), the removal was 10.39%, which allowed us to conclude that the treatment might this system can aid with the oxidation of BOD in real wastewater.

The initial phenol concentration of the real wastewater sample was 19.60 mg/L, which is close to the highest concentration evaluated in the case of biodegradation of dyes with model wastewaters. Additionally, the experiments with real wastewaters were also conducted in the presence of 5 mg of laccase-magnetite, but with a larger volume of wastewater than those carried out using dyes and phenol. Consequently, an equivalent number of cycles to treat the total volume of sample in each case can be estimated to be roughly 200.

In general, our results showed that our continuous flow treatment presented a maximum phenol oxidation of about 75% using nanocomposites, while other authors, as Hayati et al. [45], have reached phenol degradation, also using nanoparticles, between 90 and 100%. This higher phenol oxidation [45] was obtained in the presence of photocatalytic effects, and under non-continuous operation. The differences in the results might also be related to a longer reaction time and also the differences in the efficiency of a photocatalytic process, which requires also more external energy, compared to an enzymatic one. Abdollahi et al. [47] showed phenol degradation of 100% using nanoparticles and biocatalysis, a more related oxidation method to ours, however; it took between 4 and 12 h to reach those oxidation rates. Phenol oxidation was about 70 to 90% after 1 h, an oxidation rate similar to that obtained in our continuous process, which means that the efficiency might be highly related to the reaction time. Their results [47] also showed that their method lost 50% of the efficiency after 7 cycles, while our experiments with real wastewater, which represented roughly 200 cycles, reported lost 25% of the initial efficiency. Unfortunately, to our knowledge, there are no reports of microfluidic systems tested at such high number of cycles.

4. Conclusions

The removal of contaminants from industrial effluents remains a major challenge, thus, we aimed to evaluate laccase enzymes immobilized on magnetite nanoparticles (i.e., bio-nanocomposites) by controlling their interactions with real and synthetic wastewaters with the aid of low-cost microfluidic toroidal microreactors equipped with loops to accommodate permanent magnets that take advantage of the strong magnetic response of magnetite to control the residence time of laccase based bionanocomposites within the device without inducing detrimental perturbations. Multiphysics simulations indicated that the movement of the particles induced mixing patterns that can increase the interactions between the wastewater and the bionanocomposite. This is an improvement over large-scale torus reactors where mixing requires a mechanical stirring and significant energy consumption. The microreactor with the best overall performance (i.e., longest residence time and best interaction) was the two vertical loop system. This was corroborated experimentally by evaluating the biotransformation dyes in artificial wastewater, reaching efficiencies above 90%, and phenol removal from synthetic wastewater reaching efficiencies of about 80%. Real wastewater tests showed a decrease in phenol removal efficiency to about 18%, but using the same amount of bionanocomposites but with 200-fold the volume of the treated wastewater, representing the equivalent of 200 biodegradation operating cycles. Then, it would be worth exploring the idea of expanding the process for newer and more sustainable enzyme-based biodegradation routes enabled by microreactors [48]. Next steps should focus on exploring strategies for sustained and continuous operation, including recirculation and parallelization processing schemes. We also believe that it is worth exploring other microreactor geometries where a dynamic array of magnets can be tuned to maximize magnetic gradient and other geometry mixing approaches. Finally, based on the superior performance, we are exploring scaling-up routes and processing schemes and whether with such approaches key environmental indexes are positively impacted, while maintaining profitability.

Author Contributions: Conceptualization, P.A.P., M.J.N., J.F.O. and J.C.C.; methodology, P.A.P., M.J.N., J.F.O. and J.C.C.; software, S.L.F. and J.C.C.; validation, M.J.N.,P.A.P., J.F.O. and J.C.C.; formal analysis, M.J.N., P.A.P. and J.F.O.; investigation, M.J.N. and P.A.P.; resources, J.F.O., N.O.-S., and J.H.; data curation, M.J.N., P.A.P. and J.F.O.; writing—original draft preparation, P.A.P., M.J.N. and S.L.F.; writing—review and editing, M.J.N.,P.A.P., N.O.-S., J.H., J.C.C. and J.F.O.; visualization, M.J.N., P.A.P., S.L.F. and J.F.O.; supervision, J.F.O. and J.C.C.; project administration, J.F.O. All authors have read and agreed to the published version of the manuscript.

Funding: This research received no external funding.

Institutional Review Board Statement: Not applicable.

Informed Consent Statement: Not applicable.

Data Availability Statement: The data and contributions presented in the study are included in the article. Further inquiries can be directed to the corresponding author.

Acknowledgments: The authors thank the Clean Room laboratory of the Department of Electrical and Electronic Engineering, the Department of Biomedical Engineering and the Department of Civil and Environmental Engineering at Universidad de los Andes.

Conflicts of Interest: The authors declare no conflict of interest.

References

1. Hassaan, M.A.; El Nemr, A.; Hassaan, A. Health and Environmental Impacts of Dyes: Mini Review. *Am. J. Environ. Sci. Eng.* **2017**, *1*, 64–67. [CrossRef]
2. Atieh, M.A. Removal of Phenol from Water Different Types of Carbon—A Comparative Analysis. *APCBEE Procedia* **2014**, *10*, 136–141. [CrossRef]
3. Anku, W.W.; Mamo, M.A.; Govender, P.P. Phenolic Compounds in Water: Sources, Reactivity, Toxicity and Treatment Methods. In *Phenolic Compounds—Natural Sources, Importance and Applications*; Intech Open: London, UK, 2017.

4. Helmer, R.; Hespanhol, I. *Water Pollution Control—A Guide to the Use of Water Quality Management Principles*; CRC Press: Boca Raton, FL, USA, 1997.
5. Dabhade, M.A.; Saidutta, M.B.; Murthy, D.V.R. Adsorption of phenol on granular activated carbon from nutrient medium: Equilibrium and kinetic study. *Int. J. Environ. Res.* **2009**, *3*, 557–568.
6. Zou, H.; Wang, Y. Azo dyes wastewater treatment and simultaneous electricity generation in a novel process of electrolysis cell combined with microbial fuel cell. *Bioresour. Technol.* **2017**, *235*, 167–175. [CrossRef] [PubMed]
7. Kanagaraj, J.; Senthilvelan, T.; Panda, R.C. Degradation of azo dyes by laccase: Biological method to reduce pollution load in dye wastewater. *Clean Technol. Environ. Policy* **2014**, *17*, 1443–1456. [CrossRef]
8. Brüschweiler, B.J.; Merlot, C. Azo dyes in clothing textiles can be cleaved into a series of mutagenic aromatic amines which are not regulated yet. *Regul. Toxicol. Pharmacol.* **2017**, *88*, 214–226. [CrossRef] [PubMed]
9. Abadulla, E.; Tzanov, T.; Costa, S.; Robra, K.-H.; Cavaco-Paulo, A.; Guübitz, G.M. Decolorization and Detoxification of Textile Dyes with a Laccase from *Trametes hirsuta*. *Appl. Environ. Microbiol.* **2000**, *66*, 3357–3362. [CrossRef] [PubMed]
10. van der Zee, F.P.; Villaverde, S. Combined anaerobic–aerobic treatment of azo dyes—A short review of bioreactor studies. *Water Res.* **2005**, *39*, 1425–1440. [CrossRef]
11. Moilanen, U.; Osma, J.F.; Winquist, E.; Leisola, M.; Couto, S.R. Decolorization of simulated textile dye baths by crude laccases from Trametes hirsuta and Cerrena unicolor. *Eng. Life Sci.* **2010**, *10*, 242–247. [CrossRef]
12. Rodríguez-Couto, S.; Osma, J.F.; Toca-Herrera, J.L. Removal of synthetic dyes by an eco-friendly strategy. *Eng. Life Sci.* **2009**, *9*, 116–123. [CrossRef]
13. Holkar, C.R.; Jadhav, A.J.; Pinjari, D.V.; Mahamuni, N.M.; Pandit, A.B. A critical review on textile wastewater treatments: Possible approaches. *J. Environ. Manag.* **2016**, *182*, 351–366. [CrossRef] [PubMed]
14. Osma, J.F. *Production of Laccases by the White-Rot Fungus Trametes Pubescens for Their Potential Application to Synthetic Dye Treatment*; Universitat Rovira I Virgili: Tarragona, Spain, 2009.
15. Thurston, C.F. The structure and function of fungal laccases. *Microbiology* **1994**, *140*, 19–26. [CrossRef]
16. Zhang, D.; Deng, M.; Cao, H.; Zhang, S.; Zhao, H. Laccase immobilized on magnetic nanoparticles by dopamine polymerization for 4-chlorophenol removal. *Green Energy Environ.* **2017**, *2*, 393–400. [CrossRef]
17. Dong, Z.; Liu, Z.; Shi, J.; Tang, H.; Xiang, X.; Huang, F.; Zheng, M.-M. Carbon Nanoparticle-Stabilized Pickering Emulsion as a Sustainable and High-Performance Interfacial Catalysis Platform for Enzymatic Esterification/Transesterification. *ACS Sustain. Chem. Eng.* **2019**, *7*, 7619–7629. [CrossRef]
18. Huang, W.-C.; Wang, W.; Xue, C.; Mao, X. Effective Enzyme Immobilization onto a Magnetic Chitin Nanofiber Composite. *ACS Sustain. Chem. Eng.* **2018**, *6*, 8118–8124. [CrossRef]
19. Sheldon, R.A.; van Pelt, S. Enzyme immobilisation in biocatalysis: Why, what and how. *Chem. Soc. Rev.* **2013**, *42*, 6223–6235. [CrossRef]
20. Champagne, P.-P.; Ramsay, J. Dye decolorization and detoxification by laccase immobilized on porous glass beads. *Bioresour. Technol.* **2010**, *101*, 2230–2235. [CrossRef]
21. Mohajershojaei, K.; Mahmoodi, N.M.; Khosravi, A. Immobilization of laccase enzyme onto titania nanoparticle and decolorization of dyes from single and binary systems. *Biotechnol. Bioprocess Eng.* **2015**, *20*, 109–116. [CrossRef]
22. Lin, J.; Liu, Y.; Chen, S.; Le, X.; Zhou, X.; Zhao, Z.; Ou, Y.; Yang, J. Reversible immobilization of laccase onto metal-ion-chelated magnetic microspheres for bisphenol A removal. *Int. J. Biol. Macromol.* **2016**, *84*, 189–199. [CrossRef]
23. Jiang, Y.; Cui, C.; Zhou, L.; He, Y.; Gao, J. Preparation and Characterization of Porous Horseradish Peroxidase Microspheres for the Removal of Phenolic Compound and Dye. *Ind. Eng. Chem. Res.* **2014**, *53*, 7591–7597. [CrossRef]
24. Deng, Y.; Qi, D.; Deng, C.; Zhang, A.X.; Zhao, D. Superparamagnetic High-Magnetization Microspheres with an Fe_3O_4@SiO_2 Core and Perpendicularly Aligned Mesoporous SiO_2 Shell for Removal of Microcystins. *J. Am. Chem. Soc.* **2007**, *130*, 28–29. [CrossRef] [PubMed]
25. Salado, J.; Insausti, M.; Lezama, L.; Gil de Muro, I.; Goikolea, E.; Rojo, T. Preparation and Characterization of Monodisperse Fe3O4 Nanoparticles: An Electron Magnetic Resonance Study. *Chem. Mater.* **2011**, *23*, 2879–2885. [CrossRef]
26. Netto, C.; Toma, H.E.; Andrade, L.H. Superparamagnetic nanoparticles as versatile carriers and supporting materials for enzymes. *J. Mol. Catal. B Enzym.* **2013**, *85-86*, 71–92. [CrossRef]
27. Xia, T.-T.; Liu, C.-Z.; Hu, J.-H.; Guo, C. Improved performance of immobilized laccase on amine-functioned magnetic Fe_3O_4 nanoparticles modified with polyethylenimine. *Chem. Eng. J.* **2016**, *295*, 201–206. [CrossRef]
28. Rossi, L.M.; Costa, N.J.S.; Silva, F.P.; Wojcieszak, R. Magnetic nanomaterials in catalysis: Advanced catalysts for magnetic separation and beyond. *Green Chem.* **2014**, *16*, 2906–2933. [CrossRef]
29. Zhou, W.; Zhang, W.; Cai, Y. Laccase immobilization for water purification: A comprehensive review. *Chem. Eng. J.* **2020**, *403*, 126272. [CrossRef]
30. Darwesh, O.M.; Matter, I.A.; Eida, M.F. Development of peroxidase enzyme immobilized magnetic nanoparticles for bioremediation of textile wastewater dye. *J. Environ. Chem. Eng.* **2018**, *7*, 102805. [CrossRef]
31. Zhu, Y.; Qiu, F.; Rong, J.; Zhang, T.; Mao, K.; Yang, D. Covalent laccase immobilization on the surface of poly(vinylidene fluoride) polymer membrane for enhanced biocatalytic removal of dyes pollutants from aqueous environment. *Colloids Surfaces B Biointerfaces* **2020**, *191*, 111025. [CrossRef]
32. Convery, N.; Gadegaard, N. 30 years of microfluidics. *Micro Nano Eng.* **2019**, *2*, 76–91. [CrossRef]

33. Grimmer, A.; Chen, X.; Hamidović, M.; Haselmayr, W.; Ren, C.L.; Wille, R. Simulation before fabrication: A case study on the utilization of simulators for the design of droplet microfluidic networks. *RSC Adv.* **2018**, *8*, 34733–34742. [CrossRef]
34. Han, W.; Chen, X.; Wu, Z.; Zheng, Y. Three-dimensional numerical simulation of droplet formation in a microfluidic flow-focusing device. *J. Braz. Soc. Mech. Sci. Eng.* **2019**, *41*, 265. [CrossRef]
35. Pramparo, L.; Stüber, F.; Font, J.; Fortuny, A.; Fabregat, A.; Bengoa, C. Immobilisation of horseradish peroxidase on Eupergit®C for the enzymatic elimination of phenol. *J. Hazard. Mater.* **2010**, *177*, 990–1000. [CrossRef] [PubMed]
36. Ramírez-Cavazos, L.I.; Junghanns, C.; Ornelas-Soto, N.; Cárdenas-Chávez, D.L.; Luna, C.E.H.; Demarche, P.; Enaud, E.; García-Morales, R.; Agathos, S.; Parra, R. Purification and characterization of two thermostable laccases from Pycnoporus sanguineus and potential role in degradation of endocrine disrupting chemicals. *J. Mol. Catal. B Enzym.* **2014**, *108*, 32–42. [CrossRef]
37. Lopez-Barbosa, N.; Suárez-Arnedo, A.; Cifuentes, J.; Barrios, A.F.G.; Batista, C.A.S.; Osma, J.F.; Muñoz-Camargo, C.; Cruz, J.C. Magnetite–OmpA Nanobioconjugates as Cell-Penetrating Vehicles with Endosomal Escape Abilities. *ACS Biomater. Sci. Eng.* **2019**, *6*, 415–424. [CrossRef] [PubMed]
38. Schenk, O.; Gärtner, K.; Fichtner, W.; Stricker, A. PARDISO: A high-performance serial and parallel sparse linear solver in semiconductor device simulation. *Futur. Gener. Comput. Syst.* **2001**, *18*, 69–78. [CrossRef]
39. Guillén, A.; Ardila, Y.; Noguera, M.J.; Campaña, A.L.; Bejarano, M.; Akle, V.; Osma, J.F. Toxicity of Modified Magnetite-Based Nanocomposites Used for Wastewater Treatment and Evaluated on Zebrafish (*Danio rerio*) Model. *Nanomaterials* **2022**, *12*, 489. [CrossRef] [PubMed]
40. Nor, W.F.K.W.; Soh, S.K.C.; Azmi, A.A.A.R.; Yusof, M.S.M.; Shamsuddin, M. Synthesis and physicochemical properties of magnetite nanoparticles (fe3o4) as potential solid support for homogeneous catalysts. *Malays. J. Anal. Sci.* **2018**, *22*, 768–774. [CrossRef]
41. Echen, L.; Ezou, M.; Hong, F.F. Evaluation of Fungal Laccase Immobilized on Natural Nanostructured Bacterial Cellulose. *Front. Microbiol.* **2015**, *6*, 1245. [CrossRef]
42. Xuan, X. Recent Advances in Continuous-Flow Particle Manipulations Using Magnetic Fluids. *Micromachines* **2019**, *10*, 744. [CrossRef] [PubMed]
43. Agency for Toxic Substances and Disease Registry. Toxicological Profile for Phenol. In *ATSDR's Toxicological Profiles, Atlanta*; CRC Press: Boca Raton, FL, USA, 2020; pp. 1–5.
44. Garg, S.; Kumar, P.; Singh, S.; Yadav, A.; Dumée, L.F.; Sharma, R.S.; Mishra, V. Prosopis juliflora peroxidases for phenol remediation from industrial wastewater—An innovative practice for environmental sustainability. *Environ. Technol. Innov.* **2020**, *19*, 100865. [CrossRef]
45. Hayati, F.; Isari, A.A.; Fattahi, M.; Anvaripour, B.; Jorfi, S. Photocatalytic decontamination of phenol and petrochemical wastewater through ZnO/TiO_2 decorated on reduced graphene oxide nanocomposite: Influential operating factors, mechanism, and electrical energy consumption. *RSC Adv.* **2018**, *8*, 40035–40053. [CrossRef]
46. Uisce Eireann Irish Water. *Courtmacsherry-Timoleague Waste Water Discharge Licence—Technical Amendment Application D0294-0*; Irish Water's Environmental Information System: Dublin, Ireland, 2018.
47. Abdollahi, K.; Yazdani, F.; Panahi, R.; Mokhtarani, B. Biotransformation of phenol in synthetic wastewater using the functionalized magnetic nano-biocatalyst particles carrying tyrosinase. *3 Biotech* **2018**, *8*, 419. [CrossRef] [PubMed]
48. Fuentes, O.P.; Noguera, M.J.; Peñaranda, P.A.; Flores, S.L.; Cruz, J.C.; Osma, J.F. Micromixers for Wastewater Treatment and Their Life Cycle Assessment (LCA). In *Advances in Microfluidics and Nanofluids*; Intech Open: London, UK, 2021.

nanomaterials

MDPI

Article

Sustainable Synthesis of Cadmium Sulfide, with Applicability in Photocatalysis, Hydrogen Production, and as an Antibacterial Agent, Using Two Mechanochemical Protocols

Zhandos Shalabayev [1,2,*], Matej Baláž [3], Natalya Khan [1], Yelmira Nurlan [1], Adrian Augustyniak [4,5], Nina Daneu [6], Batukhan Tatykayev [1], Erika Dutková [3], Gairat Burashev [1], Mariano Casas-Luna [7,8], Róbert Džunda [9], Radovan Bureš [9], Ladislav Čelko [7], Aleksandr Ilin [2] and Mukhambetkali Burkitbayev [1]

1 General and Inorganic Chemistry Department, Al-Farabi Kazakh National University, Al-Farabi Ave. 71, Almaty 050040, Kazakhstan; natalya.khan@kaznu.edu.kz (N.K.); nurlan_yelmira1@kaznu.edu.kz (Y.N.); batukhan.tatykaev@kaznu.kz (B.T.); kairat.burashev@mail.ru (G.B.); mukhambetkali.burkitbayev@kaznu.edu.kz (M.B.)
2 Scientific Center for Anti-Infectious Drugs, Al-Farabi Ave. 75B, Almaty 050060, Kazakhstan; ilin_ai@mail.ru
3 Institute of Geotechnics, Slovak Academy of Sciences, Watsonova 45, 04001 Košice, Slovakia; balazm@saske.sk (M.B.); dutkova@saske.sk (E.D.)
4 Chair of Building Materials and Construction Chemistry, Technische Universität Berlin, Gustav-Meyer-Allee 25, 13355 Berlin, Germany; adrian.augustyniak@zut.edu.pl
5 Faculty of Chemical Technology and Engineering, West Pomeranian University of Technology in Szczecin, Piastów Ave. 42, 71-065 Szczecin, Poland
6 Jožef Stefan Institute, Jamova Cesta 39, 01000 Ljubljana, Slovenia; nina.daneu@ijs.si
7 Central European Institute of Technology, Brno University of Technology, Purkynova 123, 612 00 Brno, Czech Republic; casas@vutbr.cz (M.C.-L.); ladislav.celko@ceitec.vcutbr.cz (L.Č.)
8 Department of Physics of Materials, Charles University, 121 16 Prague, Czech Republic
9 Institute of Materials Research, Slovak Academy of Sciences, Watsonova 47, 04001 Košice, Slovakia; rdzunda@saske.sk (R.D.); rbures@saske.sk (R.B.)
* Correspondence: zhandos.shalabayev@kaznu.edu.kz; Tel.: +7-707-793-17-65

Citation: Shalabayev, Z.; Baláž, M.; Khan, N.; Nurlan, Y.; Augustyniak, A.; Daneu, N.; Tatykayev, B.; Dutková, E.; Burashev, G.; Casas-Luna, M.; et al. Sustainable Synthesis of Cadmium Sulfide, with Applicability in Photocatalysis, Hydrogen Production, and as an Antibacterial Agent, Using Two Mechanochemical Protocols. *Nanomaterials* **2022**, *12*, 1250. https://doi.org/10.3390/nano12081250

Academic Editor: Christos A. Aggelopoulos

Received: 17 February 2022
Accepted: 2 April 2022
Published: 7 April 2022

Abstract: CdS nanoparticles were successfully synthesized using cadmium acetate and sodium sulfide as Cd and S precursors, respectively. The effect of using sodium thiosulfate as an additional sulfur precursor was also investigated (combined milling). The samples were characterized by XRD, Raman spectroscopy, XPS, UV-Vis spectroscopy, PL spectroscopy, DLS, and TEM. Photocatalytic activities of both CdS samples were compared. The photocatalytic activity of CdS, which is produced by combined milling, was superior to that of CdS, and was obtained by an acetate route in the degradation of Orange II under visible light irradiation. Better results for CdS prepared using a combined approach were also evidenced in photocatalytic experiments on hydrogen generation. The antibacterial potential of mechanochemically prepared CdS nanocrystals was also tested on reference strains of *E. coli* and *S. aureus*. Susceptibility tests included a 24-h toxicity test, a disk diffusion assay, and respiration monitoring. Bacterial growth was not completely inhibited by the presence of neither nanomaterial in the growth environment. However, the experiments have confirmed that the nanoparticles have some capability to inhibit bacterial growth during the logarithmic growth phase, with a more substantial effect coming from CdS nanoparticles prepared in the absence of sodium thiosulfate. The present research demonstrated the solvent-free, facile, and sustainable character of mechanochemical synthesis to produce semiconductor nanocrystals with multidisciplinary application.

Keywords: mechanosynthesis; combined milling; semiconductor; photocatalysis; hydrogen evolution; antibacterial activity; wastewater treatment

1. Introduction

Recently, due to increasing environmental pollution, the requirements for searching for renewable energy sources to create environmentally friendly processes for decontaminating the environment are growing every day. Among all types of pollution, the most important is the discharge of wastewater from industrial enterprises [1]. The use of organic contaminants such as pesticides, herbicides, solvents, organic dyes, pharmaceuticals, cosmetics, etc., poses many problems and concerns for the scientific community and environmental regulators around the world [2]. To tackle this problem, photocatalysis is the best option among all methods, according to its simplicity and non-toxicity [3]. Consequently, the development of photocatalysts with enhanced activities for purifying wastewater from organic pollutants is hot now [4].

There are many types of photocatalysts, including metal oxides and ending metal sulfides [5,6]. Among them, photocatalysts based on CdS nanoparticles are of huge interest of scientists due to their wide bandgap energy (2.42 eV) [7] and reusability in photocatalytic processes [8]. In addition to photocatalysis, cadmium sulfide nanoparticles can be used in various fields of industries such as medicine [9], hydrogen production [10], photovoltaic devices [11], optoelectronics [12], diodes [13], gas sensors [14], etc. For example, in medicine, CdS nanoparticles reveal antibacterial [15], antimicrobial [16], and antifungal [17] activities against micro-organisms.

Recently, many methods have been developed to produce CdS nanoparticles such as microwave-assisted co-precipitation [18], the wet chemical method [19], microwave irradiation [20], reverse micellar methods [21], electrochemical deposition [22], the Langmuir–Blodgett method [23], solvothermal [24] and hydrothermal [25] processes, green synthesis [26], gamma irradiation [27], biological methods [28], and mechanochemical methods [29,30].

The last one is of particular interest in connection with the rapid supplementing of the high amount of mechanical energy to activate the chemical reaction between the reagents through the high-energy milling process [31]. This method is also environmentally friendly, since no harmful organic solvents are used [32]. Mechanochemistry is a universal tool for the synthesis of new substances, and it can be utilized in organic chemistry, supramolecular chemistry, organometallic compounds, polymers, inorganic chemistry [33], and even for waste treatment [34].

Until now, several papers reported on the synthesis of CdS nanoparticles by the mechanochemical approach. Acetate-route-based mechanosynthesis of nanocrystalline metal sulfides, in particular CdS nanoparticles, was proposed for the first time by the Slovak mechanochemists [35,36]. In this approach, the corresponding acetates and sodium sulfide were used as precursors. Later, using this approach, CdS nanoparticles with average sizes of 4–18 nm were produced in an industrial mill at ambient temperatures, in a very short processing time [29]. CdS nanoparticles also can be mechanochemically synthesized from elementary precursors. In the work [37], CdS nanoparticles with an average crystallite size of 8 nm were obtained.

In our previous work, needle-like copper sulfide (nCuS) nanocrystals have been synthesized, introducing in situ-prepared sulfur particles (using $Na_2S_2O_3$ as an additional sulfur source and crystalline acid as a catalyst) to the mechanochemical reaction media of copper acetate monohydrate and sodium sulfide nonahydrate. Produced needle-like CuS nanocrystals showed less selective antibacterial activity in comparison with spherical ones [38]. In this work, we applied a similar approach for cadmium sulfide, with the aim to improve its application potential in photocatalytic degradation of organic dyes and antibacterial treatment. The novelty of this work is in the use of additional sulfur sources, rapid mechanochemical synthesis, and enhanced photocatalytic activity of CdS samples.

2. Materials and Methods

2.1. Chemicals

Sodium thiosulfate ($Na_2S_2O_3 \cdot 5H_2O$, Sigma-Aldrich, Taufkirchen, Germany), citric acid ($C_6H_8O_7$, Sigma-Aldrich, Taufkirchen, Germany), cadmium acetate ($Cd(CH_3COO)_2 \cdot 2H_2O$, Sigma-Aldrich, Taufkirchen, Germany), sodium sulfide ($Na_2S \cdot 9H_2O$, Acros Organics, Geel, Belgium), and Orange II sodium salt ($C_{16}H_{11}N_2NaO_4S$, Sigma-Aldrich, Taufkirchen, Germany) were of analytical grade and used without further purification. Distilled and deionized water was used in all washing and cleaning procedures.

2.2. Mechanochemical Synthesis of CdS Nanoparticles

In a typical mechanochemical synthesis procedure, the reactants (in an overall mass of 1 g) in stoichiometric amounts were transferred into a 100 mL silicon nitride (Si_3N_4) milling chamber containing 20 silicon nitride balls (10 mm), and they were milled for 5 min using an Activator 2SL planetary ball mill (Activator, Novosibirsk, Russia). The rotational speed of milling chambers was 550 rpm, the ball-to-powder ratio was 37, and all milling procedures were carried out under an air atmosphere.

CdS nanoparticles have been synthesized using two mechanochemical approaches: acetate route (aCdS) [35] and combined mechanochemical synthesis (cCdS), where sulfur prepared in situ should enter the formation of cadmium sulfide [38].

For the mechanochemical synthesis of aCdS nanoparticles by a simple acetate route, 0.526 g of $Cd(CH_3COO)_2 \cdot 2H_2O$ and 0.474 g of $Na_2S \cdot 9H_2O$ was milled in a stoichiometric amount according to the chemical reaction reported in [35]:

$$Cd(CH_3COO)_2 \cdot 2H_2O + Na_2S \cdot 9H_2O \rightarrow CdS + 2CH_3COONa + 11H_2O \quad (1)$$

cCdS nanocrystals were prepared by the combined mechanosynthesis approach via co-milling reactants, mirroring the procedure described in ref. [38]. In a typical procedure, 0.4328 g of $Cd(CH_3COO)_2 \cdot 2H_2O$, 0.1950 g of $Na_2S \cdot 9H_2O$, 0.2015 g of $Na_2S_2O_3 \cdot 5H_2O$, and 0.1706 g of $C_6H_8O_7 \cdot H_2O$ are homogenized in a stoichiometric amount and milled.

2.3. Characterization Methods and Techniques

X-ray diffraction patterns were obtained on a MiniFlex 600 diffractometer (Rigaku, Tokyo, Japan) in a digital form using copper radiation. Sample analysis modes were as follows: X-ray tube voltage—40 kV, the tube current—15 mA, goniometer movement step size—0.02 2θ, and step time was 0.12 s. During shooting, the sample was rotated in its plane at a speed of 60 rpm. For phase analysis, the ICCD-PDF2 Release 2016 database and the PDXL2 software (Rigaku Corporation, Tokyo, Japan) were used.

The Raman data were obtained using a combined system Solver Spectrum (NT-MDT Spectrum Instruments, Moscow, Russia), equipped with a photomultiplier tube (PMT) detector, a high-stability fast confocal laser (Rayleigh, UK) with imaging and 600/600 grating, and 473 nm solid-state laser excitation. In all experiments, laser power at the sample was 35 mW and the exposure time was 60 s. Continuously tunable filters with ND = 0.5, which reduce the laser intensity by 30%, were also used. When using a blue laser, an error of $\pm 4\ cm^{-1}$ is provided.

The X-ray photoelectron spectroscopy (XPS) measurements were performed in the XPS by Kratos Axis Supra apparatus (Manchester, UK), with a monochromatic Al Kα X-ray radiation, an emission current of 15 mA, and a hybrid lens model. Wide and high-resolution spectra were recorded with a pass energy of 80 eV and 20 eV, using scanning steps of 1 and 0.05 eV, respectively. The obtained data were calibrated by setting the C1s emission at 284.7 eV. The deconvolution and fitting of the interesting elements were carried out using the CasaXPS software (version 2.3.22), by applying a Spine Shirley background in the high-resolution spectra and a Gaussian/Lorentzian line shape for fitting the XPS peaks.

The ultraviolet-visible (UV-Vis) spectra were obtained by a UV-Vis spectrophotometer from Helios Gamma (Thermo Electron Corporation, Rugby, UK) measuring in the range

300–800 nm, using a 1 cm path-length quartz cuvette. The samples were diluted in distilled water by ultrasonic stirring.

The photoluminescence (PL) spectra at room temperature were recorded at the right angle on a photon-counting spectrofluorometer PC1 (ISS, Champaign, IL, USA) with an excitation wavelength of 325 nm. A 300 W xenon lamp was used as the excitation source. Excitation and emission slit widths were set at 1 and 2 mm. A one cm, path-length rectangular quartz cuvette was used. The PL intensity was measured from the powder samples ultrasonically dispersed in absolute ethanol.

Imaging of the samples was performed by transmission electron microscopy (TEM). Prior to TEM analyses, the samples were ultrasonically dispersed in absolute ethanol for a few minutes, then a droplet of the suspension was applied onto a carbon-coated nickel grid. The dried grids were additionally carbon coated to prevent charging of the samples under the high-energy electron beam. TEM analyses were performed using a 200-kV microscope, JEM 2100 (JEOL, Tokyo, Japan) with a LaB6 electron source.

The morphology and size of the products was investigated by a scanning electron microscope, Tescan Vega 3 LMU (TESCAN, Brno, Czech Republic) using an accelerating voltage of 20 kV. In order for the samples to be conductive, the powder was covered by a layer of gold on a FINE COAT ION SPUTTER JFC-1100 fy (JEOL, Akishima, Japan). To obtain the information about chemical composition, the energy-dispersive X-ray spectroscopy (EDS) analyzer, Tescan: Bruker XFlash Detector 410-M (TESCAN, Brno, Czech Republic), was used (in this case, the Au coating was not applied). The same device was used to record elemental maps.

The grain size analysis (DLS) was performed using a particle size laser diffraction analyzer.

Mastersizer 2000E (Malvern Panalytical, Malvern, UK) was used in the dry mode (dry feeder Scirocco 2000M). Each sample was measured three times.

2.4. Photocatalytic Measurements

Photocatalytic experiments of CdS samples were performed by recording the UV-visible absorption of the solution after irradiation with a sunlight-simulated lamp, Osram Vita-Lux 300W, with a UV-cutter ($\lambda = 400$ nm). The organic dye Orange II was selected as a model solution. The simulated visible light intensity at the surface of the dye solution was 15 mW/cm^2.

In a typical procedure, 20 mg of a photocatalyst CdS sample was transferred to the 40 mL of Orange II aqueous solution (10 mg/L). Before irradiation, the mixture was stirred for 1 h in a dark regime to attain an adsorption–desorption equilibrium. After that, the lamp was turned on and each 30 min sample was taken for photocatalytic degradation measurements of the sample. Before measurements, the sample was centrifuged to remove the solid CdS photocatalyst from the solution. The filtrate was analyzed with the help of a UV-Vis spectrophotometer SF-56 (LOMO, St. Petersburg, Russia). All photocatalytic tests were carried out in two repetitions to confirm the accuracy of the results obtained.

2.5. Hydrogen Evolution

The photocatalytic activity of cadmium sulfide nanoparticles for hydrogen generation experiments was studied using an Osram Ultra-Vita Lux 300 W lamp with a UV cutter ($\lambda \geq 420$) as a source of visible light, as well as an aqueous solution of glycerin (10% by v/v), which uses glycerin as a sacrificial agent. In the typical photocatalytic experiment, a 30 mg catalyst and 100 mL glycerin solution were placed into a 250 mL tree naked flask and sonicated for 15 min. The photocatalytic reactor was connected to an argon gas flow to remove oxygen molecules, and the dispersed solution was kept in the dark for the first hour. The dispersed solution was magnetic and stirred during throughout experiment. An argon gas flow was fed into the solution through a needle at a rate of 100 mL/min at the degassing (1 h), and 5 mL/min during the photocatalytic experiment. Argon gas served to maintain an inert atmosphere in the reaction system and transport generated hydrogen from the reactor to the chromatograph (Chromos 1000, Dzerzhinsk, Russia).

2.6. Antibacterial Activity

Nanoparticles were tested on two bacterial models—*Escherichia coli* ATCC® 25922™ and *Staphylococcus aureus* ATCC® 33591™. The micro-organisms were kept frozen at −20 °C in trypticase soy broth (TSB) medium with 20% *v*/*v* glycerol. Before experiments, micro-organisms were revived on a trypticase soy agar (TSA) medium and incubated at 37 °C.

The nanomaterial was suspended in deionized water, heated at 100 °C for 15 min to ensure sterility, cooled down to room temperature, and sonicated in a water bath sonicator for 30 min. The concentrations used in the analyses were ranging from 6.5 µg/mL to 100 µg/mL, depending on the test.

The antibacterial activity was tested in a 24-h toxicity test, disk assay, and respiration monitoring during the logarithmic growth phase.

For a 24-h toxicity test, bacteria were inoculated at 1:500 from the overnight culture (14–16 h old) to a fresh TSB medium and incubated at 37 °C with agitation (150 rpm). At times of 0 and 24 h, optical density (λ = 600 nm) was measured on a BioTek Synergy H1 (Winooski, VT, USA) spectrophotometer. Afterward, 10% (*v*/*v*) of resazurin (1 mg/mL, Merck, Darmstadt, Germany) was added to the cultures, the samples were incubated at ambient temperature for five minutes, and fluorescence was recorded (λ_{ex} = 520 nm and λ_{em} = 590 nm).

The material for the disk assay was prepared according to EUCAST standards for antimicrobial susceptibility testing [39]. Five µL of nanomaterial (100 µg/mL) was applied to blank disks. Plates were incubated at 37 °C for 18 h and photographed. Four repetitions were prepared for each case.

The cultures for respiration monitoring during the logarithmic growth phase were prepared in the same manner as for the 24-h toxicity test. After 2 or 3 h of incubation (*E. coli* and *S. aureus*, respectively), 10% of resazurin was added. The fluorescence was measured every ten minutes for four hours.

All samples in the 24-h assay were measured in 24 repetitions, whereas each case in the respiration monitoring assay was measured in eight repetitions. The differences between samples were analyzed with One-way ANOVA performed in Origin 2021 software (OriginLab Corp., Northampton, MA, USA) which was also used to visualize the results on figures. Tukey's posthoc test was used for comparisons. Results with $p < 0.05$ were considered significantly different.

3. Results

3.1. XRD Results

The X-ray diffraction patterns of CdS obtained from different approaches of mechanosynthesis are shown in Figure 1. The diffraction peaks in the patterns of both cadmium sulfide (aCdS and cCdS) samples match those of cadmium sulfide from the database (PDF2 089-0440). The three main peaks of both samples with 2-theta values of 26.76°, 44.14°, and 52.37° correspond to the three crystal planes of (111), (220), and (311) of cubic phase β-CdS with the mineralogical name, hawleyite. The presence of another polymorph-hexagonal greenockite (PDF2 041-1049) has also been discovered. The broadness of the peaks indicates the nanocrystalline nature of the obtained CdS powders. As shown in the XRD patterns, no other clear peaks were detected, therefore suggesting a high purity of the processed compounds.

3.2. Raman Spectroscopy Results

Raman spectra of mechanochemically prepared CdS nanoparticles are presented in Figure 2. All characteristic peaks for CdS are almost the same for both samples. Usually, the Raman spectrum of the CdS exhibits a well-resolved band at 302 cm^{-1}, corresponding to the first-order scattering of the longitudinal optical (LO) phonon mode and the second-order band at approximately 599 cm^{-1}. Also, CdS can have both hexagonal wurtzite and cubic zinc blended structures. The zone-center, longitudinal–optical A1 (LO) phonon frequency for both structures are nearly 305 cm^{-1} [40]. In this paper, the Raman spectrum of cadmium sulfide prepared by mechanochemical method contains three distinguishable peaks at 300.40, 599.44, and 910.42 cm^{-1} (acetate route (aCdS)), and 300.40, 592.42, and 900.42 cm^{-1} (combined approach (cCdS)), respectively. These values are in accordance with the main wavenumber peaks of CdS [41]. As a blue laser during shooting reveals an error with a ± 4 cm^{-1} wavenumber, a negligible difference in LO2 and LO3 wavenumbers in the Raman pattern can be ignored. Detailed characterizations of CdS samples by Raman spectroscopy can be found in [42].

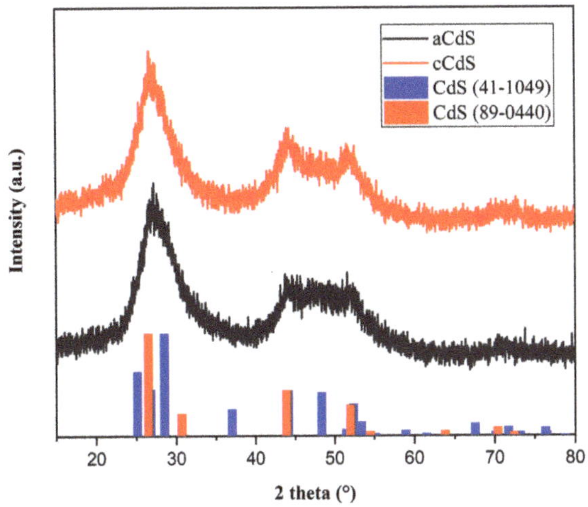

Figure 1. XRD patterns of mechanochemically synthesized aCdS and cCdS samples.

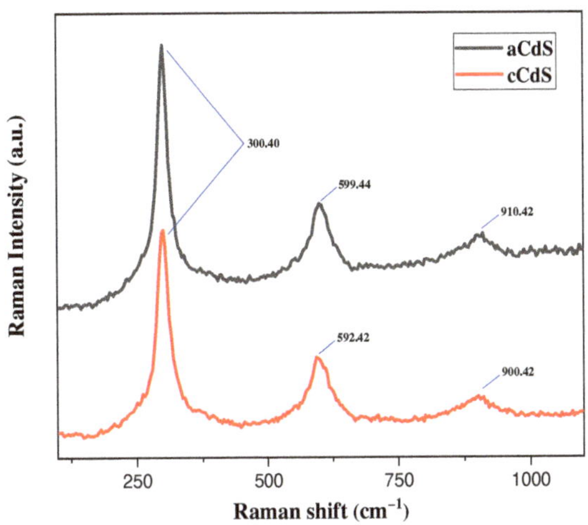

Figure 2. Raman spectra of as-synthesized aCdS and cCdS samples.

3.3. X-ray Photoelectron Spectroscopy Results

The results of XPS analysis are presented in Figure 3. The XPS measurements on the processed aCdS and cCdS compounds revealed the presence of traces from the reactants, such as sodium, and carbon or oxygen from organic compounds that could get absorbed in the surface of the fine powders during their processing. However, those impurities were very low in amount, e.g., on average less than 0.3 % in the case of Na. Besides the extra residues, the XPS spectra allowed determination of the surface chemical composition of the processed CdS compounds by analyzing the characteristic core-level emissions of sulfur and cadmium.

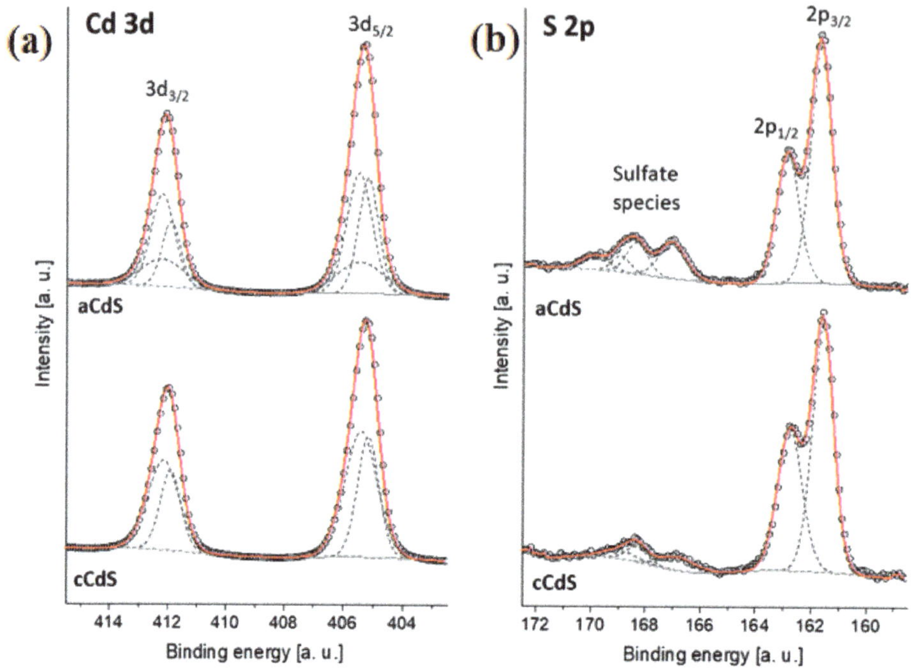

Figure 3. Core level emissions of (**a**) Cd 3d and (**b**) S 2p from the XPS spectra of mechanochemically synthesized aCdS and cCdS powders.

For the Cd, the 3d core-level spectra decomposition showed the presence of the characteristic doublet with the 3d5/2 at ~405.5 eV, and additional doublets were found at a lower binding energy (~405.2 eV), which are both related to Cd^{2+} species [43]. A load of sulfur connected to the CdS particles was evidenced by the 2p spectra, which were decomposed in two peaks positioned at 161.6 and 162.7 eV, which are connected to the S 2p3/2 and S 2p1/2 spin-orbit levels of sulfur in the CdS, respectively [44]. Additional peaks at a higher binding energy between 165–170 eV revealed a grade of oxidation in the CdS compounds. The doublets positioned at 166.8, 168.3, and 168.9 eV are connected to sulfate species formed by the oxidation of the sulfur-based compounds in air or during high-energy milling. The aCdS sample seems to be more oxidized. However, these peaks have been attributed previously to oxidized sulfur species in the form of SO_x^{2-} (x = 3 or 4), which are located on the CdS particle's surface [45].

3.4. UV-Vis Spectroscopy Results

The optical properties of aCdS and cCdS samples were investigated by ultraviolet-visible (UV-Vis) absorption spectroscopy (Figures 4 and 5). The optical bandgap energy of prepared samples was calculated from the empirical formula (Equation (2))

$$(\mathrm{ah\nu})^{\mathrm{n}} = \mathrm{A}\ (\mathrm{h\nu} - Eg) \tag{2}$$

where E_g is the optical bandgap energy, h is Planck constant, a is the absorption coefficient at ν frequency, A is constant, and n = 2, 1/2, and 3/2, correspondingly.

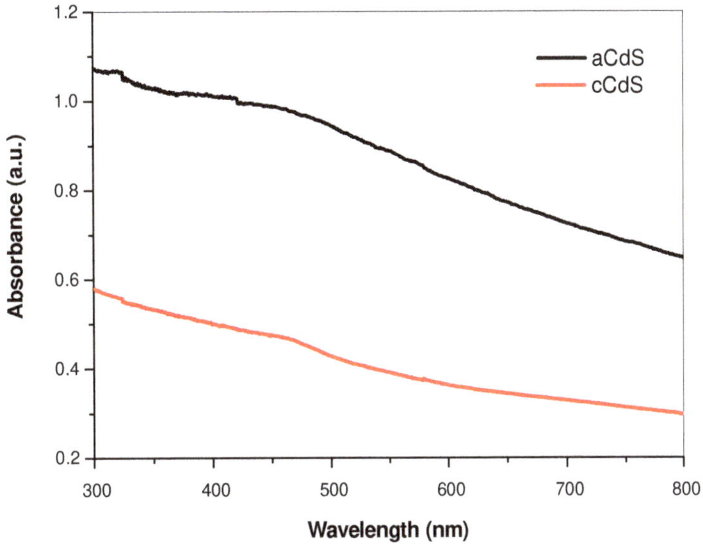

Figure 4. UV-Vis absorption spectra of aCdS and cCdS samples.

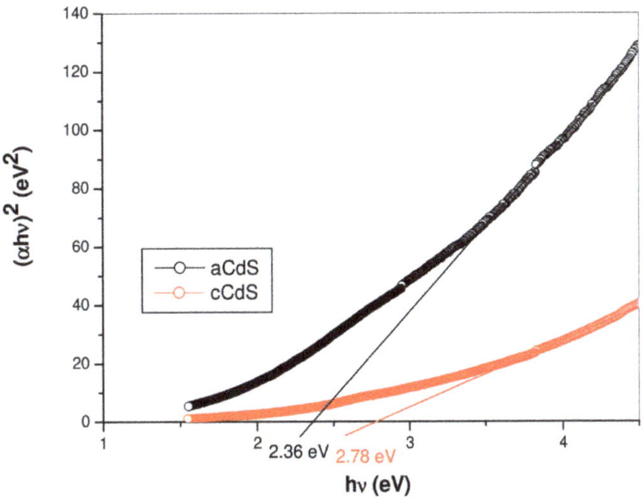

Figure 5. Tauc's plots with the determined bandgap energy of aCdS and cCdS samples.

UV-Vis absorption spectra of aCdS and cCdS samples (Figure 4) exhibit a broad absorption in the whole ultraviolet-visible region with the absorption maximum in the visible region. Figure 4 shows UV-Vis spectra with a clearly higher intensity of light absorbance evidenced in the case of the aCdS sample, in comparison with cCdS.

The optical band gap, E_g, was determined based on the Tauc Equation (2) for directly allowed optical transition, $\alpha h\nu = A\,(h\nu - E_g)^{1/2}$, by the plotting $(\alpha h\nu)^2$ against photon hν energy and extrapolating the slope in the band edge region to zero, as shown in Figure 5. The optical bandgap was determined to be 2.36 eV for aCdS and 2.78 eV for cCdS, respectively. The optical bandgap of aCdS sample is red-shifted and cCdS is blue-shifted in comparison with the bulk CdS, which has a characteristic bandgap energy of about 2.4 eV [46]. The determined bandgap for cCdS (2.78 eV) is in accordance with the results published in previous papers [47]. This blue shift indicates the presence of quantum-size effects in the cCdS sample. However, the determined bandgap for aCdS (2.36 eV) is in agreement with other reports [48,49]. The slightly red shift could originate from the surface defects of the CdS nanoparticles [50]. These shifts can be due to the influence of various factors such as grain size, structural parameters, carrier concentration, presence of impurities, deviation from stoichiometry, lattice strain, etc.

3.5. PL Spectroscopy Results

The optical properties of aCdS and cCdS samples were also analyzed by photoluminescence (PL) emission spectroscopy (Figure 6). The room-temperature PL spectra of aCdS and cCdS samples were taken using an excitation wavelength of 325 nm (3.8 eV) and are displayed in Figure 6.

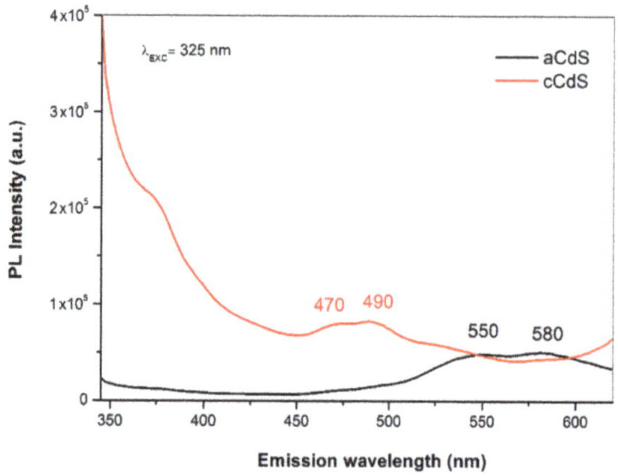

Figure 6. PL emission spectra of aCdS and cCdS samples.

The PL spectra of aCdS and cCdS samples are a little different, but both show emission spectra in the visible range. The aCdS sample has broad emission bands at 550 nm (2.2 eV) and 580 nm (2.1 eV), almost consistent with the literature [51]. However, the cCdS sample shows the broad emission bands at 470 nm (2.6 eV) and 490 nm (2.5 eV), which are in accordance with results reported in previous papers [52]. CdS nanoparticles show two types of PL emissions, identified as band-edge (400–500 nm) and surface defect, or surface state emissions (500–700). In our case, the observed green emission at 550 nm, for sample aCdS, can be assigned to charge the carrier recombination at the surface states of CdS nanoparticles. This recombination was reported as a radiative recombination of the free charge carrier and trapped charge carriers at surface defects [53]. The yellow emission observed at 580 nm (2.1 eV) for sample aCdS is less than the bandgap energy calculated (2.21 eV); the luminescence may be due to transitions involving defect states. The source of defect states may be associated with sulfur vacancies, intrinsic defects, or impurities [54]. In the case of the cCdS sample, the blue emission at around 470 nm (2.6 eV) is attributed to direct recombination of electron and hole pairs at the bandgap that cause the band edge luminescence [55]. The results for synthesized CdS nanoparticles in contrast to bulk CdS indicate that the emission depends on the particle size, shape, preparation method, etc.

3.6. SEM Results

The morphology of the prepared aCdS and cCdS samples was investigated by SEM (Figure 7). The sample prepared by a traditional acetate route contains finer particles, while in the case of cCdS, the fine particles are embedded into the matrix of large grains the size in tens of microns. It seems that the presence of an additional source of sulfur that leads to the formation of needle-like structures of CuS [38] results in the formation of large agglomerates in the present case. The distribution of Cd and S elements in both samples is homogeneous, and the locations perfectly match each other (Figure S1), thus providing another proof of successful CdS synthesis. The EDS analysis (Table S1) has shown almost ideal stoichiometry for the CdS products, albeit the content of Cd is slightly higher in all cases. The small sulfur deficiency is common for the mechanochemical synthesis of sulfides [56].

Figure 7. SEM images of the mechanochemically synthesized CdS samples: (**a**,**b**) aCdS; (**c**,**d**) cCdS.

3.7. TEM Results

Figure 8 shows the results of TEM analyses of the two samples. Low-magnification images of the samples with selected area diffraction (SAED) patterns shown in Figure 8a,d reveal that the samples are composed of agglomerated nanoparticles, which is a typical product morphology in mechanochemical synthesis. The d-values of the diffraction rings in both samples are in agreement with the results of XRD analyses, and indicate that the sphalerite-type (cubic) modification is the prevailing one in both samples. There are two main differences between the samples. According to the diffuse diffraction rings of the aCdS sample, it is composed of very small nanoparticles with an average size below 10 nm; whereas the SAED pattern of the cCdS sample contains reflections dots in addition to the diffuse rings, indicating the presence of slightly larger nanoparticles in addition to the smaller ones. The second difference between the samples is the presence of a fairly high fraction of amorphous material in addition to the crystalline CdS nanoparticles in the cCdS sample. High-resolution TEM analysis of the aCdS sample confirmed that it is composed of very small crystallites with a diameter of 10 nm or lower (Figure 8b). A more detailed analysis revealed that the grains are composed of sphalerite-type slabs that contain wurtzite-type defects, or even layers with the wurtzite-type structures (Figure 8c). We estimate that the sphalerite modification prevails in the grains of both samples; whereas the wurtzite-type sacking is mainly present at defects and thinner sections formed as a consequence of high-energy milling. The presence of both structural types is possibly due to the close structural relationship of the two modifications, as shown in the structural models in Figure 8c. A single layer is identical in both structures, but the stacking of layers is different. In sphalerite type, the subsequent layers (the $(111)_{Sp\text{-}t}$ layers) have an identical orientation, which results in the cubic close-packed stacking (ccp: -A-B-C-) of the anionic sublattice; whereas in the wurtzite-type modification, the subsequent layers (here, these are the $(001)_{Wur\text{-}t}$ layers) can be described by in-plane rotation of 180°, and this results in hexagonal close-packed stacking (hcp: -A-B-A-B-). Due to this close structural relationship between both structural types, intergrowths are common in CdS and similar compounds, e.g., ZnS, and can form for different reasons. For example, these happen as a result of crystallization near the Sph–Wur transition temperature or mechanically induced crystal slip due to high-energy milling or as observed in CdS samples synthesized in this work.

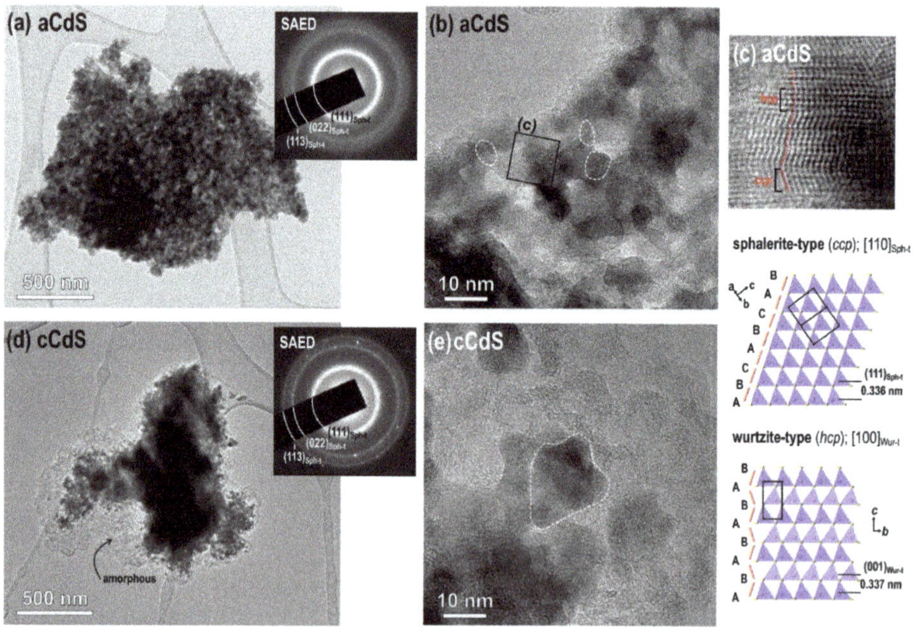

Figure 8. TEM analysis of the mechanochemically synthesized CdS samples: low-magnification images of the (**a**) aCdS and (**d**) cCdS samples with SAED; high-resolution TEM analysis of the (**b**) aCdS and (**e**) cCdS samples; (**c**) detailed analysis of aCdS sample.

3.8. DLS Analysis Results

Results of DLS analysis of mechanochemically synthesized aCdS and cCdS samples are presented in Figure 9. The grains of cCdS sample are significantly coarser, as the maximum grain size is reached around 107 μm and the overall size distribution is polymodal. On the other hand, a bimodal grain size distribution has been observed in the case of aCdS, prepared using the traditional acetate route, with the maximum being located slightly below 3 μm and being detected for the smaller fraction. A significantly minor, coarser fraction for this sample has a maximum located around 134 μm. The coarser character of the cCdS sample is in accordance with the observation of SEM analysis (Figure 7).

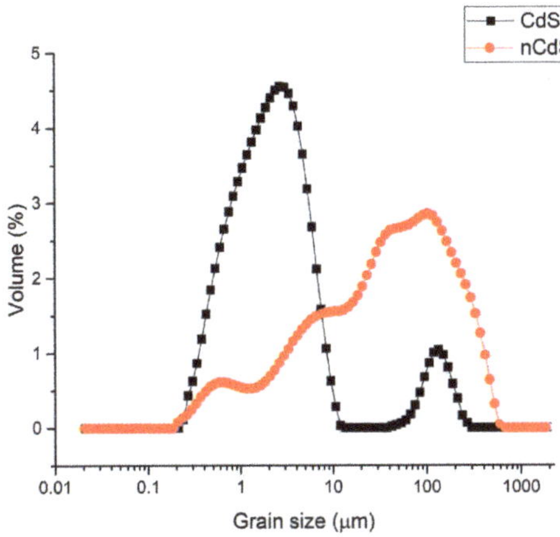

Figure 9. Grain size distribution of aCdS and cCdS samples.

3.9. Photocatalysis Results

Comparison of the photocatalytic activity of mechanochemically synthesized aCdS and cCdS (Figure 10) revealed that both substances can degrade almost all molecules of Orange II after 180 min of visible light irradiation. In general, uniform photodegradation of the model solution without sharp fluctuations is observed. However, cCdS was a more active photocatalyst than aCdS.

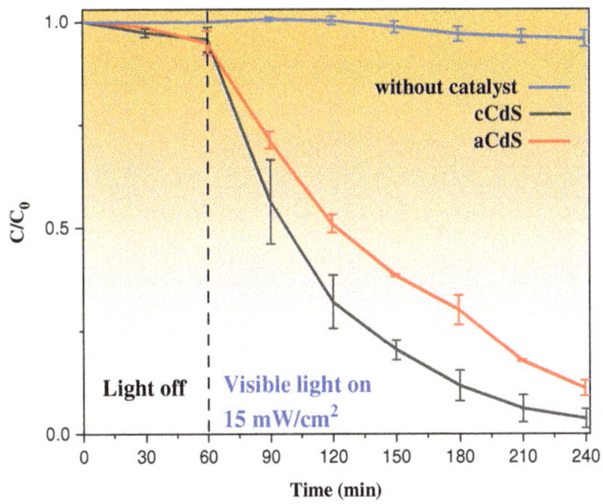

Figure 10. Photocatalytic degradation profiles of mechanochemically synthesized aCdS and cCdS catalysts under visible light irradiation.

The study of the kinetics of the photocatalytic degradation of Orange II is shown in Figure 11. The rate of the photocatalytic reaction was accepted as a pseudo-first-order reaction and described with Langmuir–Hinshelwood kinetics [57]:

$$\ln\left(\frac{C_0}{C}\right) = kt, \tag{3}$$

where C_0 and C are initial and final concentrations of Orange II solution in the moment of time, t, and k (min^{-1}) is the rate constant of the reaction. The correlation coefficient, R^2, served as proof of the kinetic order of organic dye degradation. The values of the k and R^2 are presented in Table 1. The rate constant (Table 1) of photocatalytic decomposition of Orange II by cCdS catalyst (k = 0.018 min^{-1}) is 1.5 times higher than the aCdS catalyst (k = 0.012 min^{-1}). This difference explains the effect of the shape of nanoparticles on the ability to rapidly excite an electron in a semiconductor and slow down the rate of recombination of an electron and an electron-hole, and thereby the effect on the photocatalytic activities of cadmium sulfide nanoparticles. Similarly, the higher photocatalytic activity can be compromised by the surface chemistry of the particles, and as revealed by the XPS spectra, the presence of oxidized species in the aCdS compounds was higher than in the cCdS, which increases their electrical resistivity [58]. cCdS nanoparticles have a greater propensity for photocatalytic activity on the degradation of organic compounds than usually spherical aCdS nanoparticles.

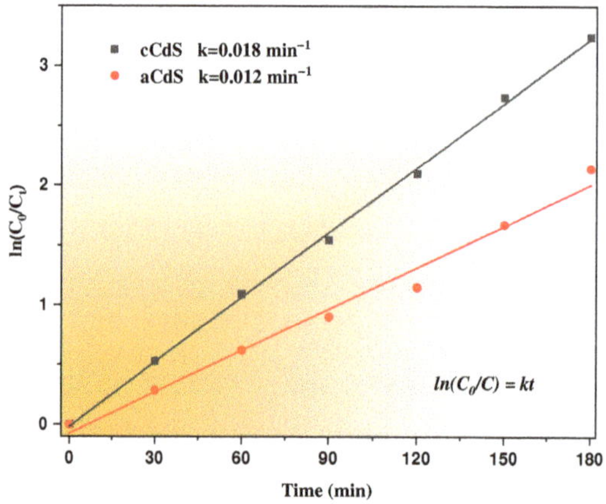

Figure 11. Kinetic linear simulation curves for Orange II photocatalytic degradation over aCdS and cCdS samples.

Table 1. The parameters for the pseudo-first-order photocatalytic reaction of aCdS and cCdS.

The Sample	k, min^{-1}	R^2
aCdS	0.012	0.98
cCdS	0.018	0.98

Figure S2 reveals the results from the cyclic test of aCdS and cCdS samples for Orange II degradation. The degradation activity of both samples for Orange II exhibits only a slight decrease after five circles. This proves that both samples have a good cyclic utilization. After checking the stability of photocatalysts, utilized aCdS and cCdS samples were centrifuged at 4000 rpm for 10 min and dried for 12 h to perform XRD analysis.

Figure S3 shows the XRD patterns of both CdS photocatalysts after performing cyclic tests. The diffraction peaks of CdS in the patterns of the samples closely match those of cadmium sulfide from the database (JCPDS 89-0444). In addition to the peaks of cadmium sulfide, both samples contain cadmium carbonate ($CdCO_3$; ICCD PDF card No. 00-001-0907), which clearly shows that some Cd(II) ions are leached out from the process. The presence of carbonate ions might come from the interaction between the solution with the photocatalayst and ambient atmosphere.

In comparison with others, photocatalysts such as TiO_2 and ZnO, and the CdS nanoparticles have a significantly narrow band gap energy and demonstrate that they can absorb a broad absorption sunlight spectrum. Moreover, the photogenerated e^- and h^+, due to the high surface-to-volume ratio, can access the particle surface more efficiently and can be easily captured by redox couples in the solution with less recombination [59].

Spectral analysis results show that the band gap of the cCdS nanoparticles was 2.78 eV, which is larger than the band gap of bulk CdS (2.42 eV) and acetate-method-synthesized aCdS nanoparticles (2.36 eV). When the visible light has a wavelength less than 448 nm, with an energy of hn that matches the energy of the band gap (Eg) of the cCdS nanoparticles, it generates electron (e^-) and hole (h^+) pairs with strong oxidizing and reducing properties:

$$2CdS + h\nu \rightarrow CdS\left(e^-\right) + CdS\left(h^+\right) \quad (4)$$

The large band gap of cCdS nanoparticles forms to a nonradiative recombination of e^- and h^+ pairs [60], which enhances photocatalytic activity. Water (H_2O) adsorbed on the surface of CdS nanoparticles captures the hole (h^+) and is oxidized to form a hydroxyl radical ($^\bullet OH$), which is expressed as:

$$H_2O + CdS\left(h^+\right) \rightarrow OH^\bullet + H^+ + CdS \quad (5)$$

Also, the presence of oxygen (O_2) prevents the recombination of pairs of electrons (e^-) and holes (h^+), and it leads a radical anion ($^\bullet O_2{}^-$), accepting electrons (e^-) from the conduction band and further combining with a proton to give $^\bullet OOH$:

$$\begin{aligned} O_2 + CdS(e^-) &\rightarrow {}^\bullet O^{2-} + CdS \\ {}^\bullet O^{2-} + H^+ &\rightarrow {}^\bullet OOH \end{aligned} \quad (6)$$

Formed radicals, $^\bullet OH$ and $^\bullet O^{2-}$, decompose organic compound Orange II during the photocatalytic process by reacting to the aromatic ring of Orange II molecules and opening it at the azo bond and the hydroxylated ring, finally giving gaseous N_2, H_2O, and CO_2 [61]:

$$\text{Orange II} + {}^\bullet OH/{}^\bullet O^{2-} \rightarrow \text{products}\ (N_2,\ H_2O \text{ and } CO_2) \quad (7)$$

In Table 2, results of photocatalytic tests of CdS nanoparticles prepared with different approaches for recent years are reviewed. According to the collected data, the photocatalytic activities of our samples are in no way inferior to those obtained by other methods. Since our approach is environmentally friendly, as no harmful organic solvents are used and only 5 min of the treatment at room temperature was necessary to obtain CdS nanoparticles, it is more beneficial than the ones reported in the Table 2.

Table 2. Comparison of photocatalytic activities of CdS nanoparticles with other methods.

№	Synthetic Method	Experimental Conditions		Precursors	Degraded Dye, Concentration	Photocatalytic Efficiency	Rate Constant (min^{-1})	[Ref]
		Time	Temperature (°C)					
1	Composite-molten-salt (CMS)	24–72 h	160–220	$Cd(NO_3)_2{\cdot}4H_2O$, $Na_2S{\cdot}9H_2O$, $LiNO_3$, KNO_3	MB, 4 mg/L	76.3%@140 min	-	[62]
					RhB, 8 mg/L	94.9%@140 min	-	
2	Solvothermal	6 h	180	$CdCl_2{\cdot}5H_2O$, $CS(NH_2)_2$	MB, 6 mg/L	95%@80 min	0.0365	[63]
3	One-step solid-state reaction	30 min	-	$Cd(CH_3COO)_2{\cdot}2H_2O$, $Na_2S_2O_3{\cdot}5H_2O$	RhB, 10 mg/L	95%@80 min	0.0429	[64]
4	Biogenic synthesis	72 h	28	Strain of T. Harzianum, $CdCl_2$, Na_2S	MB, 10 mg/L	37.15%@60 min	0.0076	[59]
5	Hydrothermal	24 h	-	$Cd(Ac)_2{\cdot}2H_2O$, PVP-K30, $CS(NH_2)_2$	MO, 20 mg/L	93.3%@240 min	-	[65]
6	Sonochemical	1 h	RT	$Cd(CH_3COO)_2$, Na_2S, tryptophan	MO, 5×10^{-6} M	75.33%@240 min	0.0062	[66]
7	Photochemical	24 h	-	$CdSO_4$, $Na_2S_2O_3$	MO, 8×10^{-6} M	26.3%@70 min	0.0058	[67]
8	Commercial CdS	-	-	-	MO, 10 mg/L	78%@90 min	-	[68]
9	Mechanochemical	5 min	RT	$Cd(CH_3COO)_2{\cdot}2H_2O$, $Na_2S{\cdot}9H_2O$, $Na_2S_2O_3{\cdot}5H_2O$, $C_6H_8O_7$	Orange II, 10 mg/L	93%@180 min	0.018	this work

3.10. Hydrogen Evolution Results

Figure 12 shows the results of a photocatalytic experiment on hydrogen generation using photocatalysts aCdS and cCdS, where the maximum values of hydrogen formation are recorded in the first hour of photocatalysis for aCdS and the second hour of photocatalysis for cCdS, and they are equal to 4.1 and 7.8 μmol g^{-1} h^{-1}, respectively. After 2 h of photocatalysis, the activity of cadmium sulfide photocatalysts drops significantly, which indicates their unstable structure. The results of photocatalytic experiments on hydrogen generation showed that cadmium sulfide obtained by combined mechanosynthesis (cCdS), once again, confirms its superiority over cadmium sulfide synthesized by the traditional acetate method (aCdS). The manifestation of a higher photocatalytic activity during hydrogen generation of cCdS nanoparticles compared to aCdS nanoparticles can be associated with the size of crystallites, morphology, and structural defects of crystal lattices of cadmium sulfide nanoparticles. From the XRD results, the diffraction peaks of cCdS nanoparticles are slightly wider, and this cCdS average crystallite size is slightly smaller than those of aCdS. A decrease in the crystallite size improves the photocatalytic activity of nanoparticles. From the results of TEM, it can be seen that there is greater content of crystal defects in cCdS nanoparticles, which should also be the reason for the higher photocatalytic activity of cCdS. The large band gap of cCdS nanoparticles (Eg = 2.78 eV) is also increased with hydrogen generation performance, due to strong oxidizing and reducing properties [59]. A decrease in hydrogen generation on the surface of both CdS nanoparticles is observed after 120 min, but the performance of cCdS remains higher than that of aCdS. The decrease in hydrogen generation could be caused by the increase in the concentration of glycerol oxidation products during photocatalysis in the solution [69]. Glycerol was used as a sacrificial agent for generating hydrogen.

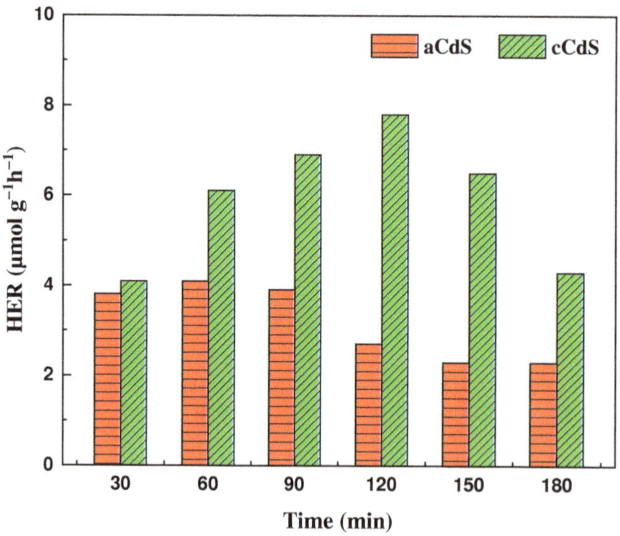

Figure 12. Photocatalytic H_2 production over aCdS and cCdS samples.

3.11. Antimicrobial Activity

To demonstrate the multidisciplinary application of the mechanochemically prepared CdS nanoparticles, their antibacterial potential was also investigated. The measurements of optical density shown in Figure 13 did not show any significant differences between the two samples in the case of *S. aureus*. On the other hand, in this test, all samples of *E. coli* had significantly higher optical density than the control sample (Supplementary Material, Figure S4). This observation confirms previous findings that bacteria could build their population after a long incubation, even if it is initially inhibited [70]. This was also confirmed in the disk diffusion test that the current study did not produce any halo zones around the nanomaterial, suggesting that ions were not being released to the agar medium (Supplementary Material, Figure S5) [71]. The respiration recorded in *E. coli* was not significantly different from the control sample (Figure 14), whereas the samples containing the highest concentrations of nanomaterials showed significantly higher respiration in *S. aureus* (Supplementary Material, Figure S6). This phenomenon could be explained by the inhibition observed during the respiration monitoring in the logarithmic phase (Figure 15). This experiment has clearly shown that nanomaterials could actively inhibit the growth of both micro-organisms, starting from the smallest concentration. Judging from the curves obtained within this step, the application of aCdS nanoparticles resulted in higher growth inhibition than in the case of cCdS. The probable cause for this effect was the difference in grain size distribution recorded for these nanoparticles. The data confirmed the concept that a nanomaterial's toxicity increases with the decrease in its size [72]. The inhibition was increasing with the concentration of the used nanomaterial. In this view, higher respiration of *S. aureus* after 24-h incubation may be a result of the retardation that occurred during the logarithmic growth. Staphylococci have a longer generation time than *E. coli*. Therefore, the effect could still be observed in this case [73,74]. Even though the inhibition was not complete in the cultures, the results have shown that the produced nanomaterials can be considered antibacterial and included in a broader group of antimicrobial agents that are used for various applications [75,76].

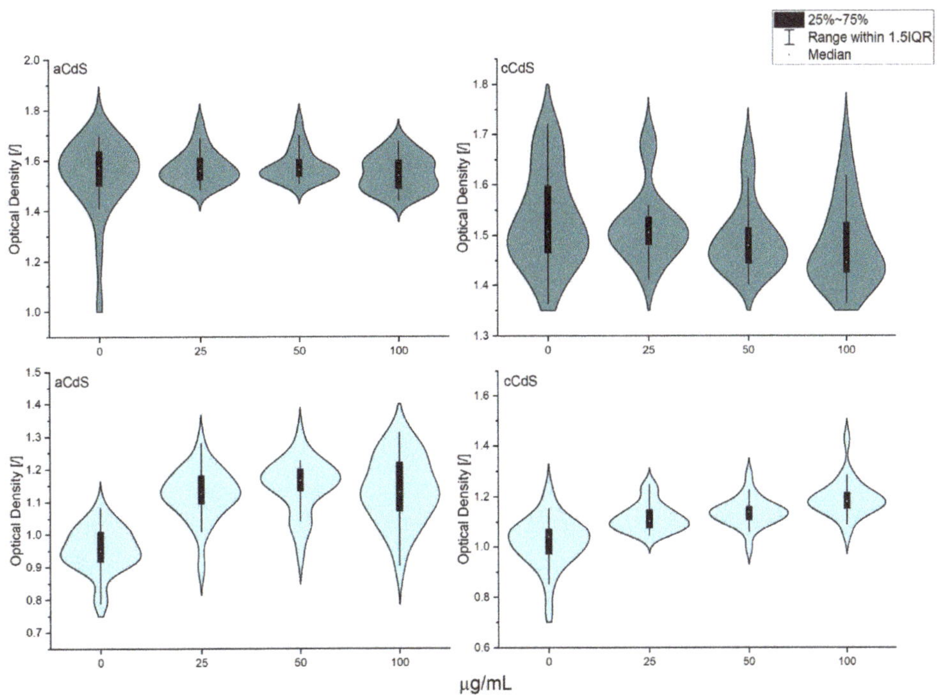

Figure 13. Optical density of 24-h cultures of *S. aureus* (first row) and *E. coli* (second row) contacted with CdS nanoparticles or deionized water.

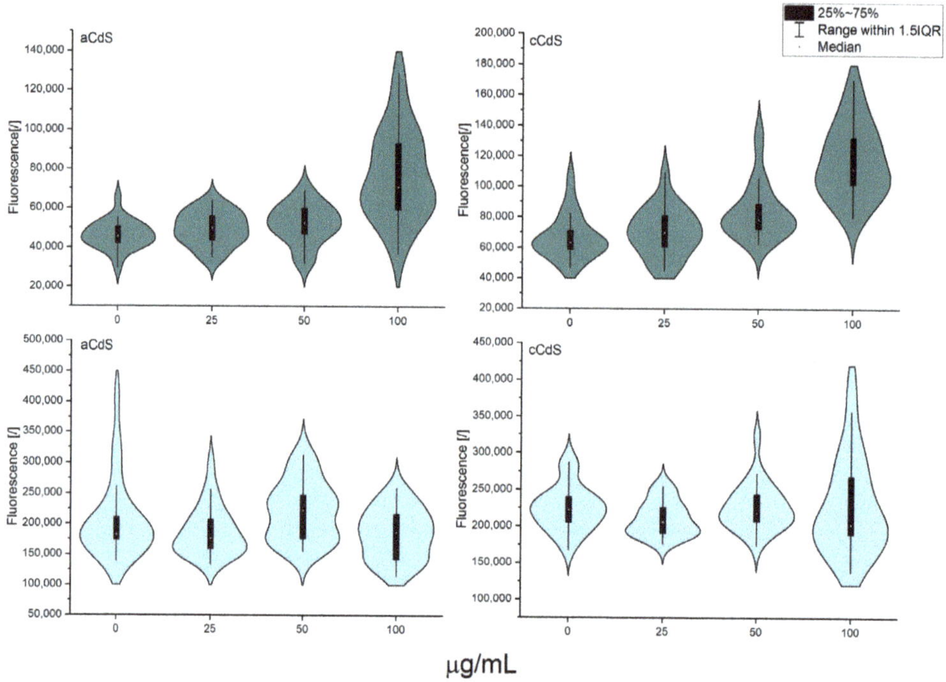

Figure 14. Respiration of cultures after 24-h incubation with nanomaterials or deionized water; *S. aureus* (first row) and *E. coli* (second row).

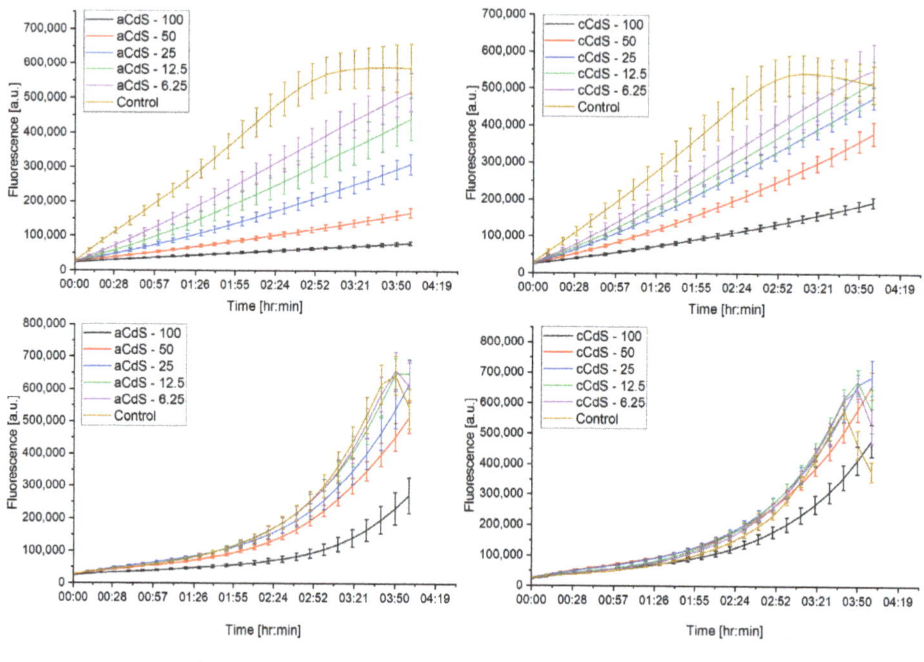

Figure 15. Respiration of cultures in the logarithmic growth phase in resazurin assay; *S. aureus* (first row) and *E. coli* (second row).

4. Conclusions

In summary, CdS nanoparticles have been successfully obtained by a mechanochemical pathway, either using the acetate route (aCdS) or its combination with sodium thiosulfate, an additional source of sulfur (cCdS). The results of UV-Vis spectroscopy revealed that the bandgap energy of cadmium sulfide (2.78 eV), produced via combined milling (cCdS), is higher than the aCdS sample prepared by acetate route (2.36 eV). The rate constant of photocatalytic degradation of dye by cCdS was 1.5 times higher than aCdS. The antibacterial properties of both CdS samples were tested on reference strains of *E. coli* and *S. aureus*. Results of experiments have confirmed that the CdS samples have some capability to inhibit bacterial growth during the logarithmic growth phase, in which the effect of aCdS was more potent than in the case of cCdS.

Supplementary Materials: The following supporting information can be downloaded at: https://www.mdpi.com/article/10.3390/nano12081250/s1. Figure S1: EDS elemental mapping of aCdS (top) and cCdS samples (bottom). (a,d) Complex image showing the analyzed data are in a green rectangle; (b,e) Cd mapping; (c,f) S mapping. Table S1: Content of Cd and S in atomic % from distinct areas in each sample, according to EDS analysis. Figure S2: Results of the cyclic test of aCdS and cCdS samples for Orange II dye. Figure S3: XRD pattern of remained cCdS and aCdS photocatalysts after performing stability tests of catalysts for Orange II dye. Figure S4: Statistical analysis of results obtained from optical density measurements after 24-h incubation with nanomaterials or deionized water. In columns, starting from the left side: *S. aureus* with aCdS, *S. aureus* with cCdS, *E. coli* with aCdS, and *E. coli* with cCdS; bottom row presents results from Tukey's test. Figure S5: Disk assays for with the produced nanomaterials; EC—*E.coli* and SA—*S. Aureus*. Figure S6: Statistical analysis of results obtained from resazurin assay (respiration) after 24-h incubation with nanomaterials or deionized water. In columns, starting from the left side: *S. aureus* with cCdS, *S. aureus* with aCdS, *E. coli* with aCdS, and *E. coli* with cCdS; bottom row presents results from Tukey's test.

Author Contributions: Z.S. and Y.N. synthesized the samples and were responsible for the interpretation of XRD and Raman results. Z.S. also wrote the first draft of the paper. M.B. (Matej Baláž) carefully reviewed the first draft and significantly helped to construct the overall conception of the paper. He partially intepreted the results of XRD analysis. Y.N., N.K. and B.T. worked on the evaluation of photocatalytic activity. G.B. tested and interpreted the hydrogen production ability of CdS samples. N.D. performed TEM analysis and interpreted the obtained results. E.D. analyzed samples using the UV-Vis and PL spectroscopies and interpreted the obtained results. A.A. tested the antimicrobial activities of the samples and interpreted the obtained results. R.D. performed the SEM-EDS analysis. R.B. analyzed particle sizes of the obtained samples by DLS analysis method. M.C.-L. and L.Č. analyzed samples by XPS analysis and interpreted the obtained results. They also partly intepreted the results of XRD analysis. M.B. (Mukhambetkali Burkitbayev) and A.I. reviewed the manuscript, provided contstructive comments, and took part in writing the introduction part of the paper. All authors have read and agreed to the published version of the manuscript.

Funding: This research was funded by the Science Committee of the Ministry of Education and Science of the Republic of Kazakhstan, grant number "AP09563428. Part of this research was made using the infrastructures of the CzechNanoLab Research Infrastructure supported by MEYS CR (LM2018110). MB acknowledges the financial support of the Slovak Research and Development Agency under contract No. APVV-18-0357 and that of The Ministry of Education, Science, Research and Sport of the Slovak Republic Grant Agency (project 2/0112/22). The support of COST Action CA18112 MechSustInd (www.mechsustind.eu (accessed on 5 April 2022), supported by the COST Association (European Cooperation in Science and Technology, www.cost.eu (accessed on 5 April 2022))) is also acknowledged. Adrian Augustyniak was supported by the German Research Foundation (DFG) as part of the Research Training Group on Urban Water Interfaces (GRK 2032).

Institutional Review Board Statement: Not applicable.

Informed Consent Statement: Not applicable.

Data Availability Statement: Not applicable.

Conflicts of Interest: The authors declare no conflict of interest.

References

1. Yadav, P.; Singh, J.; Srivastava, D.K.; Mishra, V. Environmental pollution and sustainability. In *Environmental Sustainability and Economy*; Elsevier: Amsterdam, The Netherlands, 2021; pp. 111–120.
2. Nguyen, V.-H.; Smith, S.M.; Wantala, K.; Kajitvichyanukul, P. Photocatalytic remediation of persistent organic pollutants (POPs): A review. *Arab. J. Chem.* **2020**, *13*, 8309–8337. [CrossRef]
3. Wu, Y.W.; Zhong, L.L.; Yuan, J.L.; Xiang, W.H.; Xin, X.; Liu, H.M.; Luo, H.Y.; Li, L.Y.; Chen, M.; Zhong, D.J.; et al. Photocatalytic optical fibers for degradation of organic pollutants in wastewater: A review. *Environ. Chem. Lett.* **2021**, *19*, 1335–1346. [CrossRef]
4. Rajendran, R.; Varadharajan, K.; Jayaraman, V. Fabrication of tantalum doped CdS nanoparticles for enhanced photocatalytic degradation of organic dye under visible light exposure. *Colloid Surf. A Physicochem. Eng. Asp.* **2019**, *580*, 10. [CrossRef]
5. Shume, W.M.; Murthy, H.C.A.; Zereffa, E.A. A Review on Synthesis and Characterization of Ag_2O Nanoparticles for Photocatalytic Applications. *J. Chem.* **2020**, *2020*, 5039479. [CrossRef]
6. Urakaev, F.K.; Khan, N.; Shalabaev, Z.S.; Tatykaev, B.; Nadirov, R.; Burkitbaev, M. Synthesis and Photocatalytic Properties of Silver Chloride/Silver Composite Colloidal Particles. *Colloid J.* **2020**, *82*, 76–80. [CrossRef]
7. Bhadwal, A.S.; Tripathi, R.M.; Gupta, R.K.; Kumar, N.; Singh, R.P.; Shrivastav, A. Biogenic synthesis and photocatalytic activity of CdS nanoparticles. *RSC Adv.* **2014**, *4*, 9484–9490. [CrossRef]
8. Afeesh, R.; Barakat, N.A.M.; Al-Deyab, S.S.; Yousef, A.; Kim, H.Y. Nematic shaped cadmium sulfide doped electrospun nanofiber mat: Highly efficient, reusable, solar light photocatalyst. *Colloid Surf. A Physicochem. Eng. Asp.* **2012**, *409*, 21–29. [CrossRef]
9. Harish, R.; Nisha, K.D.; Prabakaran, S.; Sridevi, B.; Harish, S.; Navaneethan, M.; Ponnusamy, S.; Hayakawa, Y.; Vinniee, C.; Ganesh, M.R. Cytotoxicity assessment of chitosan coated CdS nanoparticles for bioimaging applications. *Appl. Surf. Sci.* **2020**, *499*, 10. [CrossRef]
10. Zhang, J.; He, R.; Liu, X.H. Efficient visible light driven photocatalytic hydrogen production from water using attapulgite clay sensitized by CdS nanoparticles. *Nanotechnology* **2013**, *24*, 7. [CrossRef]
11. Veerathangam, K.; Pandian, M.S.; Ramasamy, P. Photovoltaic performance of Ag-doped CdS quantum dots for solar cell application. *Mater. Res. Bull.* **2017**, *94*, 371–377. [CrossRef]
12. Waldiya, M.; Narasimman, R.; Bhagat, D.; Vankhade, D.; Mukhopadhyay, I. Nanoparticulate CdS 2D array by chemical bath deposition: Characterization and optoelectronic study. *Mater. Chem. Phys.* **2019**, *226*, 26–33. [CrossRef]
13. Steckel, J.S.; Snee, P.; Coe-Sullivan, S.; Zimmer, J.R.; Halpert, J.E.; Anikeeva, P.; Kim, L.A.; Bulovic, V.; Bawendi, M.G. Color-saturated green-emitting QD-LEDs. *Angew. Chem. Int. Edit.* **2006**, *45*, 5796–5799. [CrossRef] [PubMed]
14. Sonker, R.K.; Yadav, B.C.; Gupta, V.; Tomar, M. Synthesis of CdS nanoparticle by sol-gel method as low temperature NO2 sensor. *Mater. Chem. Phys.* **2020**, *239*, 7. [CrossRef]
15. Haq Bhat, I.U.; Yi, Y.S. Green synthesis and antibacterial activity of cadmium sulfide nanoparticles (CdSNPs) using Panicum sarmentosum. *Asian J. Green Chem.* **2019**, *3*, 455–469.
16. Sekar, P.V.; Parvathi, V.D.; Sumitha, R. Green nanotechnology in cadmium sulphide nanoparticles and understanding its toxicity and antimicrobial properties. *Biomed. Res.* **2019**, *30*, 805–809. [CrossRef]
17. Shivashankarappa, A.; Sanjay, K.R. Escherichia coli-based synthesis of cadmium sulfide nanoparticles, characterization, antimicrobial and cytotoxicity studies. *Braz. J. Microbiol.* **2020**, *51*, 939. [CrossRef]
18. Manthrammel, M.A.; Ganesh, V.; Shkir, M.; Yahia, I.S.; Alfaify, S. Facile synthesis of La-doped CdS nanoparticles by microwave assisted co-precipitation technique for optoelectronic application. *Mater. Res. Express* **2019**, *6*, 8. [CrossRef]
19. Suresh, S. Studies on the dielectric properties of CdS nanoparticles. *Appl. Nanosci.* **2014**, *4*, 325–329. [CrossRef]
20. Khushboo; Umar, A.; Kansal, S.K.; Mehta, S.K. Highly-sensitive and selective detection of hydrazine at gold electrode modified with PEG-coated CdS nanoparticles. *Sens. Actuator B Chem.* **2013**, *188*, 372–377. [CrossRef]
21. Zhang, X.J.; Xie, Y.; Zhao, Q.R.; Tian, Y.P. 1-D coordination polymer template approach to CdS and HgS aligned-nanowire bundles. *New J. Chem.* **2003**, *27*, 827–830. [CrossRef]
22. Routkevitch, D.; Bigioni, T.; Moskovits, M.; Xu, J.M. Electrochemical fabrication of CdS nanowire arrays in porous anodic aluminum oxide templates. *J. Phys. Chem.* **1996**, *100*, 14037–14047. [CrossRef]
23. Ferreira, P.M.S.; Timmons, A.B.; Neves, M.C.; Dynarowicz, P.; Trindade, T. Langmuir-Blodgett manipulation of capped cadmium sulfide quantum dots. *Thin Solid Films* **2001**, *389*, 272–277. [CrossRef]
24. Zhang, Z.; Ren, Y.; Han, L.; Xie, G.; Zhong, B. Mixed-solvothermal synthesis of CdS micro/nanostructures with optical and ferromagnetic properties. *Phys. E Low Dimens. Syst. Nanostructures* **2017**, *92*, 30–35. [CrossRef]
25. Loudhaief, N.; Labiadh, H.; Hannachi, E.; Zouaoui, M.; Salem, M.B. Synthesis of CdS nanoparticles by hydrothermal method and their effects on the electrical properties of Bi-based superconductors. *J. Supercond. Nov. Magn.* **2018**, *31*, 2305–2312. [CrossRef]
26. Naranthatta, S.; Janardhanan, P.; Pilankatta, R.; Nair, S.S. Green Synthesis of Engineered CdS Nanoparticles with Reduced Cytotoxicity for Enhanced Bioimaging Application. *ACS Omega* **2021**, *6*, 8646–8655. [CrossRef]
27. Chatterjee, A.; Priyam, A.; Das, S.K.; Saha, A. Size tunable synthesis of cysteine-capped CdS nanoparticles by γ-irradiation. *J. Colloid Interface Sci.* **2006**, *294*, 334–342. [CrossRef]
28. Dameron, C.; Reese, R.; Mehra, R.; Kortan, A.; Carroll, P.; Steigerwald, M.; Brus, L.; Winge, D. Biosynthesis of cadmium sulphide quantum semiconductor crystallites. *Nature* **1989**, *338*, 596–597. [CrossRef]
29. Godočíková, E.; Baláž, P.; Gock, E.; Choi, W.S.; Kim, B.S. Mechanochemical synthesis of the nanocrystalline semiconductors in an industrial mill. *Powder Technol.* **2006**, *164*, 147–152. [CrossRef]

30. Tsuzuki, T.; McCormick, P.G. Mechanochemical synthesis of metal sulphide nanoparticles. *Nanostruct. Mater.* **1999**, *12*, 75–78. [CrossRef]
31. Nasser, A.; Mingelgrin, U. Mechanochemistry: A review of surface reactions and environmental applications. *Appl. Clay Sci.* **2012**, *67*, 141–150. [CrossRef]
32. Ozer, D. Mechanochemistry: A Power Tool for Green Synthesis. In *Advances in Green Synthesis*; Springer: Berlin/Heidelberg, Germany, 2021; pp. 23–39.
33. Tan, D.; García, F. Main group mechanochemistry: From curiosity to established protocols. *Chem. Soc. Rev.* **2019**, *48*, 2274–2292. [CrossRef] [PubMed]
34. Balaz, M. *Environmental Mechanochemistry: Recycling Waste into Materials Using High-Energy Ball Milling*; Springer: Berlin/Heidelberg, Germany, 2021.
35. Baláž, P.; Boldižárová, E.; Godočíková, E.; Briančin, J. Mechanochemical route for sulphide nanoparticles preparation. *Mater. Lett.* **2003**, *57*, 1585–1589. [CrossRef]
36. Dutková, E.; Baláž, P.; Pourghahramani, P. CdS nanoparticles mechanochemically synthesized in a high-energy mill. *J. Optoelectron. Adv. Mater.* **2009**, *11*, 2102–2107.
37. Kristl, M.; Ban, I.; Gyergyek, S. Preparation of nanosized copper and cadmium chalcogenides by mechanochemical synthesis. *Mater. Manuf. Process.* **2013**, *28*, 1009–1013. [CrossRef]
38. Shalabayev, Z.; Baláz, M.; Daneu, N.; Dutková, E.; Bujnáková, Z.; Kanuchová, M.; Danková, Z.; Balázová, L.; Urakaev, F.; Tkáciková, L. Sulfur-mediated mechanochemical synthesis of spherical and needle-like copper sulfide nanocrystals with antibacterial activity. *ACS Sustain. Chem. Eng.* **2019**, *7*, 12897–12909. [CrossRef]
39. Matuschek, E.; Brown, D.F.J.; Kahlmeter, G. Development of the EUCAST disk diffusion antimicrobial susceptibility testing method and its implementation in routine microbiology laboratories. *Clin. Microbiol. Infect.* **2014**, *20*, O255–O266. [CrossRef]
40. Kumar, P.; Saxena, N.; Chandra, R.; Gupta, V.; Agarwal, A.; Kanjilal, D. Nanotwinning and structural phase transition in CdS quantum dots. *Nanoscale Res. Lett.* **2012**, *7*, 584. [CrossRef]
41. Devi, R.A.; Latha, M.; Velumani, S.; Oza, G.; Reyes-Figueroa, P.; Rohini, M.; Becerril-Juarez, I.; Lee, J.-H.; Yi, J. Synthesis and characterization of cadmium sulfide nanoparticles by chemical precipitation method. *J. Nanosci. Nanotechnol.* **2015**, *15*, 8434–8439. [CrossRef]
42. Chava, R.K.; Son, N.; Kim, Y.S.; Kang, M. Controlled growth and bandstructure properties of one dimensional cadmium sulfide nanorods for visible photocatalytic hydrogen evolution reaction. *Nanomaterials* **2020**, *10*, 619. [CrossRef]
43. El Maliki, H.; Bernede, J.C.; Marsillac, S.; Pinel, J.; Castel, X.; Pouzet, J. Study of the influence of annealing on the properties of CBD-CdS thin films. *Appl. Surf. Sci.* **2003**, *205*, 65–79. [CrossRef]
44. Fedoseeva, Y.V.; Bulusheva, L.G.; Asanov, I.P.; Kurenya, A.G.; Gusel'nikov, A.V.; Maksimovskiy, E.A.; Gulyaev, D.V.; Zhuravlev, K.S.; Gutakovskii, A.K.; Okotrub, A.V. Electrically activated chemical bath deposition of CdS on carbon nanotube arrays. *Synth. Met.* **2021**, *273*, 8. [CrossRef]
45. Winkler, U.; Eich, D.; Chen, Z.H.; Fink, R.; Kulkarni, S.K.; Umbach, E. Detailed investigation of CdS nanoparticle surfaces by high-resolution photoelectron spectroscopy. *Chem. Phys. Lett.* **1999**, *306*, 95–102. [CrossRef]
46. Enriguez, J.P.; Mathew, X. Influence of the thickness on structural, optical and electrical properties of chemical bath deposited CdS thin films. *Sol. Energy Mater. Sol. Cells* **2003**, *76*, 313–322. [CrossRef]
47. Dey, P.C.; Das, R. Enhanced photocatalytic degradation of methyl orange dye on interaction with synthesized ligand free CdS nanocrystals under visible light illumination. *Spectrochim. Acta Part A Mol. Biomol. Spectrosc.* **2020**, *231*, 118122. [CrossRef] [PubMed]
48. Kamble, M.M.; Rondiya, S.R.; Bade, B.R.; Kore, K.B.; Nasane, M.P.; Dzade, N.Y.; Funde, A.M.; Jadkar, S.R. Optical, structural and morphological study of CdS nanoparticles: Role of sulfur source. *Nanomater. Energy* **2020**, *9*, 72–81. [CrossRef]
49. Rose, M.M.; Christy, R.S.; Benitta, T.A.; Kumaran, J.T.T. Phase transitions in cadmium sulfide nanoparticles. *AIP Adv.* **2021**, *11*, 085129. [CrossRef]
50. Cortes, A.; Gomez, H.; Marotti, R.E.; Riveros, G.; Dalchiele, E.A. Grain size dependence of the bandgap in chemical bath deposited CdS thin films. *Sol. Energy Mater. Sol. Cells* **2004**, *82*, 21–34. [CrossRef]
51. Javed, H.; Fatima, K.; Akhter, Z.; Nadeem, M.A.; Siddiq, M.; Iqbal, A. Fluorescence modulation of cadmium sulfide quantum dots by azobenzene photochromic switches. *Proc. R. Soc. A Math. Phys. Eng. Sci.* **2016**, *472*, 20150692. [CrossRef]
52. Alipour, A.; Lakouraj, M.M.; Tashakkorian, H. Study of the effect of band gap and photoluminescence on biological properties of polyaniline/CdS QD nanocomposites based on natural polymer. *Sci. Rep.* **2021**, *11*, 1913. [CrossRef]
53. Su, J.Z.; Zhang, T.; Li, Y.F.; Chen, Y.B.; Liu, M.C. Photocatalytic Activities of Copper Doped Cadmium Sulfide Microspheres Prepared by a Facile Ultrasonic Spray-Pyrolysis Method. *Molecules* **2016**, *21*, 735. [CrossRef]
54. Kumar, S.; Sharma, S.K. Low-temperature synthesis and characterization of cadmium sulfide nanowires grown using simple chemical reaction through the pores in an alumina template. *Phys. Scr.* **2010**, *82*, 025801. [CrossRef]
55. Dhage, S.R.; Colorado, H.A.; Hahn, T. Morphological variations in cadmium sulfide nanocrystals without phase transformation. *Nanoscale Res. Lett.* **2011**, *6*, 420. [CrossRef] [PubMed]
56. Baláž, P.; Baláž, M.; Achimovičová, M.; Bujňáková, Z.; Dutková, E. Chalcogenide mechanochemistry in materials science: Insight into synthesis and applications (a review). *J. Mater. Sci.* **2017**, *52*, 11851–11890. [CrossRef]

57. Senasu, T.; Hemavibool, K.; Nanan, S. Hydrothermally grown CdS nanoparticles for photodegradation of anionic azo dyes under UV-visible light irradiation. *RSC Adv.* **2018**, *8*, 22592–22605. [CrossRef]
58. Hernandez-Rodriguez, E.; Loeza-Poot, M.; Riech, I.; Rejon, V.; Pena, J.L. A comparative study of CdS:F and CdS:O thin films deposited by reactive RF-sputtering technique for window layer application in solar cells. *J. Phys. D Appl. Phys.* **2015**, *48*, 6. [CrossRef]
59. Banerjee, R.; Pal, A.; Ghosh, D.; Ghosh, A.B.; Nandi, M.; Biswas, P. Improved photocurrent response, photostability and photocatalytic hydrogen generation ability of CdS nanoparticles in presence of mesoporous carbon. *Materials Research Bulletin* **2021**, *134*, 111085. [CrossRef]
60. Liu, Z.; Sun, D.D.; Guo, P.; Leckie, J.O. One-step fabrication and high photocatalytic activity of porous TiO2 hollow aggregates by using a low-temperature hydrothermal method without templates. *Chem. A Eur. J.* **2007**, *13*, 1851–1855. [CrossRef]
61. Das, D.; Biswal, N.; Martha, S.; Parida, K. Solar-light induced photodegradation of organic pollutants over CdS-pillared zirconium–titanium phosphate (ZTP). *J. Mol. Catal. A Chem.* **2011**, *349*, 36–41. [CrossRef]
62. Li, X.; Hu, C.; Wang, X.; Xi, Y. Photocatalytic activity of CdS nanoparticles synthesized by a facile composite molten salt method. *Appl. Surf. Sci.* **2012**, *258*, 4370–4376. [CrossRef]
63. Jing, Z.; Tan, L.; Li, F.; Wang, J.; Fu, Y.; Qian, L. Photocatalytic and antibacterial activities of CdS nanoparticles prepared by solvothermal method. *NIScPR Online Period. Repos.* **2013**, *52*, 57–62.
64. Chen, F.; Cao, Y.; Jia, D.; Niu, X. Facile synthesis of CdS nanoparticles photocatalyst with high performance. *Ceram. Int.* **2013**, *39*, 1511–1517. [CrossRef]
65. Zhang, L.; Cheng, Z.; Wang, D.; Li, J. Preparation of popcorn-shaped CdS nanoparticles by hydrothermal method and their potent photocatalytic degradation efficiency. *Mater. Lett.* **2015**, *158*, 439–441. [CrossRef]
66. Ayodhya, D.; Venkatesham, M.; Bhagavanth Reddy, G.; Veerabhadram, G. One-pot sonochemical synthesis of CdS nanoparticles: Photocatalytic and electrical properties. *Int. J. Ind. Chem.* **2015**, *6*, 261–271. [CrossRef]
67. Zhang, L.; Sun, F.; Zuo, Y.; Fan, C.; Xu, S.; Yang, S.; Gu, F. Immobilisation of CdS nanoparticles on chitosan microspheres via a photochemical method with enhanced photocatalytic activity in the decolourisation of methyl orange. *Appl. Catal. B: Environ.* **2014**, *156*, 293–300. [CrossRef]
68. Qin, S.; Liu, Y.; Zhou, Y.; Chai, T.; Guo, J. Synthesis and photochemical performance of CdS nanoparticles photocatalysts for photodegradation of organic dye. *J. Mater. Sci. Mater. Electron.* **2017**, *28*, 7609–7614. [CrossRef]
69. Tatykayev, B.; Chouchene, B.; Balan, L.; Gries, T.; Medjahdi, G.; Girot, E.; Uralbekov, B.; Schneider, R. Heterostructured g-CN/TiO_2 photocatalysts prepared by thermolysis of g-CN/MIL-125 (Ti) composites for efficient pollutant degradation and hydrogen production. *Nanomaterials* **2020**, *10*, 1387. [CrossRef]
70. Sikora, P.; Augustyniak, A.; Cendrowski, K.; Nawrotek, P.; Mijowska, E. Antimicrobial Activity of Al_2O_3, CuO, Fe_3O_4, and ZnO Nanoparticles in Scope of Their Further Application in Cement-Based Building Materials. *Nanomaterials* **2018**, *8*, 15. [CrossRef]
71. Sanpui, P.; Chattopadhyay, A.; Ghosh, S.S. Induction of Apoptosis in Cancer Cells at Low Silver Nanoparticle Concentrations using Chitosan Nanocarrier. *ACS Appl. Mater. Interfaces* **2011**, *3*, 218–228. [CrossRef]
72. Ivask, A.; Kurvet, I.; Kasemets, K.; Blinova, I.; Aruoja, V.; Suppi, S.; Vija, H.; Käkinen, A.; Titma, T.; Heinlaan, M. Size-dependent toxicity of silver nanoparticles to bacteria, yeast, algae, crustaceans and mammalian cells in vitro. *PLoS ONE* **2014**, *9*, e102108. [CrossRef]
73. Schleifer, K.H.; Bell, J.A. Staphylococcus. In *Bergey's Manual of Systematics of Archaea and Bacteria*; Wiley: Hoboken, NJ, USA, 2015; pp. 1–43.
74. Sheutz, F.; Strockbine, N. Escherichia. In *Bergey's Manual of Systematics of Archaea and Bacteria*; Wiley: Hoboken, NJ, USA, 2015.
75. Augustyniak, A.; Jablonska, J.; Cendrowski, K.; Głowacka, A.; Stephan, D.; Mijowska, E.; Sikora, P. Investigating the release of ZnO nanoparticles from cement mortars on microbiological models. *Appl. Nanosci.* **2021**, *12*, 489–502. [CrossRef]
76. Augustyniak, A.; Cendrowski, K.; Grygorcewicz, B.; Jabłońska, J.; Nawrotek, P.; Trukawka, M.; Mijowska, E.; Popowska, M. The response of pseudomonas aeruginosa pao1 to uv-activated titanium dioxide/silica nanotubes. *Int. J. Mol. Sci.* **2020**, *21*, 7748. [CrossRef] [PubMed]

nanomaterials

Article

Preparation of Amino-Functionalized Mesoporous SBA-15 Nanoparticles and the Improved Adsorption of Tannic Acid in Wastewater

Tzong-Horng Liou [1,2,*], Guan-Wei Chen [1] and Shang Yang [1]

[1] Department of Chemical Engineering, Ming Chi University of Technology, 84 Gungjuan Road, Taishan District, New Taipei City 24301, Taiwan; u06137029@gmail.com (G.-W.C.); a0981578295@gmail.com (S.Y.)
[2] Battery Research Center of Green Energy, Ming Chi University of Technology, 84 Gungjuan Road, Taishan District, New Taipei City 24301, Taiwan
* Correspondence: thliou@mail.mcut.edu.tw; Tel.: +886-229-089-899 (ext. 4617)

Abstract: Ordered mesoporous Santa Barbara amorphous (SBA-15) materials have high surface areas and are widely used in adsorption, separation, filtration, and heterogeneous catalytic processes. However, SBA-15 surfaces contain hydroxyl groups that are unsuited to the adsorption of organic pollutants; thus, SBA-15 must be chemically modified to promote its adsorption activity. In this study, amino-functionalized nanoporous SBA-15 was fabricated by employing sodium silicate as a precursor. The structural characteristics of the prepared composites were examined using thermogravimetric analysis, X-ray diffraction, Fourier transform infrared spectrometry, field-emission scanning electron microscopy, transmission electron microscopy, and surface area analysis. The prepared SBA-15 had a large pore size (6.46–7.60 nm), large pore volume (1.037–1.105 cm^3/g), and high surface area (546–766 m^2/g). Functionalization caused a reduction in the SBA-15 pore volume and surface area, whereas amino groups that promoted an interaction between adsorbates and solids facilitated solute adsorption. The adsorption of tannic acid (TA) onto amino-modified silica composites (SBA-15 and 3-aminopropyltriethoxysilane (SBA-15/APTES) and SBA-15 and pentaethylenehexamine (SBA-15/PEHA)) was studied. Their adsorption capacities were affected by solution temperature, solution pH, agitation speed, adsorbent dosage, and initial TA concentration. The maximum adsorption capacities for SBA-15/APTES and SBA-15/PEHA were 485.18 and 413.33 mg/g, respectively, with SBA-15/APTES exhibiting ultrafast removal of TA (98.61% removal rate at 15 min). In addition, this study explored the thermodynamics, adsorption isotherms, and kinetics. A comparison of two types of amino-functionalized SBA-15 was used for the first time to adsorb TA, which providing valuable information on TA adsorption on high adsorption capacity materials in water media.

Keywords: SBA-15; amino groups; adsorption; tannic acid; mesostructure

Citation: Liou, T.-H.; Chen, G.-W.; Yang, S. Preparation of Amino-Functionalized Mesoporous SBA-15 Nanoparticles and the Improved Adsorption of Tannic Acid in Wastewater. *Nanomaterials* **2022**, *12*, 791. https://doi.org/10.3390/nano12050791

Academic Editor: Christos A. Aggelopoulos

Received: 7 February 2022
Accepted: 24 February 2022
Published: 26 February 2022

1. Introduction

Nanomaterials are widely applied in many fields such as pharmaceutics, energy, semiconductors, and the environment [1,2]. The synthesis of SBA-15 nanomaterials has attracted considerable attention since Chmelka et al. developed the process in 1998 [3]. According to the International Union of Pure and Applied Chemistry, mesoporous materials have pore sizes of 2–50 nm, and SBA-15 materials possess uniform mesopores, a high pore volume, large specific surface area, and stability, making them suitable for numerous applications related to adsorption [4], catalysis [5], tissue engineering [6], drug release [7], CO_2 capture [8], and energy cells [9]. SBA-15 is prepared in an acidic environment through the interaction between a surfactant template and silicon precursor, producing a two-dimensional and well-ordered hexagonal structure. As compared with microporous materials, such as zeolites, mesoporous SBA-15 material has large pore sizes (approximately 10–30 nm), which can decrease mass transfer resistance for large adsorbates, making it a potential material for the adsorption of wastewater. Sodium silicate (Na_2SiO_3) and tetraethyl orthosilicate (TEOS) are both typically used as a silicon source to synthesize SBA-15 materials [10,11]. However, the high-priced TEOS reagent increases the cost of raw materials. By contrast, sodium silicate is a low-cost silicon source for the mass manufacture of SBA-15. Converting sodium silicate into mesoporous silica materials can eliminate hazardous pollutants from wastewater and air.

Tannic acid (TA) is an organic compound that is frequently found in surface and drinking water [12]. TA is also a raw material used in many industries, including in the manufacturing of ink, medicines, paper, and leather products [13]. Recently, TA has raised concerns because of its toxicity to aquatic animals and plants [14]. Furthermore, in drinking water, TA can interact with chlorine in a manner that is hazardous to human health. Therefore, removing TA from water media is essential. Many methods, such as adsorption, membrane separation, biological processes, ion exchange, coagulation, and chemical oxidation, have been used to remove TA from aqueous solutions [15]. Adsorption is considered the most efficient method because it is less expensive, highly efficient, and straightforward [16].

Because it contains Si–OH and Si–O groups, the exterior of SBA-15 has a negative charge density. This makes it unsuitable for TA adsorption because organic molecules have a negative charge. Surface modification of SBA-15 with amino groups is effective in improving the adsorption activity. Castaldo et al. [17] prepared amino-functionalized hyper-crosslinked resins through a nitration-reduction process. Functionalization can promote the evolution of microporous structures, which effectively increase the adsorption of polar dyes and CO_2. Melnyk et al. [18] used a one-stage sol-gel method to synthesize amino silica spherical particles. The highest adsorption capacity values for acid red, fluorescein, and methylene blue dyes were 262, 132, and 146 mg/g, respectively. Tang et al. [19] synthesized polyethyleneimine-modified silica nanoparticles containing abundant amino groups. The maximum adsorption capability of acid organic II was as high as 705.3 mg/g. Babapour et al. [20] used a solvothermal method to fabricate a metal organic framework (MOF), which was then amino functionalized using triethylamine. The highest adsorption amount for Cr (VI) was 78.12 mg/g. Huang et al. [21] used the functionalization of chitosan by using tetraethylenepentamine to adsorb eosin Y dye. The highest amount of adsorption (292.4 mg/g) was observed at 25 °C.

In this study, two types of amino groups (aminopropyltriethoxysilane and pentaethylene hexamine) were used for the functionalization of SBA-15 to improve TA adsorption. SBA-15 was prepared with a surfactant template and sodium silicate precursor by using a hydrothermal treatment method. The effects of the calcination temperature, calcination time, hydrothermal treatment temperature, and hydrothermal treatment time on the pore textures of the SBA-15 specimens were observed. The surface characteristics and physical properties of specimens were examined. Furthermore, the adsorption of TA on SBA-15 materials after surface modification of amino groups was explored. The influences of the experimental parameters, such as solution temperature, solution pH, agitation speed, adsorbent dosage, and initial TA concentration, on the elimination of TA were also explored. Finally, the thermodynamics (entropy, enthalpy, and Gibbs free energy), isotherm adsorption (Freundlich and Langmuir equations), and kinetics data (pseudo-first-order and pseudo-second-order models) were evaluated.

2. Materials and Methods

2.1. Materials

Sodium silicate ($Na_2O(SiO_2)_x(H_2O)_x$, SiO_2 ~26.5 wt.%, Na_2O ~10.6 wt.%), 3-aminopro pyltriethoxysilane (APTES), pluronic triblock copolymer (P123, $EO_{20}PO_{70}EO_{20}$), and TA ($C_{76}H_{52}O_{46}$, 1701 g/mol) were obtained from Sigma-Aldrich (Taufkirchen, Germany). Pentaethylene hexamine (PEHA), ethanol (99.9 wt.%), hydrochloric acid, sodium hydroxide, and sulfuric acid were purchased from Merck (Gernsheim, Germany). High-purity (99.995%) mixed gas (79% N_2 and 21% O_2) was purchased from Sun Fu (Taipei, Taiwan).

2.2. Synthesis of the SBA-15 Materials

The SBA-15 specimens were fabricated utilizing sodium silicate as the silicon precursor and P123 as the template [22,23]. For a representative experiment, 2.0 g of P123 was added to 60 mL of 2.0 M HCl solution under continuous stirring at 300 rpm and 35 °C. Then, sodium silicate (5.2 g) was dissolved slowly into the P123 solution with agitation at 600 rpm and 35 °C. The suspension was stirred for 24 h, and then placed in a Teflon-lined autoclave. After heating the mixture at 100 °C for 24 h, a white precipitate was collected through filtration and washed by distilled water. The solid was dried at 60 °C for 24 h. Then, SBA-15 was acquired by calcining the solid in air at 550 °C for 6 h.

2.3. Synthesis of the Amino-Functionalized SBA-15 Specimens

Using a reflux apparatus, 2.0 g of SBA-15 and 2.0 g of APTES were mixed in 60 mL of anhydrous alcohol at 78 °C for 24 h [24]. The sediment was repeatedly rinsed with ethanol, and then dried in a vacuum oven at 60 °C. The collected sample was marked as SBA-15/APTES. In addition, 1.0 g of PEHA was dissolved in 50 mL of ethanol under agitation for 40 min [25]. After adding 2.0 g of SBA-15, the mixture was heated under reflux at 80 °C for 4 h. Finally, the specimen was dried in a vacuum oven at 60 °C. The collected specimen was marked as SBA-15/PEHA. Figure 1 shows a schematic representation for synthesis of amino-functionalized SBA-15 specimens.

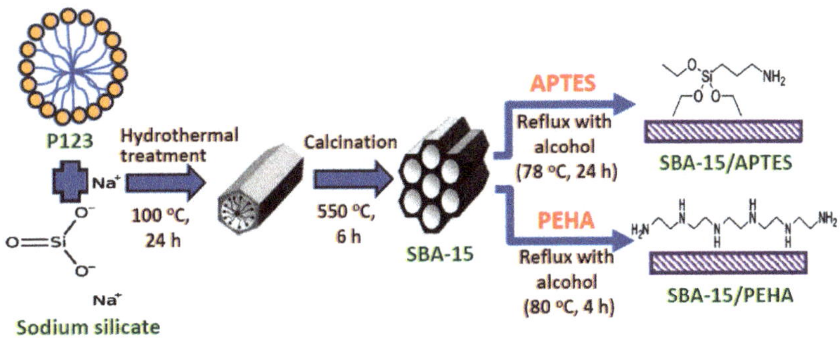

Figure 1. Schematic representation of SBA-15, SBA-15/APTES, and SBA-15/PEHA synthesis.

2.4. Characterization of the Silica Specimens

The pore characteristics and surface area of the mesoporous silica specimens were investigated using a N_2 sorption experiment at 77 K by using a Micrometrics ASAP 2020 apparatus (Norcross, GA, USA). The surface functional groups of the amino-modified SBA-15 samples were inspected by a Shimadzu FTIR-8300 analyzer (Nakagyo-ku, Kyoto, Japan). The mesophases and crystalline phases of the silica specimens were assembled on an X'pert pro system X-ray diffractometer (XRD, PANalytical, Malvern, UK) equipped with Cu-kα radiation. The mesoporous structure of the functionalized and unfunctionalized silica materials was inspected by a JEM-2100 (JEOL, Akishima, Tokyo, Japan) transmission electron microscope (TEM). The surface morphologies of the silica specimens were observed through an S-3400N (HITCHI, Chiyoda, Tokyo, Japan) field-emission scanning electron microscope (SEM). The thermal decomposition path of the uncalcined SBA-15 was observed through a thermogravimetric analyzer (TGA, Mettler Toledo, OH, USA, model TGA/SDTA851e). W/W_0 represented the remaining mass of the sample.

2.5. Adsorption Studies

The adsorption experiments were performed to observe the adsorption performance of the mesoporous silica materials by using a bath method to study changes in solution temperature, solution pH, agitation speed, initial TA concentration, and adsorbent dosage. For a typical experiment, 50 mg of adsorbent (SBA-15, SBA-15/APTES, or SBA-15/PEHA) was put into 250 mL of the TA solution, and the solution was agitated constantly. The initial TA concentrations were changed from 10 to 50 mg/L, adsorbent dosages from 10 to 100 mg, agitation speed from 100 to 250 rpm, solution pH from 2 to 5, and solution temperatures from 25 to 60 °C. After adsorption, the residual concentrations of TA at constant periods from 0 to 120 min were determined using the ultraviolet–visible Genesys spectrophotometer (Thermo Electron Corporation, Waltham, MA, USA). The maximum absorbance wavelength was 278 nm. A standard solution was used to calibrate the UV–visible spectrophotometer for every analysis. The amount of TA adsorption capacity (q_t, mg/g) and removal efficiency (R, %) of the adsorbents were evaluated using the formulas:

$$q_t = \frac{(C_0 - C_t)V}{W} \quad (1)$$

$$R = \frac{C_0 - C_t}{C_0} \times 100 \quad (2)$$

where W (g) is the mass of adsorbent, C_0 and C_t (mg/L) are the TA concentrations at the beginning and at time t, and V (L) is the volume of the TA solution.

3. Results and Discussion

3.1. Analysis of Phase Features and Surface Functional Groups

The XRD technique is frequently developed to analyze the characteristics of phase composition of silica materials in synthesis process. Figure 2 presents the XRD patterns with small-angle diffraction of silica samples obtained under different synthesis conditions. These samples exhibited three peaks at (100), (110), and (200) planes, corresponding to two-dimensional and hexagonal SBA-15 mesostructures [26]. Figure 2a displays the XRD images of the SBA-15 prepared at hydrothermal treatment time of 12–72 h. The peak intensity increased with an increase in hydrothermal time, indicating that increasing the hydrothermal time can help the formation of a well-ordered SBA-15 structure. The change in hydrothermal temperature from 25 to 130 °C (Figure 2b) exhibited the same trend as that of the change in hydrothermal time, indicating that enhancing hydrothermal temperature favors the formation of well-ordered mesoporous material. However, when the temperature was enhanced to 130 °C, the (110) and (200) peaks disappeared, indicating that violent hydrothermal reaction may destroy the micellar template, leading to a collapse of pore structure. Therefore, the optimum hydrothermal treatment temperature is 100 °C. The XRD patterns of the silica specimens synthesized at calcination time of 2–8 h are displayed in Figure 2c. The aim of calcination is to eliminate organic surfactants to obtain a porous solid. A high-ordered SBA-15 structure was observed by increasing the calcination time. The optimum calcination time was identified as 6 h. The XRD images of the silica specimens synthesized at calcination temperature of 160–760 °C are presented in Figure 2d. The peak intensity was increased with an increase in calcination temperature, suggesting that increasing the calcination temperature was useful in the synthesis of well-ordered SBA-15 materials with hexagonal symmetry. However, at a low calcination temperature of 160 °C, the three peak intensities were obviously weakened. This is probably because the surfactant template was not removed completely.

Figure 3a displays XRD images before and after amino-functionalized SBA-15 materials. The XRD data exhibit three peaks at (200), (110), and (100), implying that the pure SBA-15 sample has hexagonal symmetry. In the same situation, the amino-modified samples (SBA-15/APTES and SBA-15/PEHA) also maintained an intense (100) peak and weaker (110) and (200) peaks. The observation further confirmed that both SBA-15/APTES and SBA-15/PEHA maintained a hexagonally ordered structure after functionalization [27]. Figure 3b presents an XRD pattern with a wide-range scan ($2\Theta = 10°–80°$) of the SBA-15 sample. A broad peak presented at diffraction angles from 20° to 30°, indicating the formation of disordered cristobalite [28].

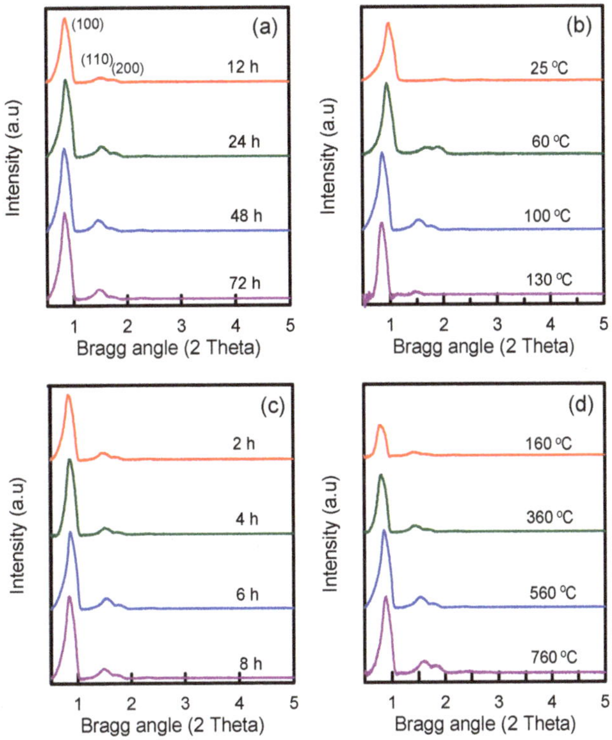

Figure 2. XRD images of SBA-15 samples prepared under various synthesis conditions: (**a**) Hydrothermal treatment time; (**b**) hydrothermal treatment temperature; (**c**) calcination time; (**d**) calcination temperature.

Qualitative identification of functional groups was accomplished by FTIR spectroscopy. The FTIR spectra of the amino-functionalized and unfunctionalized SBA-15 materials are presented in Figure 4a. In the pure SBA-15 sample, a wide peak at 3200–3700 cm^{-1} was caused by the presence of O–H groups [29]. The C=O vibrations were observed at a band of 1635 cm^{-1}. The C=C groups were identified at 1605 cm^{-1}, and the band at 1375 cm^{-1} was the –CH_3 stretching of the methyl group. The band at 980 cm^{-1} represented the appearance of the Si–OH groups, and Si–O–Si groups were located at 790 and 450 cm^{-1} [30]. In the SBA-15/APTES and SBA-15/PEHA samples, amino group stretching vibrations were found at 692, 1499, and 1572 cm^{-1} [31]. This verified that amino groups were grafted onto the silica surface. Anbia and Salehi [32] observed that broad $-NH_2$ vibrations appeared at 3250–3450 cm^{-1} in amino-modified silica samples. Nevertheless, the functional group was not easily examined due to the fact that the $-NH_2$ bands may be covered by the wide band of the –OH groups. Furthermore, SBA-15/PEHA exhibited a stronger relative amino peak intensity than SBA-15/APTES. Figure 4b displays the TG curve of the uncalcined SBA-15 sample. The mass loss occurring at temperatures lower than 200 °C was caused by water evaporation. The mass loss decreased slowly with an increase in thermal decomposition temperature because a longer heating time was required for the release of the surfactant matter from the high-porosity SBA-15 structure. The differential TG curve revealed only one peak at 195 °C. Similar results were observed by Liou [33] for the thermal decomposition of an uncalcined MCM-41 sample in air.

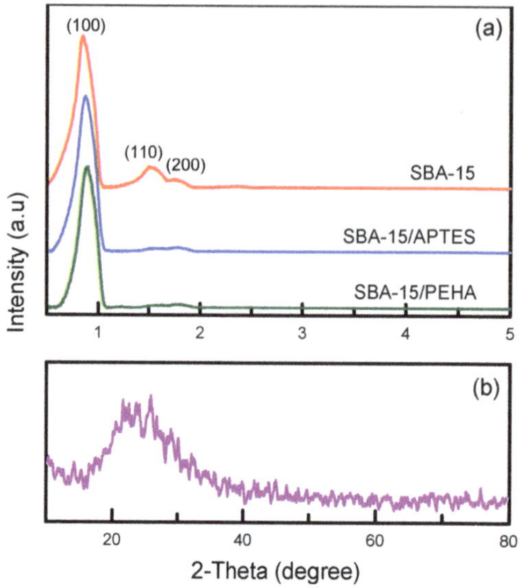

Figure 3. (**a**) XRD images of unfunctionalized and amino-functionalized silica samples; (**b**) SBA-15 sample.

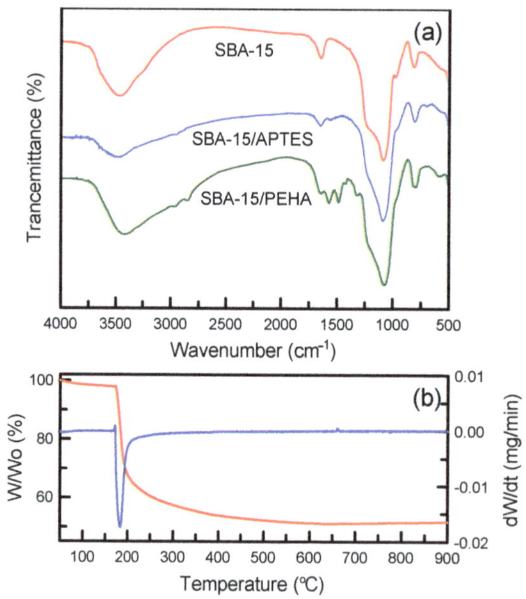

Figure 4. (**a**) FTIR spectra of unfunctionalized and amino-functionalized silica samples; (**b**) TG and DTG thermograms of uncalcined SBA-15 samples.

3.2. Analysis of the Pore Structure

The pore structure is an important physical property of mesoporous materials. Nitrogen isotherm method can standardly determine the porous structure and textural parameters. Figure 5a displays the nitrogen sorption isotherms of the SBA-15 specimens prepared at hydrothermal treatment times of 12–72 h. Pure SBA-15 exhibited a type-IV hysteresis loop with type H1 at a relative pressure of P/P_0 = 0.6–0.9, which is characteristic of a mesoporous structure [34]. By comparing the SBA-15 samples prepared at various hydrothermal treatment times, it was revealed that the isotherms were similar. This observation indicates that an increase in hydrothermal time does not affect the mesostructure of the SBA-15 framework. Figure 5b presents the features of pore size distribution. The SBA-15 samples had pore sizes of 6.62–7.60 nm. The SBA-15 obtained at hydrothermal times of 12–24 h had a smaller pore diameter than those obtained at 48–72 h. These samples displayed a narrow and uniform pore distribution, implying that the pore structure of SBA-15 was stable.

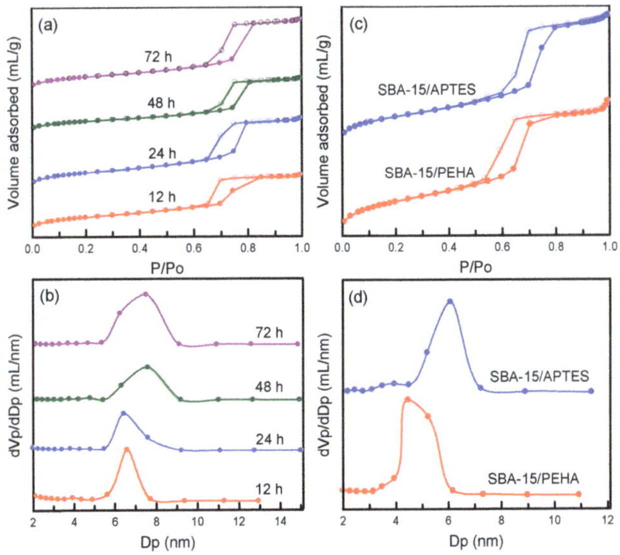

Figure 5. Nitrogen sorption isotherms and pore size distributions: (**a**,**b**) SBA-15 samples at different hydrothermal treatment times; (**c**,**d**) SBA-15/APTES and SBA-15/PEHA samples.

The isotherms of the amino-functionalized SBA-15/APTES and SBA-15/PEHA specimens are presented in Figure 5c. The two isotherms also exhibited a type-IV hysteresis loop, implying a mesoporous material. The amino-functionalized SBA-15 samples (Figure 5c) exhibited similar loops to those of pure SBA-15 (Figure 5a), indicating that the combination of SBA-15 and amino groups did not destroy the silica mesostructure. The pore size of SBA-15/APTES (6.10 nm) was larger than that of SBA-15/PEHA (4.49 nm; Figure 5d). Table 1 lists the pore volume, surface area, and mesopore fraction of the unfunctionalized and amino-functionalized silica specimens. The surface areas of the pure SBA-15 samples ranged from 546 to 766 m^2/g, which decreased with an increase in hydrothermal treatment time. The SBA-15 at a hydrothermal time of 48 h had the largest pore volume (1.105 cm^3/g). The SBA-15 samples exhibited higher surface area and pore volume as compared with the SBA-15/APTES and SBA-15/PEHA samples. The addition of amino functional groups into silica pores led to reductions in textural parameters. This is consistent with the literature [35]. Moreover, the pore volume and surface area of SBA-15/APTES (0.521 cm^3/g and 264 m^2/g) were higher than those of SBA-15/PEHA (0.272 cm^3/g and 167 m^2/g). The mesopore fractions of the pure SBA-15 samples ranged from 97.69% to 99.42%, which increased with an increase in hydrothermal time.

Table 1. Surface area, pore volume, and pore diameter of silica specimens.

Sample	S_{BET} (m^2/g)	V_t (cm^3/g)	V_{mic} (cm^3/g)	V_{meso} (cm^3/g)	V_{meso}/V_t (%)	d_p (nm)
SBA-15–12 h	766	1.038	0.024	1.014	97.69	6.62
SBA-15–24 h	655	1.044	0.014	1.030	98.66	6.46
SBA-15–48 h	604	1.105	0.009	1.096	99.19	7.60
SBA-15–72 h	546	1.037	0.006	1.031	99.42	7.55
SBA-15/APTES	264	0.521	0.005	0.516	99.04	6.10
SBA-15/PEHA	167	0.272	0.005	0.267	98.16	4.49

S_{BET}, specific surface area; V_t, total pore volume; V_{mic}, micropore volume; V_{meso}, mesopore volume; d_p, pore diameter (BJH desorption).

3.3. Surface Morphology of the Mesoporous Silica Samples

The SEM technique was utilized to observe the changes of surface morphology before and after amination of samples. The SEM pictures of SBA-15 without amination are presented in Figure 6a–d. The SBA-15 sample comprised agglomerates of rod-like particles (Figure 6a), consistent with the literature [36], whereas the SBA-15 sample acquired at a hydrothermal treatment time of 12 h exhibited irregular-shaped particles, indicating that the silica structure was not completely developed (Figure 6b). The SBA-15 samples obtained at hydrothermal times of 24 and 72 h had a glossy and regular surface (Figure 6c,d), indicating the formation of organized structures. This is consistent with the XRD measurements in Figure 2a. The width of the SBA-15 particles was in the range between 380 and 830 nm. The morphology of SBA-15/APTES and SBA-15/PEHA (Figure 6e,f) did not vary as much as that of pure SBA-15, but the average width was reduced by adding the amino groups (330–610 nm for SBA-15/APTES and 300–560 nm for SBA-15/PEHA).

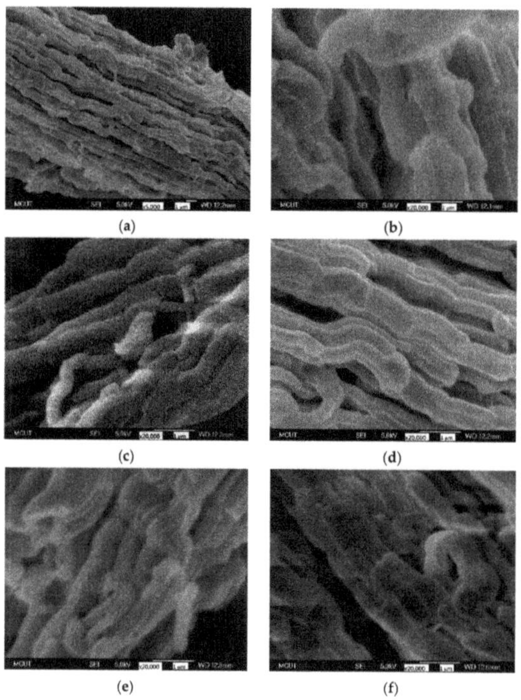

Figure 6. SEM images: (**a**) SBA-15; (**b**–**d**) SBA-15 synthesized at hydrothermal treatment times of 12, 24, and 72 h, respectively; (**e**) SBA-15/APTES; (**f**) SBA-15/PEHA.

The TEM characterization technique was utilized to identify the difference of the microstructure of the prepared silica materials. The mesostructures of the amino-functionalized and unfunctionalized SBA-15 samples were investigated using TEM. Figure 7a,b verify that SBA-15 consists of parallel pore channels and highly ordered mesopores. The silica sample also possessed a large-scale hexagonal structure (Figure 7b). The mesostructure was formed through the self-assembly of surfactant micelles and silicate species [37]. The average pore size of 7.80 nm was similar to the N_2 adsorption test in Figure 5b. After the amination reaction, the SBA-15/APTES (Figure 7c) and SBA-15/PEHA (Figure 7d) morphology was analogous to pure SBA-15. However, the pore diameters were reduced to 6.25 and 5.00 nm, which was attributed to the amino group coating on the silica surface.

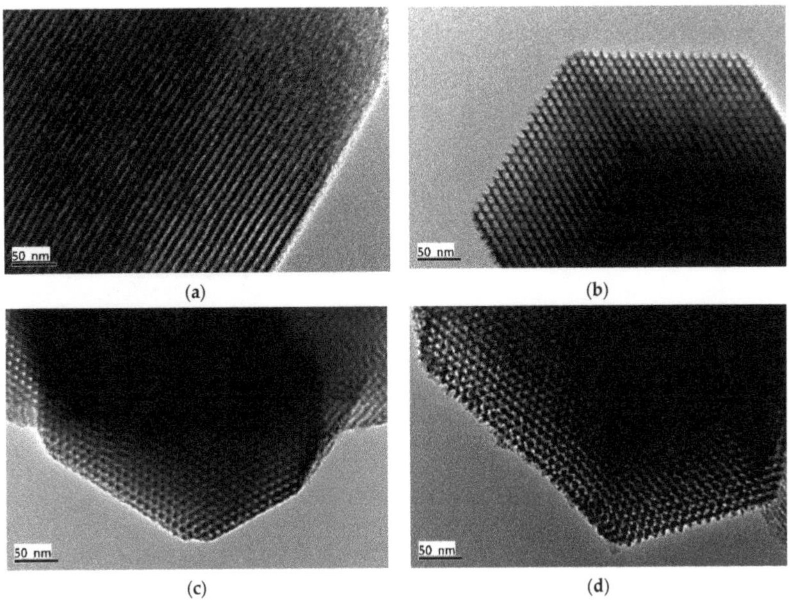

Figure 7. TEM images: (**a**,**b**) SBA-15; (**c**) SBA-15/APTES; (**d**) SBA-15/PEHA.

3.4. TA Adsorption Study

TA organic compound was utilized as the adsorbate to characterize the adsorption performance of the amino-functionalized SBA-15 materials. In order to evaluate the efficacy of the prepared adsorbents, the equilibrium adsorption of the TA was studied by changing various operating conditions. In the study, different types of amino-modified silica samples were used to remove TA from the aqueous solution. The influence of contact time on the adsorption capacities of adsorbents is exhibited in Figure 8. The adsorption capacity decreased in the following order: SBA-15/APTES > SBA-15/PEHA > pure SBA-15. The hydroxyl groups on the silica surface could not supply adsorption sites sufficiently strong to interact with the TA molecules. Therefore, the pure SBA-15 sample had the lowest adsorption performance. The amino-modified SBA-15 specimens showing the higher adsorption capacity can be explained as follows: Typically, amino groups are favorable to the adsorption of anionic species, and TA is an anionic compound. In the aqueous solution, the electrostatic attraction between the negatively charged TA molecules and positively charged amino groups on the SBA-15 surface ($-NH_3^+$) led to an increase in TA adsorption [32]. Furthermore, the SBA-15/APTES specimen had a higher surface area and pore volume than the SBA-15/PEHA specimen (Table 1) and, consequently, SBA-15/APTES exhibited more efficient adsorption than SBA-15/PEHA. The contact time curves also demonstrated that TA uptake was rapid in the first few minutes, and then proceeded at a slower rate. The rapid adsorption during the initial adsorption period was because the adsorbent surface was unoccupied. After the adsorption reached the saturation point, the slow permeation of TA into the silica pores resulted in a reduction in the adsorption rate.

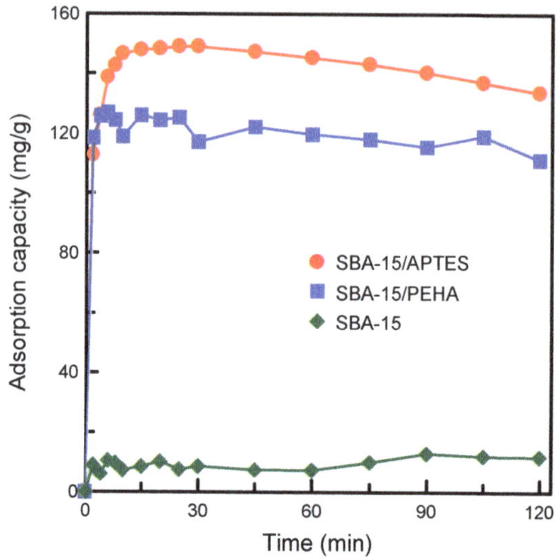

Figure 8. Effects of contact time on the adsorption performance of various kinds of adsorbents.

Figure 9a,b illustrates the influence of the initial TA concentration on adsorption capacity and the removal efficiency of amino-modified adsorbents. The adsorption capacity increased rapidly with an increase in the initial TA concentration. A higher initial TA concentration resulted in an increased mass gradient between the solute and adsorbent, which accelerated the diffusion of the TA molecules onto the SBA-15 surface. In addition, collisions between the TA and adsorbent also increased when the initial TA concentration increased. These results enhance the adsorption activity. By contrast, the removal efficiency decreased with an increase in the initial TA concentration. This was also observed in the elimination of reactive blue 21 dye from manganese oxide nanoparticles [38]. The SBA-15/APTES samples revealed a higher adsorption capacity and removal efficiency as compared with those of SBA-15/PEHA. For SBA-15/APTES specimen, a maximum removal efficiency of approximately 94% was observed at an initial concentration of 10 mg/L, suggesting that the adsorbent had sufficient ability to treat trace pollutants.

The effect of adsorbent dosage on the adsorption capacity and removal efficiency of the adsorbents is displayed in Figure 9c,d. The results indicate that increasing the adsorbent dosage from 10 to 100 mg increases the removal efficiency but decreases the adsorption capacity. Removal efficiency is increased because the available surface areas and adsorption sites are increased, whereas the reduction in adsorption capacity is because the adsorption sites are not saturated. This is consistent with a report by El-Sewify et al. on the adsorption of malachite green dye on zinc MOFs [39]. The SBA-15/APTES and SBA-15/PEHA samples displayed the maximum adsorption capacities of 485.15 and 423.33 mg/g and the highest removal efficiency of 98.61% and 87.43%, respectively.

Agitation speed can influence the dispersion of solute in the solution, which is an important factor in the adsorption process. The influence of agitation speed on adsorption capacity and removal efficiency is illustrated in Figure 9e,f. For the SBA-15/APTES specimen, the removal efficiency and adsorption capacity were enhanced with an increase in the agitation speed, whereas the adsorption capacity and removal efficiency were enhanced for the SBA-15/PEHA specimen when the agitation speed increased from 100 to 150 rpm, and then remained constant. Increasing the agitation speed can reduce the film resistance of the adsorbent, resulting in an increase in TA adsorption.

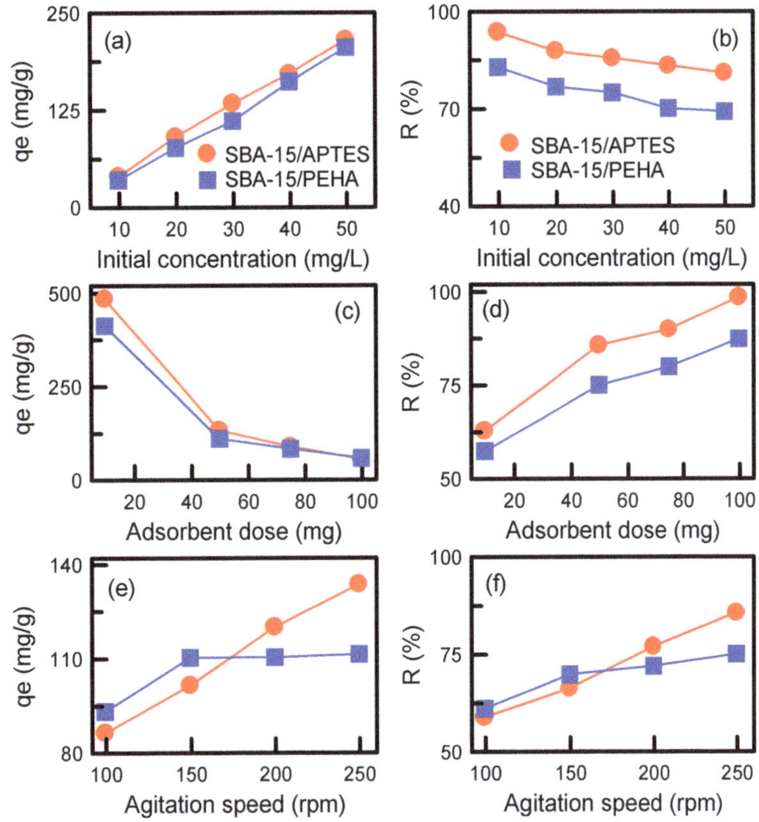

Figure 9. Effects of various adsorption conditions on the adsorption capacity and removal efficiency of adsorbents: (**a**,**b**) Initial TA concentration; (**c**,**d**) adsorbent dosage; (**e**,**f**) agitation speed.

The influence of solution pH on the adsorption capacity and removal efficiency of the adsorbents is presented in Figure 10a,b. The removal efficiency and adsorption capacity were increased with an increase in the pH values. The maximum adsorption capacities of the two adsorbents (133.76 and 111.37 mg/g) were observed at pH 5. TA exists in its molecular form at pH 4.5. The adsorption capacity at pH 5 was enhanced because of the increasing hydrogen bonding and electrostatic force attraction between the TA molecules and SBA-15 surface [31]. However, TA is an anionic molecule, and at a pH of 8–11, the negatively charged sites on the adsorbent surface increased. The existence of excess OH^- ions compete with the TA anions, and result in the dissociation of TA during the adsorption process. Hence, this study focused on TA adsorption in acidic media.

The influence of solution temperature on the adsorption of TA is presented in Figure 10c,d. The adsorption capacity and removal efficiency were increased with a decrease in the solution temperature. The maximum removal efficiency and adsorption capacity presented at 25 °C, indicating an exothermic process. Temperature was a key factor for SBA-15/APTES in the adsorption process. However, SBA-15/PEHA was not sensitive to the solution temperature. Increasing solution temperature may cause an increase in TA solubility, weakening the interaction forces between the adsorbent and adsorbate and decreasing TA adsorption. Koyuncu and Okur [40] observed similar behavior in the adsorption of acid violet 90 dye on ordered mesoporous carbon.

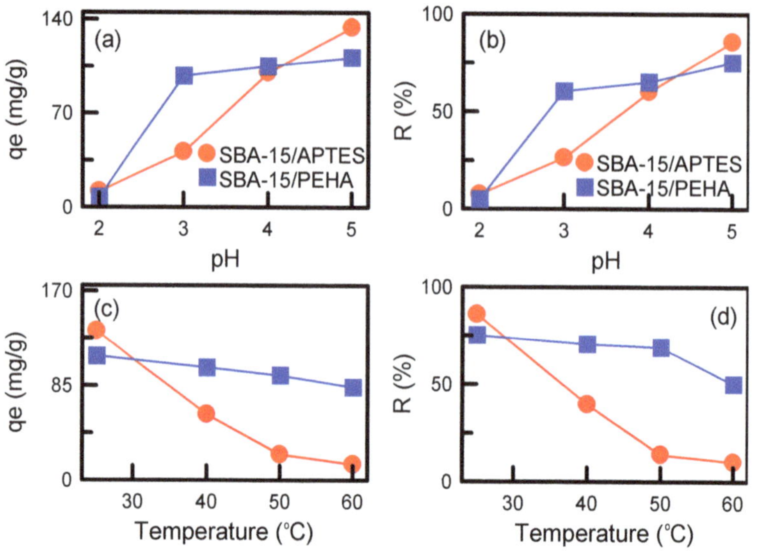

Figure 10. Effects of various adsorption conditions on the adsorption capacity and removal efficiency of adsorbents: (**a**,**b**) Solution pH; (**c**,**d**) solution temperature.

3.5. Thermodynamic Studies

Thermodynamic parameters provide in-depth information on internal energy changes associated with adsorption. The standard free energies (Δ*G*, kJ/mol) for the adsorption of TA onto SBA-15/APTES and SBA-15/PEHA were calculated using the thermodynamic equilibrium constant (K_c), which is written as below [41]:

$$\Delta G = -RT \ln K_c \quad (3)$$

$$K_c = \frac{C_s}{C_e} \quad (4)$$

where R is the gas constant (8.314 J/mol-K); T is the absolute temperature (K); and C_s and C_e (mg/L) are the equilibrium concentration of TA on the solid and in the solution, respectively.

The enthalpy (Δ*H*, kJ/mol) and entropy (Δ*S*, J/mol-K) can be estimated by the van't Hoff equation:

$$\ln K_c = \frac{\Delta S}{R} - \frac{\Delta H}{RT} \quad (5)$$

ΔS and ΔH can be acquired from the intercept and slope of Equation (5) (Figure 11), and Table 2 lists the thermodynamic parameters. The ΔG values were negative at 25 and 40 °C, indicating that adsorption was spontaneous. However, the ΔG values increased with an increase in temperature, suggesting that the adsorption process was not feasible at a high temperature. The negative values of ΔH (−100.99 and −24.58 kJ/mol) further confirmed that the adsorption process was exothermic. The negative values of ΔS (−324.99 and −72.99 J/mol-K) reflect the decrease in the freedom of mobility of TA at the liquid–solid interface of the adsorption process [42].

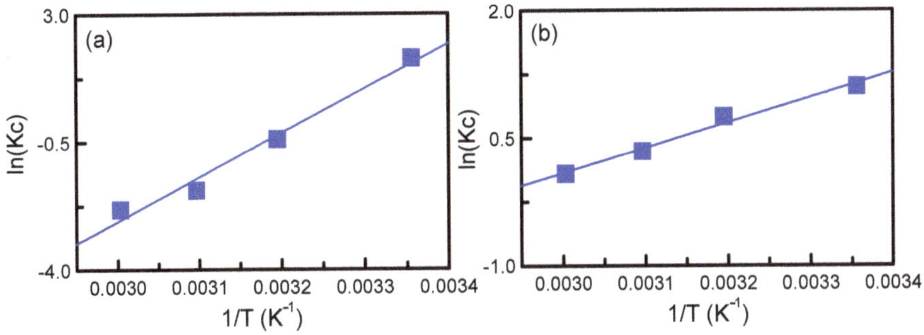

Figure 11. Plot of equilibrium constant versus temperature by using Van't Hoff equation: (**a**) SBA-15/APTES; (**b**) SBA-15/PEHA.

Table 2. Thermodynamic parameters for TA adsorption onto amino-modified silica specimens.

Sample	ΔS (J/K × mol)	ΔH (kJ/mol)	ΔG (kJ/mol)			
			25 °C	40 °C	50 °C	60 °C
SBA-15/APTES	−324.99	−100.99	−4.434	−1.096	4.909	6.507
SBA-15/PEHA	−72.99	−24.58	−2.734	−1.953	−0.926	−0.227

3.6. Adsorption Isotherm Experiments

The adsorption equilibrium between the solid and liquid surfaces in the adsorption processes can be characterized using the Langmuir and Freundlich equations. The adsorption isotherm experiments were performed using 10 mg of SBA-15/APTES or SBA-15/PEHA in 50 mL of TA solution. To reach adsorption equilibrium, the solution was stirred for 24 h. The concentrations of TA were altered from 10 to 50 mg/L. The solution pH was 5.

The linear forms of the Langmuir and Freundlich equations are [43]:

$$\frac{1}{q_e} = \frac{1}{q_L} + \frac{1}{q_L K_L C_e} \quad (6)$$

$$\log q_e = \log K_F + \frac{1}{n} \log C_e \quad (7)$$

where C_e (mg/L) is the equilibrium concentration, q_e (mg/g) is the adsorption capacity, K_F and n are the Freundlich constants, K_L (mL/mg) is the Langmuir constant, and q_L (mg/g) is the maximum adsorption capacity.

The Freundlich and Langmuir equation plots are displayed in Figure 12, and the corresponding parameters are given in Table 3. The R^2 value was employed to estimate the fit of the models to the experimental data. The adsorption of TA onto SBA-15/APTES and SBA-15/PEHA followed both Langmuir and Freundlich isotherms, as shown by R^2 values (0.98–0.99) in Table 3 [44]. The maximum adsorption capacities (q_L) of SBA-15/APTES and SBA-15/PEHA were 418.41 and 303.03 mg/g, respectively. The R_L values of the two adsorbents were within the range of 0–1, with $n > 1$ indicating an advantageous adsorption course. This result implies that adsorption was likely to occur [45].

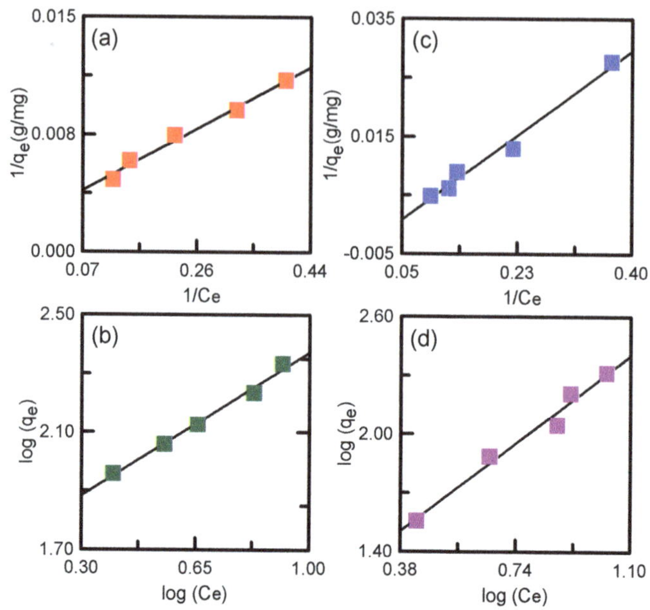

Figure 12. Plots of Langmuir model for (**a**) SBA-15/APTES and (**c**) SBA-15/PEHA; Freundlich model for (**b**) SBA-15/APTES and (**d**) SBA-15/PEHA.

Table 3. Isotherm constants for TA adsorption onto amino-modified silica specimens.

Sample	Langmuir				Freundlich		
	R_L	q_L (mg/g)	K_L	R^2	n	K_F (mg/g)	R^2
SBA-15/APTES	0.154	418.41	0.11	0.986131	1.44	47.49	0.990632
SBA-15/PEHA	0.333	303.03	0.04	0.984151	1.02	10.7	0.978013

3.7. Kinetic Studies

To further realize the adsorption mechanism, pseudo-first-order and pseudo-second-order models were examined. The equations for the two models are [46]:

$$q_t = q_e(1 - e^{-k_1 t}) \tag{8}$$

$$\frac{t}{q_t} = \frac{1}{k_2 q_2^2} + \frac{t}{q_e} \tag{9}$$

where k_1 and k_2 are the rate constants of each model, respectively; q_t and q_e (mg/g) are the adsorption capacities of adsorbents at time t and at equilibrium, respectively. The solution pH was 5.

Adsorption capacity versus time using Equations (8) and (9) is plotted in Figure 13. In Table 4, R^2 indicates that the pseudo-second-order model provides the optimum fitting for the adsorption of TA onto SBA-15/APTES and SBA-15/PEHA, suggesting that the adsorption was mainly controlled by chemisorption. Furthermore, the experimental q_e values (215.78 and 205.52 mg/g) are similar to the calculated q_e values (218.34 and 206.61 mg/g).

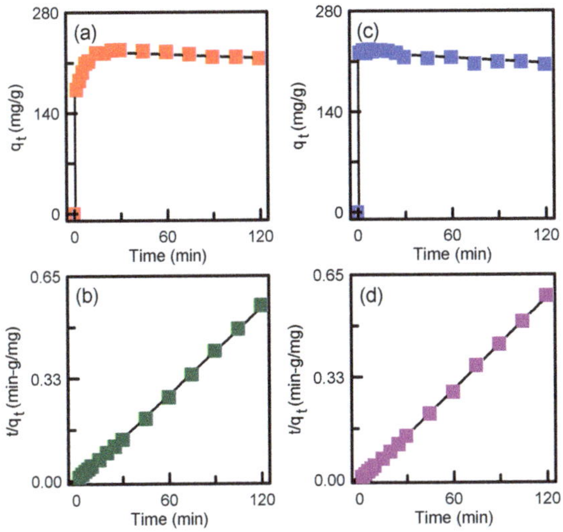

Figure 13. Plots of pseudo-first-order model for (**a**) SBA-15/APTES and (**c**) SBA-15/PEHA; pseudo-second-order model for (**b**) SBA-15/APTES and (**d**) SBA-15/PEHA.

Table 4. Kinetics constants for TA adsorption onto amino-modified silica specimens.

Model	Parameter	Value	
		SBA-15/APTES	**SBA-15/PEHA**
Pseudo-first-order adsorption kinetic	$q_{e,experiment}$ (mg/g)	215.78	205.52
	$q_{e,calculated}$ (mg/g)	219.25	218.08
	k_1 (min^{-1})	0.6473	43.49
	R^2	0.97224	0.98146
Pseudo-second-order adsorption kinetic	$q_{e,experiment}$ (mg/g)	215.78	205.52
	$q_{e,calculated}$ (mg/g)	218.34	206.61
	k_2 (min^{-1})	0.02	0.01
	R^2	0.99971	0.99978

Table 5 displays the maximum adsorption capacity of TA onto different adsorbents. The SBA-15/APTES and SBA-15/PEHA samples prepared in this study had high adsorption capacities (485 and 413 mg/g, respectively) as compared with adsorbents used in other studies. Thus, the amino-functionalized and mesoporous silica-based materials synthesized in this work could be utilized as efficient adsorbents for the elimination of TA from aqueous solutions.

Table 5. Maximum adsorption capacity for TA adsorption onto various adsorbents.

Adsorbent	q (mg/g)	Reference
Montmorillonite	219	[47]
Activated carbon	162	[48]
Clay	138	[49]
Polyaniline	117	[50]
Chitosan/NaOH/fly ash composite	244	[51]
Porous crosslinked polystyrene	248	[52]
Biochar	87	[53]
SBA-15/APTES	485	Present work
SBA-15/PEHA	413	Present work

4. Conclusions

The preparation of amino-functionalized SBA-15 composites for the adsorption of TA was investigated in the present work. The XRD observations indicate that the mesostructure of SBA-15 materials is highly sensitive to hydrothermal time, hydrothermal temperature, calcination time, and calcination temperature. The FTIR analysis revealed that amino groups were successfully grafted onto the SBA-15 surface, and TEM verified that the silica pore framework remained nearly unchanged after the amination reaction. The surface area and pore volume of the adsorbents were SBA-15 > SBA-15/APTES > SBA-15/PEHA. The adsorption efficiency of SBA-15 material was enhanced by way of surface modification with amino groups, and the adsorption capacity increased because of the formation of an electrostatic interaction and hydrogen bond between the adsorbent surface and TA molecules. SBA-15/APTES exhibited higher adsorption activity for the removal of TA than SBA-15/PEHA. The adsorption capacities were increased with increases in the initial TA concentration, agitation speed, and solution pH. However, increasing the adsorbent dosage and solution temperature resulted in a reduction in adsorption capacity. Pure SBA-15 had little affinity for TA in aqueous solution. Amino-functionalized SBA-15 presented higher TA adsorption efficiency than pure SBA-15. The results of the study identified high-efficiency adsorbents that could be applied to the removal of organic pollutants. As future perspectives, we propose to use magnetic iron compounds to facilitate the separation for larger scale application, as well as to adsorb other pollutants of interest, for example, hazardous heavy metals, and finally to a real industrial wastewater.

Author Contributions: Conceptualization, T.-H.L.; methodology, G.-W.C.; formal analysis, S.Y.; investigation, G.-W.C.; data curation, S.Y.; writing (original draft), T.-H.L.; writing (review and editing), T.-H.L. All authors have read and agreed to the published version of the manuscript.

Funding: This research received no external funding.

Institutional Review Board Statement: Not applicable.

Informed Consent Statement: Not applicable.

Data Availability Statement: Data sharing not applicable.

Acknowledgments: The authors gratefully appreciate the financial supports by the Ministry of Science and Technology (MOST) of Taiwan through project no. MOST 110-2221-E-131-008-MY2.

Conflicts of Interest: The authors declare no conflict of interest.

References

1. Zielinska, A.; Costa, B.; Ferreira, M.V.; Miguéis, D.; Louros, J.M.S.; Durazzo, A.; Lucarini, M.; Eder, P.; Chaud, M.V.; Morsink, M.; et al. Nanotoxicology and nanosafety: Safety-by-design and testing at a glance. *Int. J. Environ. Res. Public Health* **2020**, *17*, 4657. [CrossRef] [PubMed]
2. Souto, E.B.; Silva, G.F.; Dias-Ferreira, J.; Zielinska, A.; Ventura, F.; Durazzo, A.; Lucarini, M.; Novellino, E.; Santini, A. Nanopharmaceutics: Part I—Clinical trials legislation and good manufacturing practices (GMP) of nanotherapeutics in the EU. *Pharmaceutics* **2020**, *12*, 146. [CrossRef] [PubMed]
3. Zhao, D.; Feng, J.; Huo, Q.; Melosh, N.; Fredrickson, G.H.; Chmelka, B.F.; Stucky, G.D. Triblock copolymer syntheses of mesoporous silica with periodic 50 to 300 angstrom pores. *Science* **1998**, *279*, 548–552. [CrossRef]
4. Xian, Q.; Chen, L.; Fan, W.; Liu, Y.; He, X.; Dan, H.; Zhu, L.; Ding, Y.; Duan, T. Facile synthesis of novel Bi^0-SBA-15 adsorbents by an improved impregnation reduction method for highly efficient capture of iodine gas. *J. Hazard. Mater.* **2022**, *424*, 127678. [CrossRef]
5. Chongdar, S.; Bhattacharjee, S.; Azad, S.; Bal, R.; Bhaumik, A. Selective *N*-formylation of amines catalysed by Ag NPs festooned over amine functionalized SBA-15 utilizing CO_2 as C1 source. *Mol. Catal.* **2021**, *516*, 111978. [CrossRef]
6. Vargas-Osorio, Z.; Klotschan, A.; Arango-Ospina, M.; Piñeiro, Y.; Liverani, L.; Rivas, J.; Michálek, M.; Galusek, D.; Boccaccini, A.R. Effect of glycerol and H_3PO_4 on the bioactivity and degradability of rod-like SBA-15 particles with active surface for bone tissue engineering applications. *Microporous Mesoporous Mat.* **2022**, *329*, 111543. [CrossRef]
7. Su, H.L.; Xu, L.; Hu, X.J.; Chen, F.F.; Li, G.; Yang, Z.K.; Wang, L.P.; Li, H.L. Polymer grafted mesoporous SBA-15 material synthesized via metal-free ATRP as pH-sensitive drug carrier for quercetin. *Eur. Polym. J.* **2021**, *148*, 110354. [CrossRef]

8. Chen, C.; Wang, H.; Chen, Y.; Wei, X.; Zou, W.; Wan, H.; Dong, L.; Guan, G. Layer-by-layer self-assembly of hierarchical flower-like HKUST-1-based composite over amino-tethered SBA-15 with synergistic enhancement for CO_2 capture. *Chem. Eng. J.* **2021**, *413*, 127396. [CrossRef]
9. Rajabi, Z.; Javanbakht, M.; Hooshyari, K.; Adibi, M.; Badiei, A. Phosphoric acid doped polybenzimidazole based polymer electrolyte membrane and functionalized SBA-15 mesoporous for elevated temperature fuel cell. *Int. J. Hydrogen Energy* **2021**, *46*, 33241–33259. [CrossRef]
10. Chiani, E.; Azizi, S.N.; Ghasemi, S. PdCu bimetallic nanoparticles decorated on ordered mesoporous silica (SBA-15) /MWCNTs as superior electrocatalyst for hydrogen evolution reaction. *Int. J. Hydrogen Energy* **2021**, *46*, 25468–25485. [CrossRef]
11. Wang, Z.; Sun, R.; Wang, P.; Wang, W. Unit-cell wide SBA-15 type mesoporous silica nanoparticles. *Microporous Mesoporous Mat.* **2021**, *328*, 111491. [CrossRef]
12. Wei, W.; Li, J.; Han, X.; Yao, Y.; Zhao, W.; Han, R.; Li, S.; Zhang, Y.; Zheng, C. Insights into the adsorption mechanism of tannic acid by a green synthesized nano-hydroxyapatite and its effect on aqueous Cu(II) removal. *Sci. Total Environ.* **2021**, *778*, 146189. [CrossRef] [PubMed]
13. Yan, W.; Shi, M.; Dong, C.; Liu, L.; Gao, C. Applications of tannic acid in membrane technologies: A review. *Adv. Colloid Interface Sci.* **2020**, *284*, 102267. [CrossRef]
14. Kumar, R.; Barakat, M.A.; Soliman, E.M. Removal of tannic acid from aqueous solution by magnetic carbohydrate natural polymer. *J. Ind. Eng. Chem.* **2014**, *20*, 2992–2997. [CrossRef]
15. Katheresan, V.; Kansedo, J.; Lau, S.Y. Efficiency of various recent wastewater dye removal methods: A review. *J. Environ. Chem. Eng.* **2018**, *6*, 4676–4697. [CrossRef]
16. Aryee, A.A.; Dovi, E.; Li, Q.; Han, R.; Li, Z.; Qu, L. Magnetic biocomposite based on peanut husk for adsorption of hexavalent chromium, Congo red and phosphate from solution: Characterization, kinetics, equilibrium, mechanism and antibacterial studies. *Chemosphere* **2022**, *287*, 132030. [CrossRef]
17. Castaldo, R.; Avolio, R.; Cocca, M.; Errico, M.E.; Avella, M.; Gentile, G. Amino-functionalized hyper-crosslinked resins for enhanced adsorption of carbon dioxide and polar dyes. *Chem. Eng. J.* **2021**, *418*, 129463. [CrossRef]
18. Melnyk, I.V.; Tomina, V.V.; Stolyarchuk, N.V.; Seisenbaeva, G.A.; Kessler, V.G. Organic dyes (acid red, fluorescein, methylene blue) and copper(II) adsorption on amino silica spherical particles with tailored surface hydrophobicity and porosity. *J. Mol. Liq.* **2021**, *336*, 116301. [CrossRef]
19. Tang, Y.; Li, M.; Mu, C.; Zhou, J.; Shi, B. Ultrafast and efficient removal of anionic dyes from wastewater by polyethyleneimine-modified silica nanoparticles. *Chemosphere* **2019**, *229*, 570–579. [CrossRef]
20. Babapour, M.; Dehghani, M.H.; Alimohammadi, M.; Arjmand, M.M.; Salari, M.; Rasuli, L.; Mubarak, N.M.; Khan, N.A. Adsorption of Cr(VI) from aqueous solution using mesoporous metal-organic framework-5 functionalized with the amino acids: Characterization, optimization, linear and nonlinear kinetic models. *J. Mol. Liq.* **2022**, *345*, 117835. [CrossRef]
21. Huang, X.Y.; Mao, X.Y.; Bu, H.T.; Yu, X.Y.; Jiang, G.B.; Zeng, M.H. Chemical modification of chitosan by tetraethylenepentamine and adsorption study for anionic dye removal. *Carbohydr. Res.* **2011**, *346*, 1232–1240. [CrossRef] [PubMed]
22. Liou, T.H.; Liou, Y.H. Utilization of rice husk ash in the preparation of graphene-oxide-based mesoporous nanocomposites with excellent adsorption performance. *Materials* **2021**, *14*, 1214. [CrossRef] [PubMed]
23. Liou, T.H.; Hung, L.W.; Liu, C.L.; Zhang, T.Y. Direct synthesis of nano titania on highly-ordered mesoporous SBA-15 framework for enhancing adsorption and photocatalytic activity. *J. Porous Mat.* **2018**, *25*, 1337–1347. [CrossRef]
24. Nnadozie, E.C.; Ajibade, P.A. Adsorption, kinetic and mechanistic studies of Pb(II) and Cr(VI) ions using APTES functionalized magnetic biochar. *Microporous Mesoporous Mat.* **2020**, *309*, 110573. [CrossRef]
25. Zheng, W.T.; Huang, K.; Dai, S. Solvothermal and template-free synthesis of N-Functionalized mesoporous polymer for amine impregnation and CO_2 adsorption. *Microporous Mesoporous Mat.* **2019**, *290*, 109653. [CrossRef]
26. Olejnik, A.; Goscianska, J. On the importance of physicochemical parameters of copper and aminosilane functionalized mesoporous silica for hydroxychloroquine release. *Mater. Sci. Eng. C* **2021**, *130*, 112438. [CrossRef]
27. Putz, A.-M.; Ciopec, M.; Negrea, A.; Grad, O.; Ianasi, C.; Ivankov, O.I.; Milanovic, M.; Stijepovic, I.; Almásy, L. Comparison of structure and adsorption properties of mesoporous silica functionalized with aminopropyl groups by the co-condensation and the post grafting methods. *Materials* **2021**, *14*, 628. [CrossRef]
28. Liou, T.H.; Tseng, Y.K.; Liu, S.M.; Lin, Y.T.; Wang, S.Y.; Liu, R.T. Green synthesis of mesoporous graphene oxide/silica nanocomposites from rich husk ash: Characterization and adsorption performance. *Environ. Technol. Innov.* **2021**, *22*, 101424. [CrossRef]
29. Maia, D.O.; Chagas, A.M.S.; Araújo, A.M.M.; Júnior, A.V.M.; Ferreira, I.M.L.; Lemos, F.C.D.; Araujo, A.S.; Júnior, V.J.F.; Gondim, A.D. Catalytic pyrolysis of glycerol in the presence of Nickel (II) Schiff base complex supported in SBA-15: Kinetic and products (TG-FTIR and PY-CG/MS). *Thermochim. Acta* **2018**, *669*, 160–168.
30. García, A.; Rosales, M.; Thomas, M.; Golemme, G. Arsenic photocatalytic oxidation over TiO_2-loaded SBA-15. *J. Environ. Chem. Eng.* **2021**, *9*, 106443. [CrossRef]
31. Wang, J.; Zheng, C.; Ding, S.; Ma, H.; Ji, Y. Behaviors and mechanisms of tannic acid adsorption on an amino-functionalized magnetic nanoadsorbent. *Desalination* **2011**, *273*, 285–291. [CrossRef]
32. Anbia, M.; Salehi, S. Removal of acid dyes from aqueous media by adsorption onto amino-functionalized nanoporous silica SBA-3. *Dyes Pigment.* **2012**, *94*, 1–9. [CrossRef]

33. Liou, T.H. A green route to preparation of MCM-41 silicas with well-ordered mesostructure controlled in acidic and alkaline environments. *Chem. Eng. J.* **2011**, *171*, 1458–1468. [CrossRef]
34. Zhang, Y.; Qian, C.; Duan, J.; Liang, Y.; Luo, J.; Han, Y.; Hu, J.; Shi, F. Synthesis of HKUST-1 embedded in SBA-15 functionalized with carboxyl groups as a catalyst for 4-nitrophenol to 4-aminophenol. *Appl. Surf. Sci.* **2022**, *573*, 151558. [CrossRef]
35. Wu, J.; Zhu, X.; Yang, F.; Ge, T.; Wang, R. Easily-synthesized and low-cost amine-functionalized silica sol-coated structured adsorbents for CO_2 capture. *Chem. Eng. J.* **2021**, *425*, 131409. [CrossRef]
36. Parlett, C.M.A.; Arandiyan, H.; Durndell, L.J.; Isaacs, M.A.; Lopez, A.T.; Wong, R.J.; Wilson, K.; Lee, A.F. Continuous-flow synthesis of mesoporous SBA-15. *Microporous Mesoporous Mat.* **2022**, *329*, 111535. [CrossRef]
37. Akti, F.; Balci, S. Synthesis of APTES and alcohol modified Sn/SBA-15 in presence of competitive ion: Test in degradation of Remazol Yellow. *Mater. Res. Bull.* **2022**, *145*, 111496. [CrossRef]
38. Patra, T.; Mohanty, A.; Singh, L.; Muduli, S.; Parhi, P.K.; Sahoo, T.R. Effect of calcination temperature on morphology and phase transformation of MnO_2 nanoparticles: A step towards green synthesis for reactive dye adsorption. *Chemosphere* **2022**, *288*, 132472. [CrossRef]
39. El-Sewify, I.M.; Radwan, A.; Shahat, A.; El-Shahat, M.F.; Khalil, M.M.H. Superior adsorption and removal of aquaculture and bio-staining dye from industrial wastewater using microporous nanocubic Zn-MOFs. *Microporous Mesoporous Mat.* **2022**, *329*, 111506. [CrossRef]
40. Koyuncu, D.D.E.; Okur, M. Removal of AV 90 dye using ordered mesoporous carbon materials prepared via nanocasting of KIT-6: Adsorption isotherms, kinetics and thermodynamic analysis. *Sep. Purif. Technol.* **2021**, *257*, 117657. [CrossRef]
41. Alorabi, A.Q.; Hassan, M.S.; Alam, M.M.; Zabin, S.A.; Alsenani, N.I.; Baghdadi, N.E. Natural clay as a low-cost adsorbent for crystal violet dye removal and antimicrobial activity. *Nanomaterials* **2021**, *11*, 2789. [CrossRef]
42. Bachmann, S.A.L.; Calvete, T.; Féris, L.A. Caffeine removal from aqueous media by adsorption: An overview of adsorbents evolution and the kinetic, equilibrium and thermodynamic studies. *Sci. Total Environ.* **2021**, *767*, 144229. [CrossRef]
43. Elkartehi, M.E.; Mahmoud, R.; Shehata, N.; Farghali, A.; Gamil, S.; Zaher, A. LDH nanocubes synthesized with zeolite templates and their high performance as adsorbents. *Nanomaterials* **2021**, *11*, 3315. [CrossRef] [PubMed]
44. Sarıcı-Özdemir, Ç.; Önal, Y. Equilibrium, kinetic and thermodynamic adsorptions of the environmental pollutant tannic acid onto activated carbon. *Desalination* **2010**, *251*, 146–152. [CrossRef]
45. Guo, Y.; Huang, W.; Chen, B.; Zhao, Y.; Liu, D.; Sun, Y.; Gong, B. Removal of tetracycline from aqueous solution by MCM-41-zeolite A loaded nano zero valent iron: Synthesis, characteristic, adsorption performance and mechanism. *J. Hazard. Mater.* **2017**, *339*, 22–32. [CrossRef] [PubMed]
46. Hou, Y.; Liu, L.; He, Q.; Zhang, D.; Jin, J.; Jiang, B.; Zhao, L. Adsorption behaviors and kinetics studies of chitooligosaccharides with specific degree of polymerization on a novel ion-exchange resin. *Chem. Eng. J.* **2022**, *430*, 132630. [CrossRef]
47. An, J.H.; Dultz, S. Adsorption of tannic acid on chitosan-montmorillonite as a function of pH and surface charge properties. *Appl. Clay Sci.* **2007**, *36*, 256–264. [CrossRef]
48. Hsieh, C.T.; Teng, H. Influence of mesopore volume and adsorbate size on adsorption capacities of activated carbons in aqueous solutions. *Carbon* **2000**, *38*, 863–869. [CrossRef]
49. Chang, M.Y.; Juang, R.S. Adsorption of tannic acid, humic acid, and dyes from water using the composite of chitosan and activated clay. *J. Colloid Interface Sci.* **2004**, *278*, 18–25. [CrossRef]
50. Wang, J.; Ji, Y.; Ding, S.; Ma, H.; Han, X. Adsorption and Desorption Behavior of Tannic Acid in Aqueous Solution on Polyaniline Adsorbent. *Chin. J. Chem. Eng.* **2013**, *21*, 594–599. [CrossRef]
51. Agarwal, S.; Rajoria, P.; Rani, A. Adsorption of tannic acid from aqueous solution onto chitosan/NaOH/fly ash composites: Equilibrium, kinetics, thermodynamics and modeling. *J. Environ. Chem. Eng.* **2018**, *6*, 1486–1499. [CrossRef]
52. Zhang, K.; Wang, Q.; Zhou, Y.; Gao, J.; Li, C.; Jiang, X. A low-cost crosslinked polystyrene derived from environmental wastes for adsorption of phenolic compounds from aqueous solution. *J. Mol. Liq.* **2020**, *314*, 113641. [CrossRef]
53. Lawal, A.A.; Hassan, M.A.; Zakaria, M.R.; Yusoff, M.Z.M.; Norrrahim, M.N.F.; Mokhtar, M.N.; Shirai, Y. Effect of oil palm biomass cellulosic content on nanopore structure and adsorption capacity of biochar. *Bioresour. Technol.* **2021**, *332*, 125070. [CrossRef] [PubMed]

nanomaterials

Article

Adsorptive Removal of Azithromycin Antibiotic from Aqueous Solution by Azolla Filiculoides-Based Activated Porous Carbon

Davoud Balarak [1,*], Amir Hossein Mahvi [2], Saeideh Shahbaksh [3], Md A. Wahab [4,*] and Ahmed Abdala [5,*]

1 Department of Environmental Health, Health Promotion Research Center, Zahedan University of Medical Sciences, Zahedan 9816743463, Iran
2 Center for Solid Waste Research, Institute for Environmental Research, Tehran University of Medical Sciences, Tehran 1417653911, Iran; ahmahvii@gmail.com
3 Student Research Committee, Zahedan University of Medical Sciences, Zahedan 9816743463, Iran; dbchemistry2@gmail.com
4 Institute for Advanced Study, Chengdu University, Chengdu 610106, China
5 Chemical Engineering Program, Texas A&M University at Qatar, Doha P.O. Box 23874, Qatar
* Correspondence: dbalarak2@gmail.com (D.B.); mawahab@gmail.com (M.A.W.); ahmed.abdala@qatar.tamu.edu (A.A.)

Citation: Balarak, D.; Mahvi, A.H.; Shahbaksh, S.; Wahab, M.A.; Abdala, A. Adsorptive Removal of Azithromycin Antibiotic from Aqueous Solution by Azolla Filiculoides-Based Activated Porous Carbon. *Nanomaterials* **2021**, *11*, 3281. https://doi.org/10.3390/nano11123281

Academic Editors: Christos A. Aggelopoulos and Giorgio Vilardi

Received: 27 September 2021
Accepted: 26 November 2021
Published: 3 December 2021

Publisher's Note: MDPI stays neutral with regard to jurisdictional claims in published maps and institutional affiliations.

Abstract: Due to the shortage of freshwater availability, reclaimed water has become an important source of irrigation water. Nevertheless, emergent contaminants such as antibiotics in reclaimed water can cause potential health risks because antibiotics are nonbiodegradable. In this paper, we report the adsorptive removal of azithromycin (AZM) antibiotics using activated porous carbon prepared from Azolla filiculoides (AF) (AFAC). The influence of the adsorption process variables, such as temperature, pH, time, and adsorbent dosage, is investigated and described. The prepared AFAC is very effective in removing AZM with 87% and 98% removal after the treatment of 75 min, at 303 and 333 K, respectively. The Langmuir, Temkin, Freundlich, and Dubinin–Radushkevich isotherm models were used to analyze the adsorption results. The Freundlich isotherm was best to describe the adsorption isotherm. The adsorption process follows second-order pseudo kinetics. The adsorption was endothermic ($\Delta H°$= 32.25 kJ/mol) and spontaneous ($\Delta S°$ = 0.128 kJ/mol·K). Increasing the temperature from 273 to 333 K makes the process more spontaneous ($\Delta G°$ = −2.38 and −8.72 KJ/mol). The lower mean square energy of 0.07 to 0.845 kJ/mol confirms the process' physical nature. The results indicate that AFAC can be a potential low-cost adsorbent of AZM from aqueous solutions.

Keywords: Azolla filiculoides; activated porous carbon; azithromycin; adsorption equilibrium; thermodynamics; adsorption; water treatment

1. Introduction

Pharmaceuticals play a critical role in saving millions of lives from various infectious to deadly diseases and lengthening life spans. On the other hand, pharmaceutical contaminants in water sources (surfaces, lakes, rivers, and seas), effluents of wastewater treatment plants, soils, and sludges, are emerging contaminants, leading to chronic and acute effects on the environment and human life [1,2]. Wastewater effluents are widely discharged from different sectors, such as industries, households, hospitals, and pharmaceuticals factories [3,4]. These streams may contain traces of various pharmaceuticals that might lead to bioaccumulation in the environment and endanger humans and animals [5]. Pharmaceuticals are a diverse group of chemical compounds with varying physical and chemical properties [6]. Therefore, various pharmaceuticals' successful use has the drawback of their emergence as rapidly growing harmful contaminants.

Among the pharmaceutical products, antibiotics are a critically important class of pharmaceuticals extensively used to prevent and cure infectious and deadly diseases in human and veterinary applications. Among the antibiotics, azithromycin (AZM) is a relatively new macrolide employed in eliminating erythromycin shortcomings, such as intolerance and its limited antimicrobial spectrum. AZM is more widely used as an effective antibiotic for several bacterial infections, including strep throat, middle ear infections, traveler's diarrhea, pneumonia, and intestinal disorders [7]. AZM is also employed to treat several sexually transmitted diseases, including chlamydia and gonorrhea infections [8]. In addition, AZM could function along with other medications to treat malaria, with doses administered orally or intravenously [9]. Antibiotics are usually not removed by the primary and secondary water treatment steps [10]. Therefore, they can exist in reclaimed water, developing antibiotic-resistant strains and threatening human health and the ecosystem [11,12]. A large portion of antibiotics is excreted as intact from the body, as antibiotics are usually not biodegradable and can survive in aquatic environments for a long duration. As a result, antibiotics negatively impact the environment, such as creating toxicity for infection pathogens, including bacteria, even at very low concentrations, disturbing biological processes in wastewater treatment, and decreasing the degradation rate of organic compounds. The discharge of these streams raises bacterial resistance to antibiotics and negatively impacts the environment. Therefore, the removal of them has become a crucial issue.

Meanwhile, different processes on removing antibiotics from the aquatic environment, including reverse osmosis, adsorption, biodegradation, chemical oxidation, solvent extraction, and nanoemulsion [13–16], have been demonstrated. Recently, the green nanoemulsion method was used to treat polluted water with AZM [17,18]. Among the processes, adsorption as a wastewater treatment process has become an attractive method because it offers the advantages of easy preparation and handling, low cost, high efficacy, and low or no risk of producing toxic byproducts [19].

Notably, the adsorption process is primarily affected by the adsorbent type and surface-structural properties, including surface area, pore size, porosity, and pore volume. Carbonaceous material-based adsorbents are well-recognized for solving such environmental issues. Among these carbonaceous adsorbents, activated carbon (AC) was one of the effective adsorbent materials. In this context, an inexpensive AC adsorbent is demanded to reduce the adsorption process cost [18,19]. The cost of AC production depends mainly on its precursor [20]. AC production was investigated using inexpensive materials, such as waste rice hulls, waste potato residue, seaweed, and almond shells [21,22].

Azolla filiculoides (AF) have recently been explored as a candidate for activated porous carbon production due to their relatively high carbon content and meager cost [23–25]. Therefore, AF has received tremendous attention when it comes to removing different pollutants [26,27]. AF can grow on wastewater and adsorb inorganic and organic materials from waste effluents [28]. With the recent interest in the rapid growth of AF [29,30], this alga, as an inexpensive and readily available adsorbent, should significantly alleviate the severity of this problem [27]. However, the percentage of removal using alga as an adsorbent is low. Therefore, activated carbon prepared from alga can increase the removal percentage to a significant amount [31], which has hardly been explored. Therefore, this study reports on the facile preparation of porous AC from AF (AFAC) and its performance for the removal of AZM. AFAC was analyzed by scanning electron microscope (SEM), N_2 adsorption measurements, and X-ray photoelectron spectroscopy (XPS). The main characteristics of AZM adsorption, including adsorption isotherms, adsorption kinetics, and adsorption thermodynamics, are also analyzed.

2. Materials and Methods

2.1. Materials

AZM dihydrate ($C_{38}H_{72}N_2O_{12}\cdot 2H_2O$, the structure is provided in Scheme 1) (Analytical grade) from Sigma Aldrich (Darmstadt, Germany), and research-grade acetonitrile from Merck (Darmstadt, Germany), were purchased and used without further purification. The AC precursor, AF, was obtained from the Anzali wetland (Anzali City, Iran).

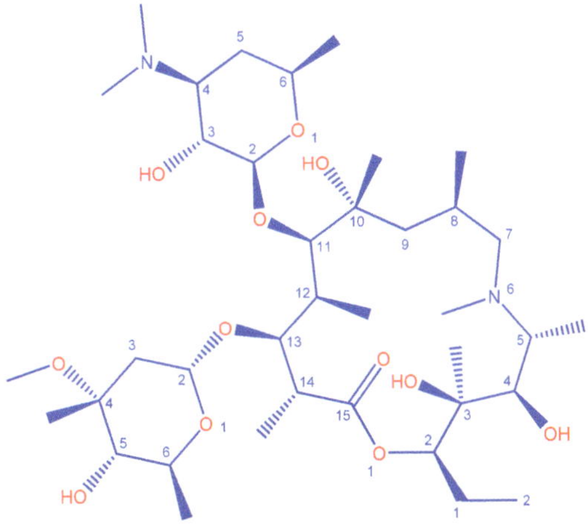

Scheme 1. Chemical Structure of azithromycin (AZM).

2.2. Preparation of Activated Porous Carbon from AF

The washed and dried AF was milled into a fine powder, sieved through 0.45 mm stainless steel mesh, and immersed in 28% $ZnCl_2$ aqueous solution for 12 h, followed by drying at 105 °C overnight. The zinc-treated AF was then pyrolyzed under N_2 atmosphere in a tube furnace at 350 °C for 2 h and then at 600 °C for 4 h. The produced porous AC (AFAC) was washed with hot 0.5 M HCl, filtered, and repeatedly rinsed with warm water until no free zinc ions were detected. The final AFAC product was dried at 105 °C for 24 h and stored in a desiccator for subsequent uses [31].

2.3. Characterization of Activated Porous Carbon

The porous morphology of the AFAC was observed using SEM (S-4800F FE-SEM, HITACHI, Japan. The surface area, pore-volume, and pore size distribution were also measured with Tristar-3000 surface area and porosity analyzer (Micromeritics, Norcross, GA, USA) at −196 °C. Before measuring the surface area, the sample was degassed at 200 °C for 12 h.

2.4. Point of Zero Charges (pH_{pzc}) Measurement

The AFAC pH_{pzc} characteristics were measured following the solid addition method. In this method, 25 mL of 0.2 M KNO_3 was placed in different flasks, and their pH was adjusted between 2 and 12 by adding 0.2 M HCl. The volume of the solution was adjusted to 50 mL by adding the KNO_3 solution. Then, 0.5 g of AFAC was added to each flask, and the suspension was placed on a shaker for 24 h. After equilibrating for 1 h, the supernatant's final pH (pH_f) was noted.

2.5. Batch Adsorption Study

A 1000 mg/L AZM stock solution was prepared and used to prepare the other concentrations, i.e., 25–200 mg/L, by dilution. Each batch adsorption experiment was carried in a 150-mL flask containing 100 mL of AZM solution and placed in a thermostatic shaker (CF0201003, Guangdong, China) (120 rpm) bath. The pH was controlled by adding 0.1 N NaOH or HCl. The adsorbent dosage was varied from 0.1 to 1.5 g/L, corresponding to an adsorbent to adsorbate ratio of 0.5 to 60. The temperature was varied from 273 to 325 K to evaluate the adsorption thermodynamics. Moreover, the adsorption kinetics were analyzed by measuring the AZM concertation at different time intervals up to 120 min. The concentration of AZM was measured using high-performance liquid chromatography (HPLC) (Agilent Technologies, Santa Clara, CA, USA) with Eclipse XDB-C18, 5 μm (4.6 × 150 mm) column, and UV detector operated at 230 nm with 1 mL/min of 40/60 acetonitrile/water as the mobile phase.

The AZM removal percentage (*R*) and adsorption capacity (q_e, mg/g) are calculated as follows [17]:

$$R = \frac{(C_0 - C_t)}{C_0} \times 100 \quad (1)$$

$$q_e = \frac{(C_0 - C_t) \times V}{m} \quad (2)$$

where C_0 is the initial AZM concentration (mg/L), C_t is AZM concentration (mg/L) at time t, *V* is the solution volume (L), and *m* is the AFAC (g). The Langmuir adsorption model, Temkin model and other associated calculations are also provided in the Supporting Information Figures S1–S4 and Table S1.

3. Results and Discussions

3.1. Characteristic of AZM and Characterization of Adsorbent AFAC

AZM is poorly soluble in water, but these antibiotics are widely used against various infections where different pathogens infect them. Azalide-type drugs, such as AZM, are recognized as pharmaceutical pollutants when the drug is drained into the water after use. Meanwhile, few approaches were employed to remove this drug from the aqueous wastewater, but are found to be very inefficient and ineffective [16–18], as the removal efficiency largely depends on the important factors, including the physicochemical characteristics of AZM, type, and nature of the solvent, and approached used for this purpose. The drug AZM anhydrous has the following physiochemical characteristics (aqueous solubility ~ 0.514 mg/mL; logP ~ 3.0 [16–18]).

The purpose of preparing a high surface area porous adsorbent is to ensure fast and efficient adsorption of the antibiotics into the porous network. Figure 1a shows the SEM image of the AFAC before adsorption, revealing the type of porous structure of AC derived from AF. As shown in Figure 1a, the particles from AFAC are in irregular pore sizes and shapes but are porous, revealing that AFAC could be used as the potential adsorbent. The structure of AFAC post-adsorption (Figure 1b) is different from the structure shown in Figure 1a, due to the loading of antibiotics into the porous network of AFAC. The hysteresis loops confirm that the N_2 adsorption-desorption isotherm of this porous structure is consistent with type IV isotherm, corresponding to mesoporous solid (Figure 1c). The inset in Figure 1c shows the pore size distribution, with an average pore size of 4.9 nm. The isotherm suggested that the adsorbent contains micropores and mesopores as the relative pressure (P/P_0) at 0.4 to 0.65, which could be ascribed to the sample's mesoporous nature [31]. In contrast, low-relative pressure (P/P_0) at 0.4 could be responsible for micropores.

The AFAC nitrogen adsorption-desorption results are also outlined in Table 1, indicating that the AFAC has a specific surface area of 484.1 m^2/g, pore volume of 0.47 cm^3/g, a porosity of 53.4%, bulk density of 0.245 $g/cm^{3,}$ and average pore size of ~ 5 nm. The results of EDX are shown in Table 1. As you can see, the elements carbon, oxygen, magnesium, potassium, and hydrogen are almost the main elements of AFAC, and the negligible amounts of sodium and iron were observed in the final composition of AFACHere; the yield (Y) of the prepared AFAC was calculated as follows [32]:

$$Y = \frac{W_{AC}}{W_{AF}} \times 100 \quad (3)$$

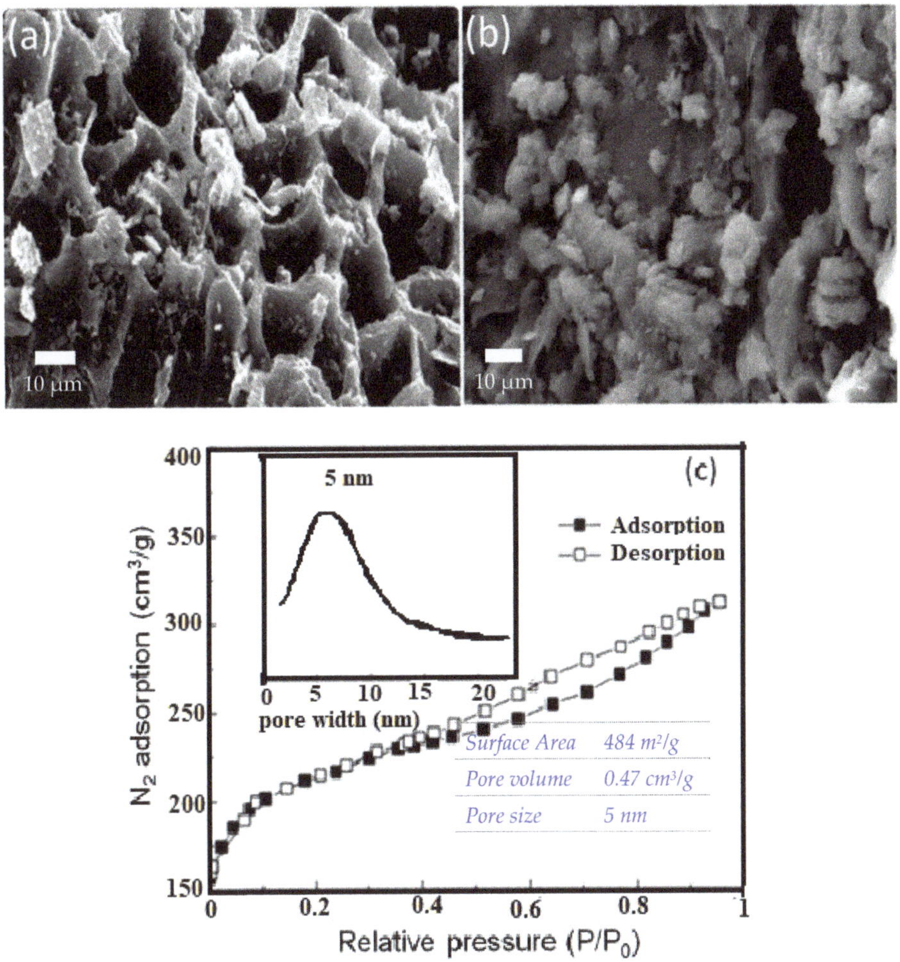

Figure 1. SEM images of (**a**) the AFAC before and (**b**) after adsorption of AZM and (**c**) the N_2 adsorption-desorption isotherm. The inset of Figure 1c shows the pore size distribution curve of AFAC.

Table 1. Surface-structural properties of the AFAC.

Porosity, %	C%	% Moisture	Y	Density, g/cm^3	%H	Mg%	Fe%	K%	O%
53.4	53.7	0.2	51.3	0.245	3.7	2.4	0.48	1.7	37.9

3.2. Adsorption of AZM on AFAC

3.2.1. Effect of Temperature and Contact Time

The adsorption time and temperature effects were investigated under the optimum pH of 9, AFAC dose of 1 g/L, and 100 mg initial AZM concentration. The adsorption capacity and removal efficiency of AZM by the AFAC adsorbent is shown in Figure 2. The adsorption process is very rapid at the initial stage (0–30 min) due to the adsorbent's high surface area and large pore volume. This initial stage is followed by a slower stage (30–60 min), and finally, an equilibrium stage is reached after ~75 min. The rapid removal rate at the first stage is due to the high concentration of the active adsorption sites on the AFAC surface in the early adsorption process [33]. These active sites become fully occupied by the adsorbed AZM molecules as the length of contact time between the adsorbent and the adsorbate increases, and the repulsive force that occurs between the AZM molecules on the surface of adsorbents and AZM molecules in the bulk liquid phase reduces the AZM adsorption rate [34,35].

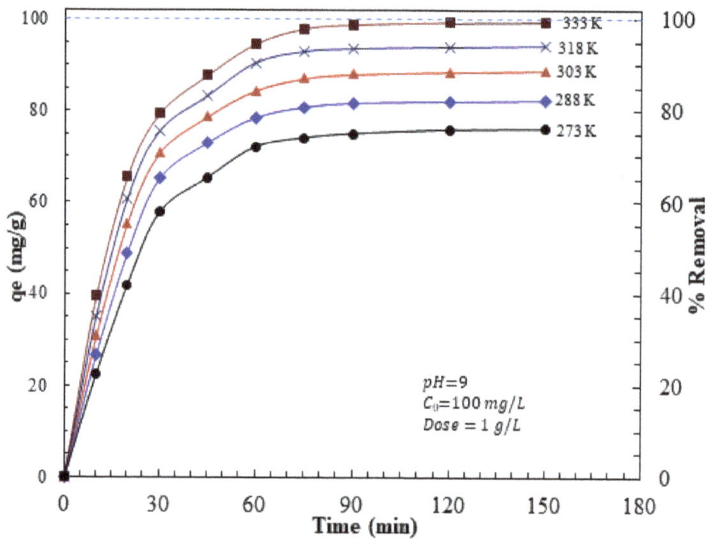

Figure 2. Influence of contact time and temperatures on the adsorption capacity and removal efficiency of AZM by AFAC.

The high AZM adsorption rate is due to the higher probability of interactions between AZM ions and AFAC porous particles [32] and the presence of more active pores. These pores can accommodate more AZM on their surface. The rapid uptake of AZM ions by AFAC is one of the parameters which could be potentially considered for economic wastewater treatment plant applications [33].

3.2.2. Effect of Adsorbent Dose

The impact of AFAC dosage on the AZM adsorption capacity and removal efficiency was investigated using aqueous by changing the AFAC adsorbent dosage between 0.1 to 1.4 g/L. All the other adsorption parameters, i.e., pH, temperature, and contact time, were constant. The results in Figure 3 show that the AZM adsorption capacity decreased from 297.4 to 63.8 mg/g, with the adsorption dosage of 0.1 and 1.4 g/L, respectively. This decrease in adsorption capacity with the AFAC dosage is attributed to the decrease in the adsorbate/adsorbent ratio with the increase in dosage consistent with previously reported results [34,35]. On the other hand, the % removal increases with the AFAC dosage and reaches a near plateau at a 1 mg/mL dosage. Therefore, 1 mg/mL AFAC was selected as the optimum dosage used in the following experiments.

Consequently, it was found that the availability of a larger surface area and more adsorption sites already accelerated with an increase in the adsorbent dose [36,37]. Increasing the adsorbent dose over 1 mg/mL resulted in a minor change in the removal percentage. The establishment of a dense screening layer at the adsorbent surface because of the accumulation of AFAC particles and a decrease in the distance between the AFAC molecules, known as the screening effect, happens with a higher adsorbent dose, could explain this phenomenon [38]. The binding sites were hidden from AZM molecules by the condensed layer on the adsorbent surface. Furthermore, because AFAC overlapped, AZM molecules competed for a limited number of accessible binding sites [39]. Agglomeration or aggregation at higher AFAC dosages lengthens the diffusion channel for AZM adsorption, lowering the adsorption rate [40].

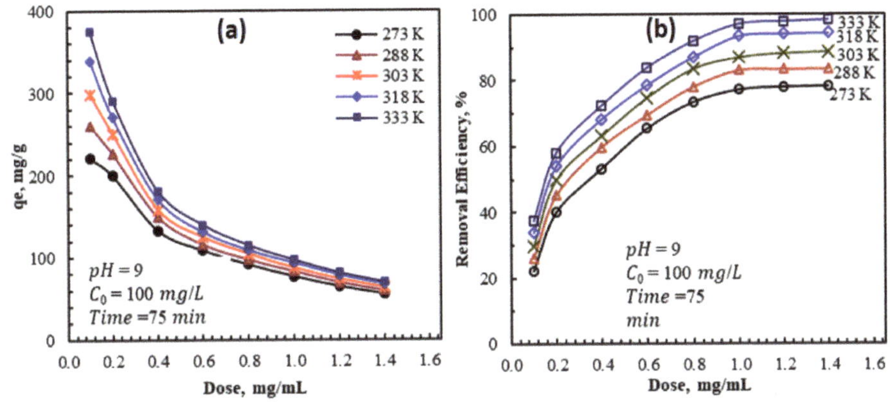

Figure 3. Effect of AFAC dosage and adsorption temperature on (**a**) the adsorption capacity and (**b**) removal efficiency of AZM by AFAC.

3.2.3. Effect of pH

The effect of solution pH on the adsorption capacity of AFAC is explained by analyzing the AZM surface charges and dissociation constant (pKa) [35]. The oxygen functional groups on the adsorbent surface can also impact the adsorbent performance related to the $pH_{(pzc)}$. In other words, the H^+ or OH^- ions in the solution change the adsorbent's surface charge and ultimately impact the antibiotics removal efficacy [36].

Figure 4a shows the effect of pH on AZM adsorption by AFAC. The pH_{pzc} of AFAC is shown in Figure 4b. At pH lower than pH_{pzc}, the surface functional groups deprotonate by the presence of OH^- ions in the solution [37]. At pH higher than the pH_{pzc}, the surface functional groups deprotonate by the H^+ ions in the solution [38]. The surface of AFAC is positively charged at pH lower than pH_{pzc}, i.e., 8.45.

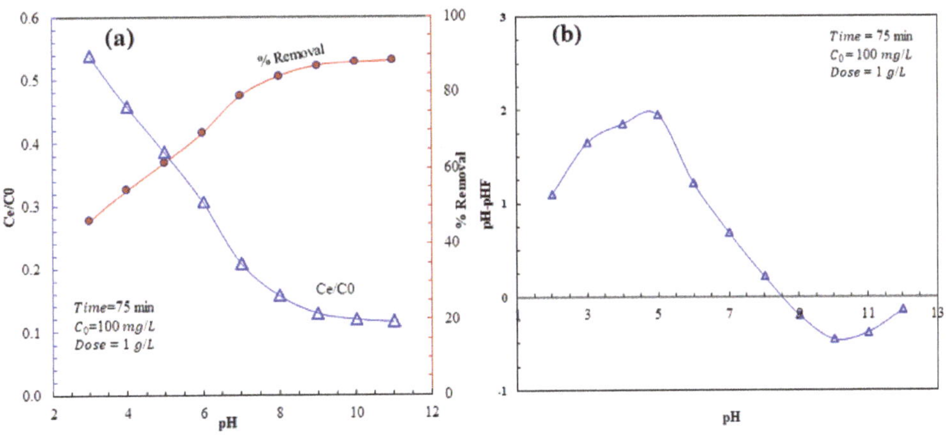

Figure 4. (**a**) Effect of pH on AZM adsorption by AFAC and (**b**) the pH_{pzc} of AFAC.

On the other hand, the pKa of AZM is ~8.6–9.5; hence, in an acidic medium, AZM is protonated. As a result, the adsorption capacity of AFAC in pH < 3 is low due to electrostatic repulsion between AZM and AFAC, and competition between H^+ with AZM for adsorption on AFAC active sites. Accordingly, the electrostatic attraction between AZM and the AFAC surface gradually increases to yield the highest adsorption at pH = 9–11.

3.2.4. Adsorption Isotherms

The adsorption isotherm for removing AZM ions from the aqueous solution was obtained at equilibrium conditions to establish the relationship between the adsorbed AZM amount onto the AFAC, and that amount remained in the aqueous solution [39]. The experimental data were examined using four different isotherm models, i.e., Langmuir, Freundlich, Temkin, and Dubinin–Radushkevich (D-R) [40–42].

The Freundlich model is an empirical isotherm that describes multilayer adsorption on a heterogeneous system. It is mathematically expressed by Equation (4) [43]:

$$ln\ q_e = ln\ K_F + \frac{1}{n} ln\ c_e \quad n > 1 \tag{4}$$

where n and K_F (mg/g) are the Freundlich Isotherm constants related to adsorption intensity and capacity. n and K_F are calculated from the slope and intercept of the linear fit to the lnq_e vs. lnc_e. A slope ranging between zero and unity indicates the heterogeneity of the surface and the adsorption intensity, as the slope is very low (near zero). A slope near unity refers to a chemisorption process, where $1/n$ higher than unity implies cooperative adsorption. The Freundlich isotherm constant relates to the favorability of the adsorption process. The adsorption process is favorable when $1 < n < 10$ [44].

The results in Figure 5 show that Freundlich isotherm reasonably describes the adsorption of antibiotics to AFAC. The value of n for AZM adsorption (Table 2) is in the range of $1 < n < 10$, indicating favorable adsorption. The Freundlich constant, K_F, signifies the strength of the interactions between the adsorbate and the adsorbent. The obtained K_F ranged from 3.08 at 273 K to 58.95 at 333 K, indicating enhanced interactions at higher temperatures. It is worth noting that the three other isotherms did not provide any reasonable fit to the adsorption results, as shown by the very low R^2 provided in Table 2.

Table 2. Adsorption isotherm constants for the adsorption of AZM onto AFAC at various temperatures.

T (K)	Freundlich			Langmuir				Temkin			D–R Model		
	n	K_F	R^2	q_m	K_L	R_L	R^2	α	β	R^2	q_m	β	R^2
273	1.00	3.08	0.96	120.3	0.0003	0.98	0.712	0.08	123	0.92	212.1	1.2×10^{-4}	0.93
288	1.10	5.43	0.97	159.7	0.0027	0.79	0.765	0.10	125	0.85	218.8	6.2×10^{-5}	0.88
303	1.28	10.57	0.97	164.2	0.0085	0.54	0.781	0.15	118	0.83	221.7	2.8×10^{-5}	0.86
318	1.72	26.11	0.95	196.1	0.0243	0.29	0.792	0.32	94	0.86	206.0	6.1×10^{-6}	0.71
333	2.43	55.42	0.93	217.4	0.0588	0.15	0.793	1.12	70	0.81	197.0	7.1×10^{-7}	0.60

The maximum measured q_m was 374 mg/g, obtained at 333 K and a dose of 0.1 g/L. This maximum adsorption capacity was much higher than the reported capacity for antibiotics, using a variety of adsorbents, as shown in Table 3. This indicates the potential of the AFAC as an adsorbent for large-scale applications.

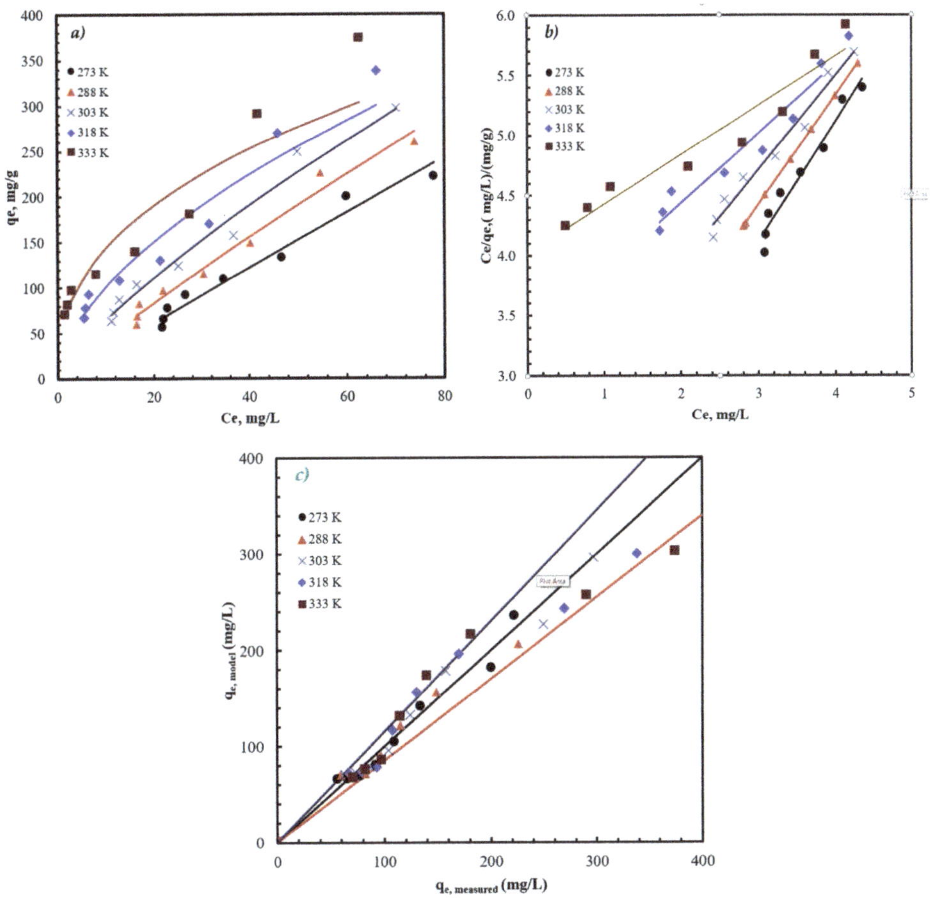

Figure 5. The adsorption data fitted to the Freundlich isotherm in (**a**) nonlinear form, (**b**) linear form, and (**c**) model predicted versus measured capacity at different temperatures. The blue and red lines in (c) represent the ±15% limit. Conditions: pH = 9, C_0 = 100 mg/L, and time = 75 min.

Table 3. The comparison of different adsorption capacities.

Antibiotics	Adsorbent	$q_{e,\,max}$ (mg/g)	Reference
Amoxicillin	Chitosan	46.21	[2]
Tetracycline	Illite	78.12	[11]
Penicillin	Chestnut Shell	27.4	[14]
Amoxicillin	AC	56.81	[16]
Tetracycline	maize stalks	71.38	[31]
SMZ	GO	54.33	[32]
Tetracycline	MWCNTS	56.5	[33]
Penicillin	TiO_2	252.6	[34]
Amoxicillin	Bentonite	46.54	[37]
Tetracycline	MIP	38.9	[38]
Cephalexin	AC	43.5	[40]
AZM	**AFAC**	**374**	**This work**

3.2.5. Adsorption Kinetic

The kinetic of the AZM adsorption on AFAC is also assessed using the pseudo-first-order kinetic model, the pseudo-second-order model, and the intra-diffusion model. The integrated form of the pseudo-first-order kinetic equation is given by [45]:

$$ln(q_e - q_t) = ln\ q_e - k_1 t \tag{5}$$

where q_e is the equilibrium sorption uptake, q_t (mg/g) is the amount of adsorbed AZM on the AFAC at time t, and k_1 (1/min) is the rate constant of the first-order adsorption. q_e is extrapolated from the experimental data at time t = 0. A straight line fitting of Ln $(q_e - q_t)$ versus t confirms the pseudo-first-order kinetic.

The analysis of the adsorption date (Figure 6) revealed that the adsorption process follows the pseudo-first-order kinetics, with a rate constant that ranges from 0.046 to 0.055 min^{-1}, indicating a weak temperature dependence. The analysis of the results using second-order and intra-particle diffusion models showed a poor correlation between the results and these models, as summarized in Table 4. Other details of the analysis are included in the Supporting Information.

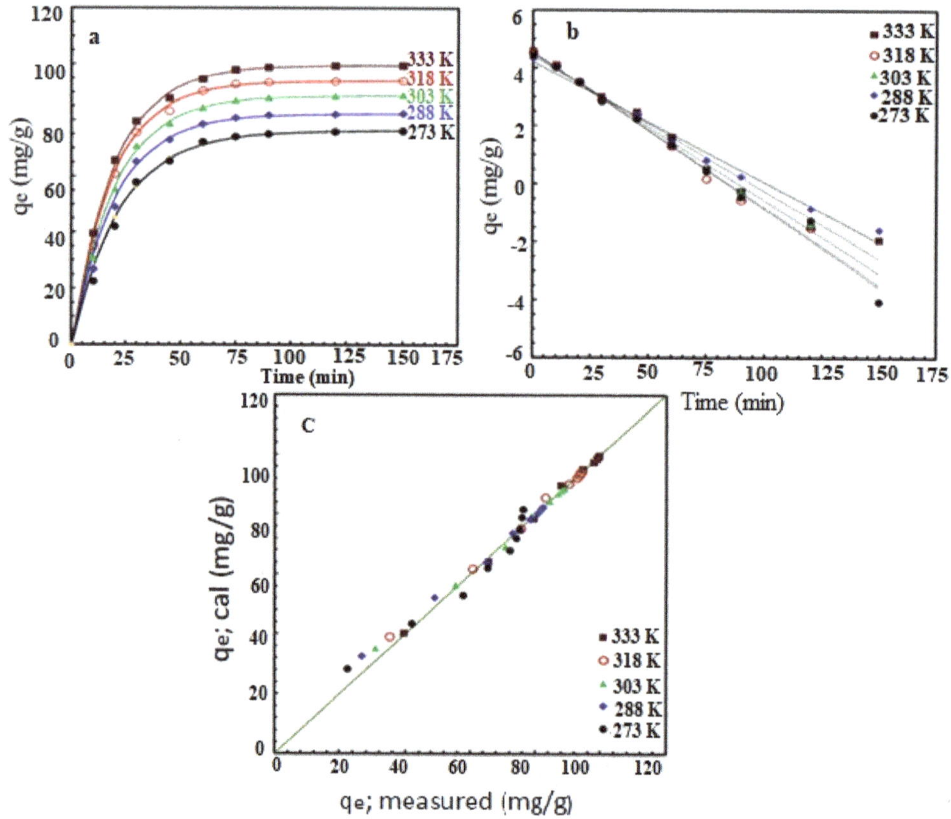

Figure 6. The adsorption kinetics for the adsorption of AZM on AFAC at various temperatures. (**a**) qe vs. time, the Symbols represent the experimental results, and the curves represent the pseudo-first-order model fitting, (**b**) the linearized (integral form) of the pseudo-first-order model, and (**c**) the model calculated adsorption capacity vs. the experimentally measured adsorption capacity shown the excellent agreement as the data points are on, or very close to, the 45° line. Conditions: pH = 9, C_0 = 100 mg/L, and dose = 1 g/L.

Table 4. Analysis of the kinetics of the adsorption of AZM on AFAC using different kinetic models.

T (K)	Pseudo-First-Order			Pseudo-Second-Order		
	k_1 min^{-1}	qe mg/g	R^2	k_2 $(mg/g)^{-1}min^{-1}$	qe, mg/g	R^2
273	0.045	76.3	0.995	4.5×10^{-4}	94.5	0.983
288	0.050	82.3	0.990	5.3×10^{-4}	98.7	0.987
303	0.051	88.7	0.996	5.5×10^{-4}	104.4	0.991
318	0.053	94.0	0.989	5.9×10^{-4}	109.2	0.993
333	0.052	99.6	0.995	5.9×10^{-4}	114.4	0.996

In the intraparticle diffusion (IPD) kinetic model is described by Equation (6) [46]:

$$q_t = k_P t^{1/2} + C \quad (6)$$

where k_p is the IPD rate constant $\left(\frac{mg}{g \cdot min^{0.5}}\right)$ and C is a constant that describes the initial adsorption capacity. When IPD governs the adsorption process, a linear plot is obtained by drawing q_t versus $t^{1/2}$ (in Figure 7), and IDP is addressed as a rate-controlling step when the obtained lines pass through the origin. In our case, the kinetic data cannot be described by a single IPD model (Figure 7a), but by the three-stage process, as shown in Figure 7b. As shown in Table 5, the first stage (t < 30 min) is very rapid, indicated by the large rate constant, k_{d_1}. This adsorption mechanism during this stage is associated with adsorption on the external surface. The second stage (t = 30–60 min) has a moderate rate constant (k_{d_2}). During this second stage, the adsorption process proceeds via adsorption on the internal mesopores. The last stage (t > 60 min) is a very slow process, as indicated by the very low rate constants (k_{d_3}). The adsorption could be attributed to the chemical adsorption step on the remaining active sites.

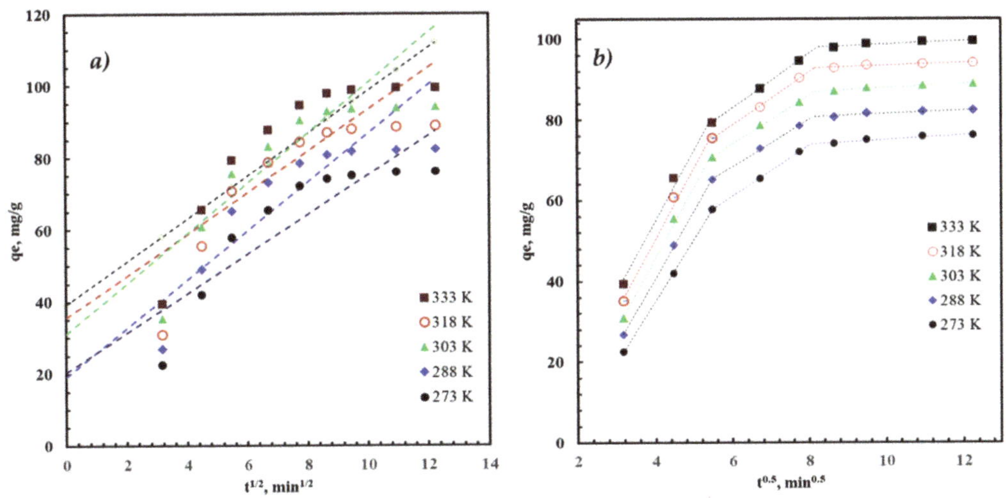

Figure 7. The adsorption capacity (symbols) of AZM using AFAC fitted to the (**a**) one stage IPD kinetic model (curves) and (**b**) three-stages IPD kinetic model (dashed-curves). Conditions: pH = 9, C_0 = 100 mg/L, and dose = 1 g/L.

Table 5. IPD parameters for the adsorption of AZM on AFAC.

T, (K)	Stage 1, (t < 30 min)			Stage 2, (30–60 min)			Stage 3, (t > 60 min)		
	k_{d1}	C	R^2	k_{d2}	C	R^2	k_{d3}	C	R^2
273	15.2	−25.8	0.912	6.3	23.2	0.915	0.56	69.5	0.884
288	16.6	−25.7	0.923	5.8	33.1	0.926	0.39	77.6	0.916
303	17.3	−23.4	0.943	6	37.9	0.973	0.47	83.2	0.898
318	17.5	−19.6	0.951	6.5	39.5	0.951	0.33	90.1	0.873
333	17.4	−14.5	0.891	6.7	42.6	0.936	0.41	94.7	0.901

3.2.6. Adsorption Thermodynamics

To study the thermodynamic of the adsorption process, batch adsorption experiments were carried at 273, 288, 303, 318, and 333 under the optimum adsorption conditions. The AZM adsorption capacity has increased with temperature. For example, the capacity has increased from 76.13 mg/g at 273 K, to 99.44 mg/g at 333 K. This is attributed to the high kinetic energy of AZM cations at higher temperatures; causing sufficient contact between the AZM and the AFAC active sites, and leading to an increase in the removal efficiency [47]. These results also indicate that the adsorption is a physical rather than a chemical process. A similar trend has been observed in other studies to remove pollutants from the aqueous phase [48,49]. Moreover, increasing the temperature may increase the pore size significantly, affecting the porous carbon uptake capacity [50].

Thermodynamic parameters such as the change in free energy ($\Delta G°$), enthalpy ($\Delta H°$), and entropy ($\Delta S°$) were calculated using Equations (7) and (8) [51,52]:

$$\Delta G_0 = -RT \ln K \tag{7}$$

$$\ln K_L = \frac{\Delta S^{\circ}}{R} - \frac{\Delta H^{\circ}}{RT} \tag{8}$$

where T (K) is the absolute temperature, R (8.314 KJ/mol·K) is the ideal gas constant, K_L (L/mol) represents the Langmuir constant. $\Delta H°$ and $\Delta S°$ are calculated from the slope and intercept of the Vant Hoff plot of Ln K_L versus (1/T). The thermodynamic parameters of the adsorption process are provided in Table 6. $\Delta G°$ was negative over the entire temperature range, revealing the feasible and spontaneous adsorption process in the 273–333 K studied temperature range. As physical adsorption process has −20 kJ/mol < $\Delta G°$ < 0 kJ/mol, while chemical adsorption has −400 kJ/mol < $\Delta G°$ < −80 kJ/mol [53]. In this study, $\Delta G°$ was between −2.38 and −8.72 kJ/mol, confirming the physical nature of the adsorption process.

Table 6. Thermodynamic parameters for the adsorption of AZM on AFAC.

T, (K)	$\Delta G°$, (kJ/mol)	$\Delta H°$, (kJ/mol)	$\Delta S°$, (kJ/mol)
273	−2.38	33.25	0.128
288	−3.24		
303	−4.31		
318	−5.81		
333	−8.72		

The positive $\Delta H°$ indicates that the adsorption process is endothermic. The diffusion coefficient of AZM molecules increases with temperature, enhancing the adsorption. The positive $\Delta S°$ confirms that the process is spontaneous, with a positive binding affinity of AZM on the AFAC active sites.

4. Conclusions

We have successfully prepared the AFAC with a specific surface area of 484.1 m^2/g, pore volume of 0.472 cm^3/g, and porosity of 53.4% using Azolla filiculoides (AF) as an activated porous carbon source for the removal of antibiotic AZM as a pollutant. The porosity of AFAC was confirmed by SEM and BET isotherms. Then, isotherms, kinetic and thermodynamic studies were conducted for the AZM antibiotic removal using AFAC. The characteristics of the AFAC and its performance for the adsorption of AZM from an aqueous solution were investigated, and the results have suggested that the percentage removal was dependent on temperature, pH, adsorbent dosage, and contact time. AFAC has exhibited a high adsorption capacity for antibiotic AZM, and the antibiotic was effectively removed from the aqueous solution used. For example, maximum removal was 97.9% at optimum pH of 9, contact time 75 min, and adsorbent dose 1 g/L. The negative ΔG° and positive ΔH° and ΔS° confirmed that the adsorption of AZM on AFAC was spontaneous and endothermic. The Freundlich isotherm has described the equilibrium adsorption results well. Our results suggested that AFAC is an efficient and low-cost adsorbent for removing antibiotics from wastewater.

Supplementary Materials: The following are available online at https://www.mdpi.com/article/10.3390/nano11123281/s1, Figure S1: The plot Ce/qe versus Ce based on Langmuir isotherm; Figure S2: The Adsorption kinetics for the adsorption of AZM on AFAC at various temperatures. (a) qe vs. time, the Symbols represent the experimental results. The curves represent the pseudo-first-order model fitting, (b) the linearized (integral form) of the pseudo-first-order model, and (c) the model calculated adsorption capacity vs. the experimentally measured adsorption capacity shown the excellent agreement as the data points are on or very near to the 45° line; Figure S3: The Adsorption kinetics for the adsorption of AZM on AFAC at various temperatures. The Symbols represent the experimental results, and the curves represent the pseudo-second-order model fitting. The inset shows the linear form of the pseudo-second-order model; Figure S4: The Adsorption kinetics for the adsorption of AZM on AFAC at various temperatures. The Symbols represent the experimental results, and the curves represent the inter-particle diffusion model fitting. The inset shows the linear form of the inter-particle diffusion model; Table S1: Kinetic parameters for the adsorption of AZM on AFAC. References [54–63] are cited in Supplementary Materials.

Author Contributions: D.B.: conceptualization, investigation data, and collecting, figure analysis, writing-original draft. A.H.M.: draft writing, correcting, and reading. A.A.: writing, reading, figure analysis, correcting, and resourcing. M.A.W.: writing—revising and figure analysis. S.S.: Data analysis, revising, and editing. All authors have read and agreed to the published version of the manuscript.

Funding: This paper is funded by Zahedan University of Medical Sciences, Zahedan, Iran.

Institutional Review Board Statement: Not applicable.

Informed Consent Statement: Not applicable.

Data Availability Statement: The data presented in this study are available on request from the corresponding authors.

Acknowledgments: The authors would like to acknowledge the Student Research Committee, Zahedan University of Medical Sciences, for this study's financial support. The publication of this article was funded by Qatar National Library.

Conflicts of Interest: The authors declare no conflict of interest.

References

1. Balarak, D.; Mostafapour, F.K. Photocatalytic Degradation of Amoxicillin Using UV/Synthesized NiO from Pharmaceutical Wastewater. *Indones. J. Chem.* **2019**, *19*, 211. [CrossRef]
2. Rostamian, R.; Behnejad, H. A comparative adsorption study of sulfamethoxazole onto graphene and graphene oxide nanosheets through equilibrium, kinetic and thermodynamic modeling. *Process. Saf. Environ. Prot.* **2016**, *102*, 20–29. [CrossRef]
3. Huang, L.; Wang, M.; Shi, C.; Huang, J.; Zhang, B. Adsorption of tetracycline and ciprofloxacin on activated carbon prepared from lignin with H3PO4activation. *Desalination Water Treat.* **2014**, *52*, 2678–2687. [CrossRef]

4. Li, M.; Shu, D.; Jiang, L. Cu (II)-influenced adsorption of ciprofloxacin from aqueous solutions by magnetic graphene oxide/nitrilotriacetic acid nanocomposite: Competition and enhancement mechanisms. *Chem. Eng. J.* **2017**, *319*, 219–228. [CrossRef]
5. Balarak, D.; Azarpira, H.; Mostafapour, F.K. Study of the Adsorption Mechanisms of Cephalexin on to Azolla Filiculoides. *Der. Pharm. Chem.* **2016**, *8*, 114–121.
6. Zhang, B.; Zhang, H.; Li, X.; Lei, X.; Li, C.; Yin, D.; Fan, X.; Zhang, Q. Synthesis of BSA/Fe3O4 magnetic composite microspheres for adsorption of antibiotics. *Mater. Sci. Eng. C* **2013**, *33*, 4401–4408. [CrossRef]
7. Yazdani, A.; Sayadi, M.H. Sonochemical degradation of azithromycin in aqueous solution. *Environ. Health Eng. Manag. J.* **2018**, *5*, 85–92. [CrossRef]
8. Homem, V.; Santos, L. Degradation and removal methods of antibiotics from aqueous matrices–A review. *J. Environ. Manag.* **2011**, *92*, 2304–2347. [CrossRef]
9. Mandić, Z.; Weitner, Z.; Ilijaš, M. Electrochemical oxidation of azithromycin and its derivatives. *J. Pharm. Biomed. Anal.* **2003**, *33*, 647–654. [CrossRef]
10. Yang, Z.Y.; Wang, L.; Tang, X. Determination of azithromycin by ion-pair HPLC with UV detection. *J. Pharm. Biomed. Anal.* **2009**, *49*, 811–815. [CrossRef]
11. Kamranifar, M.; Allahresani, A.; NaghiZadeh, A. Application of $CoFe_2O_4$@CuS magnetic nanocomposite as a novel adsorbent for removal of Penicillin G from aqueous solutions: Isotherm, kinetic and thermodynamic study. *Desalin. Water Treat.* **2019**, *148*, 263–273. [CrossRef]
12. Balarak, D.; Azarpira, H. Photocatalytic degradation of Sulfamethoxazole in water: Investigation of the effect of operational parameters. *Int. J. ChemTech Res.* **2016**, *9*, 731–738.
13. Gao, J.; Pedersen, J.A. Adsorption of Sulfonamide Antimicrobial Agents to Clay Minerals. *Environ. Sci. Technol.* **2005**, *39*, 9509–9516. [CrossRef] [PubMed]
14. Peterson, J.W.; Petrasky, L.J.; Seymourc, M.D.; Burkharta, R.S.; Schuilinga, A.B. Adsorption and breakdown of penicillin antibiotic in the presence of titanium oxide nanoparticles in water. *Chemosphere* **2012**, *87*, 911–917. [CrossRef] [PubMed]
15. Aksu, Z.; Tunç, Ö. Application of biosorption for penicillin G removal: Comparison with activated carbon. *Process. Biochem.* **2005**, *40*, 831–847. [CrossRef]
16. Putra, E.K.; Pranowoa, R.; Sunarsob, J.; Indraswatia, N.; Ismadjia, S. Performance of activated carbon and bentonite for adsorption of amoxicillin from wastewater: Mechanisms, isotherms and kinetics. *Water Res.* **2009**, *43*, 2419–2430. [CrossRef] [PubMed]
17. Mahdi, W.A.; Hussain, A.; Bukhari, S.I.; Alshehri, S.; Singh, B.; Alic, N. Removal of clarithromycin from aqueous solution using wa-ter/triton X-100/ ethanol/ olive oil green nanoemulsion method. *J. Water Proc. Eng.* **2021**, *40*, 101973. [CrossRef]
18. Hussain, A.; Afzal, O.; Altamimi, A.S.; Ali, R. Application of green nanoemulsion to treat contaminated water (bulk aqueous solution) with azithromycin. *Environ. Sci. Pollut. Res.* **2021**, *24*, 61696–61706. [CrossRef] [PubMed]
19. Ahmadi, S.; Banach, A.; Mostafapour, F.K.; Balarak, D. Study survey of cupric oxide nanoparticles in removal efficiency of ciprofloxacin antibiotic from aqueous solution: Adsorption isotherm study. *Desalin. Water Treat.* **2017**, *89*, 297–303. [CrossRef]
20. Meshkinain, A.; Davoud, B.; Nastaran, Y. Optimization of nickel oxide nanoparticle synthesis through the sol-gel method for adsorption of Penicillin G. *Res. J. Chem. Environ.* **2021**, *25*, 31–36.
21. Mahvi, A.H.; Mostafapour, F.K. Biosorption of tetracycline from aqueous solution by azolla filiculoides: Equilibrium kinetic and thermodynamics studies. *Fresen. Environ. Bull.* **2018**, *27*, 5759–5767.
22. Zazouli, M.A.; Mahdavi, Y.; Bazrafshan, E.; Balarak, D. Phytodegradation potential of bisphenolA from aqueous solution by Azolla Filiculoides. *J. Environ. Heal. Sci. Eng.* **2014**, *12*, 66. [CrossRef]
23. Balarak, D.; Mahdavi, Y.; Bazrafshan, E.; Mahvi, A.H. Kinetic, isotherms and thermodynamic modeling for adsorption of acid blue 92 from aqueous solution by modified azolla filicoloides. *Fresen. Environ. Bull.* **2016**, *25*, 1321–1330.
24. Diyanati, R.; Yazdani, J.; Belarak, D. EEffect of sorbitol on phenol removal rate by lemna minor. *J. Maz. Uni. Med. Sci.* **2013**, *22*, 58–65.
25. Diyanati, R.A.; Yousefi, Z.; Cherati, J.Y. The ability of Azolla and lemna minor biomass for adsorption of phenol from aqueous solutions. *J. Maz. Uni. Med. Sci.* **2013**, *23*, 17–23.
26. Dyanati, R.A.; Yousefi, Z.A.; Cherati, J.Y. Investigating phenol absorption from aqueous solution by dried azolla. *J. Maz. Uni. Med. Sci.* **2013**, *22* (Suppl. S2), 13–20.
27. Padmesh, T.; Vijayaraghavan, K.; Sekaran, G.; Velan, M. Application of Azolla rongpong on biosorption of acid red 88, acid green 3, acid orange 7 and acid blue 15 from synthetic solutions. *Chem. Eng. J.* **2006**, *122*, 55–63. [CrossRef]
28. Tan, C.; Li, G.; Lu, X.Q.; Chen, Z.-L. Biosorption of Basic Orange using dried A. filiculoides. *Ecol. Eng.* **2010**, *36*, 1333–1340. [CrossRef]
29. Padmesh, T.; Vijayaraghavan, K.; Sekaran, G.; Velan, M. Batch and column studies on biosorption of acid dyes on fresh. Water macro alga Azolla filiculoides. *J. Hazard. Mater.* **2005**, *125*, 121–129. [CrossRef]
30. Zazouli, M.A.; Mahvi, A.H.; Dobaradaran, S.; Barafrashtehpour, M.; Mahdavi, Y. Adsorption of fluoride from aqueous solution by modified Azolla Filiculoides. *Fluoride* **2014**, *47*, 349–358.
31. Khatibi, A.D.; Mahvi, A.H.; Mengelizadeh, N. Adsorption-desorption of tetracycline onto molecularly imprinted polymer: Isotherm, kinetics, and thermodynamics studies. *Desal. Water Treat.* **2021**, *230*, 240–251. [CrossRef]
32. Adriano, W.; Veredas, V.; Santana, C.; Gonçalves, L. Adsorption of amoxicillin on chitosan beads: Kinetics, equilibrium and validation of finite bath models. *Biochem. Eng. J.* **2005**, *27*, 132–137. [CrossRef]

33. Chang, P.-H.; Li, Z.; Jean, J.-S.; Jiang, W.-T.; Wang, C.-J.; Lin, K.-H. Adsorption of tetracycline on 2:1 layered non-swelling clay mineral illite. *Appl. Clay Sci.* **2012**, *67–68*, 158–163. [CrossRef]
34. Mohammadi, A.S.; Sardar, M. The Removal of Penicillin G from Aqueous Solutions using Chestnut Shell Modified with H2SO4: Isotherm and Kinetic Study. *J. Health Environ.* **2012**, *6*, 497–508.
35. Rahardjo, A.K.; Susanto, M.J.J.; Kurniawan, A.; Indraswati, N.; Ismadji, S. Modified Ponorogo bentonite for the removal of ampicillin from wastewater. *J. Hazard. Mater.* **2011**, *190*, 1001–1008. [CrossRef] [PubMed]
36. Li, S.; Zhang, X.; Huang, Y. Zeolitic imidazolate framework-8 derived nanoporous carbon as an effective and recyclable adsorbent for removal of ciprofloxacin antibiotics from water. *J. Hazard. Mater.* **2017**, *321*, 711–719. [CrossRef] [PubMed]
37. Yaghmaeian, K.; Moussavi, G.; Alahabadi, A. Removal of amoxicillin from contaminated water using NH 4 Cl-activated carbon: Continuous flow fixed-bed adsorption and catalytic ozonation regeneration. *Chem. Eng. J.* **2014**, *236*, 538–544. [CrossRef]
38. Balarak, D.; Azarpira, H.; Mostafapour, F.K. Adsorption isotherm studies of tetracycline antibiotics from aqueous solutions by maize stalks as a cheap biosorbent. *Int. J. Pharm. Technol.* **2016**, *8*, 16664–16675.
39. Zha, S.X.; Zhou, Y.; Jin, X.; Chen, Z. The removal of amoxicillin from wastewater using organobentonite. *J. Environ. Manag.* **2013**, *129*, 569–576. [CrossRef]
40. Liu, W.; Xie, H.; Zhang, J.; Zhang, C. Sorption removal of cephalexin by HNO3 and H2O2 oxidized activated carbons. *Sci. China Ser. B Chem.* **2012**, *55*, 1959–1967. [CrossRef]
41. Erşan, M.; Bağd, E. Investigation of kinetic and thermodynamic characteristics of removal of tetracycline with sponge like, tannin based cryogels. *Coll. Surfaces B Biointerfaces* **2013**, *104*, 75–82. [CrossRef]
42. Liu, W.; Zhang, J.; Zhang, C.; Ren, L. Sorption of norfloxacin by lotus stalk-based activated carbon and iron-doped activated alu-mina: Mechanisms, isotherms and kinetics. *Chem. Eng. J.* **2011**, *171*, 431–438. [CrossRef]
43. Wang, F.; Yang, B.; Wang, H.; Song, Q.; Tan, F.; Cao, Y. Removal of ciprofloxacin from aqueous solution by a magnetic chitosan grafted graphene oxide composite. *J. Mol. Liq.* **2016**, *222*, 188–194. [CrossRef]
44. Kyzas, G.Z.; Bikiaris, D.N.; Seredych, M.; Bandosz, T.J.; Deliyanni, E.A. Removal of dorzolamide from biomedical wastewaters with adsorption onto graphite oxide/poly(acrylic acid) grafted chitosan nanocomposite. *Bioresour. Technol.* **2014**, *152*, 399–406. [CrossRef]
45. Danalıoğlu, S.T.; Bayazit, S.; Kuyumcu, K.; Salam, M.A. Efficient removal of antibiotics by a novel magnetic adsorbent: Magnetic activated carbon/chitosan (MACC) nanocomposite. *J. Mol. Liq.* **2017**, *240*, 589–596. [CrossRef]
46. Peng, X.; Hu, F.; Dai, H.; Xiong, Q. Study of the adsorption mechanism of ciprofloxacin antibiotics onto graphitic ordered meso-porous carbons. *J. Taiwan Instit. Chem. Eng.* **2016**, *8*, 472–481. [CrossRef]
47. Balarak, D.; Baniasadi, M.; Lee, S.-M.; Shim, M.J. Ciprofloxacin adsorption onto Azolla filiculoides activated carbon from aqueous solutions. *Desal. Water Treat.* **2021**, *218*, 444–453. [CrossRef]
48. Agarwal, S.; Sadegh, H.; Monajjemi, M.; Hamdy, A.S.; Ali, G.A.; Memar, A.O.; Shahryari-Ghoshekandi, R.; Tyagi, I.; Gupta, V.K. Efficient removal of toxic bromothymol blue and methylene blue from wastewater by polyvinyl alcohol. *J. Mol. Liq.* **2016**, *218*, 191–197. [CrossRef]
49. Gupta, V.K.; Agarwal, S.; Sadegh, H.; Ali, G.; Bharti, A.K.; Makhlouf, A.S.H. Facile route synthesis of novel graphene oxide-β-cyclodextrin nanocomposite and its application as adsorbent for removal of toxic bisphenol A from the aqueous phase. *J. Mol. Liq.* **2017**, *237*, 466–472. [CrossRef]
50. Habeeb, O.A.; Ramesh, K.; Gomaa, A.M.A.; Bin, R.; Yunus, M. Low-cost and eco-friendly activated carbon from modified palm kernel shell for hydrogen sulfide removal from wastewater: Adsorption and kinetic studies. *Desal. Water Treat.* **2017**, *84*, 205–214. [CrossRef]
51. Omar, A.H.; Ramesh, K.; Gomaa, A.M.A.; Rosli, B.M.Y. Experimental design technique on removal of hydrogen sulfide using CaO-eggshells dispersed onto palm kernel shell activated carbon: Experiment, optimization, equilibrium and kinetic studies. *J. Wuhan Univ. Technol. Sci. Ed.* **2017**, *32*, 305–320. [CrossRef]
52. Habeeb, O.A.; Kanthasamy, R.; Gomaa, A.M.A. Isothermal modelling based experimental study of dissolved hydrogen sulfide ad-sorption from waste water using eggshell based activated carbon. *Malay. J. Anal. Sci.* **2017**, *21*, 334–345.
53. Balarak, D.; Taheri, Z.; Shim, M.J.; Lee, S.-M.; Jeon, C. Adsorption kinetics and thermodynamics and equilibrium of ibuprofen from aqueous solutions by activated carbon prepared from Lemna minor. *Desal. Water Treat.* **2021**, *215*, 183–193. [CrossRef]
54. Ding, R.; Zhang, P.; Seredych, M.; Bandosz, T.J. Removal of antibiotics from water using sewage sludge- and waste oil sludge-derived adsorbents. *Water Res.* **2012**, *46*, 4081–4090. [CrossRef]
55. Bui, T.X.; Choi, H. Adsorptive removal of selected pharmaceuticals by mesoporous silica SBA-15. *J. Hazard. Mater.* **2009**, *168*, 602–608. [CrossRef] [PubMed]
56. Moussavi, G.; Alahabadi, A.; Yaghmaeian, K.; Eskandari, M. Preparation, characterization and adsorption potential of the NH4Cl-induced activated carbon for the removal of amoxicillin antibiotic from water. *Chem. Eng. J.* **2013**, *217*, 119–128. [CrossRef]
57. Guler, U.A.; Sarioglu, M. Removal of tetracycline from wastewater using pumice stone: Equilibrium, kinetic and thermodynamic studies. *J. Environ. Health Sci. Eng.* **2014**, *12*, 79–87. [CrossRef]
58. Kerkez-Kuyumcu, Ö.; Bayazit, Ş.S.; Salam, M.A. Antibiotic amoxicillin removal from aqueous solution using magnetically modified graphene nanoplatelets. *J. Ind. Eng. Chem.* **2016**, *36*, 198–205. [CrossRef]
59. Kim, S.H.; Shon, H.K.; Ngo, H.H. Adsorption characteristics of antibiotics trimethoprim on powdered and granular activated carbon. *J. Ind. Eng. Chem.* **2010**, *16*, 344–349. [CrossRef]

60. Azarpira, H.; Mahdavi, Y.; Khaleghi, O. Thermodynamic Studies on the Removal of Metronidazole Antibiotic by Multi-Walled Carbon Nanotubes. *Der Pharm. Lett.* **2016**, *8*, 107–113.
61. Yu, F.; Li, Y.; Han, S.; Ma, J. Adsorptive removal of antibiotics from aqueous solution using carbon Materials. *Chemosphere* **2016**, *153*, 365–385. [CrossRef]
62. Gao, Y.; Li, Y.; Zhang, L.; Huang, H.; Hu, J.; Shah, S.M.; Su, X. Adsorption and removal of tetracycline antibiotics from aqueous solution by graphene oxide, *J. Colloids Interface Sci.* **2012**, *368*, 540–546. [CrossRef] [PubMed]
63. Hu, D.; Wang, L. Adsorption of amoxicillin onto quaternized cellulose from flax noil: Kinetic, equilibrium and thermodynamic study. *J. Taiwan Inst. Chem. Eng.* **2016**, *64*, 227–234. [CrossRef]

nanomaterials

Article

Synergetic Effect of Organic Flocculant and Montmorillonite Clay on the Removal of Nano-CuO by Coagulation-Flocculation-Sedimentation Process

Rizwan Khan [1], Muhammad Ali Inam [2], Kang-Hoon Lee [3,*], Abdul Sami Channa [1], Mukhtiar Ali Mallah [1], Young-Min Wie [4] and Mahmood Nabi Abbasi [1]

1 Department of Chemical Engineering, Quaid-e-Awam University of Engineering, Science and Technology (QUEST), Nawabshah 67480, Pakistan; rizwansoomro@quest.edu.pk (R.K.); abdul.sami@quest.edu.pk (A.S.C.); mukhtiarmallah@quest.edu.pk (M.A.M.); mahmoodabbasi73@quest.edu.pk (M.N.A.)

2 Institute of Environmental Sciences and Engineering (IESE), School of Civil and Environmental Engineering (SCEE), H-12 Campus, National University of Sciences and Technology (NUST), Islamabad 44000, Pakistan; ainam@iese.nust.edu.pk

3 Department of Civil and Environmental Engineering, Hanyang University, 222 Seongdong-gu, Seoul 04763, Korea

4 Department of Materials Engineering, Kyonggi University, Suwon 16227, Korea; supreme98@kyonggi.ac.kr

* Correspondence: diasyong@hanyang.ac.kr

Citation: Khan, R.; Inam, M.A.; Lee, K.-H.; Channa, A.S.; Mallah, M.A.; Wie, Y.-M.; Abbasi, M.N. Synergetic Effect of Organic Flocculant and Montmorillonite Clay on the Removal of Nano-CuO by Coagulation-Flocculation-Sedimentation Process. *Nanomaterials* **2021**, *11*, 2753. https://doi.org/10.3390/nano11102753

Academic Editor: Christos A. Aggelopoulos

Received: 24 September 2021
Accepted: 14 October 2021
Published: 17 October 2021

Abstract: The widespread usage of nano-copper oxide particles (nano-CuO) in several industrial products and applications raises concerns about their release into water bodies. Thus, their elimination from drinking water is essential to reduce the risk to human health. This work investigated the removal of nano-CuO from pure water and montmorillonite clay (MC) suspensions using poly aluminum ferric chloride (PAFC) as well as cationic polyacrylamide (PAM) by the coagulation-flocculation-sedimentation (C/F/S) process. Moreover, the PAFC and PAFC/PAM flocculation performance for various nano-CuO particles concentrations, dosages, pH, settling times and stirring speeds were also investigated. The findings showed that the removal of nano-CuO and turbidity in MC suspension were higher as compared to pure water. Moreover, the combined effect of PAFC/PAM on the elimination of nano-CuO and turbidity was also substantially better than the individual use of PAFC or PAM. The efficient removal of CuO was observed in the solution containing higher mass concentration in the order (10 mg/L > 2.5 mg/L > 1 mg/L) with an increased coagulant dose. The improved removal performance of nano-CuO was observed in a pH range of 7–11 under various water matrices. The C/F/S conditions of nano-CuO were further optimized by the Box–Behnken statistical experiment design and response surface methodology. The PAFC/PAM dose resulted in the maximum removal of nano-CuO (10 mg/L) in both pure water (>97%) and MC suspension (>99%). The results of particle monitoring and Fourier transform infrared of composite flocs revealed that the main removal mechanism of nano-CuO may be the combined effect of neutralization, complexation as well as adsorption.

Keywords: Box–Behnken design; coagulation-flocculation-sedimentation; clay; optimization; CuO

1. Introduction

Developments in nanotechnology have led to the widespread application of several metal-based nanoparticles (NPs) in different products and processes. The increased usage of NPs results in their release into the ecosystem, thereby affecting aquatic life and human health via the food chain [1]. Metal particles such as nano-copper oxide (nano-CuO) are typical metal-based particles, with an estimated global production of around 570 tons/year, which is projected to enhance over time [2,3]. The failure of wastewater treatment processes to effectively remove large quantities of nano-CuO can lead to the contamination of freshwater sources. The hazardous effects of nano-CuO upon many biological species, for instance, Lymphocytes, Fagopyrum esculentum, Daphnia magna and Pseudokirchneriella are well known [4–8]. Moreover, human exposure to these metal-based NPs via ingestion may cause cytotoxicity as well as genotoxicity even at a low concentration of 1 μg/mL [1,9]. Consequently, failure to remove nano-CuO effectively might increase the potential risk of exposure of these NPs to aquatic life and humans.

The mobility of metal-based NPs in the aquatic environment is controlled by different technologies such as bioremediation, phytoremediation, ion exchange, membrane separation, adsorption, electrochemical and coagulation processes. In wastewater treatment plants, biological treatments such as activated sludge can remove nano-CuO particles. However, the toxicity of NPs on the bacterial film of activated sludge interferes with the overall removal process by modifying the properties of sludge [10,11]. The membrane process can also effectively remove the NPs from water, but it is not commercially viable due to membrane fouling caused by highly concentrated NPs [12,13]. The incomplete removal of NPs through sewage treatment may eventually increase the threat of freshwater contamination. Consequently, a cost-effective method for complete nano-CuO particle removal is imperative.

Previous studies have confirmed that the conventional coagulation-flocculation-sedimentation (C/F/S) method can be used for the effective removal of various metal-based NPs such as nano-CuO, TiO_2, cadmium telluride (CdTe), multiwall carbon nanotubes (MWCNT) and zinc oxide (ZnO) from water. In alum treated water, Chalew et al. reported residual concentrations of 48–99%, 3–8% and 2–20% in spiked ZnO, TiO_2 and AgO NPs, respectively [14]. Earlier reports suggested that ferric chloride ($FeCl_3$) yielded a better coagulation efficiency of TiO_2 and ZnO NPs under a heterogeneous environment compared to polyferrous sulfate (PFS) and polyaluminum chloride (PACl) [15]. Moreover, the removal of metal-based NPs might be enhanced by the higher dosages of $FeCl_3$ and $Al_2(SO_4)_3$ [16,17]. The C/F/S process appears effective in removing metal-based NPs; however, the higher coagulant dose and unstable performance might be related to the single aluminum-based or iron-based coagulant. A recent study [18] described that the combined effect of aluminum and iron salt coagulants such as polyaluminum ferric chloride (PAFC) enhanced the coagulation performance of TiO_2 NPs at a low dosage. Moreover, high-molecular-weight flocculants such as cationic polyacrylamide (PAM) were reported to enhance the impact of removal via charge neutralization of the negatively charged flocculants [19,20]. Thus, a comprehensive exploration of the concurrent effect of inorganic coagulant and organic flocculants is essential for the removal of nano-CuO particles from aquatic environments.

Earlier studies focused on the C/F/S behavior of the primary pollutant nano-CuO; however, the interaction between several substances hindered the aggregation and removal of NPs. The interaction between clay particles such as kaolinite and NPs enhances the aggregation in ground and surface water, although the special structure of clay significantly affects the coagulation performance of NPs [21]. Wang et al. demonstrated the effect of anisotropy aggregation on the stability of NPs and clay mineral mixtures [22]. Other researchers have also shown the high adsorption capacity of the modified clay minerals of heavy metal ions under coagulation conditions [6]. Consequently, it is essential to systematically analyze the effect of clay minerals on the fate, mobility and removal behavior of nano-CuO during the conventional C/F/S process.

Many researchers have used the classical approach known as the single-factor method to investigate the coagulation behavior, although this technique fails to predict the interactive behavior under different operating parameters [23]. Amongst many statistical techniques, the Box–Behnken statistical experiment design (BBD) is a classical response surface methodology (RSM) used for modeling and analysis of experimental data with multiple operating parameters [24]. To the best of our knowledge, studies that simulate the C/F/S performance of nano-CuO removal using RSM have not been reported yet. It is crucial to systematically investigate the removal performance of nano-CuO by the C/F/S process using a mathematical modeling approach.

The present study aims to investigate the coagulation behavior of nano-CuO particles in water using PAFC and PAM via C/F/S under different experimental conditions. The effects of various mass concentrations of nano-CuO in pure water and clay suspensions were investigated. First, the optimum coagulation conditions of nano-CuO NPs were explored under PAFC, PAM and PAFC/PAM using a one factor at a time approach. Secondly, BBD and RSM were used to investigate the effect of various factors such as inorganic coagulant, organic flocculant and stirring speed. Last, the validity and reliability of the statistical analysis with various experimental data points were determined by comparing experimental and predicted nano-CuO removal efficiency response values.

2. Materials and Methods

2.1. Materials

Nano-copper oxide powder with an average <50 nm diameter, 99.8% purity, and montmorillonite clay (MC) used as the clay minerals in the current study were obtained from Sigma-Aldrich (St. Louis, MO, USA). The inorganic coagulant polyaluminum iron chloride (PAFC) (containing 26% of Al and Fe) and organic flocculant cationic polyacrylamide (PAM) with a molecular weight of 12 million were obtained from water treatment material Gongyi Tenglong Co., Ltd., Henan, China and Tianjin Chemical Reagent, Tianjin, China, respectively.

2.2. Stock Solutions

The nano-CuO stock solution was prepared by mixing various concentrations of CuO powder in pure water (18.2 MΩ). Initially, the stock solution of 100 mg/L of CuO was prepared and the final pH of the stock suspension was adjusted to 7.0 using 100 mM NaOH or HCl. The NP solution was dispersed by a probe-type ultrasonicator (Bio-safer 1200-90, Nanjing, China) for a period of 30 min at 400 W to acquire a well stable nano-CuO particles suspension. The 1 g/L MC suspension was prepared by weighing 1 g of MC powder using the microbalance (Mettler Toledo AG, Ag Model-XP26DR, Greidensee, Switzerland) and dissolved in 1 L pure water. Before the experiment, NPs' suspension was dispersed by a probe-type sonicator for 30 min. The stock solution comprising PAM 100 mg/L was prepared in pure water. The solution was stirred at a temperature of 50 °C and 250 rpm for 1 h with a lab magnetic stirrer to ensure complete dissolution. The dry weight method was used to calculate the dosage of inorganic coagulant (PAFC) and organic flocculant (PAM) in the current C/F/S experiments.

2.3. C/F/S Experiments

The experiments were conducted in the jar tester with a six-synchronous automatic lifting mixer (Young Tech Co., Ltd. Gyeongsangbuk-Do, Korea). Pure and artificial test water (MC suspension) were transferred to glass beakers. Predetermined amounts of PAFC, PAM and PAFC/PAM were dosed, and the suspension was rapidly stirred for 2 min at 250 rpm, and slow stirred for 10 min at 70 rpm. The different sets of the experiment including nano-CuO concentration (1–10 mg/L), pH (6–11), settling time (5–30 min) and stirring speed (100–350 rpm) were also investigated. The samples were collected 2 cm below the solution in the sampling vessel after the completion of the experiment for further analysis. The particle size was analyzed through the particle-size analyzer Zetasizer (Nano ZS90, Worcestershire, UK). During the coagulation, the water sample containing flocs was circulated using a silicone tube (dia 8 mm) through a peristaltic pump with controlled flow rate into the Zetasizer.

2.4. Analytical Procedure

Turbidity removal was determined with a turbidity measurement of the supernatant using a turbidimeter (Hach 2100-N, Loveland, CO, USA). Moreover, the residual concentration of nano-CuO was calculated by determining the absorbance at 254 nm with a UV-Vis spectrophotometer (Optizen, Mecasys, Daejeon, South Korea) as shown in the supplementary information (Figure S1). Additionally, a Fourier transform infrared (FTIR- JASCO, Easton, PA, USA) in the range of 400 to 4000 cm^{-1} was performed before and after the C/F/S experiments to understand possible removal mechanisms. Each test was performed three times and the relative standard deviations (STD) were reported as <5%. Moreover, the standard error was calculated through the division of STD by the square root of the number of tests performed. The popular Design Expert (version 8.0.5) software was used for the regression analysis, response-surface and contour maps of the experimental data.

3. Results

3.1. Influence of Coagulant and Flocculant Dosage on the Removal of Nano-CuO

Figure 1 shows the influence of PAFC, PAM, and PAFC/PAM on the removal performance in two water environments with varying concentrations (i.e., 1, 2.5 and 10 mg/L) of nano-CuO. The optimum dose of PAM (3 mg/L) and PAFC (50 mg/L) was observed for removing nano-CuO from pure water as well as MC containing water. Increasing the dosage of PAFC and PAFC/PAM led to the enhanced removal rate of nano-CuO and turbidity as depicted in Figure 1. The excess dose of PAM significantly decreased the removal rate after the optimal critical point; however, this effect of removal was found insignificant in the case of PAFC/PAM. It was observed that the removal efficiency of nano-CuO was enhanced with increasing mass concentration in the solution. This might be related to the fact that the higher concentration of NPs enhanced the probability of collision as well as increased the effect of co-precipitation via formation of the floc core. These findings are in good agreement with earlier work by Honda et al. [15]. The relatively better removal rate of various nano-CuO (1, 2.5 and 10 mg/L) concentrations in pure water was observed in PAFC/PAM (83.33%, 92.78% and 98.54%) rather than PAFC (72.48%, 86.23% and 93.90%) alone. The efficacy of co-precipitation and adsorption among the flocculants as well as NPs improved with the addition of PAFC/PAM in the solution. The enhanced removal efficiency of nano-CuO was found in suspension containing MC. Moreover, the addition of PAFC in both the pure and MC environments increased the removal rate of nano-CuO (10 mg/L) from 94.60% to 97.34%. As illustrated in Figure 1, the addition of MC and PAM had beneficial effects on the nano-CuO removal rates and the turbidity of the solution with similar flocculation conditions. In addition, the combined effect of PAFC/PAM improved the nano-CuO removal compared with PAFC as shown in Figure 1.

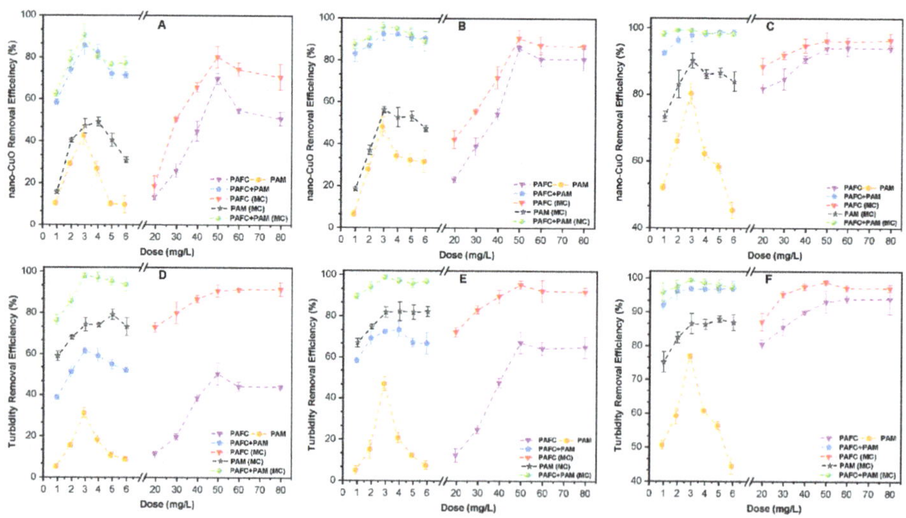

Figure 1. Influence of coagulant and flocculant dosage on the removal efficiency of nano-CuO and turbidity for various mass concentrations (**A**,**D**) 1 mg/L, (**B**,**E**) 2.5 mg/L and (**C**,**F**) 10 mg/L, respectively.

In pure water, the single system of nano-CuO (10 mg/L), the removal efficiency of PAFC before and after the addition of PAM was 94.90% and 98.54%, respectively. The PAM resulted in the formation of larger stable flocs due to increased adsorption bridging of the linear polymer, thereby improving the flocculation of nano-CuO [25]. The turbidity removal efficiency was strongly correlated with the removal of nano-CuO. It was observed that after optimal dosage, the increase of the PAM dosage decreased the rate of turbidity removal by 38%; however, this effect was found to be insignificant in the PAFC and PAFC/PAM cases. These results can be attributed to the fact that PAFC has strong potency to ionize high amounts of Fe and Al cations in the aquatic environment; thereby neutralizing the negative surface potential of nano-CuO [26]. However, the residual cations enhanced the effect of steric hindrance among colloids, resulting in particle stabilization. For the MC system, the removal efficiency was slightly higher compared with CuO in pure water. This phenomenon may be related to the fact that the addition of MC increased the floc nucleus and collision probability of the NPs, thereby resulting in adsorption onto the flocculant surface and formation of tiny dense flocs [27]. During flocculation, the simultaneous effects of PAFC and MC provided a net sweep, thus forming compact flocs and improving the overall effect of flocculation and impeded cohesion.

3.2. *Influence of pH on the Removal of Nano-CuO*

Figure 2 shows the effect of pH on the removal rates of nano-CuO and turbidity within the pH range of 6–11 at optimum doses of PAFC (50 mg/L) and PAM (3 mg/L). Fluctuation of the pH in the two systems significantly affected the removal of nano-CuO and turbidity. For instance, the removal efficiency was remarkably enhanced with the increase in pH and then declined to different levels. In the absence of MC suspension, PAM resulted in less than 30% removal of nano-CuO and turbidity at pH 6. The addition of MC enhanced the removal rates of nano-CuO and turbidity up to 60% under similar flocculation conditions as shown in Figure 2. However, the system containing MC had an insignificant effect on the nano-CuO removal by PAFC and PAFC/PAM. At pH 9 the removal efficiency of various initial concentrations of nano-CuO enhanced to the maximum level. In the MC environment, the removal efficiency of nano-CuO by PAFC/PAM was found to be more than 90%. In general, under acidic pH, the removal of both nano-CuO and turbidity was reduced compared with slightly alkaline environments and this difference was significant in the pure water system. The removal efficiency of PAFC, PAM and PAFC/PAM in the alkaline condition remained the same, suggesting that the higher pH environment had a negligible effect. The results of nano-CuO and turbidity removal curves showed similar trends, which indicate a strong correlation between the initial mass concentration of nano-CuO and turbidity. Our observation is consistent with earlier findings [28].

In the aquatic environment, the pH plays an important role in the surface potential of nano-CuO and also affects the formation of flocculant hydrolysates. The removal efficiency of PAM was mainly influenced by the solution pH due to a positive charge at acidic and alkaline conditions. Under the acidic environment, the hydroxyl groups on the surface of nano-CuO adsorbed protons and thus were positively charged, while PAM released cations in the solution, resulting in the poor removal efficiency of NPs. The surface potential of CuO above pH 7 was negatively charged owing to the loss of protons. The release of cations due to the ionization of PAM neutralized the surface charge thereby enhancing the removal due to formation of large aggregated flocs in the system [29]. However, under highly alkaline pH environments, both NPs and PAM were negatively charged thus increasing the electrostatic repulsion amongst them, and so reducing the overall flocculation and precipitation efficiency. Furthermore, under alkaline conditions, PAFC rapidly hydrolyzed and precipitated to form hydroxide, thereby reducing the overall removal efficiency of NPs. The effect of nano-CuO destabilization was enhanced by increasing the solution pH following the addition of PAFC; however, the tiny flocs remained suspended in the solution [30]. The compounding process of PAFC/PAM was relatively more stable than PAFC and PAM alone due to the wider pH adaptation range of organic flocculant (PAM) [31,32]. Moreover, the complex structure of the PAFC/PAM polymer remarkably enhanced the flocculant-specific surface area and improved the bridging effect during adsorption [32]. Consequently, the combination of inorganic and organic flocculants had a synergistic effect on the C/F/S process and on the overall nano-CuO removal performance from the aquatic environment.

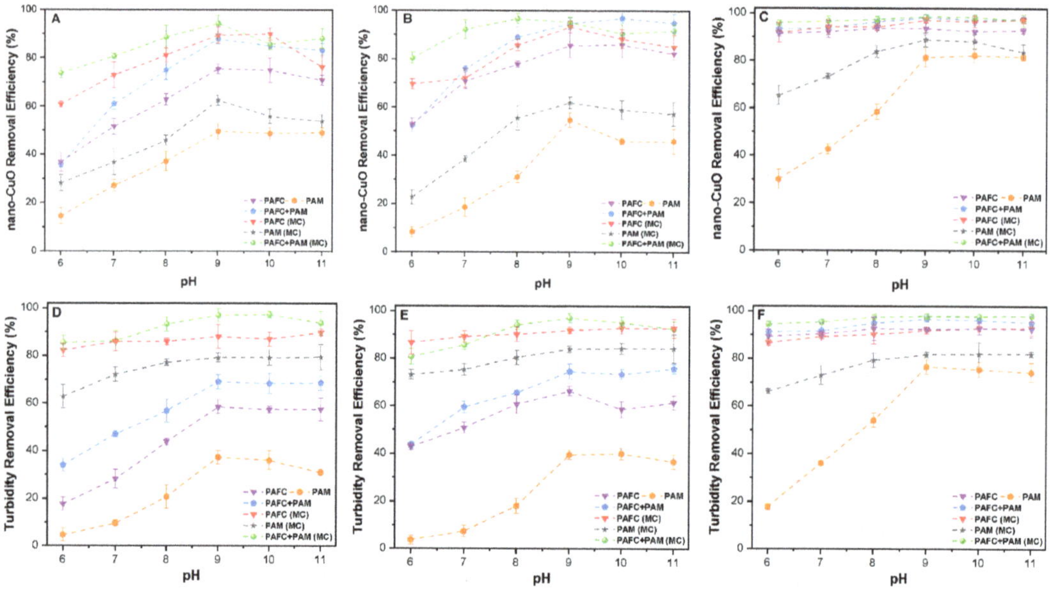

Figure 2. Influence of pH (6–11) on the removal efficiency of nano-CuO and turbidity for various mass concentrations (**A**,**D**) 1 mg/L, (**B**,**E**) 2.5 mg/L and (**C**,**F**) 10 mg/L.

3.3. Influence of Sedimentation Period on the Removal of Nano-CuO

The sedimentation period, which is directly related to the design of the settling tank, is considered as a key parameter affecting the magnitude of design, investment and operation cost. In the present study, the settling time was used as the analytical index based on the optimal dosage of coagulant. The removal rates of nano-CuO and turbidity under specific PAFC, PAM and PAFC/PAM settling time were investigated with various concentrations of nano-CuO (1, 2.5 and 10 mg/L) in two separate systems. Figure 3 shows the effect of the prolonged duration of precipitation in increasing the turbidity and removal efficacy of nano-CuO. The addition of the PAFC coagulant resulted in stable removal rates of CuO after 30 min of sedimentation time in both systems. In the single CuO system, the removal performance of PAFC and PAM for 1, 2.5 and 10 mg/L CuO under a settling time of 5 min reached 23.38, 26.04 and 79.13% and 13.87, 26.10 and 73.11%, respectively. However, the removal rates remained constant after 15 min of settling time. The precipitation period under different concentrations of CuO was substantially shorter in the composite coagulant-flocculant system (PAFC/PAM) compared to the single PAFC and PAM. The removal efficiency of nano-CuO by PAFC/PAM exceeded 75% within 5 min of settling. In the single system of CuO, the removal rate of 1, 2.5 and 10 mg/L CuO by PAFC after 5 min of settling increased from 23.38% to 65.17%, 26.04% to 78.54% and 79.13% to 92.94%, respectively. In general, the nano-CuO removal rates and turbidity stabilized with a further increase in the precipitation period up to 20 min. The optimal removal rate of nano-CuO (10 mg/L) after 30 min by PAFC, PAM and PAFC/PAM was found to be 94.14%, 83.89% and 99.24%, respectively. Moreover, the addition of MC facilitated the removal efficiency of CuO and turbidity under similar settling conditions as depicted in Figure 3.

The removal efficiencies, turbidity and settling time of nano-CuO in the three initial mass concentrations were similar in both systems. Our results are consistent with those reported in the literature [33].

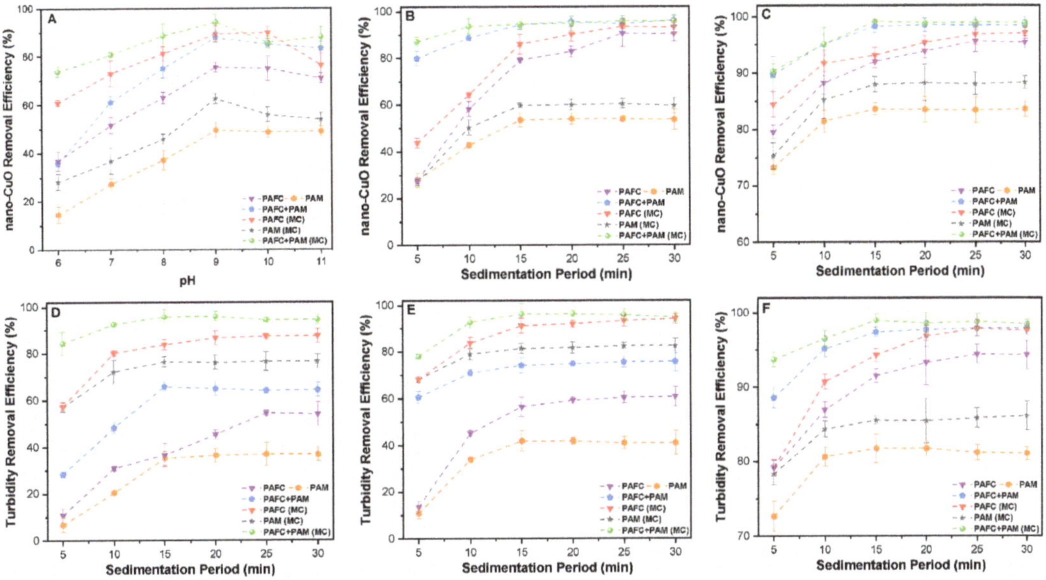

Figure 3. Influence of settling time (5–30 min) on the removal efficiency of nano-CuO and turbidity for various mass concentrations (**A**,**D**) 1 mg/L, (**B**,**E**) 2.5 mg/L and (**C**,**F**) 10 mg/L.

In the PAFC case, the optimal settling period of CuO was found to be shorter than that of PAM alone. The ionization of PAFC released Fe(III) and Al(III) ions in the solution, leading to the formation of complexes with nano-CuO. Moreover, PAM contained the reactive groups in its polymer chains which provided the favorable adsorption sites for tiny flocs of CuO. Thus, the removal of nano-CuO by PAM mainly occurs via the adsorption bridging mechanism [34]. These observations are consistent with the literature [35], which reported similar turbidity removal rates of the kaolinite−humic acid solution using PAFC and PAM−PAFC. In the PAFC/PAM case, stable nano-CuO (10 mg/L) and turbidity removal rates of up to 97.85% and 96.45%, respectively, were observed after 15 min as shown in Figure 3C,F. The enhanced removal efficiency may be related to the bridging effect of PAFC/PAM, which forms dense and stable flocs that rapidly settle down. In contrast, the flocs formed by PAFC and PAM alone were small and had longer settling times. Thus, the combination of organic and inorganic flocculants, i.e., PAFC/PAM can significantly decrease the treatment cost by reducing the settling time.

3.4. Influence of Stirring Speed on the Removal of Nano-CuO

The optimum doses of PAFC and PAM were determined at neutral pH and a precipitation duration of 30 min, at 10 min intervals under a slow stirring speed of 70 rpm and at 2 min under a fast rotation speed (FRS) of 250 rpm. Figure 4 shows the effect of different stirring speeds on the elimination of nano-CuO particles. The optimal removal of NPs under different mass concentrations (1, 2.5 and 10 mg/L) in both systems was found at an FRS of 200 rpm. Moreover, the changes in the removal pattern in both systems were consistent with the variation in the hydraulic conditions. For instance, at an initial concentration of 1 mg/L in the single system of CuO, the effect of the stirring conditions was significantly high. The removal efficiencies of 1, 2.5 and 10 mg/L of nano-CuO using PAFC were found at an FRS of 200 and 350 rpm in the ranges of 69.10% and 28.01%, 95.37%

and 38.80% and 95.39% and 98.12%, respectively. Moreover, the removal rates of 1, 2.5 and 10 mg/L of nano-CuO by PAFC/PAM at an FRS of 200 and 350 rpm were found in the ranges of 87.23% and 28.98%, 96.54% and 42.01% and 97.45% and 93.70%, respectively. The variation in stirring conditions significantly affected the removal rates of NPs at lower initial concentrations of 1 and 2.5 mg/L of CuO compared with relatively higher concentrations of 10 mg/L. The addition of MC enhanced the removal rates of nano-CuO and the suspension turbidity under similar conditions. For instance, in the MC suspension system, the removal efficiency of nano-CuO (10 mg/L) using PAFC/PAM at an FRS of 200 and 350 rpm was found as 98.54% and 92.10%, respectively.

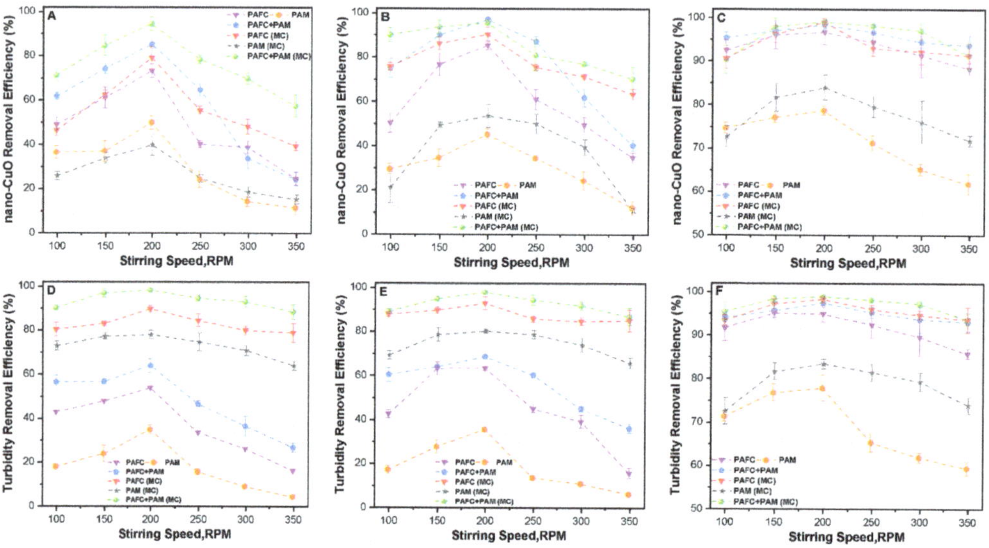

Figure 4. Influence of stirring speed (100–350 RPM) on the removal performance of nano-CuO and turbidity for various initial mass concentrations (**A**,**D**) 1 mg/L, (**B**,**E**) 2.5 mg/L and (**C**,**F**) 10 mg/L.

Hydraulic parameters, specifically the agitation time and speed, are major parameters in the formation of stable flocs. Rapid agitation is required for the homogeneous dispersion of the coagulant and flocculant in the sample, while a slower speed is used to form stable flocs. The velocity gradient (G) is another critical parameter influencing the C/F/S performance. Furthermore, mixing speeds also have a direct effect on the formation and stabilization of flocs. During the rapid mixing phase, the rotation speed around 100 rpm was too low and produced a small G value in the solution. Hence, in such systems, the low removal efficiency of nano-CuO was attributed to insignificant floc formation. Furthermore, fast speeds of about 350 rpm produced a higher G value, which was not conducive to floc formation, thereby reducing the removal efficiency of nano-CuO. The optimum G value leads to compact agglomerates with increased fractal dimension and flocs size in the suspension becoming more even [36]. The significant effect of the stirring speed was observed at low concentrations of nano-CuO (1 and 2.5 mg/L) as shown in Figure 4A,B. Such an observation might be related to the fact that the higher agitation rate leads to the formation of small aggregates during the C/F/S process. Further, the tiny and unstable flocs broke down resulting in poor removal efficiency of nano-CuO in a low mass concentration of NPs. Relatively large and stable flocs were formed during the C/F/S process at an initial mass concentration of 10 mg/L, with higher resistance to variations in hydraulic conditions. In the case of PAM alone, the high mixing speed during the C/F/S process resulted in the poor adsorption bridging effect of long-chain polymer [18,37]. Thus, the removal performance of nano-CuO and turbidity were significantly reduced in such a system.

3.5. Response Surface Methodology

Response surface methodology (RSM) based on the Box–Behnken model (BBD) design principle was used to further understand the removal behavior of nano-CuO (10 mg/L) in MC suspensions. Three independent factors, i.e., PAFC, PAM dosage and stirring conditions, were selected at three levels and the experimental design was set to simulate different C/F/S conditions as shown in Table 1. The removal rate of nano-CuO was taken as the response value and a regression model was generated, as presented in Table 2.

Table 1. Type, parameters and corresponding coded levels for RSM.

Type	Variables	Coded Level of Variables		
		−1	0	1
		Corresponding Operation Values		
A	PAFC (mg/L)	20	50	80
B	PAM (mg/L)	1	3	5
C	Stirring Speed (RPM)	100	200	350

Table 2. Test design and response values for RSM.

Experiment	A	B	C	Nano-CuO Removal (%)
1	0	1	1	92.26
2	−1	1	0	90.15
3	−1	0	1	91.15
4	0	0	0	98.82
5	0	1	−1	88.12
6	0	−1	−1	87.87
7	0	0	0	98.67
8	0	0	0	99.15
9	1	1	0	91.95
10	−1	0	−1	90.23
11	0	0	0	99.48
12	0	0	0	99.51
13	1	0	−1	87.14
14	1	0	1	93.49
15	−1	−1	0	89.37
16	1	−1	0	91.22
17	0	−1	1	91.14

The slope of the response surface plot shows the effect of different parameters on the removal efficiency of nano-CuO as illustrated in Figure 5. The higher slope value indicates the greater influence, whereas the shape of the contour map reflects the effect of the interaction of other parameters on the removal performance. The effect of linear parameters, i.e., hydraulic conditions and PAFC dosage, and all quadratic parameters on the removal efficiency of nano-CuO was found significant. However, the influence of different interactions was found insignificant (Figure 5A–C). At constant PAFC dosage, the removal efficiency of nano-CuO improved and then decreased with the increased PAM dosage (Figure 5A). Likewise, at constant PAM dosage, the increase in PAFC dosage initially increased the removal of nano-CuO and then decreased the overall C/F/S performance (Figure 5B).

The optimum influencing ranges of PAFC and PAM dosages on the removal of nano-CuO were found to be 37–65 mg/L and 2.20–4.00 mg/L, respectively. However, the model shows no significant effect of the interaction between PAFC and PAM dosages on the removal rate of the pollutant (Table 3). The optimal range of PAFC and stirring speed were found to be 37–65 mg/L and 170–270 rpm, respectively (Figure 5B). The stirring speed had a substantial influence on the removal efficiency of nano-CuO (Table 3). As indicated in Figure 5C, when the PAFC dosage was kept constant, the influence of other parameters on the elimination of nano-CuO initially increased and then decreased. The optimum ranges for the PAM dosage and stirring speed were observed as 2.20–4.00 mg/L and 170–270 rpm, respectively. The regression analysis was performed to further understand the influence of each factor. The regression equation between PAFC, PAM dosage, stirring speed and nano-CuO removal rate is presented in Equation (1).

$$\begin{aligned} nano-CuO\ removal\quad Efficiency \\ &= 98.13 + 0.36A + 0.36B + 1.83C - 0.013AB + 1.36AC \\ &+0.22BC - 3.90A^2 - 4.55B^2 - 4.72C^2 \end{aligned} \tag{1}$$

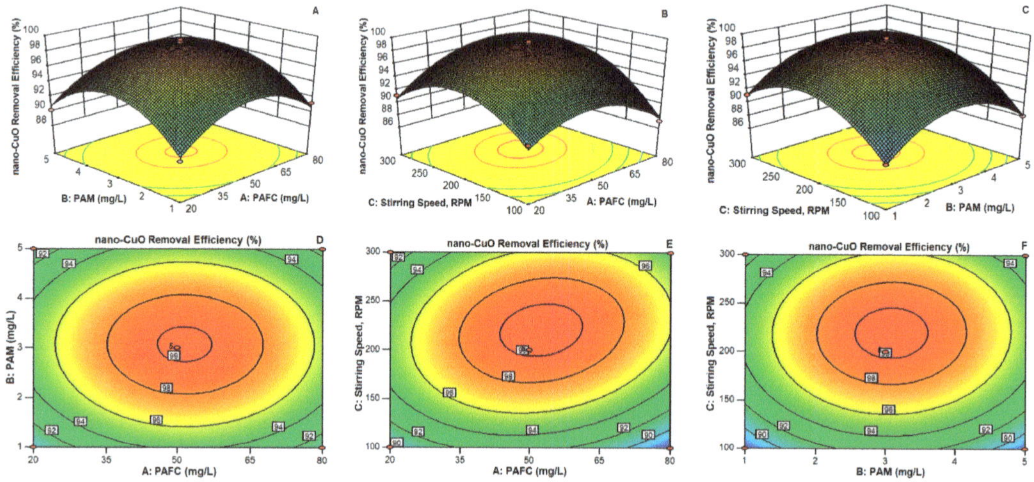

Figure 5. 3D surface plots and corresponding contour plots showing the removal rate of nano-CuO under the influences of (**A–D**) PAFC dosage and PAM dosage, (**B–E**) PAFC dosage and stirring speed and (**C–F**) PAM dosage and stirring speed.

Table 3. Analysis of variance.

Source	Sum of Squares	Df	Mean Square	F Value	p-Value Prob > F	
Model	310.5	9	34.5	80.51	<0.0001	Significant
A-(PAFC)	1.05	1	1.05	2.45	0.1613	Not significant
B-(PAM)	1.04	1	1.04	2.42	0.1638	Not significant
C-(Stirring Speed)	26.94	1	26.94	62.86	<0.0001	Significant
AB	6.250×10^{-4}	1	6.25×10^{-4}	1.458×10^{-3}	0.9706	Not significant
AC	7.37	1	7.37	17.2	0.0043	Significant
BC	0.19	1	0.19	0.44	0.5276	Not significant
A2	64.02	1	64.02	149.38	<0.0001	Significant
B2	87.33	1	87.33	203.79	<0.0001	Significant
C2	93.97	1	93.97	219.28	<0.0001	Significant
Residual	3	7	0.43			
Lack of Fit	2.42	3	0.81	5.62	0.0643	Not significant
Pure Error	0.57	4	0.14			
Cor Total	313.5	16				

The detail of the analysis of variance is shown in Table 3. The mathematical model was found significant (F = 80.51, $p < 0.0001$). Moreover, the individual effects of stirring conditions and PAFC dosage were found significant. In contrast, the interaction influence of these parameters was found insignificant. These observations are in good agreement with the results of the response surface analysis. The adjusted coefficient of determination $Radj^2$ was found to be 0.9781. Moreover, the obtained experimental results show a good correlation with the estimated value. In general, the model might be utilized to optimize the analysis and estimate the removal of nano-CuO by PAFC/PAM during the C/F/S process.

3.6. Model Validation and Monitoring of Floc

The regression equation was solved to find the optimum C/F/S condition for removal of nano-CuO. The obtained settings for NPs' removal were as follows: PAFC (~52.48 mg/L), PAM (~3.09 mg/L), and stirring conditions of 2 min FRS (220.76 rpm) and 10 min of slow rotation (80 rpm). The calculated value of nano-CuO removal under these environmental conditions was around 99.34%. The precision of the model was further verified by performing the three different sets of the experiment under the obtained model conditions. The results of three identical sets of trials presented that the treatment efficacy of nano-CuO was 97.95%, 98.76% and 99.07%. The calculated relative error among the obtained and estimated value was found to be 0.75%, which shows that the model can better reflect the optimal removal conditions for nano-CuO.

Figure 6 shows the monitoring of floc formation after the flocculation process with various conditions of PAFC, PAM, PAFC/PAM, MC and (2.5 mg/L) of nano-CuO. As demonstrated in Figure 6, the flocs formed under the condition of 50 mg/L PAFC + 3 mg/L PAM and 20 mg/L MC were larger in size as compared to those found at 50 mg/L PAFC + 3 mg/L PAM in pure water. The measured particle size (dia:0.5) of the flocs formed by 50 mg/L PAFC was around 95 μm in pure water, moreover, flocs obtained with 50 mg/L PAFC + 20 mg/L MC was around 145 μm. Thus, the addition of MC enhanced the particle size and density of flocs during flocculation. In contrast, the size of the floc formed by the combination of 50 mg/L PAFC and 3 mg/L PAM was found above 250 μm. The particle size distribution measurement of the flocs after flocculation experiments further revealed that the addition of PAM helped to enhance the floc size above 85%. The excess dosage of PAFC resulted in higher hydrolysis of Al(III) and Fe(III) in coagulation thereby incorporating cations to neutralize the negative surface potential of nano-CuO. However, excess cations in the solution resulted in the formation of tiny flocs with low compactness [38]. The addition of PAM neutralized the negative charge of NPs' surface thereby destabilizing the suspended NPs in the solution. Moreover, in the presence of PAM, the adsorption bridging occurred during the C/F/S process, which helped PAFC to form large flocs with an enhanced degree of compactness.

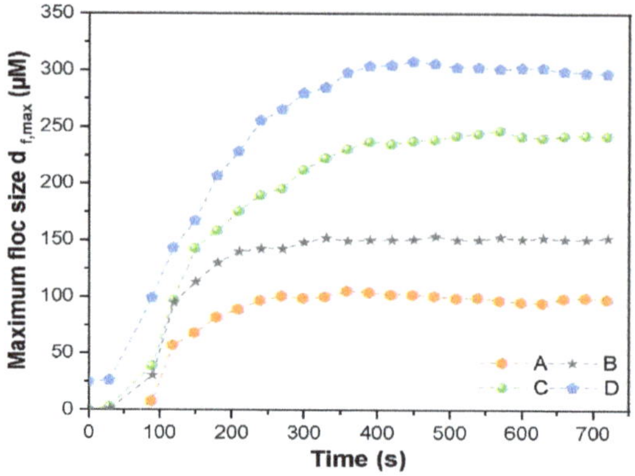

Figure 6. Size of composite floc of nano-CuO (2.5 mg/L) under various conditions at (**A**) 50 mg/L PAFC, (**B**) 50 mg/L PAFC + 20 mg/L MC, (**C**) 50 mg/L PAFC with 3 mg/L PAM and (**D**) 50 mg/L PAFC + 3 mg/L PAM and 20 mg/L MC.

3.7. Characteristics of Composite Flocs

The results of the FTIR analysis of pristine chemicals including MC, PAM, PAFC, nano-CuO and composite flocs obtained after the C/F/S experiment were used to illustrate the probable removal mechanisms, as demonstrated in Figure 7. The observed broad bands at 3300–3400 cm^{-1} due to the stretching vibration of the (^{-}OH) functional group linked to aluminum or iron (Al/Fe–OH) [39]. The spectrum peaks observed around 2972, 2887, 1653 and 1460 cm^{-1} were associated with asymmetric and symmetric stretching vibrations of C-H, CH_2, C=O and acylamino groups, respectively [40]. The absorption bands at 1381 cm^{-1} attributed to the symmetric stretching vibration of CH_3, COO^-,while the band appearing at 1254 cm^{-1} corresponded to the C–O anti-symmetric stretching [41]. In addition, few peaks were found at 1257, 1151 and 934 cm^{-1} owing to the stretching of C–OH (phenolic), C–O and carboxylic acid groups [42]. However, the peaks at 883 and 542 cm^{-1} were ascribed to the bending vibration of the Fe–OH– Fe and Cu-O bond stretching, respectively.

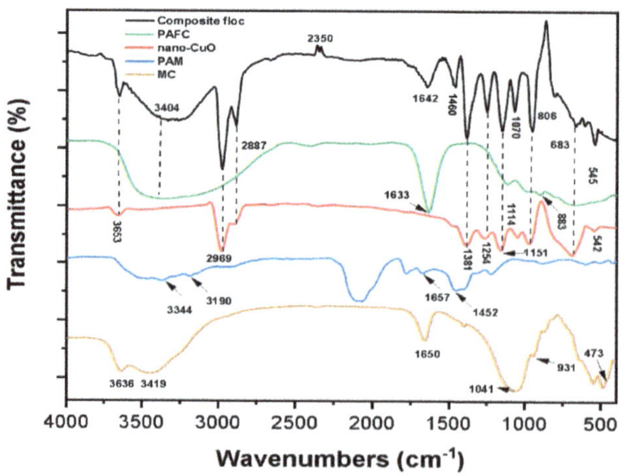

Figure 7. FT-IR spectrum of pristine chemicals, i.e., MC, PAM, nano-CuO, PAFC and composite flocs obtained at optimum C/F/S conditions.

In addition, the IR spectra of composite floc obtained by the PAFC/PAM (MC) system showed a broader absorption peak with higher intensity at 3653 cm^{-1} than that of nano-

CuO, implying some inner hydroxyl in MC bonded to the Cu-O group of nano metals [43]. The observed peak between 3250 cm^{-1} to 3500 cm^{-1} with strong intensity at 3404 cm^{-1} was larger than that of PAFC, indicating the formation of moderately strong H-bonding between inner surface hydroxyls of the MC and metals ions during the C/F/S process [44,45]. The absorption peak at 1633–1657 cm^{-1} was due to the bending vibration of H–O–H. The peak in the composite floc decreased with the significant shift, indicating that less (OH) groups were combined with Al/Fe ions to form complexes with free Cu ions [46]. The higher intensity of peak at 1460 cm^{-1} and 1254 cm^{-1} attributed to C=O and acylamino groups further clarified the complex coordination with Cu ions in the adsorption process [47]. The bands at 1114 cm^{-1} shifted to a lower frequency at 1070 cm^{-1} with enhanced intensity owing to the stretching vibration of Fe–O/Cu–O. In addition, the peak observed at 683 cm^{-1} was found, suggesting the bending vibration of Fe–O–Cu/Fe–OH–Cu [48]. The enrichment of hydroxyl (OH) bridging in PAFC/PAM (MC) suspension formed the compact structure flocs. Moreover, the flocs formed would increase adsorption and sweep coagulation. This was supported by the findings of floc monitoring, where the increase in floc size was observed in the heterogenous system. The substantial shifts and increase in the intensity of some bands in the obtained composite flocs also well supported the complexations of metal ions and PAFC/PAM in the MC environment. Consequently, it might be concluded from the improved NPs removal rate and IR spectra, that the key mechanism for nano-CuO removal may be the compound effect of charge neutralization, complexation and adsorption.

4. Conclusions

In this research, we explored the influence of coagulant dose, pH, settling time and stirring speed on CuO removal during the C/F/S process. The removal efficiency of nano-CuO and turbidity in both systems were enhanced by increasing the initial mass concentration of NPs. Moreover, the presence of MC improved the removal efficiency of CuO. The addition of both PAFC and PAM enhanced the compression and stability of flocs. The surface potential of flocs was greatly influenced in the acidic environment. In addition, the PAFC/PAM significantly increased removal efficiency and reduced the settling time in the MC containing suspension. The variation in initial mass concertation and the stirring speed affected the floc formation and removal performance of nano-CuO. The BBD and RSM techniques were applied to determine optimal C/F/S process conditions for maximizing CuO removal from water. The obtained model responses suggested the optimum C/F/S conditions as PAFC (52.48 mg/L), PAM (3.09 mg/L), and mixing at 2 min of fast rotation (220.76 rpm) and 10 min of slow rotation (80 rpm). Furthermore, the validity of the model was accessed under different environmental conditions. The FT-IR analysis of composite flocs revealed that primary mechanisms including charge neutralization, complexation and adsorption may be involved in the removal of nano-CuO by the C/F/S process. In general, the findings provide insight into enhanced flocculation and coagulation of CuO in drinking water containing clay particles.

Supplementary Materials: The following are available online at https://www.mdpi.com/article/10.3390/nano11102753/s1, Figure S1: Spectra of pristine nano-CuO, MC and the mixture of nano-CuO + MC.

Author Contributions: Conceptualization, R.K.; methodology, R.K. and K.-H.L.; software, R.K.; validation, R.K., M.A.I. and K.-H.L.; formal analysis, R.K., Y.-M.W. and K.-H.L.; investigation, R.K.; resources, R.K. and K.-H.L.; data curation, R.K.; writing—original draft preparation, R.K.; writing—review and editing, R.K., A.S.C., M.A.M., M.A.I., M.N.A. and Y.-M.W.; visualization, R.K. and M.A.I.; supervision, R.K.; project administration, R.K. and K.-H.L. All authors have read and agreed to the published version of the manuscript.

Funding: This research was supported by the Basic Science Research Program through the National Research Foundation of Korea (NRF) funded by the Ministry of Education (2021R1I1A1A01041492).

Institutional Review Board Statement: Not applicable.

Informed Consent Statement: Not applicable.

Data Availability Statement: Not applicable.

Acknowledgments: This study was jointly conducted by the Department of Chemical Engineering, Quaid-e-Awam University of Engineering, Science and Technology (QUEST), Nawabshah, Sindh, Pakistan, and Department of Civil and Environmental Engineering, Hanyang University, Seoul, Korea.

Conflicts of Interest: The authors declare no conflict of interest.

References

1. Pietroiusti, A.; Stockmann-Juvala, H.; Lucaroni, F.; Savolainen, K. Nanomaterial exposure, toxicity, and impact on human health. *Wiley Interdiscip. Rev. Nanomed. Nanobiotechnol.* **2018**, *10*, 1513. [CrossRef] [PubMed]
2. Gottschalk, F.; Sun, T.; Nowack, B. Environmental concentrations of engineered nanomaterials: Review of modeling and analytical studies. *Environ. Pollut.* **2013**, *181*, 287–300. [CrossRef]
3. Janković, N.Z.; Plata, D.L. Engineered nanomaterials in the context of global element cycles. *Environ. Sci. Nano* **2019**, *6*, 2697–2711. [CrossRef]
4. Lee, S.; Chung, H.; Kim, S.; Lee, I. The Genotoxic Effect of ZnO and CuO Nanoparticles on Early Growth of Buckwheat, Fagopyrum Esculentum. *Water Air Soil Pollut.* **2013**, *224*, 1–11. [CrossRef]
5. Bondarenko, O.; Juganson, K. Angela Ivask, Kaja Kasemets, Monika Mortimer & Anne Kahru. *Arch. Toxicol.* **2013**, *87*, 1181–1200.
6. Hussain, S.T.; Ali, S.A.K. February. Removal lead Pb (II) from wastewater using kaolin clay. In *IOP Conference Series: Materials Science and Engineering*; IOP Publishing: Bristol, UK, 2021.
7. Bour, A.; Mouchet, F.; Silvestre, J.; Gauthier, L.; Pinelli, E. Environmentally relevant approaches to assess nanoparticles ecotoxicity: A review. *J. Hazard. Mater.* **2015**, *283*, 764–777. [CrossRef]
8. Muna, M.; Blinova, I.; Kahru, A.; Vinković Vrček, I.; Pem, B.; Orupõld, K.; Heinlaan, M. Combined effects of test media and dietary algae on the toxicity of CuO and ZnO nanoparticles to freshwater microcrustaceans *Daphnia magna* and Heterocypris incongruens: Food for thought. *Nanomaterials* **2019**, *9*, 23. [CrossRef]
9. Lin, W.; Xu, Y.; Huang, C.-C.; Ma, Y.; Shannon, K.; Chen, D.-R.; Huang, Y.-W. Toxicity of nano- and micro-sized ZnO particles in human lung epithelial cells. *J. Nanoparticle Res.* **2009**, *11*, 25–39. [CrossRef]
10. Eduok, S.; Coulon, F. Engineered Nanoparticles in the environments: Interactions with microbial systems and microbial Activity. In *Microbial Ecotoxicology*; Springer: Cham, Switzerland, 2017; pp. 63–107.
11. Chen, H.; Zheng, X.; Chen, Y.; Li, M.; Liu, K.; Li, X. Influence of copper nanoparticles on the physical-chemical properties of activated sludge. *PLoS ONE* **2014**, *9*, e92871. [CrossRef]
12. Wang, Y.; Xue, N.; Chu, Y.; Sun, Y.; Yan, H.; Han, Q. CuO nanoparticle–humic acid (CuONP–HA) composite contaminant removal by coagulation/ultrafiltration process: The application of sodium alginate as coagulant aid. *Desalination* **2015**, *367*, 265–271. [CrossRef]
13. Springer, F.; Laborie, S.; Guigui, C. Removal of SiO_2 nanoparticles from industry wastewaters and subsurface waters by ultrafiltration: Investigation of process efficiency, deposit properties and fouling mechanism. *Sep. Purif. Technol.* **2013**, *108*, 6–14. [CrossRef]
14. Chalew, T.E.A.; Ajmani, G.; Huang, H.; Schwab, K.J. Evaluating nanoparticle breakthrough during drinking water treatment. *Environ. Health Perspect.* **2013**, *121*, 1161–1166. [CrossRef]
15. Honda, R.J.; Keene, V.; Daniels, L.; Walker, S.L. Removal of TiO_2 Nanoparticles during primary water treatment: Role of Coagulant type, dose, and nanoparticle concentration. *Environ. Eng. Sci.* **2014**, *31*, 127–134. [CrossRef] [PubMed]
16. Khan, R.; Inam, M.A.; Park, D.R.; Khan, S.; Akram, M.; Yeom, I.T. The Removal of CuO Nanoparticles from Water by Conventional Treatment C/F/S: The Effect of pH and Natural Organic Matter. *Molecules* **2019**, *24*, 914. [CrossRef]
17. Lv, M.; Liu, T.; Chen, F.; Zhang, Z.; Li, D.; Sun, M.; Feng, Y. Interactions between magnetic particles and polyaluminum chloride on the coagulation behavior in humic acid-kaolin synthetic water treatment. *Environ. Res.* **2021**, *197*, 111093. [CrossRef] [PubMed]
18. You, Z.; Zhuang, C.; Sun, Y.; Zhang, S.; Zheng, H. Efficient Removal of TiO_2 nanoparticles by enhanced flocculation–coagulation. *Ind. Eng. Chem. Res.* **2019**, *58*, 14528–14537. [CrossRef]
19. Xu, W.; Yue, Q.; Gao, B.; Du, B. Impacts of organic coagulant aid on purification performance and membrane fouling of coagulation/ultrafiltration hybrid process with different Al-based coagulants. *Desalination* **2015**, *363*, 126–133. [CrossRef]
20. He, M.; Zhou, R.; Guo, X. Behavior of stabilized multiwalled carbon nanotubes in a $FeCl_3$ coagulation system and the structure characteristics of the produced flocs. *J. Colloid Interface Sci.* **2012**, *366*, 173–178. [CrossRef]
21. Wang, H.; Dong, Y.-N.; Zhu, M.; Li, X.; Keller, A.A.; Wang, T.; Li, F. Heteroaggregation of engineered nanoparticles and kaolin clays in aqueous environments. *Water Res.* **2015**, *80*, 130–138. [CrossRef]
22. Wang, J.; Zhao, X.; Wu, F.; Tang, Z.; Zhao, T.; Niu, L.; Fang, M.; Wang, H.; Wang, F. Impact of montmorillonite clay on the homo-and heteroaggregation of titanium dioxide nanoparticles ($nTiO_2$) in synthetic and natural waters. *Sci. Total Environ.* **2021**, *784*, 147019. [CrossRef]
23. Neubauer, D. Process optimization: A statistical approach. *J. Qual. Technol.* **2008**, *40*, 348–352. [CrossRef]
24. Lucas, J.M. Response surface methodology: Process and product optimization using designed experiments, 3rd edition. *J. Qual. Technol.* **2010**, *42*, 228–230. [CrossRef]

25. Sang, Y.; Lin, A.; Liu, X. Population balance modeling of cationic polyacrylamide (CPAM) induced flocculation process for lignin recovery from the pre-hydrolysis liquor of kraft pulping process. *Sep. Purif. Technol.* **2019**, *221*, 152–158. [CrossRef]
26. Sun, Y.; Zhu, C.; Zheng, H.; Sun, W.; Xu, Y.; Xiao, X.; You, Z.; Liu, C. Characterization and coagulation behavior of polymeric aluminum ferric silicate for high-concentration oily wastewater treatment. *Chem. Eng. Res. Des.* **2017**, *119*, 23–32. [CrossRef]
27. Sousa, V.S.; Corniciuc, C.; Teixeira, M.R. The effect of TiO_2 nanoparticles removal on drinking water quality produced by conventional treatment C/F/S. *Water Res.* **2017**, *109*, 1–12. [CrossRef] [PubMed]
28. Wang, H.T.; Ye, Y.Y.; Qi, J.; Li, F.T.; Tang, Y.L. Removal of titanium dioxide nanoparticles by coagulation: Effects of coagulants, typical ions, alkalinity and natural organic matters. *Water Sci. Technol.* **2013**, *68*, 7–1143. [CrossRef] [PubMed]
29. Salamanca, C.H.; Yarce, C.J.; Zapata, C.A.; Giraldo, J.A. Relationship between the polymeric ionization degree and powder and surface properties in materials derived from poly (maleic anhydride-alt-octadecene). *Molecules* **2018**, *23*, 320. [CrossRef] [PubMed]
30. Zhang, Y.; Li, S.; Wang, X.; Li, X. Coagulation performance and mechanism of polyaluminum ferric chloride (PAFC) coagulant synthesized using blast furnace dust. *Sep. Purif. Technol.* **2015**, *154*, 345–350. [CrossRef]
31. Sun, Y.; Zheng, H.; Tan, M.; Wang, Y.; Tang, X.; Feng, L.; Xiang, X. Synthesis and characterization of composite flocculant PAFS-CPAM for the treatment of textile dye wastewater. *J. Appl. Polym. Sci.* **2014**. [CrossRef]
32. Sun, W.; Tang, M.; Sun, Y.; Xu, Y.; Zheng, H. Effective sludge dewatering technique using the combination of Fenton's reagent and CPAM. *Can. J. Chem. Eng.* **2017**, *96*, 1256–1263. [CrossRef]
33. He, J.; Chu, J.; Tan, S.K.; Vu, T.T.; Lam, K.P. Sedimentation behavior of flocculant-treated soil slurry. *Mar. Georesour. Geotechnol.* **2016**, *35*, 593–602. [CrossRef]
34. Sun, Y.; Chen, A.; Pan, S.-Y.; Sun, W.; Zhu, C.; Shah, K.J.; Zheng, H. Novel chitosan-based flocculants for chromium and nickle removal in wastewater via integrated chelation and flocculation. *J. Environ. Manag.* **2019**, *248*, 109241. [CrossRef]
35. Wang, X.; Jiang, S.; Tan, S.; Wang, X.; Wang, H. Preparation and coagulation performance of hybrid coagulant polyacrylamide-polymeric aluminum ferric chloride. *J. Appl. Polym. Sci.* **2018**. [CrossRef]
36. Mohammadighavam, S.; Heiderscheidt, E.; Marttila, H.; Kløve, B. Optimization of gravity-driven hydraulic flocculators to treat peat extraction runoff water. *J. Irrig. Drain. Eng.* **2016**, *142*, 04015045. [CrossRef]
37. Chi, Y.-L.; Guo, L.-F.; Xu, Y.; Liu, J.-W.; Xu, W.; Zhao, H.-Z. Rapid removal of bound water from dredged sediments using novel hybrid coagulants. *Sep. Purif. Technol.* **2018**, *205*, 169–175. [CrossRef]
38. Sun, W.; Zhou, S.; Sun, Y.; Xu, Y. Synthesis and evaluation of cationic flocculant P(DAC-PAPTAC-AM) for flocculation of coal chemical wastewater. *J. Environ. Sci.* **2021**, *99*, 239–248. [CrossRef]
39. Zhang, Y.; Li, S.; Wang, X.; Ma, X.; Wang, W.; Li, X. Synthesis, purification and characterization of polyaluminum ferric chloride (PAFC) with high (Al+Fe)b content. *Sep. Purif. Technol.* **2015**, *146*, 311–316. [CrossRef]
40. Matrajt, G.; Borg, J.; Raynal, P.I.; Djouadi, Z.; Hendecourt, L.D.; Flynn, G.; Deboffe, D. FTIR and Raman analyses of the Tagish Lake meteorite: Relationship with the aliphatic hydrocarbons observed in the Diffuse Interstellar Medium. *Astron. Astrophys.* **2004**, *416*, 983–990. [CrossRef]
41. Liao, Y.; Tang, X.; Yang, Q.; Chen, W.; Liu, B.; Zhao, C.; Zhai, J.; Zheng, H. Characterization of an inorganic polymer coagulant and coagulation behavior for humic acid/algae-polluted water treatment: Polymeric zinc–ferric–silicate–sulfate coagulant. *RSC Adv.* **2017**, *7*, 19856–19862. [CrossRef]
42. Khan, R.; Inam, M.A.; Zam, S.Z.; Park, D.R.; Yeom, I.T. Assessment of Key Environmental Factors Influencing the Sedimentation and Aggregation Behavior of Zinc Oxide Nanoparticles in Aquatic Environment. *Water* **2018**, *10*, 660. [CrossRef]
43. Zhu, Z.; Gao, C.; Wu, Y.; Sun, L.; Huang, X.; Ran, W.; Shen, Q. Removal of heavy metals from aqueous solution by lipopeptides and lipopeptides modified Na-montmorillonite. *Bioresour. Technol.* **2013**, *147*, 378–386. [CrossRef]
44. Reinert, L.; Batouche, K.; Lévêque, J.-M.; Muller, F.; Bény, J.-M.; Kebabi, B.; Duclaux, L. Adsorption of imidazolium and pyridinium ionic liquids onto montmorillonite: Characterisation and thermodynamic calculations. *Chem. Eng. J.* **2012**, *209*, 13–19. [CrossRef]
45. Hussein, S.M.; Shihab, O.H.; Ibrahim, S.S. Interaction between Kaolin and DMSO: FTIR, XRD, thermodynamic and Nano studies. *J. Univ. Anbar Pure Sci.* **2014**, *8*, 32–40.
46. Seifi, S.; Masoum, S. Preparation of copper oxide/oak-based biomass nanocomposite for electrochemical hydrogen storage. *Int. J. Hydrog. Energy* **2019**, *44*, 11979–11988. [CrossRef]
47. Schindler, P.W. Co-adsorption of metal ions and organic ligands: Formation of ternary surface complexes. *Miner. Water Interface Geochem.* **2018**, 281–308.
48. Rahimi-Nasrabadi, M.; Behpour, M.; Sobhani-Nasab, A.; Jeddy, M.R. Nanocrystalline Ce-doped copper ferrite: Synthesis, characterization, and its photocatalyst application. *J. Mater. Sci. Mater. Electron.* **2016**, *27*, 11691–11697. [CrossRef]

nanomaterials

Article

Efficiency of Wood-Dust of *Dalbergia sisoo* as Low-Cost Adsorbent for Rhodamine-B Dye Removal

Dibyashree Shrestha

Department of Chemistry, Patan Multiple Campus, Tribhuvan University, Patan Dhoka, Lalitpur 44613, Nepal; dibyashreeshrestha@gmail.com

Abstract: Wood-dust of *Dalbergia sisoo* (Sisau) derived activated carbon (AC) was successfully tested as an adsorbent material for the removal of rhodamine B dye from an aqueous solution. The AC was prepared in a laboratory by the carbonization of wood powder of *Dalbergia sisoo* at 400 °C in an inert atmosphere of N_2, which was chemically activated with H_3PO_4. Several instrumental techniques have been employed to characterize the as-prepared AC (Db-s). Thermogravimetric analysis (TGA)/differential scanning colorimetry (DSC) confirmed that 400 °C was an appropriate temperature for the carbonization of raw wood powder. The FTIR spectra clearly confirmed the presence of oxygenated functional groups such as hydroxyl (–OH), aldehyde/ketone (–CHO/C=O) and ether (C–O–C) at its surface. The XRD pattern showed the amorphous structure of carbon having the 002 and 100 planes, whereas the Raman spectra clearly displayed G and D bands that further confirmed the amorphous nature of carbon. The SEM images displayed the high porosity, and the BET analysis revealed a high surface area of 1376 $m^2\ g^{-1}$, a pore volume of 1.2 $cm^3\ g^{-1}$, and a pore size of 4.06 nm with the coexistence of micropores and mesopores. The adsorption of dyes was performed by varying the dye concentration, pH, time, and the sample dose. The maximum percent of RhB dye removal by AC (Db-s) was 98.4% at an aqueous solution of 20 ppm, pH 8.5, an adsorbent dose of 0.03 g, and a time of 5 min. This study proved to be successful in addressing the local problem of wastewater pollution of garment and textile industrial effluents using locally available agro-waste of *Dalbergia sisoo*.

Keywords: activated carbon; *Dalbergia sisoo*; rhodamine B; contact time; contact temperature; pH

Citation: Shrestha, D. Efficiency of Wood-Dust of *Dalbergia sisoo* as Low-Cost Adsorbent for Rhodamine-B Dye Removal. *Nanomaterials* **2021**, *11*, 2217. https://doi.org/10.3390/nano11092217

Academic Editor: Christos A. Aggelopoulos

Received: 11 August 2021
Accepted: 19 August 2021
Published: 28 August 2021

1. Introduction

In present days, effluents discharged from various industries, such as textiles, paints, printing, cosmetics, pulp mills, paper, rubber, pharmaceuticals, plastics, foods, and leather, have become a universal problem. These industries extensively use a large volume of dyes and generate plentiful wastewater which pollutes the environment. They have been rated as the most infamous and unadorned pollutants amongst all other sectors owing to the fact that textile industries are the major dye consumers in each of their products [1]. Within dyes, textile dyes are one of the chief components among them [2] and, unfortunately, have a harmful effect on human and animal health [3]. These pollutant dye compounds change the color of water, and it makes the degrading quality of water for both flora and fauna [4]. It is reported that about 100,000 commercial dyes are in industry and larger than 7×10^5 tons of dyestuff are produced annually. The issue of enormous amounts of synthetic dyes to the water streams has stood a serious risk to the environment [5]. A very small amounts of dyes in water (less than 1 ppm for some dyes) is highly visible and undesirable [6].

Most dyes molecules are complex and stable and can be resistant to degradation upon contact with water, detergents, or any other thing [7]. In order to remove dyes from wastewater, a wide range of techniques, such as adsorption, coagulation and flocculation, biological treatment, ion exchange, membrane filtration chemical oxidation, reverse osmosis, and electrochemical processes [8], have been employed. Among all, adsorption technology, being very easy, economical, effective and flexible, has become the most preferred methods for the elimination of toxic and hazardous dyes from wastewater.

To greatly decease the cost of activated carbon (AC), various ACs prepared from non-conventional sources, such as coir pith [9], sawdust and rice husk [10], and pinewood [11], applied in the removal of some compounds have also been investigated, and better results have been attained.

The yield and absorptive property of AC is increased by using various suitable activating agents, such as H_3PO_4, KOH, H_2SO_4, and Na_2CO_3. Among them, the AC prepared from H_3PO_4 has been reported to be the best in terms of active surface areas and porosity and functional groups. In this study, phosphoric acid (H_3PO_4) has been used as an activating agent to open pores and enlarge the surface area of AC, so that the adsorption power will increase [12]. H_3PO_4 removes impurities contained in the carbon [13], as a result of the carbonization process of biomass such as ketones, alcohols, acids, and aldehydes. Moreover, H_3PO_4 also helps to form a bridge that connects the fragments of biopolymers (layer of carbon structure) by forming phosphate bonds that make pores more open and enlarge the surface area. In the present study, basic/cationic water-soluble dyes rhodamine B (RhB), which is carcinogenic in nature, was selected for the adsorption experiment due to its presence in wastewaters from by-products of several industries such as textile, leather, jute, and food industries.

RhB is an amphoteric dye, listed in the class of the xanthene dye causing harmful effects such as acute oral toxicity. When consumed, it causes damage to the eyes or skin irritation and is hazardous to the aquatic organism with long-term effects. Therefore, the treatment of effluent containing RhB dye becomes paramount before discharging them into water streams so as to protect the aquatic organism and make the environment safer for the public [14].

In addition, AC is amongst the most used adsorbents utilized due to its high surface area, pore volumes, fast adsorption rate, and high adsorption capacity. Herein, AC from *Dalbergia sisoo* has been prepared by chemical activation with H_3PO_4 [15]. The porous properties, chemical functionality, and surface morphology of the prepared ACs are performed by N_2 adsorption, FTIR, and SEM [16].

It is found from previous research that ACs modified by different surfactants have different effects on various water qualities of dye wastewater. Therefore, it is necessary to determine the coexistence of other ions that would affect the adsorption behavior. Meanwhile, more information is still required in order to better understand the adsorption behavior of RhB cationic dyes on the modified AC. The physical and chemical characteristics of RhB is given in Table 1. The aim of this study was to examine the potential of using the as-prepared AC from wood powder of *Dalbergia sisoo* tree as a low-cost adsorbent to remove RhB dye from aqueous solutions using UV-Vis spectrophotometry. The effect of the precursor dose and time of adsorption was the central focus of this study.

Table 1. Physical and chemical characteristics of rhodamine B (RhB) dye.

Dye	IUPAC Name	Chemical Formula	Chemical Structure	Mol. Wt. ($gmol^{-1}$)	Appearance	λ_{max} (nm)
Rhodamine B	[9-(2-carboxyphenyl)-6-diethylamino3-xanthenylidene]-diethylammonium chloride	$C_{28}H_{31}ClN_2O_3$	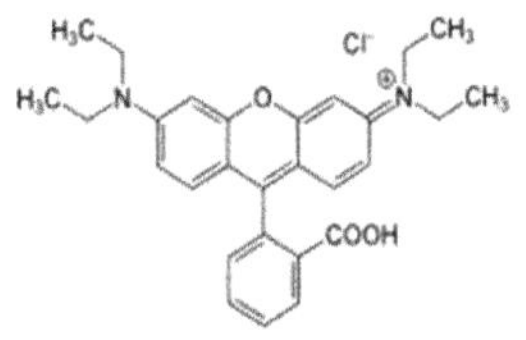	479.02	Reddish violet powder	554

2. Materials and Methods

2.1. Reagents

All the chemicals used in this study were of analytical grade and were used without further refinement. Phosphoric acid (85%) was obtained from Fischer Scientific, Pune, Maharashtra, India. Similarly, RhB (98%) was procured from Alfa Aesar A13572 (Haverhill, MA, USA), and aqueous NH_3 (30%) were from Baker (Anford, ME, USA). Double-distilled water was used throughout the experimental work. The wood dust of *Dalbergia sisoo* (Sisau) was collected from indigenous carpentry/sawmill, Kathmandu, Nepal. A commercial carbon made from coconut shells having a surface area of 876.02 $m^2\ g^{-1}$, a pore size of 4.23 nm, and a pore volume of 0.19 $cm^3\ g^{-1}$ was also used for comparing the results.

2.2. Methods

2.2.1. Preparation of Wood Powder

Wood dust was sun-dried for a few days. Then, a few initial stages were done, such as crushing, grinding, and sieving. The particle size was controlled by sieving through a 150 µm sized sieve. Thus, the obtained fine wood powder was used as a precursor for the preparation of AC.

2.2.2. Preparation of AC by the Carbonization Method

Wood dust (precursor) powder was dried in an oven to remove all volatile matters. The dried precursor was chemically activated with H_3PO_4 using the ratio of precursor to activating agent of 1:1 (*w*/*w*), left for overnight at room temperature for proper soaking and then evaporated to dry at 110 °C in an oven. The weighed amount of samples were inserted into a horizontal electric tubular furnace tube and carbonized at 400 °C for 3 h in an inert atmosphere of N_2. A continuous flow of pure N_2 was used to create an inert atmosphere. The carbonized samples were then cooled to room temperature maintaining the inert atmosphere of nitrogen and were washed with distilled water for several times until the pH of the washed water became neutral. It was dried, ground and stored in an air-tight container. The resulting material was dried in an oven at 110 °C and named as AC (Db-s). Thus, the prepared AC (Db-s) was stored in an air-tight container until further use in adsorption experiments.

A schematic diagram for the preparation of AC (Db-s) followed by carbonization is shown below in Scheme 1.

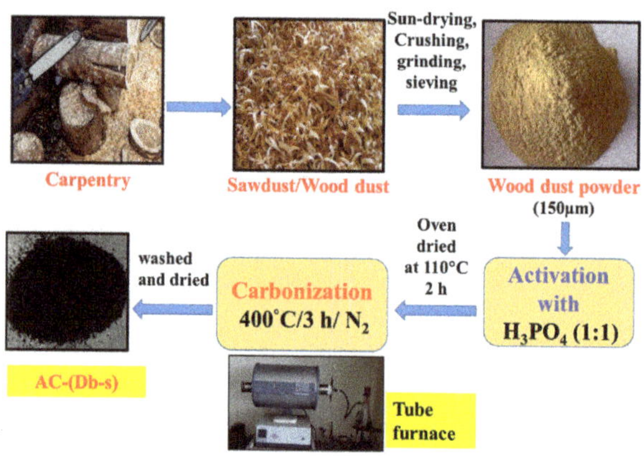

Scheme 1. Steps involved in the preparation of activated carbon from *Dalbergia sisoo*.

The descriptions of the as-prepared AC (Db-s) and conditions are given in Table 2.

Table 2. Description and preparation conditions of the sample.

Sample	Particle Size (μm)	Activating Agent	H3PO4:Wood Powder Ratio (*w/w*)	Preheating Temperature (°C)	Carbonization Temperature (°C)	Carbonization Time (h)	Atmosphere
AC (Db-s)	150	H_3PO_4	1:1	110	400	3	N_2

2.2.3. Characterization Technique of the As-Prepared AC (Db-s)

The moisture content, ash content, volatile matter, and fixed carbon of the precursor were analyzed by proximate analysis. The pyrolytic behaviors of the raw wood powder of *Dalbergia sisoo* were investigated by thermogravimetric analysis (TGA)/differential scanning colorimetry (DSC; SDT Q600 V20.9 Build 20). The morphology of AC (Db-s) was determined by a scanning electron microscope (Nanoeye, Korea). The phase state was evaluated by using XRD (Rigaku RINT 2000 diffractometer). Similarly, surface carbon-oxygen functional groups present in the samples were identified performing FTIR measurements using an FTIR instrument (Bruker, Vertex 70, Germany). The % transmission of the samples was recorded between 4000 and 400 cm^{-1}. The presence of the amorphous nature of AC (Db-s) was confirmed by Raman signals obtained by the labRAM HR800 (JOBIN YVON). The surface area, pore size, and pore volume of the AC (Db-s) were measured by BET (Micromeritics ASAP 2020 system).

2.2.4. Methodology of Dye Adsorption

A 20 ppm stock solution of RhB was been prepared by dissolving 0.02 g of the dye per liter of double-distilled water. At first, a 100 mL RhB dye solution was taken in 250 mL Erlenmeyer flasks, and 0.02 g of the as-prepared AC (Db-s) was added and stirred at 400 rpm at neutral pH for 5 min on a magnetic stirrer at room temperature. During the stirring period, about 3 mL of the mixture solution were taken in a microcentrifuge tube in every single minute time interval for 5 times. All the 5 microcentrifuge tubes were then centrifuged for 5 min to separate the insoluble particles. The agitation speed was kept constant at 400 rpm.

3. Results and Discussion

3.1. Proximate Analysis

The results obtained from the proximate analysis of precursor (wood powder of *Dalbergia sisoo*) are given in Table 3. The low ash content in the samples indicated that the precursor contained low inorganic matter, and the low moisture content exhibited the raw material is good for the preparation of AC and the efficiency of reactivation.

Table 3. Proximate analysis of the precursor (wood powder of *Dalbergia sisoo*).

S.N.	Parameter	Observed Value (%)
1.	Moisture	8.66
2.	Total ash	1.07
3.	Volatile matter	42.27
4.	Total carbon	48.0

3.2. TGA/DSC

The pyrolytic behavior of the wood powder of *Dalbergia sisoo* (Sisau) was measured by TGA/DSC. Figure 1 Illustrates the thermogravimetric (TG) and DSC curves of the raw wood powder of *Dalbergia sisoo* [(Db-s) R].

A minor mass loss was detected at around 100 °C in TG curves which might be due to dehydration, which was again confirmed by a sharp peak of the DSC curve at 100 °C. Similarly, at around 200–300 °C, there was a contracted peak, which was not clear in the TG plot. However, from the DSC curve, minor mass loss was clearly detected between 200 and 300 °C, which might be due to the breakdown of hemicellulose which was noticeably completed at 300–310 °C.

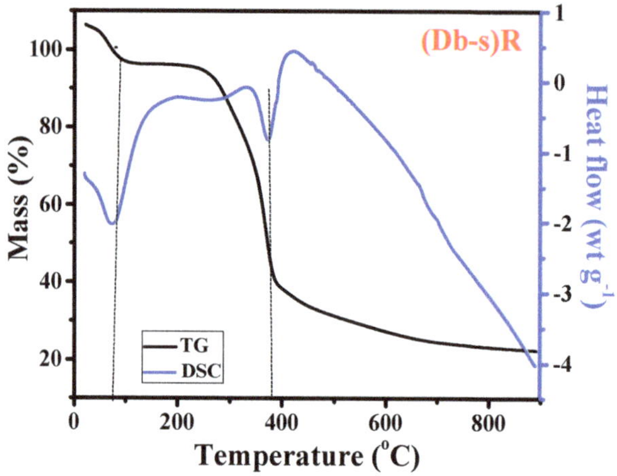

Figure 1. Thermogravimetric (TG) and differential scanning colorimetry (DSC) curves of raw wood powder of *Dalbergia sisoo* (Db-s) R.

A significant mass loss was detected between 300 and 400 °C from the TG curve, which was confirmed by peak at around 390 °C of the DSC curve. The breakdown of cellulose started at around 300 °C and by 400 °C, and all cellulose changed to organic volatile matters present in raw wood powder. It exhibited that more than 60% of weight loss occurred around 400 °C [17]. The samples were more stable beyond 400 °C. Accordingly, 400 °C is an appropriate temperature for the carbonization process [18].

3.3. XRD Measurement

XRD analysis was carried out in order to determine the degree of crystallinity or amorphous nature of the as-prepared AC (Db-s). The XRD pattern of AC (Db-s) is presented in Figure 2.

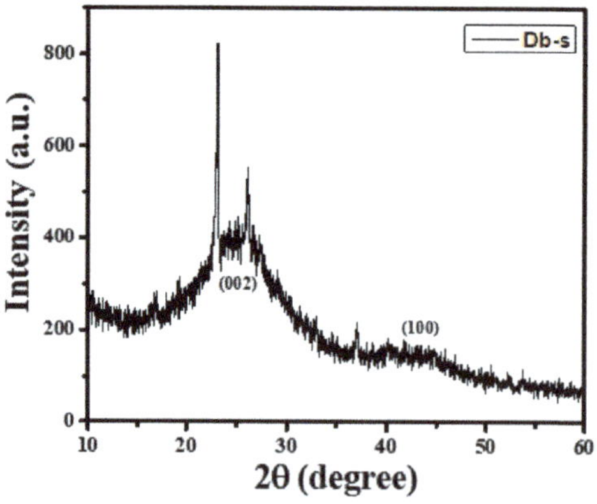

Figure 2. XRD pattern of AC (Db-s).

The broad peaks at 2θ degrees of around 24 to 26 from the (002) plane and also at 2θ degrees of around 43 to 45 from the (100) plane could be clearly seen in AC (Db-s). The absence of a sharp peak and the appearance of a wide peak indicating the amorphous nature and graphitic crystallites as in commercial AC [19], in which 2 distinct peaks at

2θ degrees of 25 and 45 coming from the (002) and (100) planes of graphitic clusters, respectively, were reported. Similarly, a wide peak of graphite clusters has been reported in porous carbon by Molina-Sabio [20]. However, a few sharp peaks could be seen in Ac (Db-s), which might be due to some impurities or moisture present in the samples.

3.4. Raman Scattering Analysis

The Raman spectra of the as-prepared AC (Db-s) are presented in Figure 3. The Raman spectra of AC (Db-s) revealed strong D and G bands at approximately 1349 and 1576 cm^{-1}, respectively.

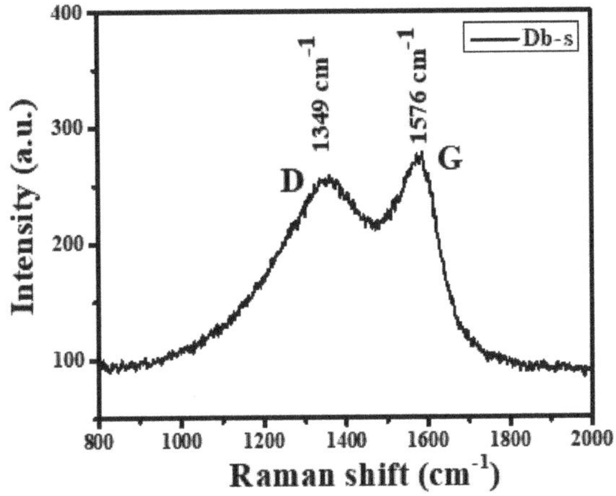

Figure 3. Raman spectra of the as-prepared AC (Db-s).

Typically, the advent of D band in Raman scattering is associated with a disordered carbon structure, and the intensity of D band is correlated with defects (amorphous). If there are a large number of defects, it will show a high intensity of D band, whereas the absence of defects correlates with the absence of D band as in case of carbon nanotubes and graphitic carbons. Hence, D band at 1349 cm^{-1} is a characteristic of disordered carbon or amorphous carbon in AC (Db-s). Therefore, this finding of Raman spectroscopy approved pleasantly with the XRD results [19].

3.5. FTIR Analysis

The FTIR spectra of AC (Db-s) are shown in Figure 4. As can be seen in Figure 4, the spectra showed a strong wide adsorption band at around 3260 cm^{-1}. The position of the band was characteristics of the stretching vibration of hydroxyl groups of carboxyl, phenol, and alcohol. –OH band can also be correlated with adsorbed water present in raw samples as indicated by TG/DSC analysis, where, at around 100 °C, mass loss was clearly seen, indicating the evaporation of water molecules. The band observed at the region around 1590 cm^{-1} is due to olefinic C=C vibrations in aromatic rings. The band at around 1189 cm^{-1} is due to the asymmetric stretching of C–O bond in acids, alcohols, phenols, and esters. The band at around 668 cm^{-1} is attributed to C–C stretching in a fingerprint region. These results are in good agreement with the findings of many investigators [21]. The FTIR results clearly expressed that the as-prepared AC (Db-s) was fully functionalized with oxygen containing groups such as carboxylic, phenolic, lactonic, and ether groups by the use of an activating agent (H_3PO_4).

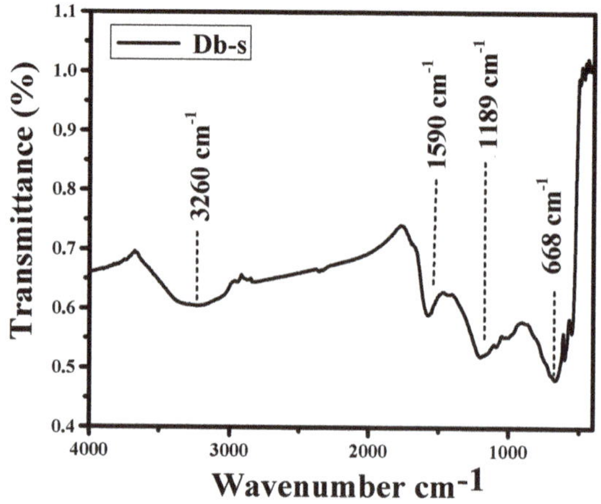

Figure 4. FTIR spectra of AC (Db-s).

3.6. N_2 Adsorption/Desorption Isotherm/Brunauer–Emmett–Teller (BET) Theory

The N_2 adsorption/desorption isotherms of AC (Db-s) at 77 K are shown in Figure 5. As can be seen in Figure 5, Type II H_3 isotherm could be seen. At relatively lower pressure ($P/P^0 < 0.1$), the isotherms showed an insignificant uptake, indicating the presence of a few micropores in AC (Db-s).

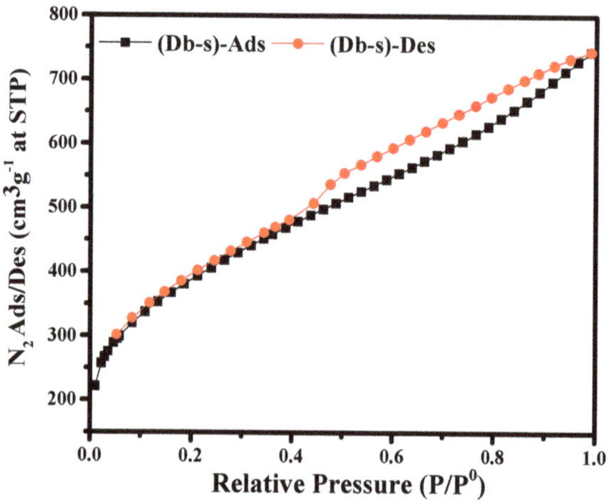

Figure 5. N_2 adsorption/desorption isotherms of AC (Db-s) at 77 K.

However, at around $P/P^0 = 0.4$, the amount of nitrogen uptake increased significantly, showing a hysteresis loop during the adsorption and desorption of nitrogen. The adsorption volume of AC (Db-s) increased with the relative pressure until $P/P^0 \cong 0.5$. The widely opened knees and the slight hysteresis loops at the relative pressure of 0.5 to 1.0 indicated the presence of a considerable amount of small mesopores in AC (Db-s). Therefore, from the isotherm of AC (Db-s), it can clearly be seen that the as-prepared AC sample consisted of an innumerable amount of mesoporores along with some micropores, since a hysteresis loop is the characteristics of mesopores [22].

Then, the BET specific surface area, pore size, and pore volume of AC (Db-s) were measured, and the results are given in Table 4. The large surface area, small pore size, and pore volume might be due to the presence of innumerable mesoporosity along with microporosity [20].

Table 4. Brunauer–Emmett–Teller (BET) specific surface area, pore size and pore volume of AC (Db-s).

Name of Sample	Specific Surface Area (m^2 g^{-1})	Pore Size (BJH ads) (nm)	Pore Volume (BJH ads) (cm^3 g^{-1})
AC (Db-s)	1376	4.06	1.2

3.7. SEM Analysis

Figure 6 illustrates the SEM micrographs of AC (Db-s), at s magnification of 1000, showing a well-developed porous morphology with a layered structure.

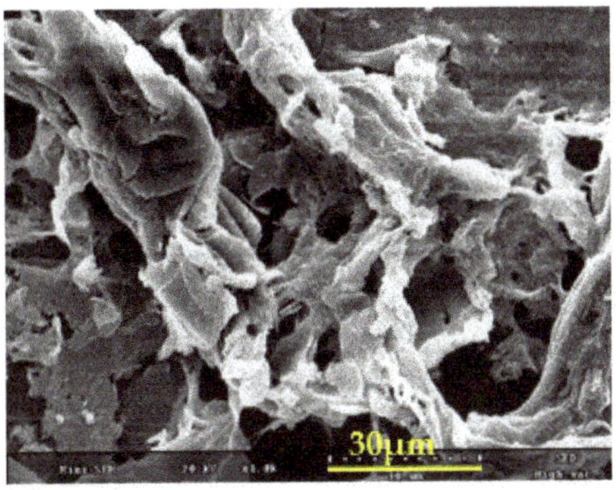

Figure 6. SEM image of AC (Db-s) showing a porous morphology.

The pore development on the surfaces of AC (Db-s) might be due to the course of dehydration of an activating agent, i.e., H_3PO_4 [19] and then react with oxygen to generate phosphoric anhydride (P_2O_5). Here, P_2O_5 sublimated from solid to gas at 360 °C, and then, gaseous P_2O_5 escaped from the surface of the AC-creating pores [23]. In addition to this, during the washing process, remaining P_2O_5 was hydrolyzed and removed, creating an empty space or pores. Pores may be micropores or mesopores with a volume corresponding to that of the removed phosphoric acid [24]. Accordingly, the presence of incredible mesoporosity along with microporosity is expected to enhance the adsorption capacity of AC (Db-s). It is also supported by the BET results of Figure 5 and Table 4.

The adsorption capacity of AC depends on surface functional groups, surface areas, and pore sizes and morphology. As reported in the relevant literatures [24], for the adsorption of dyes from effluents, ACs with oxygen containing functional groups, high surface areas (above 1000 mg^{-1}), and good morphology with extensive mesoporosity are required, because dyes are larger molecules (above 2 nm) which are fitted into the mesopores. Under the basis of these requirements, the as-prepared AC (Db-s) is considered to be well-matched for the adsorption of RhB dye.

The adsorption of dye molecules are affected by various factors such as adsorbent dose, solution pH, initial dye concentration, and contact time. The optimization of these parameters will greatly help in the research work and mainly in the development of industrial scale treatment process for the dye removal.

3.8. Effects of Variable Parameters

3.8.1. Effect of pH and Contact Time on the Adsorption of RhB

The effects of pH and contact time on the adsorption of 20 ppm RhB were studied spectrophotometrically at a fixed dose (0.030 g/100 mL) of adsorbent AC (Db-s). Figure 7a–d shows the UV-Vis spectra of RhB dyes during the adsorption at pH 3.5, 6.5, and 8.5 at time intervals of 1, 2, 3, 4, and 5 min.

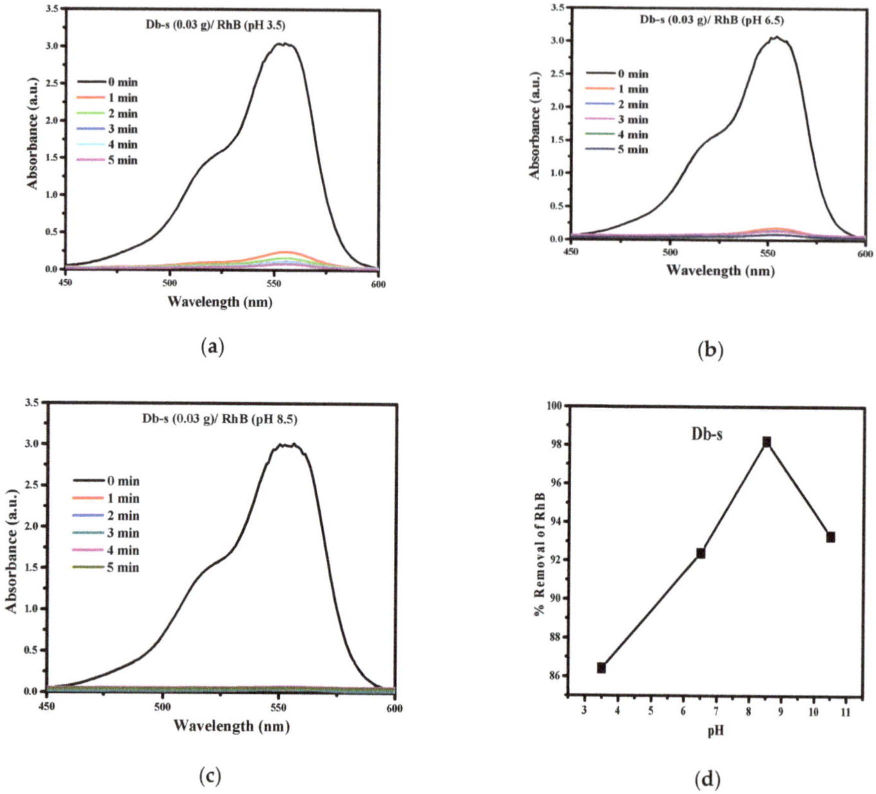

Figure 7. (**a**) Adsorption spectra of RhB during adsorption at pH 3.5 and at different times containing 0.03g of AC (Db-s). (**b**) Adsorption spectra of RhB during adsorption at pH 6.5 and at different times containing 0.030g of AC (Db-s). (**c**) Adsorption spectra of RhB during adsorption at pH 8.5 and at different times containing 0.03g of AC (Db-s). (**d**) Effect of pH on the removal of RhB by AC (Db-s).

Here, as can be seen in Figure 7a, at pH 3.5, the adsorption of RhB dye by AC (Db-s) was not completed up to 5 min.

It is obvious in Figure 7b that, at pH 6.5, the adsorption of RhB dye by AC (Db-s) was still not completed within 5 min. However, the adsorption was higher than in the case at pH 3.5.

Figure 7c shows that at pH 8.5, almost all of RhB dyes were adsorbed by AC (Db-s) and all the adsorption spectra are in a shape of a straight line for 4 and 5 min. A similar measurement carried out at pH 11 revealed a decrease in the adsorption of RhB.

The percentage removal of RhB was estimated from Figure 7a–c at different pH values, and the effect of pH on the removal of RhB is plotted in Figure 7d. The results showed a gradual increase in % removal of RhB with pH until 8.5 and afterwards a decrease in % removal of RhB.

The less % removal (low adsorption) of RhB dye in an acidic medium (pH 3.5) can be explained by the fact that at this acidic pH the adsorbent surface acquired a positive charge which repelled the cationic dye RhB [25,26]. In addition, increased hydrogen ion (H^+) at low pH may compete with dye ions for the adsorption on AC (Db-s). However, in an alkaline medium, i.e., at higher solutions at pH 8.5, AC (Db-s) may be negatively charged, which enhances the adsorption of positively charged dye cations through electrostatic forces of attraction.

RhB is an aromatic amino acid with amphoteric characteristics due to the presence of both the amino group (–NHR_2) and the carboxyl group (–COOH). Thus, the charge state of RhB is dependent on the solution pH. When the pH increases higher than 8.5, the ionization of the carboxyl group takes place which is attracted to positive charges on the xanthane group and the zwitterions form of RhB is formed. The zwitterions form of RhB in water may increase the dimerization of RhB, which makes the molecule too large to enter most of the pore structure of AC. The inaccessibility to the pore structure of AC, which is smaller than the dimer's effective size, results in a decrease in RhB removal at pH above 8.5. The variation of the RhB removal with the solution pH is similar to that reported previously [27–30].

Figure 7a–d reveals that the initial adsorption percentage of dyes was quite rapid. Almost 80% of the dyes were adsorbed within a single minute. Then, the adsorption process was more or less constant up to 5 min. This is probably due to a high surface area of 1376 m^2 g^{-1} with extensive mesoporosity in AC (Db-s). Highly developed mesoporous ACs are favorable for the adsorption of larger molecules such as dyes [27]. This might be due to the fact that initially all the sites of the AC samples were vacant and available for dye molecules to get occupied. Later, the percentage removal/adsorption rate of dye was decreased significantly, resulting from the saturation of dye molecules on the surfaces of the AC samples.

3.8.2. Optimization of the Dose of AC (Db-s)

The efficiency of adsorbent doses on the RhB dye adsorption is reported by many researchers to determine the most economical minimum dosage. In general, the dye removal percentage increases with the increase of the adsorbent dosage [28].

Here, various doses of adsorbent AC (Db-s) were mixed with a fixed amount of dye solutions, and the mixture was agitated in a mechanical shaker. The adsorption capacities for different doses were determined at 1 min interval till 5 min by keeping all other factors constant.

The percentage of adsorption increased with the increase in the dose of the adsorbent AC (Db-s) (Figure 8a–d), which are presented in Table 5. This is attributed to the increased surface area and availability of a more adsorption site [29].

The adsorption/removal capacity of the adsorbent (AC) is due to their porous structure and pore size distribution, and it depends on the polarity, solubility, and molecular size of the adsorbate [30]. From the adsorption spectra, it was observed that the optimum dose of adsorbent AC (Db-s) for the removal of RhB from an aqueous solution was 0.03 g/100 mL at a basic medium (pH 8.5) (Figure 8c).

As can be seen in Table 5, the percentage removal capacity of RhB increased from 82.6%, 91.2%, to 98.4% by the as-prepared AC (Db-s) with the increase of dose from 0.02 g, 0.025 g, to 0.030 g, respectively. When the AC (Db-s) dose was increased, free adsorption sites also increased and thus more dye molecules were adsorbed [31].

When the AC (Db-s) was increased to 0.035 g/100 mL, there was no significant difference in percentage adsorption/removal of dye from the dose of 0.030 g, and therefore, 0.030 g was suitable for the adsorption of dyes in this study. Table 6 summarizes the adsorption efficiency of Rhb dye at a dose of 0.030 g L^{-1}.

From the above results in Tables 5 and 6, it is clear that the data generated from experiments are valid and not highly deviated. Table 7 gives the summary of the analysis of the spectrometric method. Similarly, Table 8 presents the results of % removal of RhB dye by different doses of the AC (Db-s).

Furthermore, to study the efficiency of the as-prepared AC (Db-s), the % removals of RhB by using different doses of AC (Db-s) were compared with the same doses of commercial carbon, which are tabulated at Table 9. Since the best adsorption efficiency was exhibited by the dose of 0.030 g, the removal of RhB at this dose was again compared with that by commercial carbon, which is shown in Figure 8d. As can be seen in Figure 8d, the as-prepared AC (Db-s) revealed a better removal result than commercial carbon.

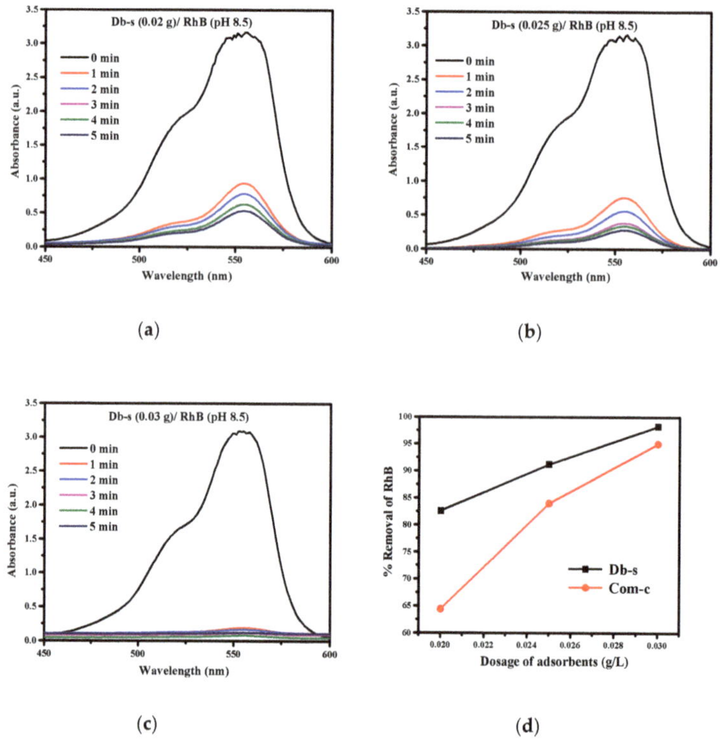

Figure 8. (**a**) RhB dye adsorption by AC (Db-s) at a dose of 0.02 g; (**b**) RhB dye adsorption by AC (Db-s) at a dose of 0.025 g; (**c**) RhB dye adsorption by AC (Db-s) at a dose of 0.03 g; (**d**) comparison of the % removals of RhB by AC (Db-s) and commercial carbon (at 0.030 g).

Table 5. Adsorption efficiency of RhB dye with the surface area, pH, and doses of the samples and adsorption capacity in 5 min.

S.N.	Surface Area (m^2 g^{-1})	Number of Adsorption	Dose (g L^{-1})	Initial Conc.(ppm)	Adsorption Capacity (%)	D@A=
1		A1	0.015		81.5	1.225
2		A2	0.020		83.1	1.662
3	1376	A3	0.025	20	91.2	2.28
4		A4	0.030		98.4	2.952
5		A5	0.035		98.4	3.444

Table 6. Adsorption efficiency of RhB dye at a dose of 0.030 g L^{-1}.

S.N.	No. of Adsorption (N)	Dose (g L^{-1})	Initial Concentration (ppm)	Adsorption Capacity (%) (X)	$X - \mu$	$(X - \mu)^2$
1	A1			98.4	$98.4 - 98.36 = -0.04$	0.0016
2	A2			98.2	$98.2 - 98.36 = -0.16$	0.0256
3	A3	0.030	20	98.5	$98.5 - 98.36 = 0.14$	0.0196
4	A4			98.4	$98.4 - 98.36 = 0.04$	0.0016
5	A5			98.3	$98.3 - 98.36 = 0.06$	0.0036

Table 7. Accuracy and precision of the spectrometric method for the determinations of RhB dye.

S.N.	Surface Area (m^2 g^{-1})	Dose Taken (μmg L^{-1})	Dose Found (μmg L^{-1})	Mean Standard Deviation (%)	Relative Standard Deviation (%)
1		15	14.78 ± 0.226	98.5	1.5
2		20	$19.\ 84 \pm 0.157$	99.2	0.8
3	1376	25	24.87 ± 0.126	99.5	0.5
4		30	29.99 ± 0.015	99.9	0.1
5		35	39.99 ± 0.012	99.9	0.1

Table 8. Percentage of RhB dye adsorbed by different doses of the as-prepared AC (Db-s).

S.N.	Dye	Adsorbent Dose				Result
		0.020 mg	0.025 mg	0.030 mg	0.035 mg	
1	RhB	83.1	91.2	98.4	98.4	No effect after 0.035 mg

Table 9. Details of the comparison of the % removal of RhB dye using different doses of AC (Db-s) and commercial carbon.

Surfactant	Surface Area (m^2 g^{-1})	Dye	pH	Initial Conc. (ppm)	Dosage (g L^{-1})	Time (min)	% Removal
Db-s	1376	RhB	8.5	20	0.020 0.025 0.030	5	83.1 91.2 98.4
Com. C	876.02	RhB	8.5	20	0.020 0.025 0.030	5	64.4 84 95

4. Conclusions

From the results on the removal of RhB dye, from agro waste of *Dalbergia sisoo* in this study, the following conclusions can be drawn:

1. AC was successfully prepared from the agro waste of *Dalbergia sisoo* by activating with H_3PO_4 followed by carbonization at 400 °C.
2. BET measurements showed the formation of a mesoporous structure with an active surface area of 1376 m^2 g^{-1}, a pore volume of 1.2 cm^3 g^{-1}, and a pore size of 4.06 nm.
3. AC (Db-s) was found to possess the best adsorption capacity at pH 8.5 and a dose of 0.030 g. Under this condition, 98.4% of 20 ppm of RhB dye was removed from an aqueous solution within 5 min.
4. The results revealed that laboratory prepared phosphoric acid AC derived from agro waste such as wood dust of *Dalbergia sisoo* can be converted into value-added materials for the removal of toxic dyes such as RhB.

Funding: This research received no external funding.

Institutional Review Board Statement: Not applicable.

Informed Consent Statement: Not applicable.

Data Availability Statement: Data can be made available upon request.

Acknowledgments: Dibyashree Shrestha is thankful to Patan Multiple Campus, Tribhuvan University, Patan Dhoka, lalitpur and Central Department of Chemistry, Tribhuvan University, Kathmandu, Nepal, and Global Research Laboratory, Sun Moon University, South Korea for their support to carry out this research.

Conflicts of Interest: The author declares no conflict of interest.

References

1. Javaid, A.; Bajwa, R.; Shafique, U.; Anwar, J. Removal of heavy metals by adsorption on Pleurotus ostreatus. *Biomass Bioenergy* **2011**, *35*, 1675–1682. [CrossRef]
2. Ding, L.; Zou, B.; Gao, W.; Liu, Q.; Wang, Z.; Guo, Y.; Wang, X.; Liu, Y. Adsorption of Rhodamine-B from aqueous solution using treated rice husk-based activated carbon. *Colloids Surf. A Physicochem. Eng. Asp.* **2014**, *446*, 1–7. [CrossRef]
3. Geçgel, Ü.; Üner, O.; Gökara, G.; Bayrak, Y. Adsorption of cationic dyes on activated carbon obtained from waste Elaeagnus stone. *Adsorpt. Sci. Technol.* **2016**, *34*, 512–525. [CrossRef]
4. Tang, L.; Cai, Y.; Yang, G.; Liu, Y.; Zeng, G.; Zhou, Y.; Li, S.; Wang, J.; Zhang, S.; Fang, Y.; et al. Cobalt nanoparticles-embedded magnetic ordered mesoporous carbon for highly effective adsorption of rhodamine B. *Appl. Surf. Sci.* **2014**, *314*, 746–753. [CrossRef]
5. Lacerda, V.D.S.; López-Sotelo, J.B.; Correa-Guimarães, A.; Hernández-Navarro, S.; Bascones, M.S.; Gracia, L.M.N.; Martín-Ramos, P.; Martín-Gil, J. Rhodamine B removal with activated carbons obtained from lignocellulosic waste. *J. Environ. Manag.* **2015**, *155*, 67–76. [CrossRef] [PubMed]
6. Laasri, L.; Elamrani, M.K.; Cherkaoui, O. Removal of two cationic dyes from a textile effluent by filtration-adsorption on wood sawdust. *Environ. Sci. Pollut. Res.* **2007**, *14*, 237–240. [CrossRef] [PubMed]

7. Adegoke, K.A.; Bello, O.S. Dye sequestration using agricultural wastes as adsorbents. *Water Resour. Ind.* **2015**, *12*, 8–24. [CrossRef]
8. Inyinbor, A.A.; Adekola, F.A.; Olatunji, A.G. Adsorption of Rhodamine B dye from aqueous solution on Irvingia gabonensis biomass: Kinetics and thermodynamics studies. *S. Afr. J. Chem.* **2015**, *68*, 115–125. [CrossRef]
9. Dahri, M.K.; Kooh, M.R.R.; Lim, L.B.L. Remediation of Rhodamine B Dye from Aqueous Solution Using Casuarina equisetifolia Cone Powder as a Low-Cost Adsorbent. *Adv. Phys. Chem.* **2016**, *2016*, 9497378. [CrossRef]
10. Lim, L.B.L.; Priyantha, N.; Fang, X.Y.; Mohamad Zaidi, N.A.H. Artocarpusodoratissimus peel as a potential adsorbent in environmental remediation to remove toxic Rhodamine B dye. *J. Mater. Environ. Sci.* **2017**, *8*, 494–502.
11. Bello, O.S.; Lasisi, B.M.; Adigun, O.J.; Ephraim, V. Scavenging Rhodamine B dye using moringa oleifera seed pod. *Chem. Speciat. Bioavailab.* **2017**, *29*, 120–134. [CrossRef]
12. Rasapoor, M.; Young, B.; Asadov, A.; Brar, R.; Sarmah, A.K.; Zhuang, W.-Q.; Baroutian, S. Effects of biochar and activated carbon on biogas generation: A thermogravimetric and chemical analysis approach. *Energy Convers. Manag.* **2020**, *203*, 112221. [CrossRef]
13. Suiuay, C.; Sudajan, S.; Katekaew, S.; Senawong, K.; Laloon, K. Production of gasoline-like-fuel and diesel-like-fuel from hard-resin of Yang (Dipterocarpus alatus) using a fast pyrolysis process. *Energy* **2019**, *187*, 115967. [CrossRef]
14. Senturk, H.B.; Ozdes, D.; Duran, C. Biosorption of Rhodamine 6G from aqueous solutions onto almond shell (Prunus dulcis) as a low cost biosorbent. *Desalination* **2010**, *252*, 81–87. [CrossRef]
15. Guo, Y.; Tan, C.; Sun, J.; Li, W.; Zhang, J.; Zhao, C. Porous activated carbons derived from waste sugarcane bagasse for CO_2 adsorption. *Chem. Eng. J.* **2020**, *381*, 122736. [CrossRef]
16. Yagmur, E.; Gokce, Y.; Tekin, S.; Semerci, N.I.; Aktas, Z. Characteristics and comparison of activated carbons prepared from oleaster (Elaeagnus angustifolia) fruit using KOH and $ZnCl_2$. *Fuel* **2020**, *267*, 117232. [CrossRef]
17. Gonzalez-Serrano, E.; Cordero, T.; Rodriguez-Mirasol, J.; Cotoruelo, L.; Rodriguez, J. Removal of water pollutants with activated carbons prepared from H3PO4 activation of lignin from kraft black liquors. *Water Res.* **2004**, *38*, 3043–3050. [CrossRef]
18. Wang, J.-G.; Yang, Y.; Huang, Z.-H.; Kang, F. Incorporation of nanostructured manganese dioxide into carbon nanofibers and its electrochemical performance. *Mater. Lett.* **2012**, *72*, 18–21. [CrossRef]
19. Shrestha, D.; Maensiri, S.; Wongpratat, U.; Lee, S.W.; Nyachhyon, A.R. Shorea robusta derived activated carbon decorated with manganese dioxide hybrid composite for improved capacitive behaviors. *J. Environ. Chem. Eng.* **2019**, *7*, 103227. [CrossRef]
20. Molina-Sabio, M.; Rodriguez-Reinoso, F. Role of chemical activation in the development of carbon porosity. *J. Collids Surf. Physicochem. Eng.* **2004**, *241*, 15–25. [CrossRef]
21. Shi, K.; Ren, M.; Zhitomirsky, I. Activated carbon-coated carbon nanotubes for energy storage in supercapacitors and capacitive water purification. *ACS Sustain. Chem. Eng.* **2014**, *2*, 1289–1298. [CrossRef]
22. Kondrat, S.; Perez, C.R.; Presser, C.R.; Gogotsi, Y.; Kornyshev, A.A. Effect of pore size and its dispersity on the energy storage in nanoporous supercapacitors. *Energy Environ. Sci.* **2012**, *5*, 6474. [CrossRef]
23. Sarkar, A.; Singh, A.K.; Sarkar, D.; Khan, G.G.; Mandal, K. Three-Dimensional Nano-Architecture of $BiFeO_3$ anchored TiO_2 Nanotube Arrays for Electrochemical Energy Storage and Solar Energy Conversion. *J. ACS Sustain. Chem. Eng.* **2015**, *3*, 2254–2263. [CrossRef]
24. Tawfik, A.S.; Gaddafi, I.D. Adsorptive desulfurization of dibenzothiophene from fuels by rubber tyres-derived carbons: Kinetics and isotherms evaluation. *Proc. Safe. Environ. Prot.* **2016**, *102*, 9–19.
25. Arivoli, S.; Viji, J.M.; Rajachandrasekar, T. Cobalt Adsorption on a Low Cost Carbon-Kinetic, Equilibrium and Mechanistic Studies. *Mat. Sci. Res. India* **2006**, *3*, 241–250.
26. Doumic, L.I.; Soares, P.A.; Ayude, M.A.; Cassanello, M.; Boaventura, R.A.; Vilar, V.J. Enhancement of a solar photo-Fenton reaction by using ferrioxalate complexes for the treatment of a synthetic cotton-textile dyeing wastewater. *Chem. Eng. J.* **2015**, *277*, 86–96. [CrossRef]
27. Stavropoulos, G.; Zabaniotou, A. Production and characterization of activated carbons from olive-seed waste residue. *Microporous Mesoporous Mater.* **2005**, *82*, 79–85. [CrossRef]
28. Sen, T.K.; Dawood, S. Review on Dye Removal from Its Aqueous Solution into Alternative Cost Effective and Non-Conventional Adsorbents. *J. Chem. Process. Eng.* **2013**, *1*, 104. [CrossRef]
29. Namasivayam, C.; Radhika, R.; Suba, S. Uptake of dyes by a promising locally available agricultural solid waste: Coir pith. *Waste Manag.* **2001**, *21*, 381–387. [CrossRef]
30. Bedmohata, M.A.; Chaudhari, A.R.; Singh, S.P.; Choudhary, M.D. Adsorption Capacity of Activated Carbon prepared by Chemical Activation of Lignin for the Removal of Methylene Blue Dye. *Int. J. Adv. Res. Chem. Sci.* **2015**, *2*, 1–13.
31. Chen, Y.; Zhu, Y.; Wang, Z.; Li, Y.; Wang, L.; Ding, L.; Gao, X.; Ma, Y.; Guo, Y. Application studies of activated carbon derived from rice husks produced by chemical-thermal process—A review. *Adv. Colloid Interface Sci.* **2011**, *163*, 39–52. [CrossRef] [PubMed]

nanomaterials

MDPI

Article

Cinnamomum tamala Leaf Extract Stabilized Zinc Oxide Nanoparticles: A Promising Photocatalyst for Methylene Blue Degradation

Sajina Narath [1], Supin Karonnan Koroth [1], Sarojini Sharath Shankar [2,3,*], Bini George [1,*], Vasundhara Mutta [4], Stanisław Wacławek [5,*], Miroslav Černík [5,*], Vinod Vellora Thekkae Padil [5,*] and Rajender S. Varma [6,*]

1 Department of Chemistry, School of Physical Sciences, Central University of Kerala, Kasaragod 671316, Kerala, India; sajina@cukerala.ac.in (S.N.); supinkk09@gmail.com (S.K.K.)
2 Department of Biochemistry and Molecular Biology, School of Biological Sciences, Central University of Kerala, Kasaragod 671316, Kerala, India
3 Department of Medicine, Thomas Jefferson University, Jefferson Alumni Hall, 1020 Locust Street, Philadelphia, PA 19107, USA
4 Polymer and Functional Materials Department, CSIR-Indian Institute of Chemical Technology, Tarnaka 500007, Hyderabad, India; mvas@iict.res.in
5 Institute for Nanomaterials, Advanced Technologies and Innovation (CXI), Technical University of Liberec (TUL), Studentská 1402/2, 461 17 Liberec, Czech Republic
6 Regional Centre of Advanced Technologies and Materials, Czech Advanced Technology and Research Institute, Palacky University, Šlechtitelů 27, 783 71 Olomouc, Czech Republic
* Correspondence: sharathshankar82@gmail.com (S.S.S.); binigeorgek@cukerala.ac.in (B.G.); Stanislaw.waclawek@tul.cz (S.W.); miroslav.cernik@tul.cz (M.Č.); vinod.padil@tul.cz (V.V.T.P.); Varma.Rajender@epa.gov (R.S.V.)

Citation: Narath, S.; Koroth, S.K.; Shankar, S.S.; George, B.; Mutta, V.; Wacławek, S.; Černík, M.; Padil, V.V.T.; Varma, R.S. *Cinnamomum tamala* Leaf Extract Stabilized Zinc Oxide Nanoparticles: A Promising Photocatalyst for Methylene Blue Degradation. *Nanomaterials* **2021**, *11*, 1558. https://doi.org/10.3390/nano11061558

Academic Editor: Christos A. Aggelopoulos

Received: 24 May 2021
Accepted: 11 June 2021
Published: 13 June 2021

Abstract: A facile green synthetic method is proposed for the synthesis of zinc oxide nanoparticles (ZnO NPs) using the bio-template *Cinnamomum tamala* (*C. tamala*) leaves extract. The morphological, functional, and structural characterization of synthesized ZnO NPs were studied by adopting different techniques such as energy dispersive X-ray analysis (EDX), high-resolution transmission electron microscopy (HR-TEM), scanning electron microscopy (SEM), X-ray diffraction (XRD), UV-Visible spectroscopy, fourier transform infrared (FTIR) spectroscopy, raman spectroscopy, and X-ray photoelectron spectroscopy (XPS). The fabricated ZnO NPs exhibit an average size of 35 nm, with a hexagonal nanostructure. Further, the well-characterized ZnO NPs were employed for the photocatalytic degradation of methylene blue (MB) in an aqueous solution. The photocatalytic activity was analyzed by changing the various physicochemical factors such as reaction time, amount of photocatalyst, precursor concentration, and calcination temperature of the ZnO NPs. All the studies suggest that the ZnO synthesized through the green protocol exhibits excellent photocatalytic potency against the dye molecules.

Keywords: green synthesis; zinc oxide nanoparticles; photocatalysis; methylene blue

1. Introduction

Environmental nanotechnology has garnered immense significance for removing hazardous chemicals, with remarkable potential to promote economically viable synthesis. Nanoparticles (NPs), the building blocks of nanotechnology, are categorized into different classes based on their morphology, structure, size, and chemical properties [1,2]. The synthesis of metal oxide NPs has been widely employed using physical and chemical methods, including electro-explosion, solvothermal, hydrothermal, microwave, spray pyrolysis, vapour deposition, microemulsion, coprecipitation, and the wet-chemical method [3–6]. However, these conventional methods of the synthesis of NPs limit their application due to the issues of toxicity, costliness, and the formation of diverse nanostructures [3]. In recent times, the emergence of various synthesis methods for semiconductor nanoparticles gained much attention [7,8]. The synthesis of NPs using green chemistry principles provides a solution to ecological concerns by minimizing toxic chemicals and the subsequent conversion of products without much process investment [9,10]. The greener approach to the synthesis of NPs came into the limelight to restrain the ecosystem with a naturally available biodegradable matter for its production [11,12]. The plant extract-mediated synthesis is a simple, cost-effective, efficient, and feasible process because of the availability of effective phytochemicals and functional groups present in the extractives [13].

C. tamala is an evergreen tree belonging to the Lauraceae family, commonly known as Tejpat in India [14]. It is medicinally used as a carminative, anti-flatulent, diuretic, and for cardiac disorders. Alkaloids, terpenes, flavonoids, tannins, polyphenols, and saponins are present in *C. tamala* [15], which can facilitate the synthesis of nanoparticles by serving as a reducing, as well as a capping or stabilizing, agent.

Zinc oxide (ZnO) is a promising metal oxide that can eliminate most of the ecological concerns, as it is relatively nontoxic [16–18] and can be used for the removal of environmental pollutants via photodegradation (especially wastewater treatment); ZnO NPs have acquired attention due to their extraordinary property of degrading contaminants [19]. One of the most important applications of semiconductor nanoparticles is their photocatalytic activity, which is crucial in removing or degrading pollutants. The industrially essential organic dyes extensively deployed in textile industries releases nearly 146,000 tonnes of dyes per year, alongside other associated pollutants [1,20,21]. The large number of dyes discharged in water bodies are difficult to degrade because of their non-biodegradable nature, large size, and complex nature; they often threaten humans and the environment [22]. The commonly used techniques are efficient in degradation, but they cause the generation of absorbent material, thus needing the post-treatment of waste materials [21,23–25]. Extensive research has been conducted to replace these methods, and semiconductor photocatalysis is emerging as the best choice for this purpose, as it has remarkable potential to destroy many toxic compounds in an aqueous medium, with fewer or no harmful products. The solar-driven photocatalytic processes has recently gained special attention in solving wastewater problems, as it relies on a natural process of energy conversion [26].

The present study has been focused on the fabrication of green ZnO NPs using *C. tamala* leaf extract and evaluating its ability towards methylene blue (MB) photodegradation. The photocatalytic potency or effective degradation of MB dye was studied using various physicochemical parameters such as reaction time, the concentration of precursor, calcination temperature, and amount of catalyst.

2. Materials and Methods

2.1. Materials

Zinc nitrate hexahydrate (98% purity) and methylene blue were procured from Sigma Aldrich and Himedia chemicals, Mumbai, India. *C. tamala* leaves were collected from the Kannur district, Kerala, India. All other chemicals and solvents used were of analytical grade and were used without further purification.

2.2. Preparation of *C. tamala* Aqueous Extract

Fresh *C. tamala* leaves were collected, cleaned well, and shade dried for a month. The wholly dried leaves were ground to a fine powder using a mixer grinder.

Ten g of fine powder was weighed and transferred to a 500 mL stoppered bottle. Two hundred and fifty mL of distilled water was added, and the mixture was heated at 80 °C on a magnetic stirrer for 1 h. The temperature was maintained constantly and, subsequently, the solution was filtered using a cotton cloth. The filtrate solution was used for the synthesis of ZnO NPs, and it was stored in the refrigerator for further use.

2.3. Synthesis of Zinc Oxide Nanoparticles (ZnO NPs)

ZnO NPs were synthesized using *C. tamala* extract by a simple procedure. Briefly, *C. tamala* extract (5 mL) was added to 0.1 M of zinc nitrate ($Zn(NO_3)_2{\cdot}6H_2O$) (25 mL) and stirred at 60 °C for 3 h. The product obtained was dried and calcined at 500 °C for 2 h.

2.4. Characterization of ZnO NPs

2.4.1. X-ray Diffraction (XRD) Analysis

To study the crystal structure of ZnO NPs, the calcined sample was subjected to X-ray diffraction analysis with a Rigaku Miniflex 600 (Tokyo, Japan) diffractometer with nickel filtered Cu Kα (λ = 1.54 Å) radiation. The sample scanned between 20° and 90° (diffraction angle 2θ) with a scan speed of 0.05°. The Joint Committee on Powder Diffraction Standards (JCPDS)'s data files were used for matching with the spectrum obtained from ZnO NPs.

2.4.2. FT-IR Analysis

Fourier Transform Infrared analysis was performed using a Perkin-Elmer FTIR Spectrum Two spectrometer (Singapore, Singapore) in attenuated total reflection (ATR) mode with 30 scans in a spectral range set between 4000 and 400 cm^{-1}.

2.4.3. UV-Visible Analysis

Optical properties were obtained using an ultraviolet (UV)-visible spectrophotometer (Perkin Elmer Lambda 35, Singapore) over the spectral region 200–700 nm. The present spectrophotometer uses a deuterium lamp for UV and a tungsten lamp for the visible region.

2.4.4. Scanning Electron Microscope-Energy-Dispersive X-ray Spectroscopy (SEM-EDX) Analysis

The synthesized ZnO NPs were analyzed with a scanning electron microscope (TECNAI), Model: AIS2100 from M/S. Mirero Inc, Seongam, Korea, to determine the surface morphology. The elemental composition of the sample was acquired from energy-dispersive X-ray spectroscopy (EDX) measurements, carried out using a scanning electron microscope equipped with an EDX: Model No. INCAE350 from M/S, Oxford Instruments, Abingdon, UK.

2.4.5. TEM and High-Resolution Transmission Electron Microscopy (HR-TEM) Analysis

The nanoparticles size and particle distributions were attained from transmission electron microscopy analysis (TEM) (JEOL, JEM-2100, Tokyo, Japan).

2.4.6. Raman Spectroscopy

The Micro Raman scattering study was performed using LabRam HR Evolution Raman Spectrometer (Horiba Scientific, France with 633 nm Laser in backscattering geometry).

2.4.7. XPS

The Thermo Scientific K-Alpha instrument (monochromatic Al Kα radiation, Ephoton = 1486.6 eV) was used for X-ray photoelectron spectroscopy (XPS) studies. The internal reference peak of C 1s peak centred at 284.8 eV is used for binding energy (B.E) corrections. A dual-beam charge neutralization method is used in the instrument, where a low energy ion beam is used to eliminate the samples static charge, allowing the low energy electron beam to reach the sample and neutralize the localized positive charge created by the X-ray beam. The specimens were analyzed at an electron take-off angle of 70°, and measured with respect to the surface plane. The monochromatic X-ray source is located perpendicular to the analyzer axis, and the standard X-ray source is located at 54.7° relative to the analyzer axis. For all of the spectra, the spectrometer was operated in a standard mode. All survey spectra scans were taken at a pass energy of 58.7 eV. The narrow scans of strong lines were, in most cases, just wide enough to encompass the peaks of interest and were obtained with a pass energy of 23.5 eV.

2.5. Photocatalytic Studies on Methylene Blue Degradation

The photocatalytic activity of green synthesized ZnO NPs was evaluated by MB degradation under direct sunlight. The experiment setup is made in such a way that the reaction mixture was exposed to solar radiation. The intensity of solar light was measured using a lux meter. The degrading capacity was assessed by taking out 5 mL of the reaction mixture at a regular interval of time. The samples were centrifuged at 9000 rpm for 10 min and were analyzed for the photocatalytic degradation of MB using a UV-visible spectrometer over the wavelength 200–700 nm. The percent of photodegradation of MB (photocatalytic efficiency of ZnO) in aqueous media was calculated by the following equations: $\eta = (C_0 - C)/C_0$ or $\eta = (A_0 - A)/A_0$ [21], where C_0 is the initial dye concentration without catalyst and C is the final dye concentration with catalyst after 90 min, and A_0 and A show initial and final absorbance of MB dye without and with ZnO photocatalyst.

2.5.1. Effect of Solar Irradiation Time on Photocatalytic Studies

To evaluate the influence of solar irradiation time on MB degradation, 5 mg of green ZnO NPs was added to 50 mL of dye solution. The resultant reaction mixture was kept for continuous stirring at 700 rpm under solar irradiation. The percentage degradation of MB was evaluated by the absorbance measurements of the solution using a UV-visible spectrophotometer by eluting and measuring the absorbance of 5 mL supernatant (centrifuged at 9000 rpm for 10 min) at regular time intervals of 15 min. The process was continued for 90 min.

2.5.2. Effect of Amount of Catalyst on Photocatalytic Studies

The influence of the amount of catalyst on photocatalytic degradation of MB was evaluated by adding different amounts (5–15 mg) of ZnO NPs to 50 mL of MB dye solution (10 μM). The reaction mixture was kept for stirring on a magnetic stirrer for 90 min under sunlight. The absorbance measurements were done with a UV-visible spectrometer and calculated the photodegradation capability of ZnO NPs.

2.5.3. Effect of Calcination Temperature on Photocatalytic Studies

Green synthesized ZnO NPs calcined at three different temperatures (300 °C, 500 °C, and 700 °C) was used for MB photodegradation to study the influence of calcination temperature of the sample on photocatalytic efficiency. Five mg of each of them was added to 50 mL of MB dye solution. The reaction mixture was kept on a magnetic stirrer for 90 min under sunlight. The supernatant solution was subjected to absorbance measurements.

2.5.4. Effect of Concentration of $Zn(NO_3)_2$ on Photocatalytic Studies

To study the effect of precursor concentration on photocatalytic degradation capability, ZnO NPs were synthesized from three different precursor concentrations, i.e., 0.025, 0.05, and 0.1 M of $Zn(NO_3)_2{\cdot}6H_2O$. Synthesized nanoparticles were calcined at 500 °C. The photocatalytic study was monitored by adding 5 mg of catalyst to 50 mL MB dye solution using a UV-visible spectrophotometer.

2.6. Effect of Solar Irradiation on Photodegradation of MB

The effect of solar irradiation on the percentage degradation of MB dye solution was evaluated by loading ZnO catalyst in the presence and absence of sunlight. Fifteen mg of ZnO NPs was added to 50 mL of MB dye solution, and the sample was mixed thoroughly for 90 min on a magnetic stirrer. The reaction was performed under sunlight and in a dark chamber to study the effect of solar irradiation on MB degradation.

2.7. Studies on the Effect of Concentration of $Zn(NO_3)_2$ on Crystallite Size of ZnO NPs

To study the relationship between the concentration of precursor and the crystallite size, ZnO NPs were synthesized from 0.025, 0.05, and 0.1 M $Zn(NO_3)_2{\cdot}6H_2O$. All samples were calcined at 500 °C. From the XRD spectrum obtained, the crystallite size of NPs can be calculated using Debye–Scherrer's Equation:

$$D = 0.9\lambda/(\beta\cos\theta)$$

where D is the crystallite size, λ is the X-ray wavelength, θ is Bragg's angle in radians, and β is the full width half maximum [27].

3. Results and Discussion

3.1. Mechanism of Formation of ZnO NPs via Biosynthesis

The environment is rich in plant resources which contain many phytochemicals present in roots, leaves, flowers, etc. The functional groups, such as hydroxyl, carbonyl, carboxylic acid, etc., present in these phytochemicals can facilitate the formation of nanoparticles by inducing chemical reduction. The current work is dealing with biosynthesizing ZnO NPs using a green template, *C. tamala* leaf extract, which is a rich source of polyphenols, flavonoids, and alkaloids [15].

When a metal solution, i.e., $Zn(NO_3)_2{\cdot}6H_2O$, is introduced into a well synthesized homogenous leaf extract solution, leaf extract will try to form the matrix in which zinc in the +2 oxidation state gets absorbed. The stability of Zn^{2+} on the matrix is due to the chelating effect, or a type of interaction between functional groups and Zn^{2+}', which may be due to the transfer of a lone pair of electrons (available on functional groups) to the empty orbital of zinc. Therefore, the quarantined cations M^{n+} or hydroxylated cation $[M(OH)]^{m+}$ can go through nucleation of the growth process and are accelerated by functional groups present in the leaf extract [26]. The dried mixture obtained was yellow-coloured, which may be attributed to the phytochemicals present. The sample was kept for calcination to remove the hydroxides and other impurities that rise to the formation of required oxide [28] and, finally, results in ZnO NPs. Various analytical techniques confirmed the formation of nanoparticles.

3.2. Characterization of ZnO NPs

3.2.1. X-ray Diffraction Studies

The X-ray diffraction pattern of ZnO NPs calcined at 500 °C is represented in Figure 1. A definite number of sharp, intense diffraction peaks were observed, corresponding to (100), (002), (101), (102), (110), (103), (200), (112), and (201) crystallographic planes at characteristic peak angles 31.6°, 34.31°, 36.15°, 47.44°, 56.51°, 62.81°, 66.36°, 67.85° and 69.02°, which have been in agreement with the hexagonal wurtzite crystalline structure of ZnO NPs as per Joint Committee on Powder Diffraction Studies Standards (JCPDS card no. 36–1451) [27,28], thus validating the formation of required ZnO NPs. The intense and narrow peaks obtained described the crystalline nature and high purity of ZnO. In contrast, the absence of additional peaks further confirms the higher purity of ZnO NPs, since they are free of impurities. The Debye–Scherrer equation is employed to find out the average crystallite size or diameter of green synthesized ZnO NPs, $D = 0.9\lambda/(\beta\cos\theta)$, corresponding to the maximum intense peak of the (101) plane. The average crystallite size is estimated to be approximately 35 nm.

3.2.2. FTIR Analysis

FTIR spectroscopy was employed to recognize the functionalities present in the *C. tamala* leaf extract and characterize the ZnO NPs, and the obtained spectra are presented in Figure 2a. In the mid-IR region, peaks reveal the comparison of two plots that guide the confirmation of ZnO NPs. The leaf extract spectrum was observed to have significant peaks at 3500–3100 cm^{-1}, 2924 cm^{-1}, 1603 cm^{-1}, and 1021 cm^{-1}. The broad peak obtained at 3500–3100 cm^{-1} was attributed to the OH stretching frequency. Similarly, the bands registered at 2924 cm^{-1}, 1603 cm^{-1}, and 1021 cm^{-1} were due to the C–H, C=C, and C–O stretching vibrations, respectively [29]. These peaks shown by the leaf extract were due to the phytochemicals present in it, i.e., glycosides, flavonoids, polyphenols, terpenoids, tannins, carbohydrates, and reducing sugar. These phytochemicals have shown other additional small peaks that are due to overtones. The absence of these specified peaks in the FTIR spectrum of ZnO nanopowder results from the higher purity of green synthesized ZnO NPs calcined at 500 °C. However, the ZnO NPs showed a unique sharp signal in the fingerprint region below 1000 cm^{-1}, which may be attributed to the stretching frequency of the Zn–O bond [30]. Besides, ZnO NPs also showed small intense peaks which may be corresponding to the complication of fundamental bands, i.e., overtones, combination bands, or different bands.

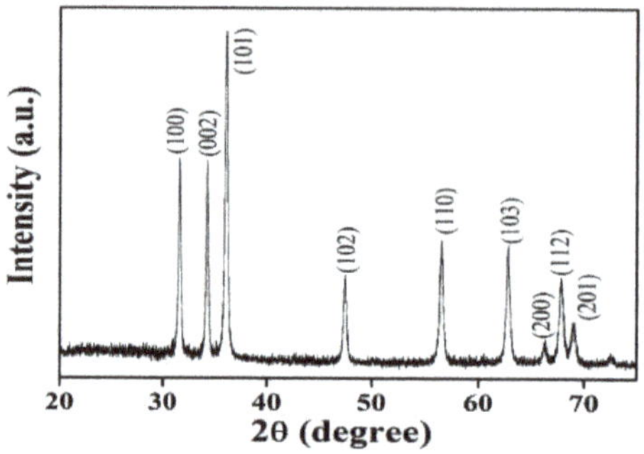

Figure 1. XRD spectrum of ZnO NPs.

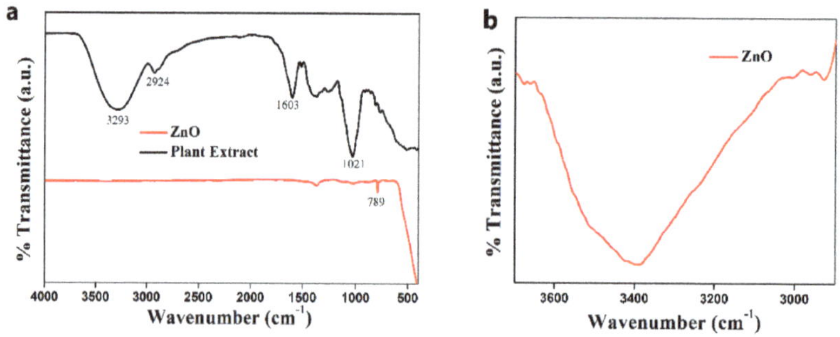

Figure 2. (**a**) FT-IR spectra of leaf extract and ZnO NPs; (**b**) The expanded region of FT-IR spectra around 3100 to 3500 cm^{-1} of the ZnO sample.

An absorption peak corresponding to the O–H frequency is barely seen in ZnO NPs. However, a broad peak at 3293 cm^{-1} is noticed in the enlarged view of the figure, as depicted in Figure 2b, suggesting the O–H band is present in the ZnO sample. The FT-IR analysis manifests the role of organic substances present in *C. tamala* leaf extract to stabilize biosynthesized ZnO NPs. Thus, the formation of ZnO NPs is accomplished due to the interaction between oxygen in functional groups involved in *C. tamala* leaf extract and zinc molecules in the precursors.

3.2.3. Optical Property Studies

The bio-synthesized ZnO NPs were further analyzed using a UV-visible spectrophotometer in the range of 200–700 nm, and the resulting optical plot is shown in Figure 3a,b. The observed absorption spectrum of synthesized nanoparticles matches the absorption spectrum of previously reported ZnO NPs, as mentioned in Kayani et al. [31]. The spectrum is also in agreement with those obtained for ZnO NPs produced by synthetic methods [32–37], exhibiting a sharp, intense peak in the UV region at 376 nm and a small number of little, extreme, sharp peaks in between 250–350 nm. This is the characteristic absorption peak of ZnO. The absorption peak obtained may be either due to the surface plasmon resonance (SPR) involving the quantum size effect [38,39] or the semiconductor band gap transition [27]. Moreover, a small number of weak intense sharp peaks between 250–350 nm indicates the surface defects due to the recombination of electrons in the conduction band and holes in the valence band, which also specify the monodisperse nature of ZnO NPs distribution [27]. The optical bandgap of bio-synthesized ZnO NPs can be calculated from the Tauc plot, and it was estimated to be 3.24 eV.

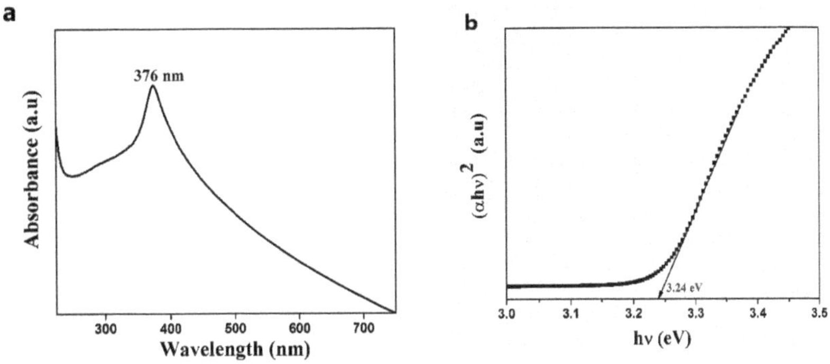

Figure 3. (**a**) The absorption spectrum of ZnO NPs; (**b**) Tauc plot of ZnO NPs.

3.2.4. FESEM-EDX Analysis

The FESEM analysis was performed to identify the surface morphology of synthesized ZnO NPs and is displayed in Figure 4a–d. The resulting images have a particle size in the nanometre range, observable at higher magnification. The ZnO NPs show aggregation to some extent under lower magnification due to weak physical force. On increasing the magnification, the particles were good enough and well separated, and it is concluded that the particles are in the nanometre range. The chemical composition was analyzed using EDX spectra (Figure 4e), which showed four absorption peaks identified as Zinc (54.21%) and Oxygen (45.79%) in their atomic percentage, as well as Zinc (82.87%) and Oxygen (17.13%) in their weight percentage (Table 1), confirming the high purity nature of synthesized ZnO NPs.

Table 1. Elemental Composition of ZnO.

Element	Weight%	Atomic%
Zn	82.87	54.21
O	17.13	45.79
Total	100	100

3.2.5. TEM and HR-TEM Analysis

ZnO NPs were analyzed by TEM analysis to understand the nanoparticles' morphology further, this analysis is shown in Figure 5. Figure 5a shows that the ZnO NPs showed polygonal morphology and were non-uniform in size. The NPs also appeared to be monodispersed with an excellent crystalline structure, and agreed with the XRD results. The high-resolution TEM images (shown in Figure 5b) show clear lattice fringes without any distortion, designating the high crystallinity of ZnO NPs. The estimated interplanar spacing of adjacent lattice fringes is estimated to be 0.24 nm. Selected area electron diffraction (SAED) patterns of the sample have been captured and displayed in Figure 5c. The rings labelled in the SAED pattern represent the lattice planes and are identical to that known for the hexagonal wurtzite crystallite structure. To check the sample size distribution, 100 NPs were selected from different sample images. The particle size distribution of the sample is acquired from the average diameters, and the obtained histogram is shown in Figure 5d. The estimated average particle sizes of the sample were found to be 35 to 40 nm.

The particle size determined from the TEM study is keenly matched with the crystallite size of ZnO acquired via XRD analysis.

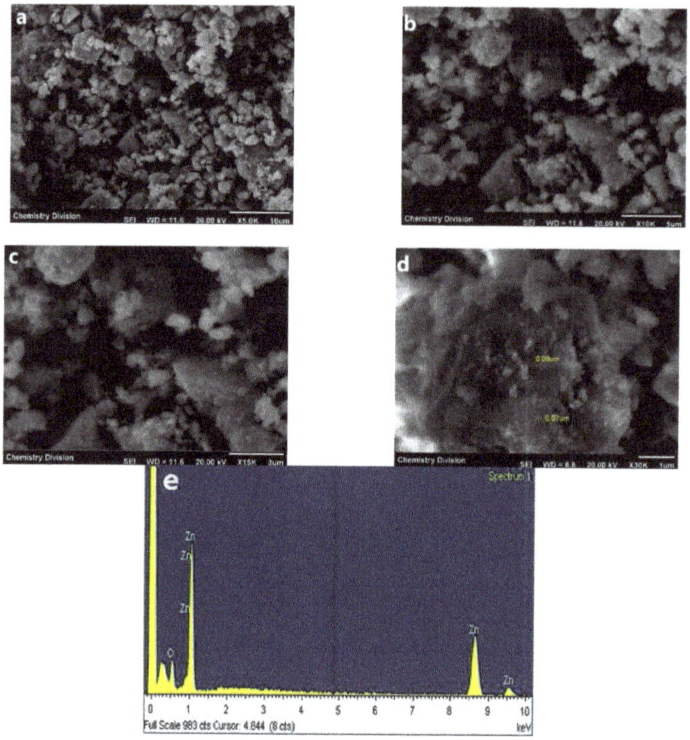

Figure 4. Field Emission Scanning electron microscopy (FESEM) images of ZnO NPs in magnification: (**a**) ×5.0 k; (**b**) ×10 k; (**c**) ×15 k; (**d**) ×30 k; (**e**) EDX (Energy Dispersive X-ray) Analysis of ZnO NPs.

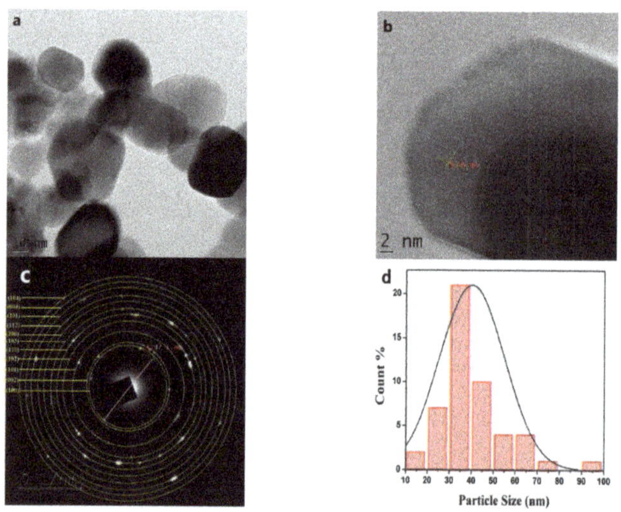

Figure 5. TEM data of ZnO NPs: (**a**)TEM image; (**b**) HR-TEM image; (**c**) SAED pattern; (**d**) Histogram showing particle size distribution in the sample.

3.2.6. Raman Spectroscopy

The micro-Raman spectra, along with peak fitting using the Gaussian function for ZnO NPs in the spectral range 60–700 cm^{-1} is shown in Figure 6. The hexagonal wurtzite structure of ZnO can be represented by $A_1 + 2B_1 + E_1 + 2E_2$, as it belongs to the C6v point group. The polar modes, A_1 and E_1, can be split into transverse optical (TO) and longitudinal optical (LO) modes. B_1 modes are Raman inactive. Raman active E_2 modes are non-polar, comprise of $E_{2\,(High)}$ (437 cm^{-1}) and $E_{2\,(Low)}$ (99 cm^{-1}), which correlated with the vibration of the oxygen atom and heavier Zn-atom, respectively. The Raman bands appear at 332 and 380 cm^{-1}, and are assigned to the $2E_{2(M)}$ and A_1(TO) mode, respectively. The peak at 156 cm^{-1} may be related to a defect induced mode. The peaks at 284 and 580 cm^{-1} are attributed to $B_{1(Low)}$ and $B_{1(High)}$ modes, respectively [40]. The blurring of peaks can be seen in the spectra, suggesting disorders present in ZnO.

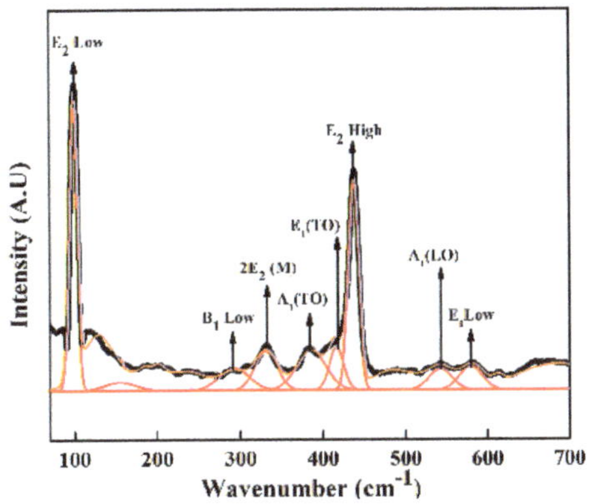

Figure 6. Raman spectrum of ZnO NPs.

3.2.7. XPS Analysis

The XPS spectra of ZnO was measured to explore the chemical states of the elements present in the sample. The broad range XPS spectrum of the sample is manifested in Figure 7a, and the high-resolution spectra of Zn 2p and O 1s were recorded and are depicted in Figure 7b,c, along with the deconvolution. All measured values are given with respect to the C 1s peak located at 284.6 eV. Figure 7b, the high-resolution XPS spectrum of Zn 2p, shows two strong peaks appearing at 1021.5 and 1044.6 eV, which are ascribed to the binding energies of Zn 2p3/2 and Zn 2p1/2, respectively. These values are in agreement with the previous reports on Zn^{2+} ion [41]. The spin−orbit splitting energy for the Zn 2p peaks was around 23.1 eV. The high-resolution O 1s spectra show an asymmetric peak, and the peak is fitted with two peaks centred at 530.2 eV and 531.5 eV, and correspond to the lattice oxygen of ZnO and hydroxyl group (–OH) oxygen.

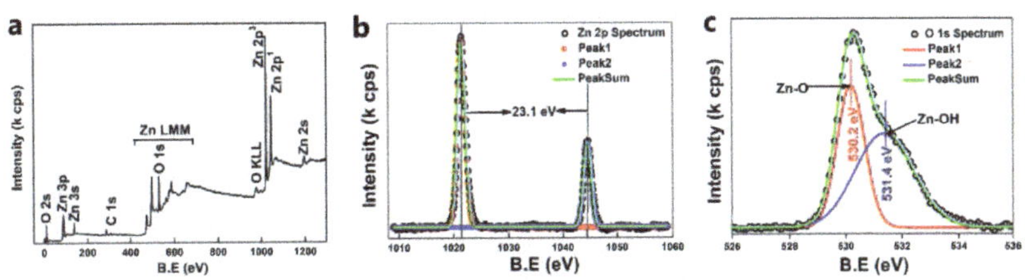

Figure 7. (**a**) Wide range XPS spectra of ZnO NPs; (**b**) high-resolution spectra of Zn 2p; (**c**) high-resolution spectra of O1s.

3.3. Mechanism of Photo-Degradability of MB by ZnO Catalyst

The photocatalytic potency of green-synthesized ZnO NPs was illustrated with an example of MB dye under solar irradiation at an average intensity of 100,000 Lux (Figure 8). The MB photodegradation occurs due to the electrons, holes, and free radicals generated, and undergoes specific reactions to destroy the structure of MB. The catalyst, upon irradiation with sunlight, generates electrons in the conduction band and holes in the valence band. The generated electrons react with adsorbed oxygen on the catalyst to give $O_2^{\cdot -}$ radical anion, and the water from the solution combines with the holes to produce OH· radicals. These generated radicals undergo the reduction and oxidation reaction while reacting with dyes [32–37]. However, $O_2^{\cdot -}$ radical anion reacts further to generate HOO·, OH^- radicals, and OH^- anions, and these active species of oxygen are responsible for the degradation of MB dye.

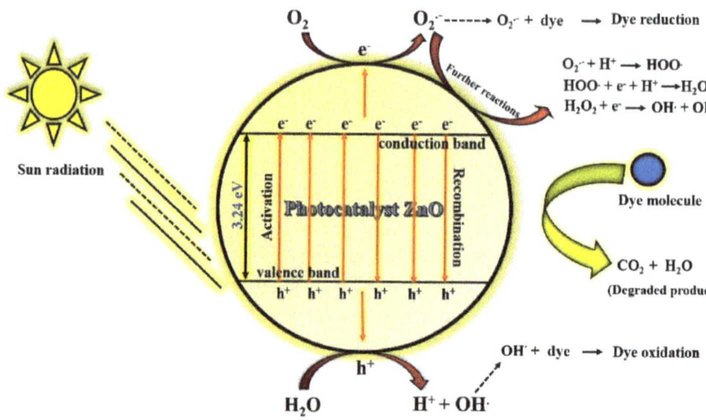

Figure 8. Schematic diagram of photocatalysis mechanism of ZnO NPs.

As a result, the prime factor in the degradation of the dye is mainly originated from the generation of electrons and holes. These electron–hole pairs react further to generate active oxygen species radicals ($O_2^{\cdot -}$, HOO$^{\cdot}$, OH$^{\cdot}$, and OH^-), which are responsible for destroying the structure of MB dye and transforming it into a less harmful or degraded product. Hence, pollutant content in the water bodies can be minimized by using such a catalyst [42,43]. The photocatalytic reaction taking place within the ZnO catalyst can be summarised as shown below,

1. $\text{ZnO (catalyst)} + \text{h}\nu\ \text{(light)} \rightarrow e^-\ \text{(conduction band)} + h^+\ \text{(valence band)}$
2. $O_2 + e^- \rightarrow O_2^{\cdot -}\ \text{(radical anion)}\ H_2O + h^+ \rightarrow H^+ + OH\cdot\ \text{(hydroxyl radical)}$
3. $O_2^{\cdot -} + \text{MB dye} \rightarrow \text{dye (reduction)}\ OH\cdot + \text{MB dye} \rightarrow \text{dye (oxidation)}$

Further reaction,

$$O_2^{\cdot -} + H^+ \rightarrow HOO\cdot$$

$$HOO\cdot + e^- + H^+ \rightarrow H_2O_2$$

$$H_2O_2 + e^- \rightarrow OH\cdot + OH\cdot$$

3.4. Effect of Time of Solar Irradiation on Photodegradation of MB

The influence of the extent of solar irradiation on the percentage dye degradation of MB, evaluated by loading ZnO catalyst in the presence and absence of sunlight, is shown in Figure 9, which provides the absorption band for each sample at a definite regular time interval. The dye exhibited a characteristic absorption, with maximum absorbance at 664 nm in UV-visible spectra. By analyzing the absorption of the mixture for overall time, the efficiency calculated was found to be 98.07% upon solar irradiation and 8.57% without solar irradiation. This result implied that the biosynthesized ZnO photocatalyst exhibited an excellent photocatalytic activity only in the presence of sunlight.

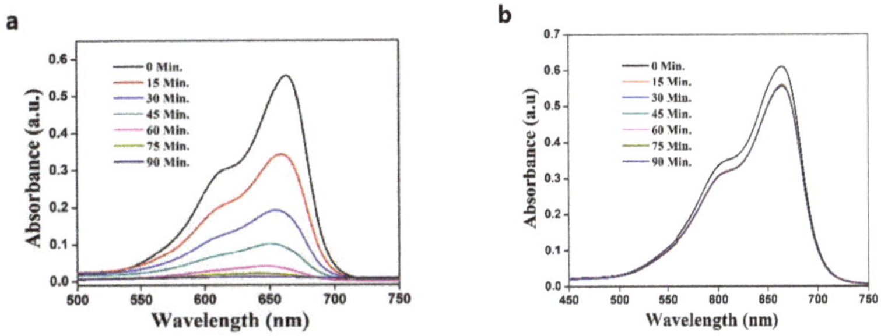

Figure 9. The photo-degradability of MB dye by ZnO catalyst (**a**) in the presence of sunlight; (**b**) in the absence of sunlight.

3.5. Effect of ZnO Catalyst Loading on Photodegradation of MB

The impact of the catalyst concentration on photocatalytic degradation of MB dye was investigated by adding 5–15 mg/50 mL of ZnO catalyst into the MB solution, and the UV-visible spectra of the resultant solution are presented in Figure 10. The catalyst concentration has a noticeable impact on MB degradation, as evidenced by the graph.

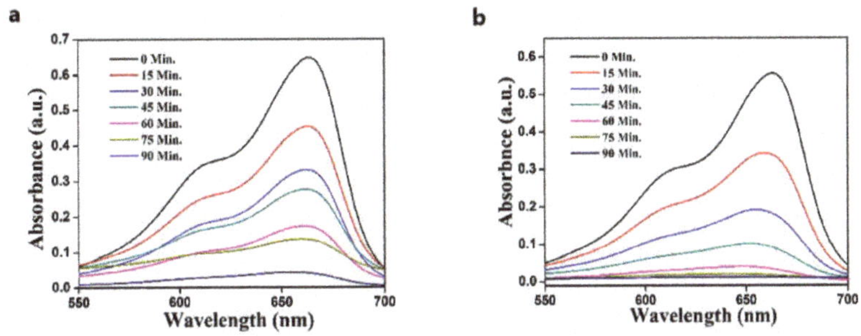

Figure 10. The effect of ZnO photocatalyst on % degradation of MB dye (**a**) with 5 mg ZnO; (**b**) with 15 mg ZnO.

The percentage of degradation of MB dye increased from 93.89% to 98.07% as the catalytic dosage increased from 5 mg to 15 mg. The effect of increment in the percentage degradation is probably due to more catalyst active sites and a higher adsorption area along with a higher specific surface area [44,45]. As a result of surface defects, a higher specific surface area is available for the origination of active radicals to degrade the dye. However, only a 5% increase in the photocatalytic efficiency was obtained with a higher amount of catalyst. This may be associated with the aggregation of particles, which increases the turbidity of the suspension, resulting from the high doses of the photocatalyst [46]. Because of it, the penetration of sunlight into the solution decreases, which results in a reduction in the photoactivated volume of suspension. Briefly, we can state that an increase in the catalyst concentration results in a higher adsorption area (more active site) as well as the generation of turbidity in the reaction mixture. Consequently, the photocatalytic efficiency concerning the amount of catalyst will be a combined impression of specific surface area and the solar light penetration. Hence, if one factor dominates over the other, the ultimate result will be produced in the extent of photocatalytic degradation of the MB dye.

3.6. The Effect of Calcination Temperature on the Photodegradation of MB

The relation between the calcination temperature of the ZnO photocatalyst and photocatalytic efficiency towards the MB degradation is shown in Figure 11. The percentage of degradation was found to be 82.41%, 87.45%, and 93.87% for the ZnO sample calcined at 300 °C, 500 °C, and 700 °C, respectively; the calcination temperature of the catalyst highly influenced its photodegradation capability. The increment of efficiency may be attributed to surface crystallinity, which relates to the specific crystallite area of the catalyst. The highest photocatalytic activity is observed for catalyst with a higher specific surface area at an elevated temperature, and it may be ascribed to more photogeneration of holes (Vb) and electrons (Cb) in the ZnO crystal, contributing to more defects as well [47,48]. The surface imperfections present in the ZnO lattice may evolve into the formation of small pores, which causes an increment in the surface area. These tiny pores benefit oxygen adsorption and thereby enhances the MB degradation rate. Hence, the higher specific surface area with an increasing calcination temperature helps in improved adsorption of MB dye on the ZnO surface and thus performs as a better photocatalyst for photodegradation of MB.

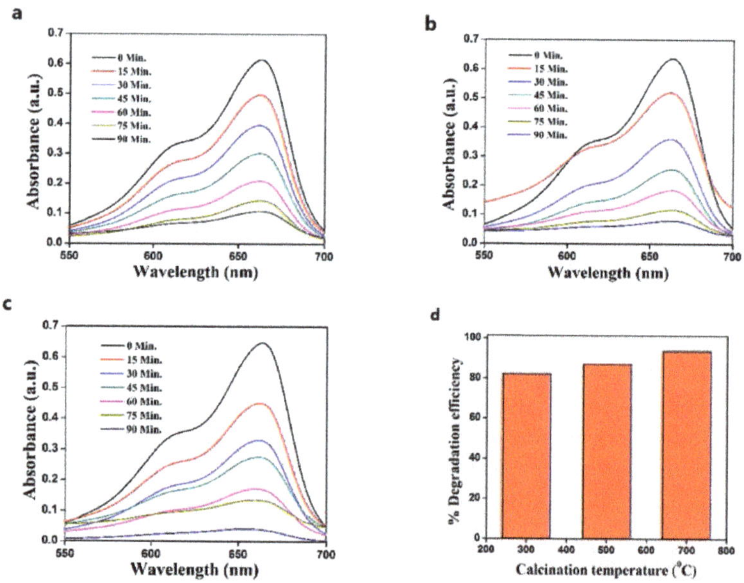

Figure 11. The photocatalytic degradation of MB after treatment with ZnO calcined at (**a**) 300 °C; (**b**) 500 °C; (**c**) 700 °C; (**d**) the effect of calcination temperature of ZnO on MB photodegradation.

3.7. Effect of Precursor Concentration on Photocatalytic Efficiency

The effect of the concentration of the precursor on the photocatalytic efficiency of the photocatalyst is presented in Figure 12. The photodegradation ability of the ZnO catalyst is in a linear relation with the $Zn(NO_3)_2{\cdot}6H_2O$ concentration, which was utilized in the synthesis process and attained a maximum of 87.45% for 0.1 M. The higher degradation capability can be ascribed to the variation in average crystallite size of the synthesized ZnO nanostructure. The crystallite size calculated from the XRD data shows that the average crystallite size of ZnO increases in the precursor concentration. A detailed description is given in Section 3.8. The increase in crystallite size results in a higher available surface area for the adsorption of more oxygen and photogeneration of more electrons and holes [41,46], i.e., if the crystallite size is increasing, crystallinity may increase. This may be due to the agglomeration of crystallite size particles which may result in the long-range of orderly arrangement of particles. Moreover, the crystallinity and larger crystallite size may enhance the reduction in the bandgap, and may generate more e-h pairs upon irradiation with solar light, which helps in producing radicals. Hence, the increase in the precursor concentration benefits the photodegradation of dyes. The photocatalytic activity depends on several factors, including surface area, particle size, crystallinity, defect (pores), adsorption (active sites), catalyst concentration, and calcination temperature. It does not rely on one factor; dominance of one factor over the other will decide the photocatalytic degradation of dyes by the ZnO photocatalyst.

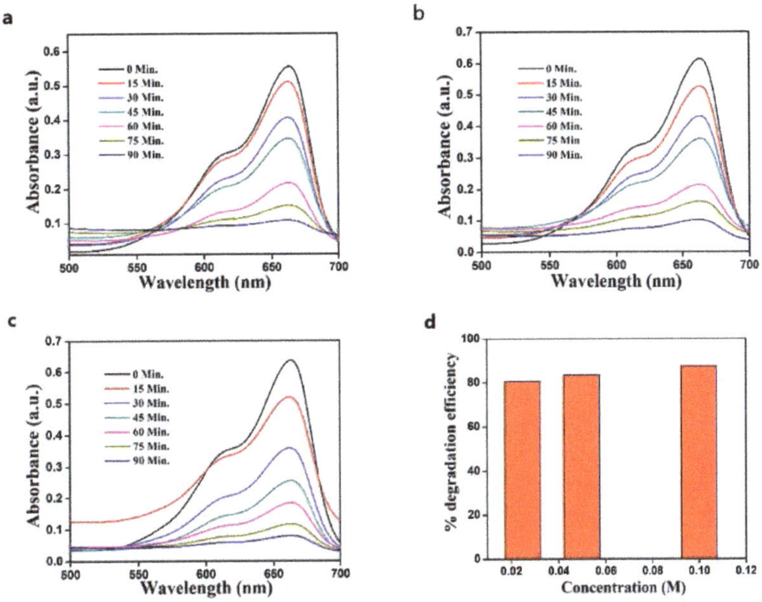

Figure 12. The photocatalytic degradation of MB after treatment with ZnO synthesized from (**a**) 0.025 M $Zn(NO_3)_2$; (**b**) 0.05 M $Zn(NO_3)_2$; (**c**) 0.1 M $Zn(NO_3)_2$; (**d**) the effect of $Zn(NO_3)_2{\cdot}6H_2O$ concentration on photodegradation of MB.

3.8. The Effect of Concentration of $Zn(NO_3)_2$ on Crystallite Size of ZnO NPs

To study the effect of $Zn(NO_3)_2$ concentration on the average crystallite size, ZnO NPs were synthesized from three different concentrations of $Zn(NO_3)_2{\cdot}6H_2O$ solution (0.025 M, 0.05 M, and 0.1 M) and the samples were calcined at 500 °C. The XRD spectra of ZnO NPs synthesized from the previously mentioned concentration are depicted in Figure 13. The Debye–Scherrer equation was used to determine the average crystallite size of green ZnO from the XRD pattern. It was observed that the intensity of every diffraction plane is increasing, but the peak of the (101) plane is of particular interest since it is the most intense peak. Hence, the crystallite size calculations were done with the (101) diffraction peak using the Scherrer formula.

The increase in precursor concentration leads to the enhancement in the preferential orientation of the intense peak, which specified the improvement in crystallinity [35,36]. The crystallite size found a maximum for 0.1 M precursor solution, which is around 49.25 nm. Similarly, the crystallite size obtained for 0.05 M and 0.025 M are 30.44 nm and 25.51 nm, respectively. That the average crystallite size decreases with a decrease in concentration is due to the increment in the FWHM value of (101) diffraction peak, as evidenced in Table 2. The growth in the crystallite size is due to the crystal agglomeration with an increase in the concentration.

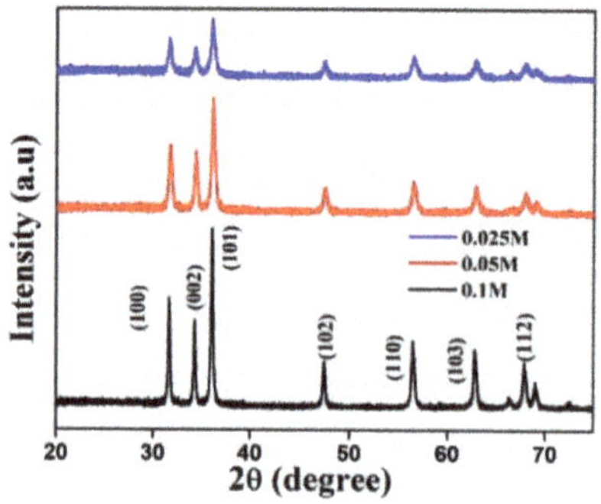

Figure 13. XRD patterns of ZnO NPs synthesized from the different molar concentrations of precursor.

Table 2. Crystallite size of ZnO NPs by varying $Zn(NO_3)_2$ concentration.

Sl. No.	Concentration (M)	2θ (Degree)	FWHM	Crystallite Size (nm)
1	0.025	36.28	0.56569	25.51
2	0.05	36.20	0.47367	30.44
3	0.1	36.15	0.28872	49.95

Moreover, in a highly concentrated solution, the rate of reduction and oxidation of $Zn(NO_3)_2$ to $Zn(OH)_2$ will be higher, as the collision between highly concentrated salt with phytochemicals will be more. Therefore, the chance of lower concentration reduction will be minimal as there are a small number of salt ions present. This increase in crystallite size is also reflected in photocatalytic studies, as discussed earlier, in which the photocatalytic activity is enhanced with the increase in precursor concentration.

The photocatalytic efficiency of *C. tamala* stabilized ZnO NPs is compared with recent studies from the literature and is shown in Table 3.

Table 3. A comparative table showing the photocatalytic efficiency of *C. tamala* stabilized ZnO NPs with other phtotcatalytic system based on ZnO NPs for the removal of MB.

Sl. No.	ZnO Synthesis Method	% Efficiency (Time)	Light Source	MB Concentration	Reference
1	Precipitation Sol-gel Method	81% (180 min) 92.5% (180 min)	UV light	20 mg/L	[49]
2	Sol-gel	86% (140 min)	UV light	10 mg/L	[50]
3	Green synthesis-*Becium grandiflorum* aqueous leaf extract	69% (200 min)	UV light	100 ppm	[51]
4	Green synthesis-Durian waste	84% (40 min)	Solar light	10mg/L	[52]
5	Ultra sound assisted green synthesis	93.25% (70 min)	Solar light	1 mg/mL	[53]
6	Green synthesis-*Kalopanax septemlobus*	97.5% (30 min)	UV light	50 μM	[54]
7	Green synthesis-*Sambucus ebulus*	80% (200 min)	UV light	50 ppm	[55]
8	Green synthesis-*Cinnamomum tamala* aqueous leaf extract	98% (90 min)	Solar light	10 μM	Present study

4. Conclusions

The green synthesis of ZnO NPs using the *C. tamala* template offers a simple, stable, sustainable, economic, and eco-friendly approach compared to other conventional methods that have been commonly deployed. The phytochemicals present in the *C. tamala* extract played a significant role in serving both as capping and reducing agents during the synthesis. The synthesized ZnO NPs were hexagonal crystalline in nature, having an average size of 35 nm. The bandgap obtained from the Tauc plot was 3.24 eV. It is noticeable that the photocatalytic activity of *C. tamala* stabilized ZnO NPs was exclusively dependent on various factors and the conditions optimized for the best catalytic performance. It was shown to have higher photocatalytic activity for the catalyst loading, 5 mg/50 mL. ZnO synthesized from 0.1 M $Zn(NO_3)_2 \cdot 6H_2O$ and ensuing ZnO calcined at 700 °C displayed higher photocatalytic efficiency towards MB degradation. The ZnO microstructure formation process and the crystallite size have a prominent impact on the photocatalytic performance of ZnO NPs. The present study has outstandingly suggested that green synthesized ZnO NPs display an excellent photocatalytic degradation capability of a maximum of 98% towards the degradation of organic dye. Hence, the presently adopted method provides a better alternative over other methods and can be developed for a large-scale operation to be employed in wastewater treatment and water purification.

Author Contributions: Investigation, writing—original draft, S.N. and S.K.K.; supervision, project administration, B.G., S.S.S. and V.V.T.P.; formal analysis, methodology, S.W., S.S.S., and V.M.; resources, validation, funding acquisition, writing—review and editing, V.V.T.P. and M.Č.; validation, writing—review and editing, R.S.V. All authors have read and agreed to the published version of the manuscript.

Funding: This research work was financially supported by the Ministry of Education, Youth and Sports of the Czech Republic and the European Union (European Structural and Investment Funds -Operational Programme Research, Development and Education) in the frames of the project "International Research Laboratories", Reg. No. CZ.02.2.69/0.0/0.0/18_054/0014685. This work also was supported by the Department of Science & Technology (DST)-Senior Research Fellowship. (Sajina No. IF160445). The authors gratefully acknowledge the financial support under Project No. LM2018124 supported by the Minis-try of Education, Youth and Sports of the Czech Republic and the European Union—European Structural and Investment Funds in the Operational Programme Research, Development, and Education framework and also supported by the Ministry of Education, Youth and Sports in the Czech Republic under the "Inter Excellence—Action Programme" within the framework of project "Bio-based Porous 2D Membranes and 3D Sponges Based on Functional-ized Tree Gum Polysaccharides and their Environmental Application" (registration No. LTAUSA19091)—TUL internal No.: 18309/136.

Data Availability Statement: The data presented in this study are available on request from the corresponding author.

Acknowledgments: The authors gratefully acknowledge financial support from the Department of Science & Technology (DST). The authors extend their thanks to Jitendra Nuwad (Bhabha Atomic Research Centre-BARC) and Sophisticated Test and Instrumentation Centre (Cochin University of Science & Technology) instrumentation facilities for nanoparticle characterization.

Conflicts of Interest: The authors declare no conflict of interest.

References

1. Zare, E.N.; Padil, V.V.T.; Mokhtari, B.; Venkateshaiah, A.; Wacławek, S.; Černík, M.; Tay, F.R.; Varma, R.S.; Makvandi, P. Advances in Biogenically Synthesized Shaped Metal- and Carbon-Based Nanoarchitectures and Their Medicinal Applications. *Adv. Colloid Interface Sci.* **2020**, *283*, 1–17. [CrossRef]
2. Hebbalalu, D.; Lalley, J.; Nadagouda, M.N.; Varma, R.S. Greener Techniques for the Synthesis of Silver Nanoparticles Using Plant Extracts, Enzymes, Bacteria, Biodegradable Polymers, and Microwaves. *ACS Sustain. Chem. Eng.* **2013**, *1*, 703–712. [CrossRef]
3. Mohammadinejad, R.; Karimi, S.; Iravani, S.; Varma, R.S. Plant-Derived Nanostructures: Types and Applications. *Green Chem.* **2015**, *18*, 20–52. [CrossRef]
4. Kim, C.S.; Moon, B.K.; Park, J.H.; Choi, B.C.; Seo, H.J. Solvothermal Synthesis of Nanocrystalline TiO_2 in Toluene with Surfactant. *J. Cryst. Growth* **2003**, *257*, 309–315. [CrossRef]
5. Gotić, M.; Musić, S. Synthesis of Nanocrystalline Iron Oxide Particles in the Iron(III) Acetate/Alcohol/Acetic Acid System. *Eur. J. Inorg. Chem.* **2008**, *6*, 966–973. [CrossRef]
6. Hayashi, H.; Hakuta, Y. Hydrothermal Synthesis of Metal Oxide Nanoparticles in Supercritical Water. *Materials* **2010**, *3*, 3794–3817. [CrossRef]
7. Yadav, A.; Burak, G.; Ahmadivand, A.; Kaushik, A.; Cheng, G.J.; Ouyang, Z.; Wang, Q.; Yadav, V.S.; Mishra, Y.K.; Wu, Y.; et al. Controlled Self-Assembly of Plasmon-Based Photonic Nanocrystals for High Performance Photonic Technologies. *Nano Today* **2021**, *37*, 101072. [CrossRef]
8. Vabbina, P.K.; Sinha, R.; Ahmadivand, A.; Karabiyik, M.; Gerislioglu, B.; Awadallah, O.; Pala, N. Sonochemical Synthesis of a Zinc Oxide Core-Shell Nanorod Radial p-n Homojunction Ultraviolet Photodetector. *ACS Appl. Mater. Interfaces* **2017**, *9*, 19791–19799. [CrossRef] [PubMed]

9. Devatha, C.P.; Thalla, A.K. Green Synthesis of Nanomaterials. In *Synthesis of Inorganic Nanomaterials*; Elsevier: Amsterdam, The Netherlands, 2018; pp. 169–184.
10. Padil, V.V.T.; Wacławek, S.; Černík, M.; Varma, R.S. Tree Gum-Based Renewable Materials: Sustainable Applications in Nanotechnology, Biomedical and Environmental Fields. *Biotechnol. Adv.* **2018**, *36*, 1984–2016. [CrossRef]
11. Machado, S.; Pinto, S.L.; Grosso, J.P.; Nouws, H.P.A.; Albergaria, J.T.; Delerue-Matos, C. Green Production of Zero-Valent Iron Nanoparticles Using Tree Leaf Extracts. *Sci. Total Environ.* **2013**, *445–446*, 1–8. [CrossRef]
12. Huang, L.; Weng, X.; Chen, Z.; Megharaj, M.; Naidu, R. Green Synthesis of Iron Nanoparticles by Various Tea Extracts: Comparative Study of the Reactivity. *Spectrochim. Acta Part A Mol. Biomol. Spectrosc.* **2014**, *130*, 295–301. [CrossRef]
13. Kumar, M.D.K.R.A. *Green Chemistry and Engineering*, 1st ed.; Academic Press: Cambridge, MA, USA, 2007.
14. Yadav, S.; Khurana, J.M. Cinnamomum Tamala Leaf Extract-Mediated Green Synthesis of Ag Nanoparticles and Their Use in Pyranopyrazles Synthesis. *Cuihua Xuebao/Chin. J. Catal.* **2015**, *36*, 1042–1046. [CrossRef]
15. Hassan, W.; Zainab Kazmi, S.N. Antimicrobial Activity of Cinnamomum Tamala Leaves. *J. Nutr. Disord. Ther.* **2015**, *6*. [CrossRef]
16. Rastogi, A.; Zivcak, M.; Sytar, O.; Kalaji, H.M.; He, X.; Mbarki, S.; Brestic, M. Impact of Metal and Metal Oxide Nanoparticles on Plant: A Critical Review. *Front. Chem.* **2017**, *5*, 1–16. [CrossRef]
17. Sabir, S.; Arshad, M.; Chaudhari, S.K. Zinc Oxide Nanoparticles for Revolutionizing Agriculture: Synthesis and Applications. *Sci. World J.* **2014**. [CrossRef]
18. Khalafi, T.; Buazar, F.; Ghanemi, K. Phycosynthesis and Enhanced Photocatalytic Activity of Zinc Oxide Nanoparticles toward Organosulfur Pollutants. *Sci. Rep.* **2019**. [CrossRef]
19. Widiyandari, H.; Ketut Umiati, N.A.; Dwi Herdianti, R. Synthesis and Photocatalytic Property of Zinc Oxide (ZnO) Fine Particle Using Flame Spray Pyrolysis Method. *J. Phys. Conf. Ser.* **2018**, *1025*. [CrossRef]
20. Akshay Kumar, K.P.; Zare, E.N.; Torres-Mendieta, R.; Wacławek, S.; Makvandi, P.; Černík, M.; Padil, V.V.T.; Varma, R.S. Electrospun Fibers Based on Botanical, Seaweed, Microbial, and Animal Sourced Biomacromolecules and Their Multidimensional Applications. *Int. J. Biol. Macromol.* **2021**, *171*, 130–149. [CrossRef]
21. Saeed, K.; Khan, I.; Shah, T.; Park, S.Y. Synthesis, Characterization and Photocatalytic Activity of Silver Nanoparticles/Amidoxime-Modified Polyacrylonitrile Nanofibers. *Fibers Polym.* **2015**, *16*, 1870–1875. [CrossRef]
22. Venkateshaiah, A.; Cheong, J.Y.; Shin, S.H.; Akshaykumar, K.P.; Yun, T.G.; Bae, J.; Wacławek, S.; Černík, M.; Agarwal, S.; Greiner, A.; et al. Recycling Non-Food-Grade Tree Gum Wastes into Nanoporous Carbon for Sustainable Energy Harvesting. *Green Chem.* **2020**, *22*, 1198–1208. [CrossRef]
23. Stoyanova, M.; Christoskova, S. Catalytic Degradation of Methylene Blue in Aqueous Solutions over Ni- and Co-Oxide Systems. *Cent. Eur. J. Chem.* **2011**, *9*, 1000–1007. [CrossRef]
24. Ali, R.; Ooi, B.S. Photodegradation of New Methylene Blue N in Aqueous Solution Using Zinc Oxide and Titanium Dioxide as Catalyst. *J. Teknol.* **2012**, *45*, 31–41. [CrossRef]
25. Ledakowicz, S.; Solecka, M.; Zylla, R. Biodegradation, Decolourisation and Detoxification of Textile Wastewater Enhanced by Advanced Oxidation Processes. *J. Biotechnol.* **2001**, *89*, 175–184. [CrossRef]
26. Saranya, K.S.; Vellora, V.; Padil, T.; Senan, C.; Pilankatta, R.; Saranya, K.; George, B.; Wacławek, S.; Černík, M. Green Synthesis of High Temperature Stable Anatase Titanium Dioxide Nanoparticles Using Gum Kondagogu: Characterization and Solar Driven Photocatalytic Degradation of Organic Dye. *Nanomaterials* **2018**, *8*, 1002. [CrossRef] [PubMed]
27. Talam, S.; Karumuri, S.R.; Gunnam, N. Synthesis, Characterization, and Spectroscopic Properties of ZnO Nanoparticles. *ISRN Nanotechnol.* **2012**, *2012*, 1–6. [CrossRef]
28. Varma, R.S. Journey on Greener Pathways: From the Use of Alternate Energy Inputs and Benign Reaction Media to Sustainable Applications of Nano-Catalysts in Synthesis and Environmental Remediation. *Green Chem.* **2014**, *16*, 2027–2041. [CrossRef]
29. Reddy, G.B.; Madhusudhan, A.; Ramakrishna, D.; Ayodhya, D.; Venkatesham, M.; Veerabhadram, G. Green Chemistry Approach for the Synthesis of Gold Nanoparticles with Gum Kondagogu: Characterization, Catalytic and Antibacterial Activity. *J. Nanostruct. Chem.* **2015**, *5*, 185–193. [CrossRef]
30. Winiarski, J.; Tylus, W.; Winiarska, K.; Szczygieł, I.; Szczygieł, B. Characterization of Selected Synthetic Corrosion Products of Zinc Expected in Neutral Environment Containing Chloride Ions. *J. Spectrosc.* **2018**. [CrossRef]
31. Kayani, Z.N.; Saleemi, F.; Batool, I. Synthesis and Characterization of ZnO Nanoparticles. *Mater. Today Proc.* **2015**, *2*, 5619–5621. [CrossRef]
32. Shi, S.; Xu, J.; Li, L. Preparation and Photocatalytic Activity of ZnO Nanorods and ZnO/Cu_2O Nanocomposites. *Main Group Chem.* **2017**, *16*, 47–55. [CrossRef]
33. Li, Z.; Zhang, Y.; Zhang, Y.; Chen, S.; Ren, Q. Facile Synthesis of ZnO/Ag Composites with the Enhanced Absorption and Degradation of MB and MO under Visible Light. *Ferroelectrics* **2018**, *526*, 152–160. [CrossRef]
34. Chikkanna, M.M.; Neelagund, S.E.; Rajashekarappa, K.K. Green Synthesis of Zinc Oxide Nanoparticles (ZnO NPs) and Their Biological Activity. *SN Appl. Sci.* 2019. [CrossRef]
35. Yang, H.Y.; Lee, H.; Kim, W. Effect of Zinc Nitrate Concentration on the Structural and the Optical Properties of ZnO Nanostructures. *Appl. Surf. Sci.* **2010**, *256*, 6117–6120. [CrossRef]
36. Alami, Z.Y.; Salem, M.; Gaidi, M. Effect of Zn Concentration On Structural and Optical Proprieties of ZNO Thin Films Deposited By Spray Pyrolysis. *Adv. Energy Int. J.* **2015**, *2*, 11–24. [CrossRef]

37. Rahman, Q.I.; Ahmad, M.; Misra, S.K.; Lohani, M. Effective Photocatalytic Degradation of Rhodamine B Dye by ZnO Nanoparticles. *Mater. Lett.* **2013**, *91*, 170–174. [CrossRef]
38. Al-Kordy, H.M.H.; Sabry, S.A.; Mabrouk, M.E.M. Statistical Optimization of Experimental Parameters for Extracellular Synthesis of Zinc Oxide Nanoparticles by a Novel Haloalaliphilic Alkalibacillus Sp.W7. *Sci. Rep.* **2021**, *11*, 1–14. [CrossRef] [PubMed]
39. Sharmila, G.; Muthukumaran, C.; Sandiya, K.; Santhiya, S.; Pradeep, R.S.; Kumar, N.M.; Suriyanarayanan, N.; Thirumarimurugan, M. Biosynthesis, Characterization, and Antibacterial Activity of Zinc Oxide Nanoparticles Derived from Bauhinia Tomentosa Leaf Extract. *J. Nanostruct. Chem.* **2018**, *8*, 293–299. [CrossRef]
40. Yang, R.D.; Tripathy, S.; Li, Y.; Sue, H.J. Photoluminescence and Micro-Raman Scattering in ZnO Nanoparticles: The Influence of Acetate Adsorption. *Chem. Phys. Lett.* **2005**, *411*, 150–154. [CrossRef]
41. Steffy, K.; Shanthi, G.; Maroky, A.S.; Selvakumar, S. Synthesis and Characterization of ZnO Phytonanocomposite Using Strychnos Nux-Vomica L. (Loganiaceae) and Antimicrobial Activity against Multidrug-Resistant Bacterial Strains from Diabetic Foot Ulcer. *J. Adv. Res.* **2018**, *9*, 69–77. [CrossRef]
42. Vasiljevic, Z.Z.; Dojcinovic, M.P.; Vujancevic, J.D.; Jankovic-Castvan, I.; Ognjanovic, M.; Tadic, N.B.; Stojadinovic, S.; Brankovic, G.O.; Nikolic, M.V. Research Photocatalytic Degradation of Methylene Blue under Natural Sunlight Using Iron Titanate Nanoparticles Prepared by a Modified Sol-Gel Method. *R. Soc. Open Sci.* **2020**. [CrossRef] [PubMed]
43. Kumar, S.; Ae, K.; Ae, N.K.; Singh, S. Photocatalytic Degradation of Two Commercial Reactive Dyes in Aqueous Phase Using Nanophotocatalysts. *Nanoscale Res. Lett.* **2009**. [CrossRef]
44. Chijioke-Okere, M.O.; Okorocha, N.J.; Anukam, B.N.; Oguzie, E.E. Photocatalytic Degradation of a Basic Dye Using Zinc Oxide Nanocatalyst. *Int. Lett. Chem. Phys. Astron.* **2019**, *81*, 18–26. [CrossRef]
45. Ram, C.; Pareek, R.K.; Singh, V. Photocatalytic Degradation of Textile Dye by Using Titanium Dioxide Nanocatalyst. *Int. J. Theor. Appl. Sci.* **2012**, *4*, 82–88.
46. Bansal, P.; Sud, D. Photodegradation of Commercial Dye, Procion Blue HERD from Real Textile Wastewater Using Nanocatalysts. *Desalination* **2011**, *267*, 244–249. [CrossRef]
47. Nadarajan, R.; Wan Abu Bakar, W.A.; Ali, R. Effect of Calcination Temperature on Metal Oxides and Their Photocatalytic Activity. *Adv. Mater. Res.* **2015**, *1107*, 73–78. [CrossRef]
48. Lv, K.; Xiang, Q.; Yu, J. Effect of Calcination Temperature on Morphology and Photocatalytic Activity. *Appl. Catal. B Environ.* **2011**, *104*, 275–281. [CrossRef]
49. Balcha, A.; Prakash Yadav, O.; Dey, T. Photocatalytic Degradation of Methylene Blue Dye by Zinc Oxide Nanoparticles Obtained from Precipitation and Sol-Gel Methods. *Environ. Sci. Pollut. Res.* **2016**. [CrossRef] [PubMed]
50. Isai, K.A.; Shrivastava, V.S. Photocatalytic Degradation of Methylene Blue Using ZnO and 2%Fe–ZnO Semiconductor Nanomaterials Synthesized by Sol–Gel Method: A Comparative Study. *SN Appl. Sci.* **2019**, *1*. [CrossRef]
51. Kahsay, M.H. Synthesis and Characterization of ZnO Nanoparticles Using Aqueous Extract of Becium Grandiflorum for Antimicrobial Activity and Adsorption of Methylene Blue. *Appl. Water Sci.* **2021**, *11*, 1–12. [CrossRef]
52. Ravichandran, V.; Sumitha, S.; Ning, C.Y.; Xian, O.Y.; Kiew Yu, U.; Paliwal, N.; Shah, S.A.A.; Tripathy, M. Durian Waste Mediated Green Synthesis of Zinc Oxide Nanoparticles and Evaluation of Their Antibacterial, Antioxidant, Cytotoxicity and Photocatalytic Activity. *Green Chem. Lett. Rev.* **2020**, *13*, 102–116. [CrossRef]
53. Khan, M.; Ware, P.; Shimpi, N. Synthesis of ZnO Nanoparticles Using Peels of Passiflora Foetida and Study of Its Activity as an Efficient Catalyst for the Degradation of Hazardous Organic Dye. *SN Appl. Sci.* **2021**, *3*. [CrossRef]
54. Lu, J.; Batjikh, I.; Hurh, J.; Han, Y.; Ali, H.; Mathiyalagan, R.; Ling, C.; Ahn, J.C.; Yang, D.C. Photocatalytic Degradation of Methylene Blue Using Biosynthesized Zinc Oxide Nanoparticles from Bark Extract of Kalopanax Septemlobus. *Optik* **2019**, *182*, 980–985. [CrossRef]
55. Alamdari, S.; Ghamsari, M.S.; Lee, C.; Han, W.; Park, H.; Tafreshi, M.J.; Afarideh, H. Preparation and Characterization of Zinc Oxide Nanoparticles Using Leaf Extract of Sambucus ebulus. *Appl. Sci.* **2020**, *10*, 3620. [CrossRef]

nanomaterials

Communication

Removal of Radioactive Iodine Using Silver/Iron Oxide Composite Nanoadsorbents

Mah Rukh Zia [1,†], Muhammad Asim Raza [2,3,†], Sang Hyun Park [2,3], Naseem Irfan [1], Rizwan Ahmed [1], Jung Eun Park [4], Jongho Jeon [4,*] and Sajid Mushtaq [1,2,3,*]

1 Department of Nuclear Engineering, Pakistan Institute of Engineering and Applied Sciences, P. O. Nilore, Islamabad 45650, Pakistan; Mahrukhzia14@gmail.com (M.R.Z.); naseem@pieas.edu.pk (N.I.); Rizwanahmed@pieas.edu.pk (R.A.)
2 Advanced Radiation Technology Institute, Korea Atomic Energy Research Institute, Jeongeup 56212, Korea; masimraza@ust.ac.kr (M.A.R.); parksh@kaeri.re.kr (S.H.P.)
3 Radiation Science and Technology, University of Science and Technology, Daejeon 34113, Korea
4 Department of Applied Chemistry, College of Engineering, Kyungpook National University, Daegu 41566, Korea; pje1204@knu.ac.kr
* Correspondence: jeonj@knu.ac.kr (J.J.); sajidmushtaq@pieas.edu.pk (S.M.); Tel.: +82-53-950-5584 (J.J.); +92-51-9248611-3716 (S.M.)
† These Authors contributed equally to this study.

Abstract: Efficient and cost-effective removal of radioactive iodine (radioiodine) from radioactive contaminated water has become a crucial task, following nuclear power plant disasters. Several materials for removing radioiodine have been reported in the literature. However, most of these materials exhibit some limitations, such as high production cost, slow adsorption kinetics, and poor adsorption capacity. Herein, we present silver/iron oxide nanocomposites (Ag/Fe_3O_4) for the efficient and specific removal of iodine anions from contaminated water. The Ag/Fe_3O_4 were synthesized using a modified method and characterized via scanning electron microscopy, transmission electron microscopy, and X-ray diffraction analyses. This adsorbent showed a high adsorption capacity for iodine anions (847 mg/g of the adsorbent) in pure water. Next, Ag/Fe_3O_4 was applied to the removal of radioiodine, and high removal efficiencies were observed in water. In addition, its desalination capacity was retained in the presence of competitive ions and varied pH. After the adsorption process, Ag/Fe_3O_4 was easily removed from the water by applying an external magnetic field. Moreover, the same operation can be repeated several times without a significant decrease in the performance of Ag/Fe_3O_4. Therefore, it is expected that the findings presented in this study will offer a new method for desalinating radioiodine in various aqueous media.

Keywords: adsorbents; radioactive wastes; radioactive iodine; desalination; nanocomposites

Citation: Zia, M.R.; Raza, M.A.; Park, S.H.; Irfan, N.; Ahmed, R.; Park, J.E.; Jeon, J.; Mushtaq, S. Removal of Radioactive Iodine Using Silver/Iron Oxide Composite Nanoadsorbents. *Nanomaterials* **2021**, *11*, 588. https://doi.org/10.3390/nano11030588

Academic Editor: Christos A. Aggelopoulos

Received: 8 February 2021
Accepted: 22 February 2021
Published: 26 February 2021

1. Introduction

The safe and reliable treatment of radioactive waste is inevitably linked to the safe production of nuclear energy [1,2]. Environmental damage caused by radioactive waste has attracted global attention. Radioisotopes, such as radioactive iodine (radioiodine), which exhibit a high degree of dispersion in water and air, are produced by nuclear fission. These can exert long-term adverse effects on human lives [3–5]. Notably, the global concern regarding nuclear waste leakage was kindled by the Fukushima accident in 2011 [6]. Further, the Chernobyl disaster in Ukraine occurred in 1986 wherein iodine radioisotopes were a major component of the radioactive contamination [7]. Moreover, the continuous operation of nuclear power plants can produce and introduce trace amounts of radioiodine into the environment [8]. Radioiodine has also been extensively used in the diagnosis of diseases and treatment of thyroid cancer on the basis of the selective uptake of iodine into the thyroid [9,10]. Consequently, the introduction of trace amounts of radioiodine from nuclear medicine research institutes also needs to be considered. For example, the medical applications of iodine-131 (^{131}I; half-life: 8 days) and iodine-129 (^{129}I; half-life: 15.7×10^6 years) are considered to be the main generators of radioiodine waste [11–13]. The short-lived as well as long-lived radioisotopes of iodine can accumulate and cause serious damage to the human body. Therefore, the efficient treatment of radioactive iodine in nuclear wastes and contaminated water is an essential area of research. In past decades, various adsorbents such as graphene-based sorbents [14,15], deep eutectic solvents [16,17], hydrogelators [18], nanoporous carbons [19,20], polyacrylonitrile–chalcogel [21], microporous polymers [22–24], metal–organic frameworks (MOFs) [25,26], and functionalized zeolites [27,28] were employed to remove radioiodine that was dissolved in solutions and/or gaseous radioiodine. However, these materials exhibited several drawbacks, including low removal efficiency, slow adsorption kinetics, and high production cost. Furthermore, layered bismuth–iodine–oxide [29], titanate nanolamina [30], Mg–Al(NO_3) layered double hydroxide (LDH) [31], and magnetite composites [32,33] have been employed to remove iodine. However, setbacks ranging from their poor reusability to their low adsorption capacities have limited the application of these methods. In previous studies, we reported that gold nanoparticles (AuNPs) immobilized adsorbents for the removal of radioiodine anions in aqueous media [34–37]. The method exhibited efficient and ion-selective desalination; however, the high cost of AuNPs-based systems hampered their large-scale syntheses and remediation applications.

Silver-based materials have also demonstrated a great potential for removing iodine owing to the high affinity of iodine toward silver [3,38,39]. In a typical procedure in the previous studies, silver nanoparticles or silver-based composite materials were immersed in the contaminated water to remove radioiodine. Thereafter, radioiodine containing solid waste was separated from the water via filtration or centrifugation. However, most of these methods require further steps to separate solid radioactive wastes from water after the desalination procedure. Moreover, the separation of nanosized adsorbents via these methods is time-consuming and non-applicable at an industrial scale. Thus, the development of additional cost-effective, efficient remediation procedures for radioactive wastes is still desired. Here, we designed a stable and efficient silver/iron oxide (Ag/Fe_3O_4) nanocomposite-based desalination system for the efficient removal of radioiodine from water. Compared with nonmagnetic silver composites, the magnetic nature of the Ag/Fe_3O_4 adsorbent is advantageous as it can be harvested by an external magnetic field. Therefore, the desalination procedure and recovery of radioisotope-containing adsorbents from treated water can be easy, rapid, and cost-effective (Figure 1).

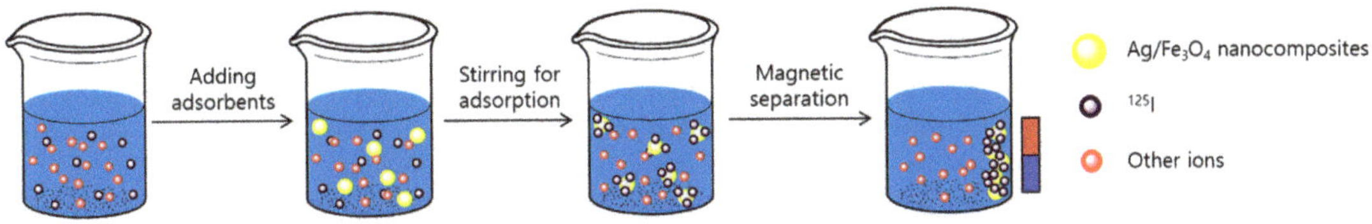

Figure 1. Schematic illustration of the desalination process using Ag/Fe_3O_4 composite nanoadsorbents.

2. Experimental Procedures

2.1. Materials

Radioiodine ([^{125}I]NaI) was supplied by PerkinElmer in an aqueous sodium hydroxide (NaOH) solution. All the chemicals, such as iron(III)chloride hexahydrate ($FeCl_3{\cdot}6H_2O$), iron(II)chloride tetrahydrate ($FeCl_2{\cdot}4H_2O$), silver nitrate ($AgNO_3$), NaOH, ethanol (C_2H_5OH), (3-aminopropyl)trimethoxysilane ($H_2N(CH_2)_3Si(OCH_3)_3$, APTMS), hydroxylamine hydrochloride ($NH_2OH{\cdot}HCl$), sodium iodide (NaI), hydrochloric acid (HCl), sodium chloride (NaCl), sodium bromide (NaBr), sodium fluoride (NaF), sodium chlorate ($NaClO_3$), sodium bromate ($NaBrCO_3$), and potassium iodide (KI), were purchased from Sigma Aldrich Korea and utilized without further purification. The radioiodine removal experiments were performed using a radio-thin-layer chromatograph (TLC, AR-2000, Bioscan, USA) that was equipped with a dose calibrator (CRC-25PET) or automatic gamma counter (2480 automatic gamma counter, PerkinElmer, UK)Further, the nonradioactive iodine removal experiments were performed via ultraviolet–visible (UV–Vis) spectroscopy (UV–Vis spectrophotometer, Evolution™ 201/220, Thermo Scientific™, USA). The Fe_3O_4 and Ag/Fe_3O_4 composite nanoparticles were characterized via transmission electron microscopy (TEM; H-7650, Hitachi, Japan), field-emission scanning electron microscopy (FE-SEM; FEI Verios 460L, Philips, USA), and X-ray diffraction (XRD, Bruker, D2 PHASER). Magnetization measurements of nanomaterials were performed by a vibrating sample magnetometer (VSM JDM-13, Lake Shore, USA) at room temperature.

2.2. Synthesis of the Fe_3O_4 Nanoparticles

The Fe_3O_4 nanoparticles were prepared using a slightly modified, coprecipitation method [38]. Briefly, $FeCl_2{\cdot}4H_2O$ (1.99 g, 0.01 mol) and $FeCl_3{\cdot}6H_2O$ (5.41 g, 0.02 mol) were dissolved in water and sonicated for 30 min. The $FeCl_2{\cdot}4H_2O$ and $FeCl_3{\cdot}6H_2O$ solutions were mixed and placed in a three-neck bottle. The resulting solution was heated at 90 °C in a nitrogen stream, after which it was vigorously mixed in a deoxygenated atmosphere. Next, the aqueous NaOH solution was added dropwise, and the Fe_3O_4 nanoparticles were obtained as dark-brown precipitates. The precipitated mixture was stirred for an additional 6 h to achieve complete conversion. The Fe_3O_4 nanoparticles were extracted from the solution by applying an external magnetic field, washed several times with water and ethanol, and vacuum-dried for 2 h at 80 °C.

2.3. Synthesis of the Ag/Fe_3O_4 Composite Nanoadsorbents

The Ag/Fe_3O_4 composite materials were prepared via APTMS [40]. Briefly, the Fe_3O_4 nanoparticles (0.30 g) were dispersed in 150 mL of ethanol and sonicated for 45 min. Next, APTMS (1 mL) was injected into the reaction mixture and stirred for 10 h. The APTMS-functionalized Fe_3O_4 nanoparticles were extracted by applying an external magnetic field, washed several times with ethanol, and vacuum-dried for 2 h at 80 °C. Further, the APTMS-functionalized Fe_3O_4 nanoparticles (0.25 g) were dispersed in 100 mL of water, after which $AgNO_3$ (0.30 wt.%) was added to the reaction mixture, followed by sonication for 1 h. For the preparation of the silver nanoparticles, NaOH (50 mL, 0.1 M) and hydroxylamine hydrochloride (45 mL, 0.05 M) were added to the reaction mixture and stirred for an additional 2 h. Finally, the Ag/Fe_3O_4 composites were extracted by applying a magnetic field, washed several times with water and ethanol, and vacuum-dried at 80 °C.

2.4. Adsorption of Nonradioactive Iodine($^{127}I^-$) Using Ag/Fe_3O_4 Composite Nanoadsorbents

Adsorption efficiency of the Ag/Fe_3O_4 nanocomposites was determined by measuring the absorbance variation of nonradioactive NaI/KI via UV–Vis spectroscopy at a maximum wavelength, λ_{max} = 225 nm. Briefly, 100 ppm stock solution was prepared by dissolving KI in water, and the pH was maintained at 7. Further, the desired low concentrations were prepared by diluting the stock solution. In the adsorption experiment, the adsorbent, the Ag/Fe_3O_4 composites, were shaken with an aqueous KI solution of a given concentration at a different time interval. After the experiment, the adsorbent was removed by an external magnet. The iodine concentration of the treated solution was measured via UV–Vis spectroscopy.

The percentage removal efficiency of Ag/Fe_3O_4 nanocomposites was measured using Equation (1):

$$\text{Removal efficiency } (\%) = \frac{(C_0 - C_e)}{C_0} \times 100 \quad (1)$$

Equilibrium adsorption capacity of Ag/Fe_3O_4 nanocomposites, Q_e (mg/g), was determined using Equation (2):

$$Q_e = \frac{(C_0 - C_e)}{M} \times V \quad (2)$$

where Q_e (mg/g) is the quantity of I^- that was adsorbed on the adsorbent at equilibrium time, C_0 (mg/L) is the initial concentration of I^- in the aqueous solution, C_e (mg/L) is the final concentration of I^- in the aqueous solution at time t, V (L) is the volume of the solution, and M (g) represents the mass of the adsorbents (Ag/Fe_3O_4).

2.5. *Determination of Removal Efficiency in the Presence of Competitive Ions*

The removal efficiency of Ag/Fe_3O_4 nanocomposites was investigated in the presence of competitive ions. Radioiodine, $[^{125}I]NaI$ (150 μCi), was diluted in an aqueous solution of NaCl, NaBr, NaF, $NaClO_3$, $NaBrCO_3$, or nonradioactive NaI (10 mL, 1.0 M). The Ag/Fe_3O_4 nanoparticles were stirred with the $[^{125}I]NaI$ solution of given radioactivity for 60 min. Next, the adsorbent was removed by an external magnet. The radioactivities of the supernatant and adsorbent material were measured using the radio-TLC system or dose calibrator.

2.6. *Removal Efficiency in Different Aqueous Media*

To investigate the removal efficiency of the adsorbents, $[^{125}I]NaI$ (150 μCi) was diluted in 10 mL of different aqueous media (pure water, 1× PBS, water at 80 °C, river water, 0.1 M NaOH, 0.1 M HCl, or 1.0 M NaI). Ag/Fe_3O_4 nanocomposites were stirred with the $[^{125}I]NaI$ solution of given radioactivity for 60 min. Subsequently, the adsorbent was removed by an external magnet. The radioactivities of the supernatant and adsorbent materials were measured using the radio-TLC system or dose calibrator.

2.7. *Reusability of the Composite Nanoadsorbents*

To investigate the reusability of Ag/Fe_3O_4 nanocomposites, $[^{125}I]NaI$ (150 μCi) was diluted in 10 mL of pure water. Ag/Fe_3O_4 (10 mg) was shaken with the $[^{125}I]NaI$ solution of given radioactivity for 60 min. Subsequently, the adsorbent was removed by an external magnet. The radioactivity in the supernatant and adsorbent materials was measured using a gamma counter. The experiment was repeated for up to seven cycles.

2.8. *Adsorption Isotherm Studies*

The adsorption isotherm was measured using the KI solution at an ambient temperature and neutral pH. Briefly, 5 mg of the Ag/Fe_3O_4 nanoparticles was treated with 100 mL of KI at different initial concentrations (100–200 ppm) with a constant increment of 10 ppm. The final concentration of iodine after adsorption procedure was determined via UV–Vis spectroscopy at different intervals. The adsorption of I^- (Q_e) was calculated using Equation (2). The Langmuir and Freundlich isotherm models were applied to describe the equilibrium adsorption:

$$\text{Langmuir equation}: \frac{C_e}{Q_e} = \frac{C_e}{Q_{max}} + \frac{1}{Q_{max}\,K_L} \quad (3)$$

$$\text{Freundlich equation}: \ln Q_e = \ln K_F + \frac{1}{n}\ln C_e \quad (4)$$

where C_e (mg/L) are concentrations of I^- at the initial and equilibrium times, respectively. Q_e (mg/g) is the quantity of I^- that was adsorbed on the adsorbing medium at the equilibrium time, and Q_{max} (mg/g) is the maximum adsorption capacity of the adsorbent. K_L and K_F are the Langmuir and Freundlich adsorption constants, respectively.

2.9. *Adsorption Kinetics of I^- on the Adsorbents*

The adsorption kinetics of I^- on Ag/Fe_3O_4 nanoparticles was determined using 100 ppm KI at pH 7 and room temperature. Briefly, 100 mL of KI (100 ppm) solution was shaken with 5 mg of Ag/Fe_3O_4 nanocomposites. At different times, the absorbent was separated from the solution by applying an external magnet and the concentration of I^- was determined via UV–Vis spectroscopy by measuring the absorbance variation at the maximum wavelength, λ_{max} = 225 nm. The adsorption capacity was fitted into the pseudo-first-order (PFO) and pseudo-second-order (PSO) kinetics equations with respect to time, as expressed in Equations (3) and (4), respectively.

$$\text{PFO kinetic model}: \ln(Q_e - Q_t) = \ln Q_e - \frac{k_1 t}{2.303} \quad (5)$$

$$\text{PSO kinetic model}: \frac{t}{Q_t} = \frac{1}{k_2 Q_e^2} + \frac{t}{Q_e} \quad (6)$$

where Q_e and Q_t are the quantities of I^- (mg/g) at equilibrium and time t, respectively. The Fe_3O_4 nanoparticles were used in the control experiment under similar conditions. k_1 (min^{-1}) and k_2 (g mg^{-1} min^{-1}) are the PFO and PSO adsorption rate constants, respectively.

3. Results and Discussion

The adsorbent, Ag/Fe_3O_4 nanoparticles, were synthesized in two steps using a modified procedure, as shown in Figure S1a. To prepare Fe_3O_4 nanoparticles, a mixture of $FeCl_2 \cdot 4H_2O$ and $FeCl_3 \cdot 6H_2O$ was treated with NaOH at 90 °C, and the product was washed several times with water and ethanol, after which it was dried at a high temperature. In the next step, APTMS was coated on the surface of the Fe_3O_4 nanoparticles. Further, the silver layer was formed using $AgNO_3$ in the presence of hydroxylamine hydrochloride and a base.

The particle morphology and size of the prepared nanoparticles were observed via SEM and TEM, respectively. SEM images of the bare Fe_3O_4 exhibited the nearly spherical shapes of the particles (Figure S2a,b), and a significant uniform particle-size distribution was observed. The observed average size of the Fe_3O_4 particle was ~27 nm (Figure S2c–e). The crystal structure and phase of the prepared nanoparticles were determined via X-ray diffraction (XRD). The strong Bragg peaks of (220), (311), (400), (422), (511), and (440) corresponded to the diffractions from the inverse spinel structure of Fe_3O_4 (Figure S2f). The energy-dispersive X-ray spectroscopy (EDS) analysis of the Fe_3O_4 nanoparticles revealed a set of peaks, which corresponded to iron as well as oxygen (Figure S3). The characterization data of Ag/Fe_3O_4 are shown in Figure 2. The TEM and SEM images of Ag/Fe_3O_4 nanocomposites showed the nearly spherical-shaped agglomerates, respectively (Figure 2a,b). The additional TEM data are shown in Figure S4. The elemental mapping patterns revealed the presence of the main elements, including O, Fe, Si, and Ag (Figure 2c–f). A new peak, which corresponded to silver, is evident compared with the Fe_3O_4 nanoparticles, thus confirming the presence of the silver coating on the Fe_3O_4 nanoparticles. In addition, the EDS data of the Ag/Fe_3O_4 nanoparticles revealed the presence of Fe, O, C, Si, and Ag atoms (Figure 3a). The presence of Si was observed after the surface modification of the Fe_3O_4 particles via APTMS. The observed particle size was 35.9 nm with a standard deviation of 2.2 nm (Figure 3b). The size of nanomaterials was further analyzed by the Scherrer equation (Tables S1 and S2). The XRD peaks revealed the crystalline nature of the nanoparticles. The peaks indicated the ultrafine nature and small crystallite size of the nanoparticles. The strong Bragg peaks of (111), (200), (220), and (311) corresponded to the diffractions from the FCC (Face-centered Cubic) structure of silver nanoparticles. The absence of spurious diffractions indicated the absence of significant impurities in the sample (Figure 3c). The magnetic properties of the nanoadsorbent were assessed by applying a magnetic field in the range of −30,000 to 30,000 Oe via vibrational sample magnetometry. The bare Fe_3O_4 nanoparticles exhibited the highest saturation magnetization (M_s) value (67.84 emu/g). However, the value decreased to 49.48 and 40.34 emu/g because of the APTMS coating and combination of silver with the APTMS coating, respectively (Figure 3d). The surface modification and formation of the silver nanoparticles on the surface of the magnetite nanoparticles caused a decrease in the M_s value. This result can be attributed to the presence of more diamagnetic material per gram of the material. Zero remanence and coercivity of the magnetization curve suggested that the nanoadsorbents possessed superparamagnetic properties. As shown in Figure S1c–e, Ag/Fe_3O_4 responded immediately to the external magnetic field, and the collected particles could be dispersed again by gentle shaking after removing the magnet. This result indicates that the adsorbents can be easily removed from wastewater via a simple separation procedure.

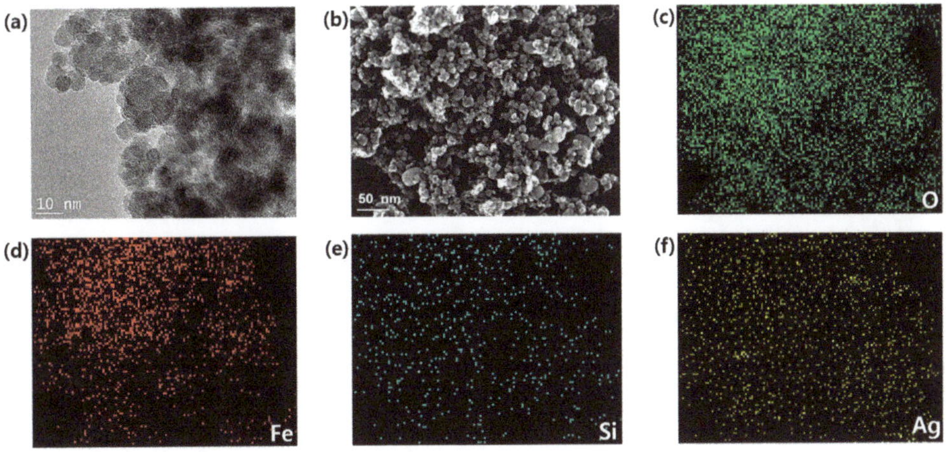

Figure 2. (**a**) TEM images of the Ag/Fe_3O_4 composite nanoparticles, (**b**) SEM image of the Ag/Fe_3O_4 composite nanoparticles, (**c–f**) EDS elemental mapping patterns of the Ag/Fe_3O_4 nanocomposites from (**b**).

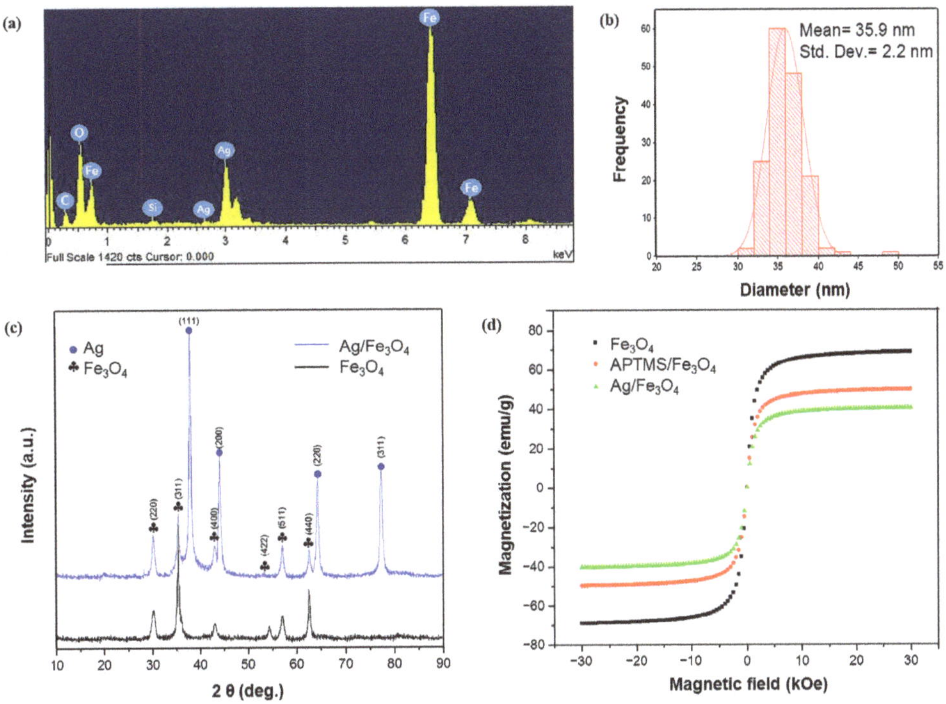

Figure 3. Characterization of the nanomaterials. (**a**) EDS spectrum of the Ag/Fe_3O_4 nanocomposites; (**b**) size-distribution histogram of the Ag/Fe_3O_4 with a standard deviation of 2.2 nm; (**c**) XRD analysis of the Fe_3O_4 and Ag/Fe_3O_4; (**d**) magnetic hysteresis loops of Fe_3O_4, APTMS/Fe_3O_4, and Ag/Fe_3O_4 nanocomposites at room temperature.

To perform the iodine adsorption experiments, nonradioactive iodide anion (I^-) or radioiodine ($^{125}I^-$) were used. The calibration curve was plotted to determine the concentrations of I^- (λ_{max} = 225 nm) in aqueous media (Figure S5) via UV–Vis spectroscopy. First, the adsorbent was added to a 100 ppm KI solution in pure water or a NaCl solution, after which the amount of I^- absorbed on Ag/Fe_3O_4 was determined by comparing the UV absorbance at 225 nm for 1 h. Figure 4a reveals that I^- in aqueous media could not be captured by the unmodified Fe_3O_4. Conversely, silver-coated adsorbents efficiently removed I^- with a removal efficiency of ~100% in 1 h (Figure 4b). Interestingly, the excellent removal efficiency was also observed using 1.0 M NaCl solution with a Cl^- to I^- anion ratio ([Cl^-]:[I^-]) exceeding 10^3:1, thereby indicating the ion-selective adsorption performance of the silver layer.

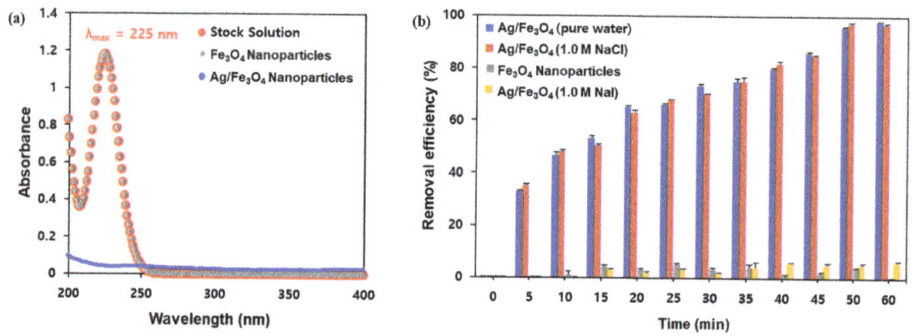

Figure 4. (**a**) UV absorption spectra of iodine after treatment with Fe_3O_4 or Ag/Fe_3O_4 composite nanoparticles in water; (**b**) removal efficiency of Ag/Fe_3O_4 nanocomposites at high salt concentrations, and the control study using Fe_3O_4 nanoparticles.

Linear fitting of the observed data according to the Langmuir (3) and Freundlich isotherm models (4) revealed that the adsorption mechanism of Ag/Fe_3O_4 was better described by the Langmuir equation with a correlation factor (R^2) of 0.995 (Figure 5a). This result also indicated the monolayer adsorption mechanism, and the observed maximum adsorption capacity (Q_{max}) obtained using Equation (3) was 847 mg/g. The corresponding parameters for these models are summarized in Table 1. The kinetic parameters of the adsorption are also pivotal to the practical application of the nanoadsorbent. The removal efficiency of I^- was measured as a function of time (5–180 min) to determine the optimum time for the desalination experiments (Figure 5b). The adsorption of I^- was rapid in the first 60 min, after which it became slower, before finally reaching a plateau after 180 min. The fitting results of the PFO and PSO kinetic models are shown in Figure S6a,b and Table 2. Based on the calculated kinetic parameters, it is clear that the PSO kinetic model fitted better with the kinetic results.

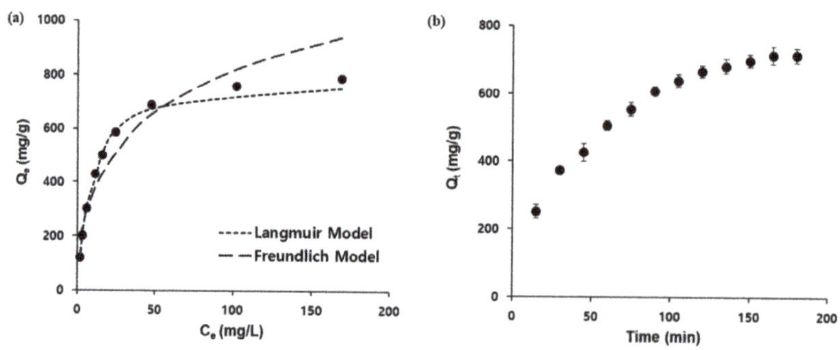

Figure 5. (**a**) Adsorption isotherm study employing the Langmuir and Freundlich models, (**b**) absorption kinetics as a function of time.

Table 1. Calculated parameters for the adsorption isotherm fittings for the Ag/Fe_3O_4 nanocomposites. (R^2 = coefficient of determination)

Isotherm Type	Parameters	R^2
Langmuir Model	Q_{max} = 846.860 mg/g K_L = 0.09162 L/mg	0.995
Freundlich Model	n = 2.6 $K_F = 148.4\ L^{-1/n}\ mg^{(1-1/n)}$	0.883

Table 2. Calculated parameters for fitting the chemical kinetics of the Ag/Fe_3O_4 nanocomposites.

Type of Chemical Kinetics	Parameters	R^2
First-order	Q_e = 613 mg/g $k_1 = 9.0 \times 10^{-3}\ min^{-1}$	0.976
Second-order	Q_e = 897 mg/g $k_2 = 6.03 \times 10^{-8}\ gmg^{-1}min^{-1}$	0.996

Using these results, we investigated the desalination of radioiodine by Ag/Fe_3O_4. For this study, the same adsorbents were immerged into aqueous solutions containing 150 μCi of [^{125}I]NaI. Afterward, the adsorbents were collected by an external magnet. The removal efficiency was determined by measuring the residual radioactivities in the solution and Ag/Fe_3O_4 nanocomposite. As shown in Figure 6a, high removal efficiencies were observed in the presence of other competing anions as well (e.g., Cl^-, Br^-, and phosphate). By contrast, the adsorption of radioiodine was completely inhibited in an aqueous solution of nonradioactive NaI, which might be due to the surface area of the adsorbents getting covered by an excess amount of I^-. The desalination performance was also evaluated in other environments, such as varied pH values, river waters, and elevated temperatures. More than 99% of the radioactive iodines were captured by Ag/Fe_3O_4 in 1 h in these environments as well. These results demonstrated that Ag/Fe_3O_4 successfully and selectively captures I^- in the presence of mixed ion species. Next, the reusability of Ag/Fe_3O_4 was explored via the repetitive adsorption of radioiodine (150 μCi) from water. As shown in Figure 6b, a high removal efficiency (>99%) was observed in seven consecutive processes, suggesting that the adsorbent retained its stability. Moreover, the adsorbed radioiodine anions on the adsorbents were not readily released during the repeated operations.

Several studies have described various silver metal or silver oxide composite nanomaterials that can remove radioactive iodines [3,41]. Removal of unsettled adsorbents after water treatment requires further separation processes. Compared with the previous studies, this method offers a simpler and more efficient method for capturing radioactive materials from different aqueous solutions. Through a single operation for 1 h, most of the radioactive components were selectively captured by Ag/Fe_3O_4. Thereafter, the magnetic separation successfully recovered the I^--loaded adsorbent from the aqueous solvent. Additionally, the observed Q_{max} value was favorable compared with those obtained in previous studies (Table S3). As a result of the large-scale synthesis and characterization of Fe_3O_4 as well as the establishment of the formation of the silver layer, a large number of adsorbents that were used in this study can be easily prepared in a short time. These advantages strongly demonstrate that desalination using Ag/Fe_3O_4 would be beneficial for the efficient treatment of radioiodine waste. To date, different silver-coated magnetic nanomaterials have been employed in a wide range of applications for specific purposes, including catalysis [42], antibacterial agents [43], imaging [44], and biosensing [45], because of their unique physical and chemical properties. This study would widen the scope of engineered nanomaterials in the field of environmental remediation. However, further optimization and validation of the process is required to investigate the industrial-scale remediation process of radioactive waste.

(a)

(b)

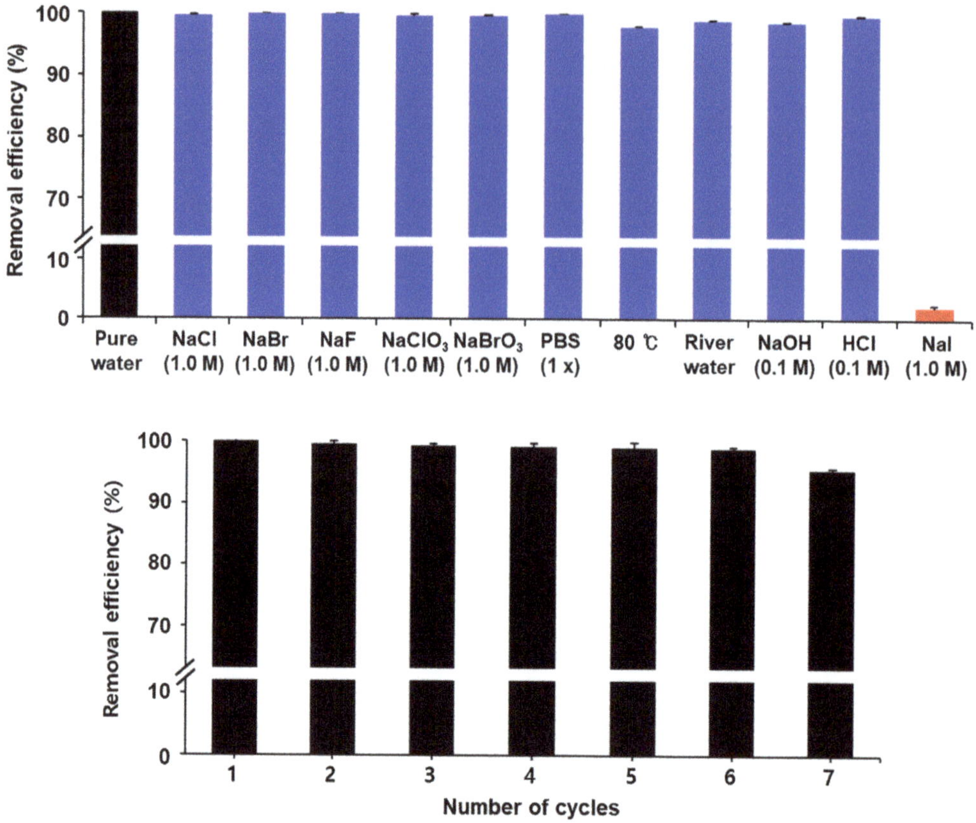

Figure 6. (**a**) Removal efficiency of the Ag/Fe_3O_4 nanocomposites in the presence of competitive ions, different aqueous solutions, and different pH conditions using radioiodine (^{125}I); (**b**) removal efficiency of the Ag/Fe_3O_4 nanocomposites using radioiodine (^{125}I) as a function of the number cycles.

4. Conclusions

In this study, we synthesized an Ag/Fe_3O_4 nanoadsorbent for the desalination of radioiodine. The synthesized nanomaterials were characterized via SEM, TEM, EDS, and XRD. The composite material exhibited high adsorption capacity for I^- (847 mg/g) in water. The Ag/Fe_3O_4 composite nanoadsorbents exhibited high removal efficiency as well as ion-selective desalination in the presence of several competing ions. The material was easily recovered from the treated water by applying an external magnetic field without the significant desorption of radioactivity. Moreover, the adsorbent maintained good desalination performance during seven consecutive remediations. Consequently, it is expected that Ag/Fe_3O_4-based desalination will present a promising direction and can be developed as a practical method for wastewater treatment.

Supplementary Materials: The following are available online at https://www.mdpi.com/2079-4991/11/3/588/s1, Figure S1: Figure S1. (a) Schematic route for the synthesis of Fe_3O_4 and Ag/Fe_3O_4 nanocomposites, (b) Experimental setup for the synthesis of nanoparticles and (c-e) Steps to collect Ag/Fe_3O_4 nanocomposites by using an external magnet; Figure S2: Figure S2. (a,b) SEM images of Fe_3O_4 nanoparticles, (c,d) TEM images of Fe_3O_4 nanoparticles, (e) Size distribution histogram of Fe_3O_4 nanoparticles with a standard deviation of 1.15 nm, (f) Powder XRD analysis of Fe_3O_4 nanoparticles.; Figure S3: Figure S3. EDS analysis of iron oxide nanoparticles; Figure S4: Figure S4. TEM images of Ag/Fe_3O_4 nanocomposite; Figure S5: Calibration curve to determine the unknown concentration using UV-Visible Spectrometer at 226 nm; Figure S6: (a) Pseudo-second-order kinetics study for Ag/Fe_3O_4, (b) Pseudo-first-order kinetics study for Ag/Fe_3O_4; Table. S1. Scherrer equation based crystallite size Fe_3O_4 nanoparticles; Table. S2. Scherrer equation based crystallite size Ag/Fe_3O_4 composite nanoparticles; Table. S3. Nanomaterials used for iodine removal from aqueous solutions.

Author Contributions: Conceptualization, S.M. and J.J.; methodology, S.M.; formal analysis, J.E.P.; investigation, M.R.Z., M.A.R., N.I., and R.A.; resources, S.H.P.; writing—original draft preparation, S.M.; writing—review and editing, S.M., J.E.P., and J.J.; supervision, S.M.; project administration, S.M.; funding acquisition, J.J. All authors have read and agreed to the published version of the manuscript.

Funding: This work was supported by the Korea Institute of Energy Technology Evaluation and Planning (KETEP) and the Ministry of Trade, Industry and Energy (MOTIE) of the Republic of Korea (No. 20191510301310).

Data Availability Statement: The data presented in this article are available on request from the corresponding authors.

Conflicts of Interest: The authors declare no conflict of interest.

References

1. Chapman, K.W.; Chupas, P.J.; Nenoff, T.M. Radioactive iodine capture in silver-containing mordenites through nanoscale silver iodide formation. *J. Am. Chem. Soc.* **2010**, *132*, 8897–8899. [CrossRef]
2. Riley, B.J.; Vienna, J.D.; Strachan, D.M.; McCloy, J.S.; Jerden, J.L., Jr. Materials and processes for the effective capture and immobilization of radioiodine: A review. *J. Nucl. Mater.* **2016**, *470*, 307–326. [CrossRef]
3. Huve, J.; Ryzhikov, A.; Nouali, H.; Lalia, V.; Augé, G.; Daou, T.J. Porous sorbents for the capture of radioactive iodine compounds: A review. *RSC Adv.* **2018**, *8*, 29248–29273. [CrossRef]
4. Liu, Y.; Gu, P.; Yang, Y.; Jia, L.; Zhang, M.; Zhang, G. Removal of radioactive iodide from simulated liquid waste in an integrated precipitation reactor and membrane separator (PR-MS) system. *Sep. Purif. Technol.* **2016**, *171*, 221–228. [CrossRef]
5. Choung, S.; Um, W.; Kim, M.; Kim, M.G. Uptake mechanism for iodine species to black carbon. *Environ. Sci. Technol.* **2013**, *47*, 10349–10355. [CrossRef]
6. Tsuruta, H.; Moriguchi, Y.; Nakajima, T. Dynamics of atmospheric 131 I in radioactive plumes in eastern Japan immediately after the Fukushima accident by analysing published data. *Sci. Rep.* **2019**, *9*, 1–15.
7. Ernst, T.; Rinke, J.; Hagen, J.; Dmytrenko, I.; Hochhaus, A.; Dyagil, I. Molecular-defined clonal evolution in patients with chronic myeloid leukemia who were exposed to ionizing radiation following the Chernobyl nuclear disaster. *Leukemia* **2020**, *34*, 645–650. [CrossRef] [PubMed]
8. Nandanwar, S.U.; Coldsnow, K.; Utgikar, V.; Sabharwall, P.; Aston, D.E. Capture of harmful radioactive contaminants from off-gas stream using porous solid sorbents for clean environment—A review. *Chem. Eng. J.* **2016**, *306*, 369–381. [CrossRef]
9. Shobab, L.; Gomes-Lima, C.; Zeymo, A.; Feldman, R.; Jonklaas, J.; Wartofsky, L.; Burman, K.D. Clinical, pathological, and molecular profiling of radioactive iodine refractory differentiated thyroid cancer. *Thyroid* **2019**, *29*, 1262–1268. [CrossRef]
10. Wallner, L.P.; Reyes-Gastelum, D.; Hamilton, A.S.; Ward, K.C.; Hawley, S.T.; Haymart, M.R. Patient-perceived lack of choice in receipt of radioactive iodine for treatment of differentiated thyroid cancer. *J. Clin. Oncol.* **2019**, *37*, 2152–2161. [CrossRef]
11. Liu, B.; Ren, X.; Chen, L.; Ma, X.; Chen, Q.; Sun, Q.; Lin, Z.; Pengchao, S.; Lijie, C. High efficient adsorption and storage of iodine on S, N co-doped graphene aerogel. *J. Hazard. Mater.* **2019**, *373*, 705–715. [CrossRef] [PubMed]
12. Guo, B.; Xiong, Y.; Chen, W.; Saslow, S.A.; Kozai, N.; Ohnuki, T.; Dabo, I.; Sasaki, K. Spectroscopic and first-principles investigations of iodine species incorporation into ettringite: Implications for iodine migration in cement waste forms. *J. Hazard. Mater.* **2020**, *389*, 121880. [CrossRef]
13. Attallah, M.F.; Rizk, S.E.; El Afifi, E.M. Efficient removal of iodine and chromium as anionic species from radioactive liquid waste using prepared iron oxide nanofibers. *J. Radioanal. Nucl.* **2018**, *317*, 933–945. [CrossRef]
14. Scott, S.M.; Hu, T.; Yao, T.; Xin, G.; Lian, J. Graphene-based sorbents for iodine-129 capture and sequestration. *Carbon* **2015**, *90*, 1–8. [CrossRef]
15. Sun, H.; Mu, P.; Xie, H.; Zhu, Z.; Liang, W.; Zhou, Z.; Li, A. Efficient capture and reversible storage of radioactive iodine by porous graphene with high uptake. *Chem. Sel.* **2018**, *3*, 10147–10152. [CrossRef]
16. Li, G.; Yan, C.; Cao, B.; Jiang, J.; Zhao, W.; Wang, J.; Mu, T. Highly efficient I_2 capture by simple and low-cost deep eutectic solvents. *Green Chem.* **2016**, *18*, 2522–2527. [CrossRef]
17. Li, R.; Zhao, Y.; Chen, Y.; Liu, Z.; Han, B.; Li, Z.; Wang, J. Imidazolate ionic liquids for high-capacity capture and reliable storage of iodine. *Commun. Chem.* **2018**, *1*, 1–8. [CrossRef]
18. Sarkar, S.; Dutta, S.; Ray, C.; Dutta, B.; Chowdhury, J.; Pal, T. A two-component hydrogelator from citrazinic acid and melamine: Synthesis, intriguing role of reaction parameters and iodine adsorption study. *Cryst. Eng. Comm.* **2015**, *17*, 8119–8129. [CrossRef]
19. Muhammad, R.; Attia, N.F.; Cho, S.; Park, J.; Jung, M.; Chung, J.; Oh, H. Exploitation of surface heterogeneity and textural properties in nanoporous carbon fabrics for efficient iodine capture. *Thin Solid Films* **2020**, *706*, 138049. [CrossRef]

20. Sun, H.; La, P.; Yang, R.; Zhu, Z.; Liang, W.; Yang, B.; Deng, W. Innovative nanoporous carbons with ultrahigh uptakes for capture and reversible storage of CO_2 and volatile iodine. *J. Hazard. Mater.* **2017**, *321*, 210–217. [CrossRef]
21. Riley, B.J.; Pierce, D.A.; Chun, J.; Matyas, J.; Lepry, W.C.; Garn, T.G.; Kanatzidis, M.G. Polyacrylonitrile-chalcogel hybrid sorbents for radioiodine capture. *Environ. Sci. Technol.* **2014**, *48*, 5832–5839. [CrossRef] [PubMed]
22. Qian, X.; Zhu, Z.Q.; Sun, H.X.; Ren, F.; Mu, P.; Liang, W.; Li, A. Capture and reversible storage of volatile iodine by novel conjugated microporous polymers containing thiophene units. *ACS Appl. Mater. Interfaces* **2016**, *8*, 21063–21069. [CrossRef] [PubMed]
23. Liao, Y.; Weber, J.; Mills, B.M.; Ren, Z.; Faul, C.F. Highly efficient and reversible iodine capture in hexaphenylbenzene-based conjugated microporous polymers. *Macromolecules* **2016**, *49*, 6322–6333. [CrossRef]
24. Sigen, A.; Zhang, Y.; Li, Z.; Xia, H.; Xue, M.; Liu, X.; Mu, Y. Highly efficient and reversible iodine capture using a metalloporphyrin-based conjugated microporous polymer. *Chem. Commun.* **2014**, *50*, 8495–8498.
25. Sava, D.F.; Chapman, K.W.; Rodriguez, M.A.; Greathouse, J.A.; Crozier, P.S.; Zhao, H.; Nenoff, T.M. Competitive I_2 sorption by Cu-BTC from humid gas streams. *Chem. Mater.* **2013**, *25*, 2591–2596. [CrossRef]
26. Katsoulidis, A.P.; He, J.; Kanatzidis, M.G. Functional monolithic polymeric organic framework aerogel as reducing and hosting media for Ag nanoparticles and application in capturing of iodine vapors. *Chem. Mater.* **2012**, *24*, 1937–1943. [CrossRef]
27. Lee, Y.R.; Do, X.H.; Cho, K.Y.; Jeong, K.; Baek, K.Y. Amine-Functionalized Zeolitic Imidazolate Framework-8 (ZIF-8) Nanocrystals for Adsorption of Radioactive Iodine. *ACS Appl. Nano Mater.* **2020**, *3*, 9852–9861. [CrossRef]
28. Inglezakis, V.J.; Satayeva, A.; Yagofarova, A.; Tauanov, Z.; Meiramkulova, K.; Farrando-Pérez, J.; Bear, J.C. Surface Interactions and Mechanisms Study on the Removal of Iodide from Water by Use of Natural Zeolite-Based Silver Nanocomposites. *Nanomaterials* **2020**, *10*, 1156. [CrossRef] [PubMed]
29. Krumhansl, J.L.; Nenoff, T.M. Hydrotalcite-like layered bismuth–iodine–oxides as waste forms. *J. Appl. Geochem.* **2011**, *26*, 57–64. [CrossRef]
30. Bo, A.; Sarina, S.; Zheng, Z.; Yang, D.; Liu, H.; Zhu, H. Removal of radioactive iodine from water using Ag_2O grafted titanate nanolamina as efficient adsorbent. *J. Hazard. Mater.* **2013**, *246*, 199–205. [CrossRef]
31. Kentjono, L.; Liu, J.C.; Chang, W.C.; Irawan, C. Removal of boron and iodine from optoelectronic wastewater using Mg–Al (NO_3) layered double hydroxide. *Desalination* **2010**, *262*, 280–283. [CrossRef]
32. Li, J.; Dou, X.; Qin, H.; Sun, Y.; Yin, D.; Guan, X. Characterization methods of zerovalent iron for water treatment and remediation. *Water Res.* **2019**, *148*, 70–85. [CrossRef] [PubMed]
33. Madrakian, T.; Afkhami, A.; Zolfigol, M.A.; Ahmadi, M.; Koukabi, N. Application of modified silica coated magnetite nanoparticles for removal of iodine from water samples. *Nano-Micro Lett.* **2012**, *4*, 57–63. [CrossRef]
34. Choi, M.H.; Shim, H.E.; Yun, S.J.; Park, S.H.; Choi, D.S.; Jang, B.S.; Jeon, J. Gold-nanoparticle-immobilized desalting columns for highly efficient and specific removal of radioactive iodine in aqueous media. *ACS Appl. Mater. Interfaces* **2016**, *8*, 29227–29231. [CrossRef] [PubMed]
35. Mushtaq, S.; Yun, S.J.; Yang, J.E.; Jeong, S.W.; Shim, H.E.; Choi, M.H.; Park, S.H.; Choi, Y.J.; Jeon, J. Efficient and selective removal of radioactive iodine anions using engineered nanocomposite membranes. *Environ. Sci. Nano* **2017**, *4*, 2157–2163. [CrossRef]
36. Choi, M.H.; Jeong, S.W.; Shim, H.E.; Yun, S.J.; Mushtaq, S.; Choi, D.S.; Jang, B.S.; Yang, J.E.; Choi, Y.J.; Jeon, J. Efficient bioremediation of radioactive iodine using biogenic gold nanomaterial-containing radiation-resistant bacterium, Deinococcus radiodurans R1. *Chem. Commun.* **2017**, *53*, 3937–3940. [CrossRef]
37. Park, J.E.; Shim, H.E.; Mushtaq, S.; Choi, Y.J.; Jeon, J. A functionalized nanocomposite adsorbent for the sequential removal of radioactive iodine and cobalt ions in aqueous media. *Korean J. Chem. Eng.* **2020**, *37*, 2209–2215. [CrossRef]
38. Shim, H.E.; Yang, J.E.; Jeong, S.W.; Lee, C.H.; Song, L.; Mushtaq, S.; Jeon, J. Silver nanomaterial-immobilized desalination systems for efficient removal of radioactive iodine species in water. *Nanomaterials* **2018**, *8*, 660. [CrossRef]
39. Huang, H.; Sha, X.; Cui, Y.; Sun, S.; Huang, H.; He, Z.; Wei, Y. Highly efficient removal of iodine ions using MXene-PDA-Ag2Ox composites synthesized by mussel-inspired chemistry. *J. Colloid Interface Sci.* **2020**, *567*, 190–201. [CrossRef]
40. Yu, W.; Huang, Y.; Pei, L.; Fan, Y.; Wang, X.; Lai, K. Magnetic Fe_3O_4/Ag hybrid nanoparticles as surface-enhanced Raman scattering substrate for trace analysis of furazolidone in fish feeds. *J. Nanomater.* **2014**, *6*, 1–8. [CrossRef]
41. Zhang, X.; Liu, Y. Nanomaterials for radioactive wastewater decontamination. *Environ. Sci. Nano* **2020**, *7*, 1008–1040. [CrossRef]
42. Shama, G.; Jeevanadam, P. A Facile Synthesis of Multifunctional Iron Oxide@Ag Core–Shell Nanoparticles and Their Catalytic Applications. *Eur. J. Inorg. Chem.* **2013**, *2013*, 6126–6136. [CrossRef]
43. Khalid, H.F.; Tehseen, B.; Sarwar, Y.; Hussain, S.Z.; Khan, W.S.; Raza, Z.A.; Bajwa, S.Z.; Kanaras, A.G.; Hussain, I.; Rehman, A. Biosurfactant coated silver and iron oxide nanoparticles with enhanced anti-biofilm and anti-adhesive properties. *J. Hazard. Mater.* **2019**, *364*, 441–448. [CrossRef] [PubMed]
44. Wang, H.; Shen, J.; Li, Y.; Wei, Z.; Gai, Z.; Hong, K.; Banerjee, P.; Zhou, S. Porous Carbon Protected Magnetite and Silver Hybrid Nanoparticles: Morphological Control, Recyclable Catalysts, and Multicolor Cell Imaging. *ACS Appl. Mater. Interfaces* **2013**, *5*, 9446–9453. [CrossRef] [PubMed]
45. Sangili, A.; Annalakshmi, M.; Chen, S.; Balasubramanian, P.; Sundrarajan, M. Synthesis of silver nanoparticles decorated on core-shell structured tannic acid-coated iron oxide nanospheres for excellent electrochemical detection and efficient catalytic reduction of hazardous 4-nitrophenol. *Compos. B Eng.* **2019**, *162*, 33–42. [CrossRef]

nanomaterials

Article

Fe_3O_4@C Nanoparticles Synthesized by In Situ Solid-Phase Method for Removal of Methylene Blue

Hengli Xiang [1], Genkuan Ren [1,2], Yanjun Zhong [1], Dehua Xu [1], Zhiye Zhang [1], Xinlong Wang [1,*] and Xiushan Yang [1,*]

[1] School of Chemical Engineering, Sichuan University, Ministry of Education Research Center for Comprehensive Utilization and Clean Processing Engineering of Phosphorus Resources, Chengdu 610065, China; xianghl0908@163.com (H.X.); rgk2000@163.com (G.R.); yjzhong@scu.edu.cn (Y.Z.); dhxu@scu.edu.cn (D.X.); zhiyezhang@scu.edu.cn (Z.Z.)

[2] College of Chemistry and Chemical Engineering, Yibin University, Yibin 644000, China

* Correspondence: wangxl@scu.edu.cn (X.W.); yangxs@scu.edu.cn (X.Y.)

Abstract: Fe_3O_4@C nanoparticles were prepared by an in situ, solid-phase reaction, without any precursor, using $FeSO_4$, FeS_2, and PVP K30 as raw materials. The nanoparticles were utilized to decolorize high concentrations methylene blue (MB). The results indicated that the maximum adsorption capacity of the Fe3O4@C nanoparticles was 18.52 mg/g, and that the adsorption process was exothermic. Additionally, by employing H_2O_2 as the initiator of a Fenton-like reaction, the removal efficiency of 100 mg/L MB reached ~99% with Fe_3O_4@C nanoparticles, while that of MB was only ~34% using pure Fe_3O_4 nanoparticles. The mechanism of H_2O_2 activated on the Fe_3O_4@C nanoparticles and the possible degradation pathways of MB are discussed. The Fe_3O_4@C nanoparticles retained high catalytic activity after five usage cycles. This work describes a facile method for producing Fe_3O_4@C nanoparticles with excellent catalytic reactivity, and therefore, represents a promising approach for the industrial production of Fe_3O_4@C nanoparticles for the treatment of high concentrations of dyes in wastewater.

Keywords: Fe_3O_4@C nanoparticles; solid-phase method; adsorption; Fenton-like reaction; methylene blue

Citation: Xiang, H.; Ren, G.; Zhong, Y.; Xu, D.; Zhang, Z.; Wang, X.; Yang, X. Fe_3O_4@C Nanoparticles Synthesized by In Situ Solid-Phase Method for Removal of Methylene Blue. *Nanomaterials* **2021**, *11*, 330. https://doi.org/10.3390/nano11020330

Received: 2 January 2021
Accepted: 22 January 2021
Published: 27 January 2021

1. Introduction

Dye pollution is one of the most severe environmental concerns nowadays. Most industrial dyes contain complex components which are highly toxic, teratogenic, and carcinogenic [1–4]. A host of technologies has been applied for the degradation of dye pollutants including biological, physical, and chemical approaches [5–8]. Among the various treatments, adsorption and advanced oxidation processes (AOPs) have been shown to be highly efficient methods for the removal of dye from wastewater [9,10]. Adsorption is low-cost and free of intermediates. The Fenton-like system has drawn much attention because of its ability to cleanly and efficiently remove dyes from wastewater. Thus, the development of adsorbents with good performance coupled with Fenton-like reactions has been the focal point of a great deal of recent research.

Because Fe_3O_4 has the characteristics of easy magnetic separation, stable properties, and low toxicity, Fe^{2+} and Fe^{3+} can be safely reacted with H_2O_2 to trigger the Fenton reaction; in such cases, Fe_3O_4 is a potential adsorbent and Fenton catalyst [11,12]. However, the H_2O_2-activating ability of pure Fe_3O_4 is not strong, and the compound tends to agglomerate in the presence of strong magnetism, which inevitably results in a reduction of the adsorption capacity and catalytic activity [13]. To overcome the shortcomings of single-phase materials [14,15], the design of composite materials has become a focus of today's research. For example, core-shell structure Fe_3O_4/TiO_2 nanoparticles were successfully manufactured to enhance photocatalytic performance [16–18], Glutathione-coated

Fe_3O_4 was applied in an enhanced photo-Fenton system [19], a novel composite material of g-$C_3N_4/Fe_2O_3/Fe_3O_4$ was used to degrade Orange II via a visible-light Fenton system [20], hydrothermally synthesized C/Fe_3O_4 nanoparticles were used as Fenton-like catalysts with high-performance for dye decolorization [21], a Fe_3O_4/WO_3 core-shell photocatalyst loaded on UiO-66(Zr/Ti) nanoflakes was successfully synthesized to enhance photo-oxidation capacity [22], and a Fe_3O_4/CuO@C composite from MOF-based materials was used as a magnetic separation photocatalyst for the degradation of antibiotics [23]. Compared with combinations of Fe_3O_4 and metal oxides or polymer shells, the combination with carbon shells has a broader range of practical applications thanks to its stability in acid-base solutions and in high-temperature and high-pressure conditions [24]. Additonally, carbon-based materials are complementary to magnetite because of their conjugated π-electron effect, high porosity, and large specific surface area [25]. Owing to the unique properties of carbon-modified magnetite nanoparticles, such as ease of separation, nontoxicity, and convenient regeneration, these nanoparticles represent a promising method for dye wastewater treatment.

Synthesis methods of carbon-modified magnetite nanoparticles mainly include co-precipitation [26–28], the impregnation method [29,30], and hydrothermal carbonization processes [6,15,31], where high production costs and complicated preparation routes hamper industrial production and reduce the feasibility of using these materials to treat dye wastewater on an industrial scale. Hence, the development of a promising method for the preparation of high-performance Fe_3O_4@C particles for industrial production is significant. Recently, the synthesisis of nanoparticles by the solid-phase method has drawn much attention because of its low-cost and ease of industrial production. Peng Wang et al. [32] employed the solid-phase method to manufacture Fe_3O_4@C nanoparticles using α-Fe_2O_3 nanoparticles and acetylene black as raw materials. Kan Wang et al. [33] calcined α-Fe_2O_3 nanoparticles in an acetylene atmosphere to obtain γ-Fe_2O_3@C nanoparticles, and then calcined the γ-Fe_2O_3@C nanoparticles in an N_2/H_2 (5% H_2) atmosphere to acquire Fe_3O_4@C nanoparticles. Zhang et al. [34] manufactured porous carbon/Fe_3O_4 by the calcination of waste cigarette filters immersed in ferric nitrate solution. However, these solid-phase methods contain a two-step reaction, where the involvement of the solvent and the synthesis of precursors cannot be avoided [35,36], which represents an obstacle to large-scale industrial production. In our previous research, we [37] put forward a new solid-phase reduction process for the synthesis of pure Fe_3O_4 nanoparticles with a relatively large particle size, i.e., ~50 nm; however, severe agglomeration caused a low surface area, i.e., 10.6 m^2/g, which represented one aspect requiring improvement.

In this study, we propose an in situ, solid-phase method to fabricate Fe_3O_4@C nanoparticles with core-shell structure using $FeSO_4$, FeS_2, and PVP K30 as raw materials. Characterization of X-ray powder diffraction (XRD), Fourier transform infrared (FT-IR) and Raman spectra, High-resolution transmission electron microscopy (HRTEM), Brunauer-Emmett-Teller (BET) method, Vibrating Sample Magnetometer (VSM), and X-ray photoelectron spectroscopy (XPS) were applied to explore properties of the Fe_3O_4@C nanoparticles. The target dye was methylene blue (MB), as it is widely used. The effects of the experimental conditions, adsorption kinetics, and isothermal adsorption were investigated. Furthermore, a Fenton-like reaction was conducted to synergistically degrade high concentration MB, and the synergism between adsorption and the Fenton-like reaction was evaluated. This study presents a facile, in situ, solid-phase method to synthesize Fe_3O_4@C nanoparticles for potential industrial-scale production and high-concentration dye wastewater treatment.

2. Materials and Methods

2.1. Materials

Analytical reagents of $FeSO_4 \cdot 7H_2O$ (99%, Chengdu Kelong Chemical Co, Ltd. Chengdu, China), FeS_2, (98%, Beijing Hawk Science and Technology Co., Ltd. Bejing, China), PVP K30 ($(C_6H_9NO)_n$, 99%, Shanghai Yuanye Biotechnology Co., Ltd. Shanghai, China), tert-butanol

(99%, Chengdu Kelong Chemical Co, Ltd. Co., Ltd. Chengdu, China), and H_2O_2 (30%, Chengdu Kelong Chemical Co, Ltd. Chengdu, China) were used without further purification.

2.2. Fabrication of Fe_3O_4@C Nanoparticles

The fabrication procedure of the Fe_3O_4@C nanoparticles comprised the following steps: (i) $FeSO_4 \cdot 7H_2O$ was dried at 180 °C for 360 min to remove water, yielding $FeSO_4 \cdot H_2O$, FeS_2, and PVP K30, which were also dried at 80 °C until constant weight was achieved. (ii) Then, 15 g $FeSO_4 \cdot H_2O$, 1.0 g FeS_2, and 0.8 g PVP K30 were put into an omnidirectional planetary ball mill and ground for 30 min to obtain a homogeneous mix. (iii) The mixture was placed in a tube furnace at a given heating program (reaction temperature: 500 °C, heating rate: 10 °C/min) under a nitrogen atmosphere. (iv) After calcining, the product was cooled to ambient temperature under nitrogen. (v) The product was removed and washed with deionized water two times, before vacuum drying at 80 °C until constant weight was achieved.

2.3. Experimental Procedure for Decolorization of MB

Adsorption:

(1) A certain number of Fe_3O_4@C nanoparticles were added to 50 mL MB solutions with various concentrations; these mixtures were then placed on a thermostatic shaker. (2) After starting the thermostatic shaker, the Fe_3O_4@C nanoparticles were extracted by a magnet from the heterogeneous solution at selected interval times. (3) A UV-vis spectrophotometer was then used to determine the concentration of methylene blue.

Fenton-like reaction:

(1) A certain number of Fe_3O_4@C nanoparticles were added to 50 mL MB solutions with various concentrations; these mixtures were then placed on a thermostatic shaker (2) After running the thermostatic shaker for 60 min to achieve adsorption equilibrium, a certain amount of H_2O_2 (30%) was quickly added to the heterogeneous solution to initiate a Fenton-like reaction. (3) Then, 5 mL of the tert-butanol solution (AR) was put into the heterogeneous solution at selected times to inhibit the Fenton-like reaction; the mixture was then centrifuged to remove solid particles. (4) The supernatant liquid obtained from the serum by centrifuge was subjected to a UV-vis spectrophotometer analysis to determine the concentration of methylene blue. Furthermore, pure Fe_3O_4 nanoparticles synthesized according to our previous study [37] were used to compare the decolorization efficiency.

2.4. Adsorption Kinetic and Interparticle Diffusion Study

The adsorption kinetic models applied in this study were the pseudo-first-order model (Equation (1)), pseudo-second-order model (Equation (2)) [6], and the Elovich model (Equation (3)) [38].

$$\text{In}\,(q_e - q_t) = \text{In}\,(q_e) - k_1 t \tag{1}$$

$$\frac{t}{q_t} = \frac{1}{k_2 q_e^2} + \frac{t}{q_e} \tag{2}$$

$$q_t = \frac{1}{\beta}\,\text{In}\,(1 + \alpha\beta t) \tag{3}$$

where q_t (mg/g) is the adsorption capacity at adsorption time t, k_1 (min^{-1}) and k_2 (min^{-1}) are the rate constants of the pseudo-first-order and pseudo-second-order models, respectively, α is the initial adsorption rate (mg/g·min^{-1}), and β is the desorption constant (g/mg)

The adsorption rate (mg/g min) at the beginning of adsorption was calculated as follows:

$$h = k_2 q_e^2 \tag{4}$$

The control-step of the MB adsorption by the Fe_3O_4@C nanoparticles was determined by the Weber and Morris model [10]. The expression of the Weber and Morris model may be depicted as followed:

$$q_t = K_{dif} t^{1/2} + \varepsilon \quad (5)$$

where K_{dif} (mg/g min$^{1/2}$) is the diffusion rate constant within the adsorption process and ε (mg/g) is the dimensionless constant.

2.5. Adsorption Isotherm Study

The relevance between the equilibrium adsorption capacity of Fe_3O_4@C nanoparticles at different adsorption temperatures and the remaining MB concentration was investigated using adsorption isotherms models. Adsorption isotherm experiments were conducted at 25 °C, 35 °C, and 45 °C. Langmuir, Freundlich, Redlich-Peterson, and Temkin models were introduced to describe the adsorption category [39–41]; expressions are shown as Equations (6)–(9).

$$q_e = \frac{q_{\max} k_{LC} C_{equ}}{(1 + k_{LC} C_{equ})} \quad (6)$$

$$q_e = K_{FC}\, C_{equ}^{1/n} \quad (7)$$

$$q_e = \frac{K_R C_{equ}}{(1 + a_R C_{equ}^{\alpha})} \quad (8)$$

$$q_e = B\,\mathrm{In}(A C_{equ}) \quad (9)$$

where k_{LC} and K_{FC} are the Langmuir and Freundlich constant, respectively, K_R and a_R are the characteristics of the R-P isotherm model, *B is* the Temkin constant, A is the equilibrium binding constant, and C_{equ} is the MB concentration at adsorption equilibrium.

2.6. Characteristic Methods

The crystalline structure of the products was verified by XRD (Empyrean, PANalytical, Alemlo, The Netherlands). FT-IR spectroscopy (PerkinElmer Frontier, Waltham, MA, USA) and the Raman spectroscopy (LabRAM HR, Horiba Scientific, Paris, France) were utilized to determine the surface functional radicals of the products. The HRTEM (FEI Talos F200x, Hillsboro, OR, USA) measured the particle morphology, primary particle size, and lattice of the products. The N_2 adsorption/desorption curves were analyzed by the standard BET method (77 K, NOVA1000e analyzer) to estimate the specific surface area and the corresponding pore structure of the products. The MB solution concentration was measured with a spectrophotometer (664 nm, V-5800, Metash instrument, Shanghai, China), and the intermediates produced under the process of Fenton-like reaction were determined using the LC-MS (Thermo Scientific TSQ Quantum, Waltham, MA, USA). The concentration of leaching iron ions after decolorization was measured using the ICP-AES (DV 7000, Waltham, MA, USA).

3. Results

3.1. Characterization of the Fe_3O_4@C Nanoparticles

The XRD pattern of the Fe_3O_4@C nanoparticles shown in Figure 1 indicated that the characteristic peaks of Fe_3O_4@C nanoparticles were consistent with the crystal planes of the PDF standard card (JCPDS 00-019-0629) of magnetite [42]. The broad peak found at $2\theta = 21.6°$ was the characteristic reflection of carbon [43]. Furthermore, the crystallite sizes measured by Debye-Scherrer's equation [44] in the light of the strongest diffraction peak (311) was 20.6 nm.

The FT-IR and Raman spectra of the Fe_3O_4@C nanoparticles are shown in Figure S1. As shown in Figure S1a, the peaks emerging at 3400 cm^{-1}, 1636 cm^{-1}, 1123 cm^{-1}, and 565 cm^{-1} corresponded to the stretching vibration of the -OH bond [37], the stretching vibration of C=O in amide bond derived from the pyrolysis of the PVP 30 [45], the stretching vibration of the SO_4^{2-} of the residual ferrous sulfate [46], and the stretching vibration of the Fe^{3+}-O [47], respectively. Figure S1b demonstrates that the G band (Graphite) and D band (disordered) of the carbon-carbon bonds could be seen at 1580 cm^{-1} and 1350 cm^{-1}, respectively [48,49]. The peak found at 1180 cm^{-1} was the A_{1g} symmetry vibration of the disordered graphitic lattice [50]. Also, the observed diffraction peak at about 670 cm^{-1} was indexed to the A_{1g} mode of magnetite [51]. Therefore, the FT-IR spectroscopy and Raman spectroscopy further demonstrated that Fe_3O_4@C nanoparticles had been successfully obtained.

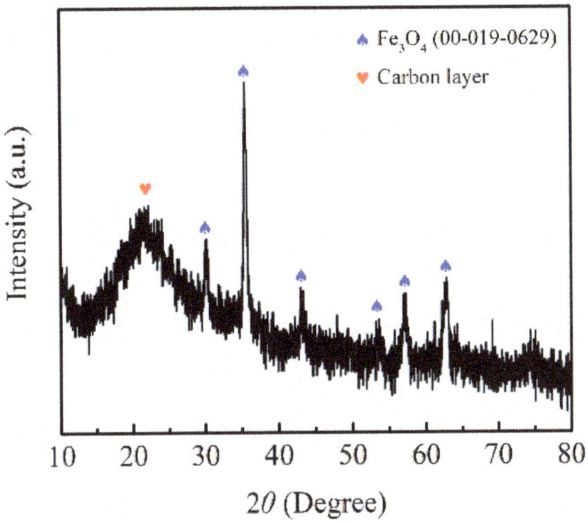

Figure 1. The XRD pattern of the Fe_3O_4@C nanoparticles.

The HRTEM graphs presented in Figure 2a,b show that the morphology of the Fe_3O_4@C nanoparticles was spherical, with a core-shell structure. The primary particle diameter was ~30 nm and the thickness of the carbon-shell was ~2 nm. Figure 2c demonstrates that the interlayer spacing of the lattice fringes was 0.26 nm, which closely matched the d-spacing of the (311) plane in cubic Fe_3O_4. The SAED micrograph shown in Figure 2d further verified the polycrystalline structure of the Fe_3O_4@C nanoparticles, where the diffraction rings could be ascribed to the (220), (311), (400), (511), and (440) planes of Fe_3O_4.

The specific surface area of the Fe_3O_4@C nanoparticles, calculated by the BET method, according to the N_2 adsorption/desorption isotherm curves was 37.74 m^2/g; see Figure S2a. The average pore diameters and pore volume estimated by the BJH method according to the pore distribution, displayed in the inset of Figure S2a, was 3.78 nm and 0.227 cm^3/g, respectively. The magnetic property of the Fe_3O_4@C nanoparticles determined the ease of the separation of the particles in a heterogeneous solution. The magnetic hysteresis loops of the Fe_3O_4@C nanoparticles measured by the VSM at 298 K are displayed in Figure S2b, and showed a superparamagnetic feature. The saturation magnetization was 77 emu/g, which was lower than the values reported in the literature [52,53]. The lower saturation magnetization of the as-synthesized Fe_3O_4@C nanoparticles might have been due to the coating of the carbon-shell. The coercivity value was found to be only 0.16 kOe and the remnant magnetization was 12.8 emu/g.

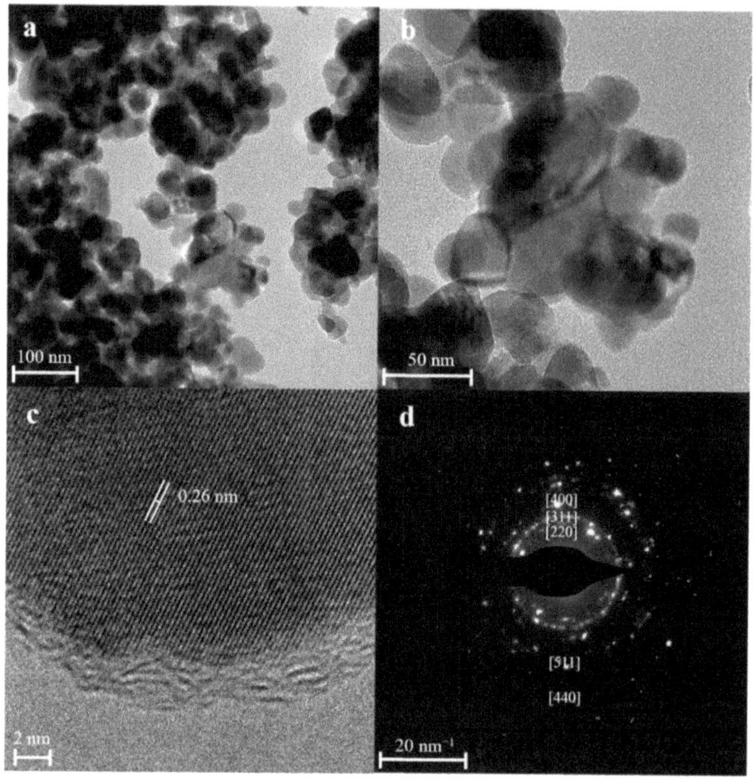

Figure 2. (**a–c**) The HRTEM graphs of the Fe_3O_4@C, and (**d**) the SAED micrograph of the Fe_3O_4@C.

3.2. *Adsorption Studies*

The adsorption of MB was the first step of the decolorization of MB. Figure 3 shows the effect of the adsorbent dosages, initial MB concentrations, temperatures, and initial pH values on the adsorption of MB. Figure 3a illustrates that when the adsorbent dosage increased to 2.0 g/L, the adsorption capacity decreased and the adsorption efficiency increased; this was due to the increase of adsorbent dosage that likely increased the adsorption activate site, thereby increasing the adsorption efficiency and decreasing the MB adsorbed per unit mass. Figure 3b shows that as the initial MB concentration increased, the adsorption efficiency gradually declined and the adsorption capacity increased; this was because the addition of the initial MB concentration likely raised the concentration gradient between the MB solution and the Fe_3O_4@C nanoparticles. Additionally, the increase of initial MB concentration increased the probability of the MB molecules coming into contact with the active sites on the surface of the adsorbent, so the adsorption capacity for MB improved. Figure 3c implies that the higher the adsorption temperature, the lower the adsorption efficiency, suggesting that the adsorption process was exothermic. Figure 3d reveals that the adsorption capacity and efficiency increased with the initial increase in pH. This was due to the fact that MB is a kind of cationic dye, and a basic solution would reduce the competition between the H^+ and MB ions, thereby offering more active sites for MB ions [54].

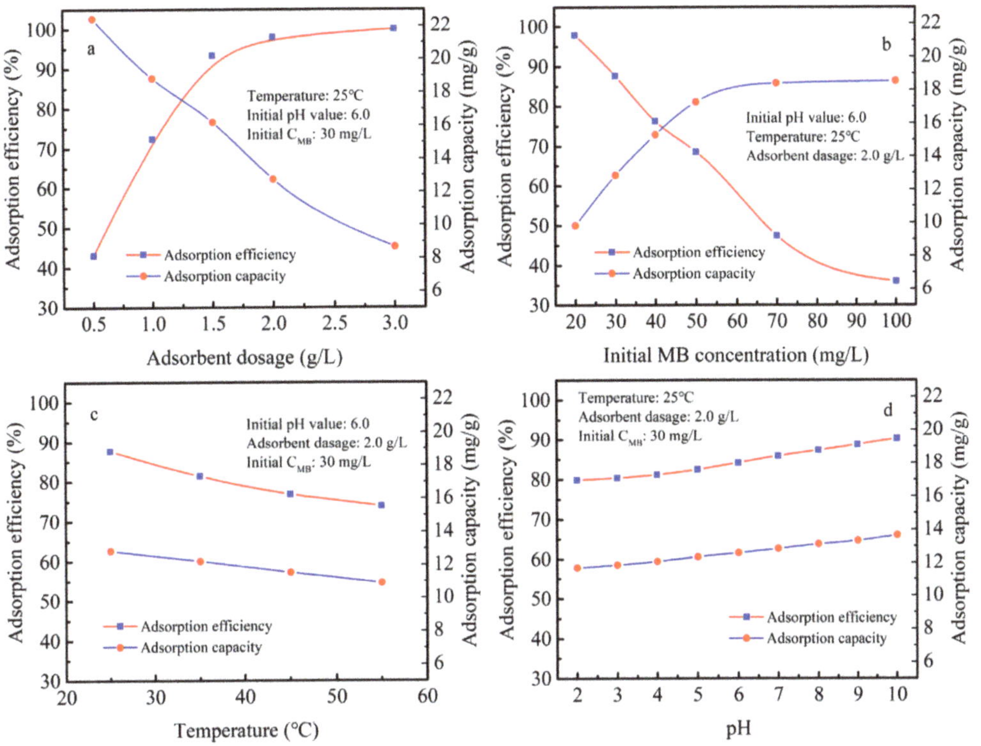

Figure 3. The (**a**) adsorbent dosages, (**b**) initial MB concentrations, (**c**) temperatures, and (**d**) effect of the initial pH values on the adsorption of MB.

3.3. Adsorption Kinetics and Interparticle Diffusion Analysis

The adsorption mechanism of MB by the Fe_3O_4@C nanoparticles synthesized by an in situ, solid-phase method is described by adsorption kinetics and interparticle diffusion analyses. It may be seen in Figure 4a that at different initial MB concentrations, the adsorption capacity curve followed the same tendency, and the adsorption capacity increased with the increasing initial C_{MB}. Figure 4b,c displays the linear forms of the kinetics model. The correlation coefficients (R^2) and the rate parameters fitted by the kinetic models are given in Table 1. By comparing Figure 4b,c, it is obvious that the pseudo-second-order model better describes the kinetic behavior than the pseudo-first-order model. Additionally, Table 1 also clearly shows that the theoretical adsorption capacity calculated from the pseudo-second-order model provided more accurate results in comparison to the values of the actual adsorption capacity. Furthermore, due to the fact that traditional linear transformation techniques used in adsorption study often misinterpret adsorption processes [55], nonlinear kinetics models were also applied to interpret the adsorption process; the results were shown in Figure S5 and Table S2. As shown in Figure S5 and Table S2, regardless of the linear or nonlinear forms of the kinetics model, the pseudo-second-order model fitted better with the adsorption behavior, while the theoretical adsorption capacity calculated from the pseudo-second-order model was more accurate regarding the values of the actual adsorption capacity. The Weber and Morris model determined the rate-limiting steps of the adsorption process. Figure 4d shows a graph of q_t versus $t^{1/2}$, while the parameters of the interparticle diffusion model of MB adsorption are illustrated in Table 2. The results of the plots shown in Figure 4d presented multilinearity and the intercept was not 0; as such, the adsorption process of MB on Fe_3O_4@C nanoparticles included interparticle diffusion and boundary layer diffusion [6]. Additionally, the heterogeneous diffusion process controlled by the reaction rate and diffusion factor, including a series of reaction mechanisms such as the diffusion of solutes at the liquid phase or interface, surface activation, and deactivation, was described by the Elovich model; the results are shown in Figure S6 and Table S3. The α values of the Elovich model were much higher than the β values, indicating that the adsorption rate was much higher than the desorption rate [38]. In addition, the R^2 values of the Elovich model were higher than those of both the pseudo-first- and pseudo-second-order models, suggesting that the Elovich model best represents the experimental kinetic data.

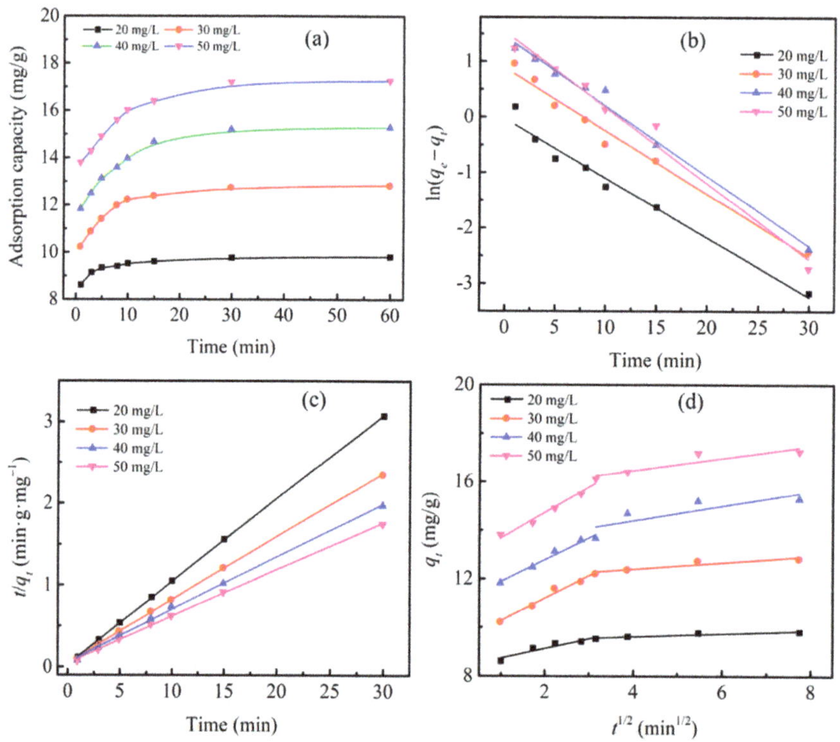

Figure 4. Adsorption kinetics of MB by the as-synthesized Fe_3O_4@C nanoparticles: (**a**) effect of adsorption time on adsorption capacity; (**b**) pseudo-first-order; (**c**) pseudo-second-order; (**d**) interparticle diffusion model. (T = 25 °C, initial pH value = 6.0, adsorbent dosage: 2.0 g/L).

Table 1. Kinetic parameters for adsorption of methyl blue on Fe_3O_4@C nanoparticles.

C_{ini} g·L^{-1}	$q_{e(expe)}$ mg/g	Pseudo-First-Order Kinetics			Pseudo-Second-Order Kinetics			
		k_1 min^{-1}	$q_{e(calc)}$ mg/g	R_1^2	k_2 g/mg·min	$q_{e(calc)}$ mg/g	R_2^2	h mg/mg·min
20	9.809	0.1018	0.9962	0.9704	0.3924	9.827	0.9992	37.89
30	12.84	0.1133	2.445	0.9795	0.1464	12.93	0.9997	24.47
40	15.29	0.1262	4.333	0.9862	0.0794	15.51	0.9990	19.10
50	17.26	0.1369	4.718	0.9784	0.0775	17.48	0.9991	23.68

Table 2. The parameters of the interparticle diffusion model of methyl blue adsorption.

C_{int} mg/L^{-1}	$K_{dif,1}$ mg/g·min$^{0.5}$	$\varepsilon_{dif,1}$ mg/g	$R^2_{dif,1}$	$K_{dif,2}$ mg/g·min$^{0.5}$	$\varepsilon_{dif,2}$ mg/g	$R^2_{dif,2}$
20	0.3964	8.336	0.8735	0.0610	9.369	0.8126
30	0.9254	9.336	0.9735	0.1353	11.86	0.8196
40	0.8963	10.98	0.9699	0.3039	13.17	0.6355
50	1.0562	12.60	0.9671	0.2554	15.45	0.7411

3.4. Adsorption Isotherm Study

An adsorption isotherm study was used to analyze the relationship between the equilibrium adsorption capacity of the Fe_3O_4@C nanoparticles and the remaining concentration of the MB solution at a selected temperature [56]. Figure 5 and Table 3 display the Langmuir, Freundlich, Redlich-Peterson, and Temkin models and the fitting parameters of the MB adsorption by the Fe_3O_4@C nanoparticles. It may be seen in Figure 5 and Table 3 that the R^2 of the Langmuir model was lower than those of the Freundlich, Redlich-Peterson, and Temkin models, indicating that the adsorption of MB on the Fe_3O_4@C nanoparticles was not a single-layer adsorption on a uniform surface [3]. The Freundlich model was based on multilayer adsorption on a reversible heterogeneous surface, considering that the Fe_3O_4@C nanoparticles tend to agglomerate because of their magnetic characteristics. As such, the Freundlich model may be more appropriate for describing the adsorption behaviour [4]. However, the Redlich-Peterson model showed a higher value of R^2 compared to the Freundlich model, which implied that the adsorption behaviour of MB on the Fe_3O_4@C nanoparticles possessed a hybrid characteristic of the traditional Langmuir and Freundlich models. Therefore, the Redlich-Peterson model could be used to describe the relationship between the equilibrium adsorption capacity of the Fe_3O_4@C nanoparticles and the remaining concentration of the MB solution under a selected temperature. The Temkin model considered the effects of the indirect adsorbate/adsorbent interactions on the distribution of adsorption heat and binding energies. The R^2 values were all above 0.98, which further demonstrated that the adsorption active sites on the Fe_3O_4@C nanoparticle surfaces were not uniform, and the increase of temperature likely reduced the binding ability between the MB molecules and the Fe_3O_4@C nanoparticles.

3.5. Fenton-Like Reaction

By only carrying out adsorption, the adsorption efficiency of Fe_3O_4@C nanoparticles on MB of 100 mg/L was lower than 40%. In this case, H_2O_2 was used as an additive to activate a Fenton-like reaction in order to achieve synergetic degradation of a high concentration of MB. The influence of the operational parameters on MB decolorization is displayed in Figure 6. Figure 6a depicts the effect of Fe_3O_4@C nanoparticle dosage on MB decolorization by Fenton-like reaction. With an increase of Fe_3O_4@C nanoparticle dosage from 1.0 g/L to 3.0 g/L, the decolorization efficiency rose from ~50% to ~99%. However, with a further increase to 4.0 g/L, the decolorization efficiency decreased; this result was caused by the excessive iron species, which led to competitive scavenging of radicals with MB molecules, thereby decreasing the decolorization efficiency [57]. Figure 6b shows that as the concentration of H_2O_2 increased from 15 mM to 30 mM, the decolorization rate and efficiency increased significantly, and when the concentration of H_2O_2 reached 45 mM, the decolorization rate and efficiency exhibited no difference compared with 30 mM of H_2O_2. Normally, more H_2O_2 led to a higher decolorization efficiency, but excess H_2O_2 also resulted in a reaction with •OH to form HO_2•, which delayed MB decolorization [58]. From Figure 6c, it may be seen that although higher temperatures were not favorable to the adsorption reaction, for the Fenton reaction, the decolorization rate and efficiency were higher at high temperatures. The initial pH value was an important parameter for the Fenton-like reaction, as this could strongly influence the amount of •OH radicals produced by the leaching Fe^{2+} and H_2O_2. As shown in Figure 6d, the higher pH values resulted in higher adsorption efficiency, while lower pH values were more favorable for the leaching of iron ions, which accounted for higher decolorization efficiency and rate.

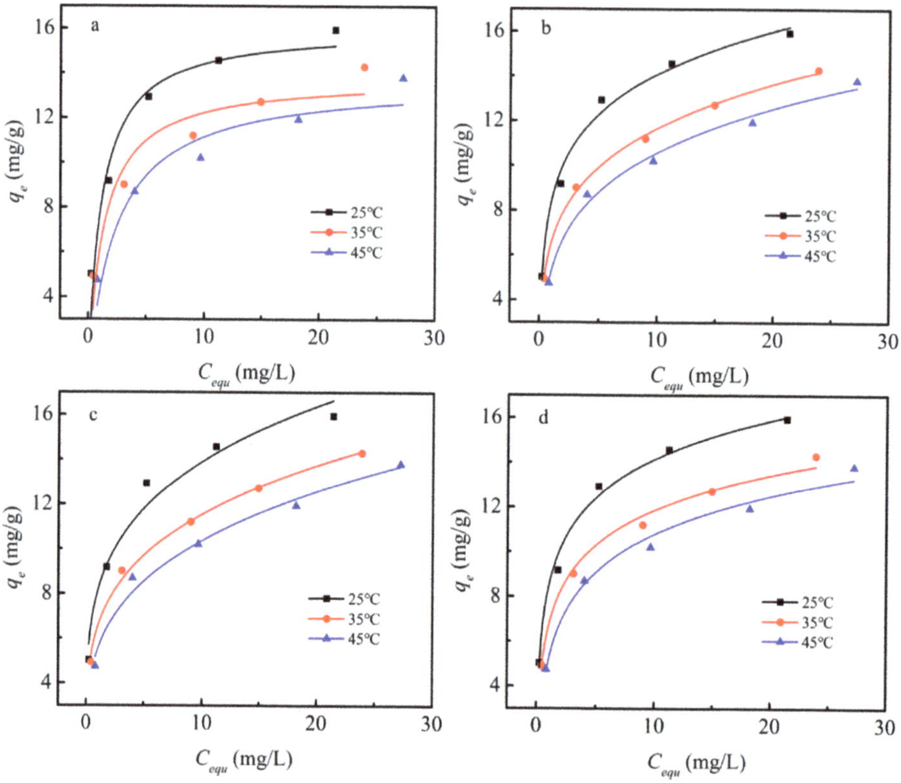

Figure 5. Adsorption isotherms of (**a**) Langmuir model, (**b**) Freundlich model, (**c**) Redlich-Peterson model, and (**d**) Temkin model for methylene blue adsorption by Fe_3O_4@C nanoparticles. (Time: 1 h, initial pH: 7.0, the adsorbent dosage of 2.0 g/L).

Table 3. Parameters of the Langmuir and Freundlich adsorption isotherm models for MB adsorbed on Fe_3O_4@C nanoparticles.

Temp. °C	Langmuir			Freundlich		
	q_{max} mg/g	k_{LC} L/mg	R_L^2	K_{FC} g/mg·min	$1/n$	R_{FC}^2
25	16.00	0.8955	0.8954	8.060	0.2365	0.9643
35	14.19	0.7804	0.8711	6.486	0.2503	0.9939
45	13.69	0.4383	0.9021	5.468	0.2779	0.9800
Temp. °C	**Redlich-Peterson**			**Temkin**		
	a_R (L/mg)$^\alpha$	K_R L/mg	R_R^2	B	A L/mg	R_T^2
25	5.963	58.11	0.9830	2.488	29.36	0.9873
35	5.496	43.11	0.9971	2.450	19.26	0.9874
45	3.762	25.03	0.9878	2.254	8.159	0.9804

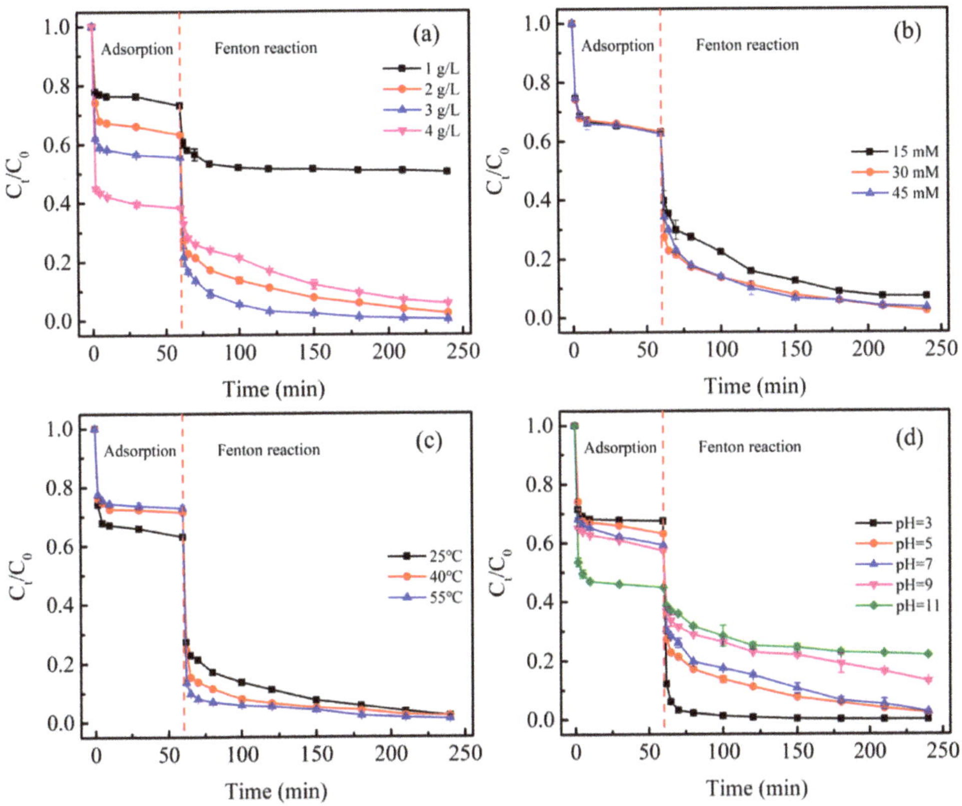

Figure 6. (**a**) Adsorbent dosage, (**b**) H_2O_2 concentration, (**c**) temperature, and (**d**) initial pH affections on the decolorization efficiency of MB. Except for the studied parameters, other parameters were fixed at pH = 5.0, Fe_3O_4@C = 2.0 g/L, $[H_2O_2]$ = 30 mM, [MB] = 100 mg/L, and temperature 25 °C.

In this study, the concentration of MB solution was much higher than that in most of the literature, and the dosage of the catalysts and H_2O_2 was not excessive. A comparison of different catalysts for MB degradation through the Fenton-like reaction is shown in Table 4.

Table 4. The comparison of different catalysts for MB degradation through the Fenton-like reaction.

Catalysts	[Catalysts] (g/L)	$[H_2O_2]$ (mM)	[MB] (mg/L)	Removal Efficiency	Ref.
MnMg/Fe LDH	1.0	10	20	93% in 300 min	[59]
Fe_3O_4/CuO	1.6	32	10	95% in 120 min	[60]
MPCMSs	2.0	16	40	~100% in 25 min	[61]
Fe_3O_4/rGO	0.3	60	10	~99% in 120 min	[62]
Fe_3O_4/C	0.5	90	10	~100% in 60 min	[63]
Fe_3O_4/rGO aerogel	0.3	20	50	~100% in 360 min	[64]
Fe_2O_3	0.5	30	10	~70% in 420 min	[65]
MIL-68(Fe)	0.2	50	20	~100% in 40 min	[66]
$CuCr_2O_4/CeO_2$	1.0	4	10	~80% in 20 min	[67]
$CuFe_2O_4$/Cu@C	0.5	16	20	~100% in 30 min	[68]
Fe_3O_4@C	2.0	30	100	~99% in 180 min	This study

3.6. The Mechanism of the Decolorization of MB

A comparison of the Fenton-like catalytic activity between Fe_3O_4@C nanoparticles and pure Fe_3O_4 nanoparticles was made for the removal of 100 mg/L MB for 180 min with a Fe_3O_4@C dosage of 2.0 g/L, an initial pH value of 3.0, and a temperature of 25 °C; see Figure 7. According to Figure 7, the pure Fe_3O_4 nanoparticles exhibited much lower decolorization of MB (~34%) than the Fe_3O_4@C nanoparticles (~99%).

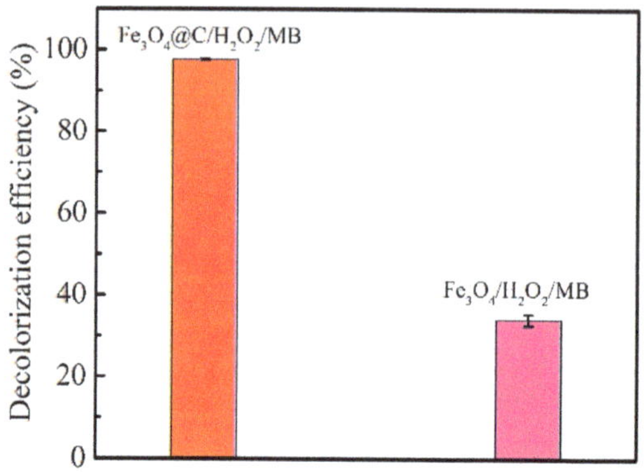

Figure 7. Comparison of Fenton-like catalytic activity between Fe_3O_4@C nanoparticles and pure Fe_3O_4 nanoparticles.

This remarkable difference in decolorization efficiency was because the carbon-shell could adsorb more MB molecules and H_2O_2 molecules to produce more radicals (•OH), thereby accelerating the removal of the MB in the solution. Based on the results of the Raman spectrum and FT-IR spectrum, the carbon-shell on the Fe_3O_4 nanoparticles was mostly amorphous (I_D/I_G = 1.2) and the main bonds on the surface were C=O and C-N. The carbon-shell, rich in the functional groups, could easily adsorb the MB molecules and H_2O_2 molecules to initiate the Fenton-like reaction on the particle surface. Additionally, the carbon-shell also inhibited particle growth, thereby increasing the specific surface area and providing more adsorption active sites. The Zeta potential of the Fe_3O_4@C nanoparticles, displayed in Figure 8a, suggested that the pH_{ZPC} of the Fe_3O_4@C nanoparticles was 9.26, which meant that the surface of Fe_3O_4@C nanoparticles was negatively charged. As MB is a cationic dye with a positive charge, the electrostatic attraction and conjugated π-electron effect were the primary mechanisms for the adsorption of MB on the Fe_3O_4@C nanoparticles.

The FT-IR spectra of Fe_3O_4@C nanoparticles, Fe_3O_4@C nanoparticles after adsorption of MB, and Fe_3O_4@C nanoparticles after Fenton-like reaction, as shown in Figure 8b, further confirmed this hypothesis. Figure 8b shows that after adsorption of MB, new peaks emerged at 878 cm^{-1} and 1046 cm^{-1}, which were assigned to the C-H in the benzene ring and the rocking vibration of -CH_3, while the peaks at 2977 cm^{-1} and 2922 cm^{-1} were ascribed to the stretching vibration of -CH_3, indicating that the MB had been adsorbed by the nanoparticles [69]. However, after the Fenton-like reaction was complete, the FT-IR spectra showed no difference compared to Fe_3O_4@C nanoparticles, indicating that the MB had been successfully degraded. In addition, XPS was used to investigate the valence changes of the Fe_3O_4@C nanoparticles before and after adsorption and Fenton-like reaction. As displayed in Figure 8c, the peak of C1s could be deconvoluted to the peaks of C-C/C=C, C-N/C-O, and C=O [70]. The ratio of C-N/C-O after adsorption of MB increased from 27.15% to 45.05%, suggesting that the MB molecules had been successfully adsorbed on the surface of the Fe_3O_4@C nanoparticles. However, when the Fenton-like reaction was complete, the ratio of C-N/C-O decreased from 45.05% to 40.42% and the ratio of C=O increased from 14.48% to 18.11%, suggesting that part of the carbon-shell had been oxidized during the Fenton-like reaction [63]. Furthermore, the peak of C-C/C=C of the Fe_3O_4@C nanoparticles at the binding energy of 284.60 eV shifted to 284.30 eV, which was caused by the oxidation of unstable amorphous carbon and the exposure of graphite. The peaks of Fe2p, as shown in Figure 8d, indicated that the Fenton-like reaction triggered on the surface of nanoparticles resulted in a change of the ratio of Fe(III)/Fe(II) from 2.01 to 2.38, because a small amount of Fe(II) was oxidized to Fe(III).

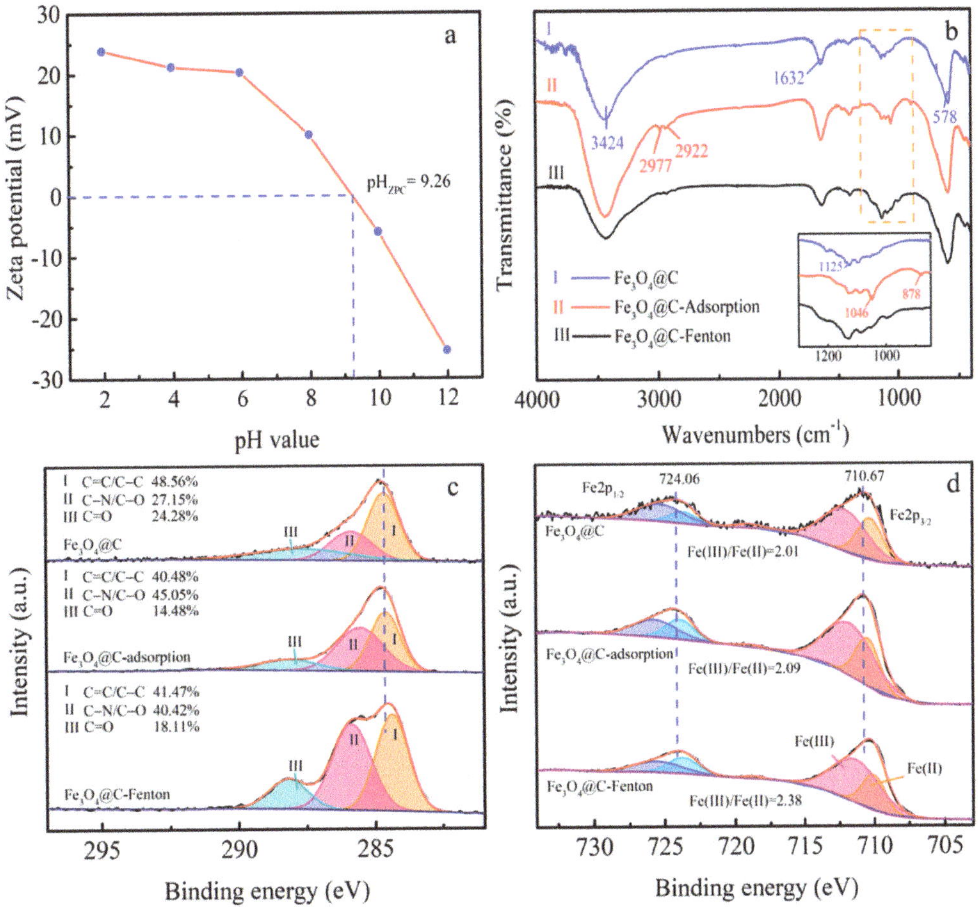

Figure 8. (**a**) The Zeta potential of the Fe_3O_4@C nanoparticles, the (**b**) FT-IR spectra, (**c**) C1s spectra, and (**d**) Fe2p spectra of the Fe_3O_4@C nanoparticles, Fe_3O_4@C nanoparticles after adsorption, and Fe_3O_4@C nanoparticles after Fenton-like reaction.

The proposed mechanism of degradation of MB is shown in Figure 9. Firstly, the Fe_3O_4@C nanoparticles could easily adsorb MB and H_2O_2 molecules due to electrostatic attraction and the conjugated π-electron effect. Then, the H_2O_2 molecules could react with the Fe ions to generate •OH to degrade MB molecules. The excessive •OH would diffuse into the solution to degrade more MB, and the Fe_3O_4@C nanoparticles would continuously adsorb MB molecules for synergetic degradation. Thus, by employing the Fenton-like reaction to promote adsorption, the degradation ability for high concentration MB was significantly enhanced.

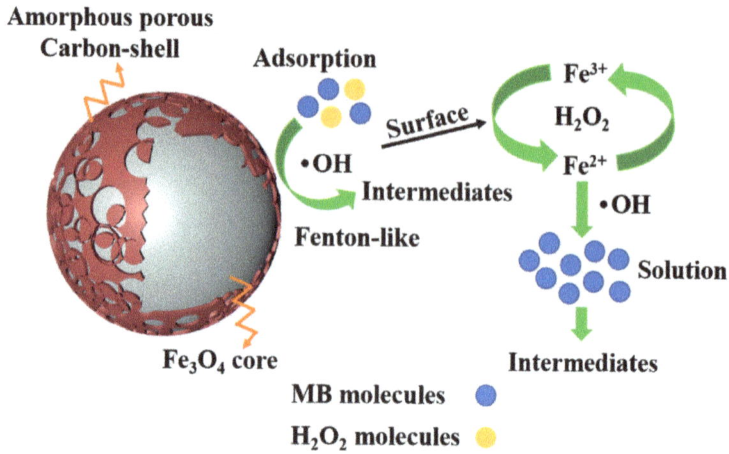

Figure 9. The activation mechanism of H_2O_2 on Fe_3O_4@C nanoparticles.

3.7. Possible Degradation Pathways of MB

The possible degradation pathways of MB were determined by LC-MS analysis; the ESI mass spectra results at reaction times of 1 h and 3 h are presented in Figures S3 and S4, and the possible intermediate degradation products of MB are presented in Table S1. From Figure S3, it may be seen that at a reaction time of 1 h, the initial reaction step was the demethylation of MB to form Azure A (m/z = 270), B (m/z = 256), and C (m/z = 241). Meanwhile, the degradation of the chromophoric group was detected where the electronic reorganization led to a change of $C\text{-}S^+\text{=}C$ to C-(S=O)-C and the break of C-N-C to form $C\text{-}NH_2$ [71,72]. Then, under the attack of •OH, the aromatic groups were continuously destroyed, giving rise to smaller intermediate products at m/z = 173, 158, 149, 136, 117, 103, and 85 [73–75]. Figure S4 indicates that the main structure of the MB molecule was destroyed at a reaction time of 3 h, and only smaller intermediate products were detected by the ESI mass spectra, indicating that the MB molecule had been successfully degraded. The possible degradation pathways of MB are displayed in Scheme 1. In addition, after 3 h of the MB removal (conditions: 100 mM MB, 30 mM H_2O_2, 2 g/L Fe_3O_4@C nanoparticles, 40 °C, and initial pH value of 3.0), the removal efficiency of TOC reached 82.38%, where the remaining TOC in water was 14.90 mg/L, indicating that most of the methylene blue molecules had been completely mineralized, and that those that had not had at least been decomposed into small molecular intermediates.

3.8. The Recyclability Tests of the Fe_3O_4@C Nanoparticles

Recycling experiments were implemented to examine the stability of the Fe_3O_4@C nanoparticles. As shown in Figure 10, the decolorization ratio remained ~97% after five cycles of use under the optimum conditions, which indicated excellent stability. Furthermore, the leaching of iron ions, as determined by ICP-OES, was 1.14 mg/L after the decolorization process, suggesting that ions leached from Fe_3O_4@C nanoparticles would not give rise to secondary pollution. The stability of the carbon-shell with respect to reaction with •OH radicals was determined using a carbon-sulfur analyzer. The results shown in Figure S7 suggested that the carbon contents in the Fe_3O_4@C nanoparticles reduced from 1.47 wt.% to 1.17 wt.% after five usage cycles, indicating the relatively high stability of the carbon-shell. Figure S8 also demonstrates that the reduction of the carbon contents in the Fe_3O_4@C nanoparticles resulted in a reduction of the adsorption efficiency from 32.5% to 24.1% after five usage cycles; however, after employing the Fenton-like reaction, the final decolorization of the MB remained ~97% after five usage cycles, indicating excellent stability of the Fe_3O_4@C nanoparticles.

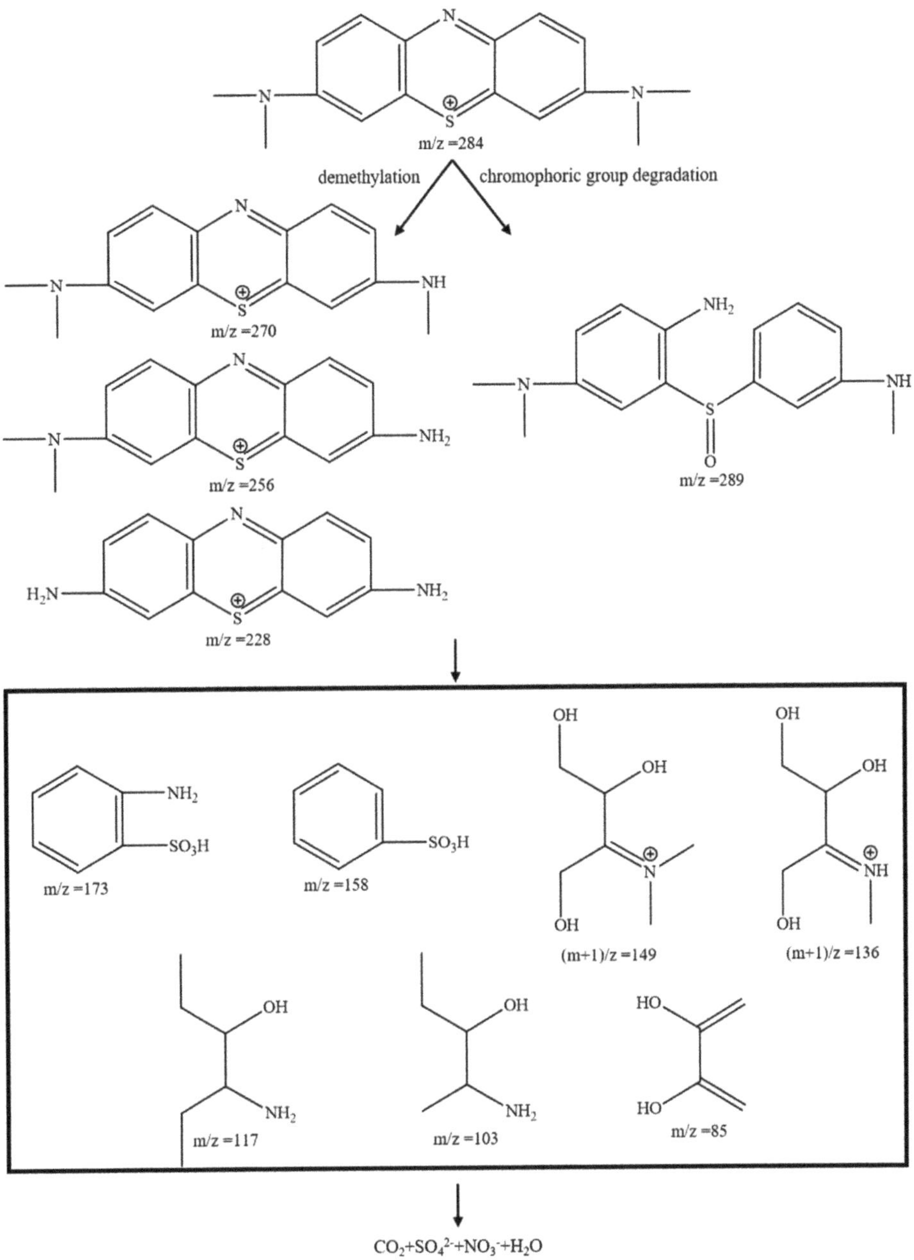

Scheme 1. The possible degradation pathway of MB.

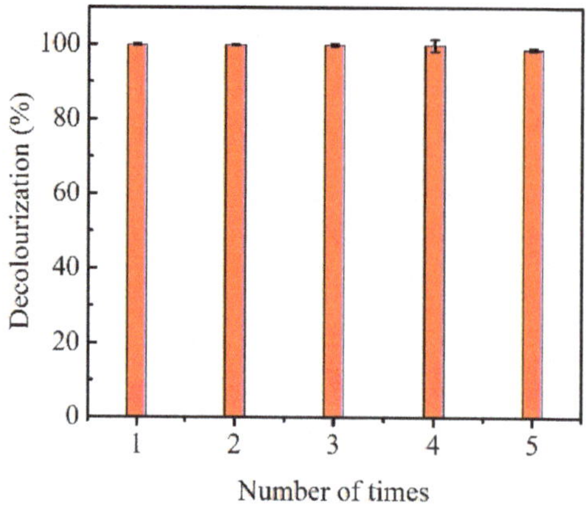

Figure 10. The decolorization efficiency of the Fe_3O_4@C nanoparticles. (Conditions: 100 mM MB, 30 mM H_2O_2, 2 g/L Fe_3O_4@C nanoparticles, the temperature of 40 °C, and initial pH value of 3.0).

4. Conclusions

Fe_3O_4@C nanoparticles were fabricated by an in situ, solid-phase method without any precursors, and were employed for the decolorization of MB. Characterization showed that the Fe_3O_4@C nanoparticles had been successfully prepared with a primary particle size of ~30 nm and a carbon-shell with a thickness of ~2 nm. The XPS and FT-IR spectra demonstrated that the carbon-shell mainly comprised C=O and C-N bonds. The specific surface area, average pore diameters, and pore volume of the Fe_3O_4@C nanoparticles was 37.74 m^2/g, 3.78 nm, and 0.227 cm^3/g, respectively. The saturation magnetization, coercivity, and remnant magnetization of the Fe_3O_4@C nanoparticles was 77 emu/g, 0.16 kOe, and 12.8 emu/g, respectively. The as Fe_3O_4@C nanoparticles showed a much lower average particle size and much higher specific surface area compared to pure Fe_3O_4 nanoparticles synthesized by the solid-phase method, demonstrating that the adjunction of PVP K30 in the solid-phase method significantly improved the particle properties. Moreover, it was shown that the Fe_3O_4@C/H_2O_2 system could effectively decolorize MB through the simultaneous involvement of the adsorption and Fenton-like process, where the carbon-shell provided adsorption active sites for MB and H_2O_2 molecules, while the core Fe_3O_4 provided Fe ions to stimulate the Fenton-like reaction. The maximum adsorption capacity of Fe_3O_4@C nanoparticles for MB was 18.52 mg/g, and the adsorption kinetic was well-fitted by the Elovich model, indicating that the adsorption process was a heterogeneous diffusion process. Additionally, the Redlich-Peterson adsorption isotherm model could better describe the adsorption behavior, implying that the adsorption active sites on the surface of the Fe3O4@C nanoparticles were not uniform, and that the increase of temperature reduced the binding ability between the MB molecules and the Fe_3O_4@C nanoparticles. Also, to degrade higher concentrations of methylene blue solution, H_2O_2 was added after the adsorption equilibrium to stimulate the Fenton reaction. The removal efficiency of 100 mg/L MB reached ~99% by Fe_3O_4@C nanoparticles after 3 h, and the maximum decolorization of the MB was still more than 97% after five usage cycles. Compared to the efficiency of different catalysts for MB degradation described in the literature, the Fe_3O_4@C nanoparticles proposed in this paper could degrade MB of much higher concentration without excessive usage of catalyst or H_2O_2. The leaching iron ions in the solution, as determined by ICP-OES, constituted 1.14 mg/L, suggesting that the Fe_3O_4@C nanoparticles had good stability. The possible degradation pathways of the MB molecule were determined by LC-MS: demethylation, chromophoric group crack, and aromatic ring opening to form smaller fragments. The primary limitations of this study are that the raw materials to synthesize Fe_3O_4@C nanoparticles were all analytical, pure reagents, and that application in real dye wastewater has not been studied. In future research, we will focus on the use of titanium dioxide waste residue (~90% $FeSO_4 \cdot 7H_2O$) and pyrite (~75% FeS_2) to synthesize the Fe_3O_4@C nanoparticles by the in situ, solid-phase method, and its application in real dye wastewater. In conclusion, this study describes a promising method for the industrial production of Fe_3O_4@C nanoparticles and their potential for industrial treatment of high concentration dye wastewater.

Supplementary Materials: The following are available online at https://www.mdpi.com/2079-4991/11/2/330/s1, Figure S1: The (**a**) FT-IR spectroscopy and (**b**) Raman spectroscopy of the as-synthesized Fe_3O_4@C nanoparticles, Figure S2: (**a**) The N_2 adsorption/desorption isotherm curves and (**b**) the magnetic property of the Fe_3O_4@C nanoparticles, Figure S3: The non-linear forms of the kinetics model. (**a**) pseudo-first-order model, (**b**) pseudo-second-order model, Figure S4: The Elovich kinetics model of the adsorption, Figure S5: ESI mass spectra of different retention time at the reaction time of 1 h, Figure S6: ESI mass spectra of different retention time at the reaction time of 3 h, Figure S7: The carbon contents of the Fe_3O_4@C nanoparticles after repeated use, Figure S8: The recyclabitily test of the Fe_3O_4@C nanoparticles (Conditions: 100 mM MB, 30 mM H_2O_2, 2 g/L Fe_3O_4@C nanoparticles, the temperature of 40°C, and initial pH value of 3.0), Table S1: Kinetic parameters for adsorption of methyl blue on Fe_3O_4@C nanoparticles, Table S2: The Elovich kinetic parameters for adsorption of methyl blue on Fe_3O_4@C nanoparticles, Table S3: The possible intermediate degradation products of MB.

Author Contributions: Conceptualization, H.X. and G.R.; Data curation, G.R.; Formal analysis, H.X.; Funding acquisition, Y.Z. and X.Y.; Investigation, H.X.; Methodology, H.X.; Project administration, Z.Z. and X.W.; Supervision, X.W.; Validation, D.X.; Writing–original draft, H.X.; Writing–review & editing, Y.Z. and X.Y. All authors have read and agreed to the published version of the manuscript.

Funding: This research was funded by [Sichuan University-Panzhihua City Science and Technology Cooperation Special Fund for Titanium White by-product Ferrous Sulfate Preparation 500 tons/year Nano-iron Red Pigment and Co-production Sulfuric Acid Pilot Study] grant number [2018CDPZH-5], [Sichuan Science and Technology Planning Project] grant number [2019YFH0149].

Institutional Review Board Statement: Not applicable.

Informed Consent Statement: Not applicable.

Data Availability Statement: The data presented in this study are available on request from the corresponding author.

Acknowledgments: The authors are grateful for the financial support provided by Sichuan University-Panzhihua City Science and Technology Cooperation Special Fund for Titanium White by-product Ferrous Sulfate Preparation 500 tons/year Nano-iron Red Pigment and Co-production Sulfuric Acid Pilot Study (Project No.:2018CDPZH-5), Sichuan Science and Technology Planning Project (Project No.: 2019YFH0149) and Sichuan University "Chemical Star" Excellent Young Talents Training Program. The authors appreciate Bo Gao from the Analytical & Testing Center of Sichuan University for help the LC-MS characterization.

Conflicts of Interest: The authors declare no conflict of interest.

References

1. Qadri, S.; Ganoe, A.; Haik, Y. Removal and recovery of acridine orange from solutions by use of magnetic nanoparticles. *J. Hazard. Mater.* **2009**, *169*, 318–323. [CrossRef]
2. Karri, R.R.; Tanzifi, M.; Tavakkoli Yaraki, M.; Sahu, J.N. Optimization and modeling of methyl orange adsorption onto polyaniline nano-adsorbent through response surface methodology and differential evolution embedded neural network. *J. Environ. Manag.* **2018**, *223*, 517–529. [CrossRef] [PubMed]
3. Lau, Y.J.; Karri, R.R.; Mubarak, N.M.; Lau, S.Y.; Chua, H.B.; Khalid, M.; Jagadish, P.; Abdullah, E.C. Removal of dye using peroxidase-immobilized Buckypaper/polyvinyl alcohol membrane in a multi-stage filtration column via RSM and ANFIS. *Environ. Sci Pollut. Res. Int.* **2020**, *27*, 40121–40134. [CrossRef] [PubMed]
4. Gill, G.K.; Mubarak, N.M.; Nizamuddin, S.; Al-Salim, H.S.; Sahu, J.N. Column performance of carbon nanotube packed bed for methylene blue and orange red dye removal from waste water. *IOP Conf. Ser. Mater. Sci. Eng.* **2017**, *206*, 012081. [CrossRef]
5. Cui, D.; Zhang, H.; He, R.; Zhao, M. The Comparative Study on the Rapid Decolorization of Azo, Anthraquinone and Triphenylmethane Dyes by Anaerobic Sludge. *Int. J. Environ. Res. Public Health* **2016**, *13*. [CrossRef] [PubMed]

6. Qu, L.; Han, T.; Luo, Z.; Liu, C.; Mei, Y.; Zhu, T. One-step fabricated Fe_3O_4@C core–shell composites for dye removal: Kinetics, equilibrium and thermodynamics. *J. Phys. Chem. Solids* **2015**, *78*, 20–27. [CrossRef]
7. Tang, X.; Li, Z.; Liu, K.; Luo, X.; He, D.; Ao, M.; Peng, Q. Sulfidation modified Fe_3O_4 nanoparticles as an efficient Fenton-like catalyst for azo dyes degradation at wide pH range. *Powder Technol.* **2020**, *376*, 42–51. [CrossRef]
8. Mubarak, N.M.; Fo, Y.T.; Al-Salim, H.S.; Sahu, J.N.; Abdullah, E.C.; Nizamuddin, S.; Jayakumar, N.S.; Ganesan, P. Removal of Methylene Blue and Orange-G from Waste Water Using Magnetic Biochar. *Int. J. Nanosci.* **2015**, *14*, 1550009. [CrossRef]
9. Yao, H.; Xie, Y.; Jing, Y.; Wang, Y.; Luo, G. Controllable Preparation and Catalytic Performance of Heterogeneous Fenton-like α-Fe_2O_3/Crystalline Glass Microsphere Catalysts. *Ind. Eng. Chem. Res.* **2017**, *56*, 13751–13759. [CrossRef]
10. Hu, L.; Guang, C.; Liu, Y.; Su, Z.; Gong, S.; Yao, Y.; Wang, Y. Adsorption behavior of dyes from an aqueous solution onto composite magnetic lignin adsorbent. *Chemosphere* **2020**, *246*, 125757. [CrossRef]
11. Zhang, Y.; Zhang, B.T.; Teng, Y.; Zhao, J.; Sun, X. Heterogeneous activation of persulfate by carbon nanofiber supported Fe3O4@carbon composites for efficient ibuprofen degradation. *J. Hazard. Mater.* **2020**, *401*, 123428. [CrossRef] [PubMed]
12. Jiao, Y.; Wan, C.; Bao, W.; Gao, H.; Liang, D.; Li, J. Facile hydrothermal synthesis of Fe_3O_4@cellulose aerogel nanocomposite and its application in Fenton-like degradation of Rhodamine B. *Carbohydr. Polym.* **2018**, *189*, 371–378. [CrossRef] [PubMed]
13. Niu, H.; Zhang, D.; Zhang, S.; Zhang, X.; Meng, Z.; Cai, Y. Humic acid coated Fe_3O_4 magnetic nanoparticles as highly efficient Fenton-like catalyst for complete mineralization of sulfathiazole. *J. Hazard. Mater.* **2011**, *190*, 559–565. [CrossRef] [PubMed]
14. Li, H.; Xu, R.; Wang, Y.; Qian, B.; Wang, H.; Chen, L.; Jiang, H.; Yang, Y.; Xu, Y. In situ synthesis of hierarchical mesoporous Fe_3O_4@C nanowires derived from coordination polymers for high-performance lithium-ion batteries. *RSC Adv.* **2014**, *4*, 51960–51965. [CrossRef]
15. Wu, R.; Liu, J.-H.; Zhao, L.; Zhang, X.; Xie, J.; Yu, B.; Ma, X.; Yang, S.-T.; Wang, H.; Liu, Y. Hydrothermal preparation of magnetic Fe_3O_4@C nanoparticles for dye adsorption. *J. Environ. Chem. Eng.* **2014**, *2*, 907–913. [CrossRef]
16. Villa, S.; Caratto, V.; Locardi, F.; Alberti, S.; Canepa, F.; Ferretti, M.; Sturini, M.; Speltini, A.; Maraschi, F. Enhancement of TiO_2 NPs Activity by Fe_3O_4 Nano-Seeds for Removal of Organic Pollutants in Water. *Materials* **2016**, *9*, 771. [CrossRef]
17. Li, Z.-D.; Wang, H.-L.; Wei, X.-N.; Liu, X.-Y.; Yang, Y.-F.; Jiang, W.-F. Preparation and photocatalytic performance of magnetic Fe_3O_4@TiO_2 core-shell microspheres supported by silica aerogels from industrial fly ash. *J. Alloy. Compd.* **2016**, *659*, 240–247. [CrossRef]
18. Rezaei, S.S.; Kakavandi, B.; Noorisepehr, M.; Isari, A.A.; Zabih, S.; Bashardoust, P. Photocatalytic oxidation of tetracycline by magnetic carbon-supported TiO_2 nanoparticles catalyzed peroxydisulfate: Performance, synergy and reaction mechanism studies. *Sep. Purif. Technol.* **2021**, *258*, 117936. [CrossRef]
19. Zhou, R.; Shen, N.; Zhao, J.; Su, Y.; Ren, H. Glutathione-coated Fe_3O_4 nanoparticles with enhanced Fenton-like activity at neutral pH for degrading 2,4-dichlorophenol. *J. Mater. Chem. A* **2018**, *6*, 1275–1283. [CrossRef]
20. Wang, Z.; Fan, Y.; Wu, R.; Huo, Y.; Wu, H.; Wang, F.; Xu, X. Novel magnetic g-C_3N_4/α-Fe_2O_3/Fe_3O_4 composite for the very effective visible-light-Fenton degradation of Orange II. *RSC Adv.* **2018**, *8*, 5180–5188. [CrossRef]
21. Zhang, X.; He, M.; Liu, J.-H.; Liao, R.; Zhao, L.; Xie, J.; Wang, R.; Yang, S.-T.; Wang, H.; Liu, Y. Fe_3O_4@C nanoparticles as high-performance Fenton-like catalyst for dye decoloration. *Chin. Sci. Bull.* **2014**, *59*, 3406–3412. [CrossRef]
22. Bahmani, M.; Dashtian, K.; Mowla, D.; Esmaeilzadeh, F.; Ghaedi, M. Robust charge carrier by Fe_3O_4 in Fe_3O_4/WO_3 core-shell photocatalyst loaded on UiO-66(Ti) for urea photo-oxidation. *Chemosphere* **2020**, *267*, 129206. [CrossRef] [PubMed]
23. Le, V.T.; Tran, V.A.; Tran, D.L.; Nguyen, T.L.H.; Doan, V.-D. Fabrication of Fe_3O_4/CuO@C composite from MOF-based materials as an efficient and magnetically separable photocatalyst for degradation of ciprofloxacin antibiotic. *Chemosphere* **2021**, *270*, 129417. [CrossRef] [PubMed]
24. Chen, T.; Wang, Q.; Lyu, J.; Bai, P.; Guo, X. Boron removal and reclamation by magnetic magnetite (Fe3O4) nanoparticle: An adsorption and isotopic separation study. *Sep. Purif. Technol.* **2020**, *231*, 115930. [CrossRef]
25. Chen, T.; Zhang, X.; Qian, J.; Li, S.; Jia, X.; Song, H.-J. One-step hydrothermal synthesis of carbon@Fe_3O_4 nanoparticles with high adsorption capacity. *J. Mater. Sci. Mater. Electron.* **2014**, *25*, 1381–1387. [CrossRef]
26. Kong, D.; Liu, Y.; Li, Y.; Chi, W.; Huang, Q.; Yu, C. Facile preparation and dye removal properties of Fe_3O_4@carbon nanocomposite. *Micro Nano Lett.* **2018**, *13*, 219–222. [CrossRef]
27. Zhang, L.; Han, Y.; Yang, Z.; Su, B. Application of multi-functional chestnut shell in one-step preparing Fe_3O_4@C magnetic nanocomposite with high adsorption performance. *Full. Nanotub. Carbon Nanostructures* **2018**, *26*, 471–478. [CrossRef]
28. Lestari, I.; Kurniawan, E.; Gusti, D.R.; Yusnelti. Magnetite Fe_3O_4-activated carbon composite as adsorbent of rhodamine B dye. *IOP Conf. Ser. Earth Environ. Sci.* **2020**, *483*, 012046. [CrossRef]
29. Du, C.; Song, Y.; Shi, S.; Jiang, B.; Yang, J.; Xiao, S. Preparation and characterization of a novel Fe_3O_4-graphene-biochar composite for crystal violet adsorption. *Sci. Total Environ.* **2020**, *711*, 134662. [CrossRef]
30. Karamipour, A.; Khadiv Parsi, P.; Zahedi, P.; Moosavian, S.M.A. Using Fe_3O_4-coated nanofibers based on cellulose acetate/chitosan for adsorption of Cr(VI), Ni(II) and phenol from aqueous solutions. *Int. J. Biol. Macromol.* **2020**, *154*, 1132–1139. [CrossRef]
31. Zhang, W.; Zhang, L.Y.; Zhao, X.J.; Zhou, Z. Citrus pectin derived ultrasmall Fe_3O_4@C nanoparticles as a high-performance adsorbent toward removal of methylene blue. *J. Mol. Liq.* **2016**, *222*, 995–1002. [CrossRef]
32. Wang, P.; Gao, M.; Pan, H.; Zhang, J.; Liang, C.; Wang, J.; Zhou, P.; Liu, Y. A facile synthesis of Fe_3O_4/C composite with high cycle stability as anode material for lithium-ion batteries. *J. Power Sources* **2013**, *239*, 466–474. [CrossRef]

33. Wang, K.; Wan, G.; Wang, G.; He, Z.; Shi, S.; Wu, L.; Wang, G. The construction of carbon-coated Fe_3O_4 yolk-shell nanocomposites based on volume shrinkage from the release of oxygen anions for wide-band electromagnetic wave absorption. *J. Colloid Interface Sci.* **2018**, *511*, 307–317. [CrossRef] [PubMed]
34. Zhang, Q.; Cheng, Y.; Fang, C.; Chen, J.; Chen, H.; Li, H.; Yao, Y. Facile synthesis of porous carbon/Fe_3O_4 composites derived from waste cellulose acetate by one-step carbothermal method as a recyclable adsorbent for dyes. *J. Mater. Res. Technol.* **2020**, *9*, 3384–3393. [CrossRef]
35. Geng, H.; Zhou, Q.; Zheng, J.; Gu, H. Preparation of porous and hollow Fe_3O_4@C spheres as an efficient anode material for a high-performance Li-ion battery. *RSC Adv.* **2014**, *4*, 6430–6434. [CrossRef]
36. Yuan, S.; Zhou, Z.; Li, G. Structural evolution from mesoporous α-Fe_2O_3 to Fe_3O_4@C and γ-Fe_2O_3 nanospheres and their lithium storage performances. *CrystEngComm* **2011**, *13*, 4709–4713. [CrossRef]
37. Ren, G.; Yang, L.; Zhang, Z.; Zhong, B.; Yang, X.; Wang, X. A new green synthesis of porous magnetite nanoparticles from waste ferrous sulfate by solid-phase reduction reaction. *J. Alloys Compd.* **2017**, *710*, 875–879. [CrossRef]
38. de la Luz-Asunción, M.; Pérez-Ramírez, E.E.; Martínez-Hernández, A.L.; García-Casillas, P.E.; Luna-Bárcenas, J.G.; Velasco-Santos, C. Adsorption and kinetic study of Reactive Red 2 dye onto graphene oxides and graphene quantum dots. *Diamond Relat. Mater.* **2020**, *109*, 108002. [CrossRef]
39. Schneider, M.; Ballweg, T.; Groß, L.; Gellermann, C.; Sanchez-Sanchez, A.; Fierro, V.; Celzard, A.; Mandel, K. Magnetic Carbon Composite Particles for Dye Adsorption from Water and their Electrochemical Regeneration. *Part. Part. Syst. Charact.* **2019**, *36*. [CrossRef]
40. Tempkin, M.I.; Pyzhev, V. Kinetics of Ammonia Synthesis on Promoted Iron Catalyst. *Acta Physicochim. URSS* **1940**, *12*, 327–356.
41. Redlich, O.; Peterson, D.L. A Useful Adsorption Isotherm. *J. Phys. Chem.* **1959**, *63*, 1024–1026. [CrossRef]
42. Ren, G.; Wang, X.; Zhang, Z.; Zhong, B.; Yang, L.; Yang, X. Characterization and synthesis of nanometer magnetite black pigment from titanium slag by microwave-assisted reduction method. *Dye. Pigment.* **2017**, *147*, 24–30. [CrossRef]
43. Bao, X.; Qiang, Z.; Chang, J.-H.; Ben, W.; Qu, J. Synthesis of carbon-coated magnetic nanocomposite (Fe_3O_4@C) and its application for sulfonamide antibiotics removal from water. *J. Environ. Sci.* **2014**, *26*, 962–969. [CrossRef]
44. Moghtada, A.; Shahrouzianfar, A.; Ashiri, R. Facile synthesis of $NiTiO_3$ yellow nano-pigments with enhanced solar radiation reflection efficiency by an innovative one-step method at low temperature. *Dye. Pigment.* **2017**, *139*, 388–396. [CrossRef]
45. Borodko, Y.; Habas, S.E.; Koebel, M.; Yang, P.; Frei, H.; Somorjai, G.A. Probing the Interaction of Poly(vinylpyrrolidone)with Platinum Nanocrystals by UV- Raman and FTIR. *J. Phys. Chem. B* **2006**, *110*, 23052–23059. [CrossRef] [PubMed]
46. Suppiah, D.D.; Hamid, S.B.A. One step facile synthesis of ferromagnetic magnetite nanoparticles. *J. Magn. Magn. Mater.* **2016**, *414*, 204–208. [CrossRef]
47. Chae, H.S.; Piao, S.H.; Choi, H.J. Fabrication of spherical Fe_3O_4 particles with a solvothermal method and their magnetorheological characteristics. *J. Ind. Eng. Chem.* **2015**, *29*, 129–133. [CrossRef]
48. Tu, J.; Yuan, J.; Kang, S.; Xu, Y.; Wang, T. One-pot synthesis of carbon-coated Fe_3O_4 nanoparticles with tunable size for production of gasoline fuels. *New J. Chem.* **2018**, *42*, 10861–10867. [CrossRef]
49. Spivakov, A.A.; Lin, C.-R.; Jhang, C.-J.; Tsai, Y.-J.; Tseng, Y.-T. Synthesis and characterization of carbon-coated wustite nanoparticles. *Mater. Lett.* **2019**, *249*, 147–150. [CrossRef]
50. Sadezky, A.; Muckenhuber, H.; Grothe, H.; Niessner, R.; Pöschl, U. Raman microspectroscopy of soot and related carbonaceous materials: Spectral analysis and structural information. *Carbon* **2005**, *43*, 1731–1742. [CrossRef]
51. Shebanova, O.N.; Lazor, P. Raman spectroscopic study of magnetite ($FeFe_2O_4$): A new assignment for the vibrational spectrum. *J. Solid State Chem.* **2003**, *174*, 424–430. [CrossRef]
52. Salazar-Camacho, C.; Villalobos, M.; Rivas-Sánchez, M.d.l.L.; Arenas-Alatorre, J.; Alcaraz-Cienfuegos, J.; Gutiérrez-Ruiz, M.E. Characterization and surface reactivity of natural and synthetic magnetites. *Chem. Geol.* **2013**, *347*, 233–245. [CrossRef]
53. Rajput, S.; Pittman, C.U., Jr.; Mohan, D. Magnetic magnetite (Fe_3O_4) nanoparticle synthesis and applications for lead (Pb2+) and chromium (Cr6+) removal from water. *J. Colloid Interface Sci.* **2016**, *468*, 334–346. [CrossRef]
54. Li, Z.; Sun, Y.; Xing, J.; Meng, A. Fast Removal of Methylene Blue by Fe(3)O(4) Magnetic Nanoparticles and Their Cycling Property. *J. Nanosci. Nanotechnol.* **2019**, *19*, 2116–2123. [CrossRef] [PubMed]
55. Karri, R.R.; Sahu, J.N.; Jayakumar, N.S. Optimal isotherm parameters for phenol adsorption from aqueous solutions onto coconut shell based activated carbon: Error analysis of linear and non-linear methods. *J. Taiwan Inst. Chem. Eng.* **2017**, *80*, 472–487. [CrossRef]
56. Ren, G.; Wang, X.; Huang, P.; Zhong, B.; Zhang, Z.; Yang, L.; Yang, X. Chromium (VI) adsorption from wastewater using porous magnetite nanoparticles prepared from titanium residue by a novel solid-phase reduction method. *Sci. Total Environ.* **2017**, *607-608*, 900–910. [CrossRef]
57. Xiang, H.; Ren, G.; Yang, X.; Xu, D.; Zhang, Z.; Wang, X. A low-cost solvent-free method to synthesize α-Fe_2O_3 nanoparticles with applications to degrade methyl orange in photo-fenton system. *Ecotoxicol. Environ. Saf.* **2020**, *200*. [CrossRef]
58. Ouyang, J.; Zhao, Z.; Suib, S.L.; Yang, H. Degradation of Congo Red dye by a Fe_2O_3@CeO_2-ZrO_2/Palygorskite composite catalyst: Synergetic effects of Fe_2O_3. *J. Colloid Interface Sci.* **2019**, *539*, 135–145. [CrossRef]
59. Gonçalves, R.G.L.; Mendes, H.M.; Bastos, S.L.; D'Agostino, L.C.; Tronto, J.; Pulcinelli, S.H.; Santilli, C.V.; Neto, J.L. Fenton-like degradation of methylene blue using Mg/Fe and MnMg/Fe layered double hydroxides as reusable catalysts. *Appl. Clay Sci.* **2020**, *187*, 105477. [CrossRef]

60. Ghasemi, H.; Aghabarari, B.; Alizadeh, M.; Khanlarkhani, A.; Abu-Zahra, N. High efficiency decolorization of wastewater by Fenton catalyst: Magnetic iron-copper hybrid oxides. *J. Water Process. Eng.* **2020**, *37*, 101540. [CrossRef]
61. Zhou, L.; Shao, Y.; Liu, J.; Ye, Z.; Zhang, H.; Ma, J.; Jia, Y.; Gao, W.; Li, Y. Preparation and Characterization of Magnetic Porous Carbon Microspheres for Removal of Methylene Blue by a Heterogeneous Fenton Reaction. *ACS Appl. Mater. Interfaces* **2014**, *6*, 7275–7285. [CrossRef]
62. Liu, W.; Qian, J.; Wang, K.; Xu, H.; Jiang, D.; Liu, Q.; Yang, X.; Li, H. Magnetically Separable Fe_3O_4 Nanoparticles-Decorated Reduced Graphene Oxide Nanocomposite for Catalytic Wet Hydrogen Peroxide Oxidation. *J. Inorg. Organomet. Polym. Mater.* **2013**, *23*, 907–916. [CrossRef]
63. Li, W.; Wu, X.; Li, S.; Tang, W.; Chen, Y. Magnetic porous Fe_3O_4/carbon octahedra derived from iron-based metal-organic framework as heterogeneous Fenton-like catalyst. *Appl. Surf. Sci.* **2018**, *436*, 252–262. [CrossRef]
64. Zhang, F.; Xue, X.; Huang, X.; Yang, H. Adsorption and heterogeneous Fenton catalytic performance for magnetic Fe_3O_4/reduced graphene oxide aerogel. *J. Mater. Sci.* **2020**, *55*, 15695–15708. [CrossRef]
65. Zong, M.; Song, D.; Zhang, X.; Huang, X.; Lu, X.; Rosso, K.M. Facet-Dependent Photodegradation of Methylene Blue by Hematite Nanoplates in Visible Light. *Environ. Sci. Technol.* **2021**, *55*, 677–688. [CrossRef]
66. Lu, Z.; Cao, X.; Wei, H.; Huo, W.; Wang, Q.; Li, K. Strong enhancement effect of bisulfite on MIL-68(Fe)-catalyzed Fenton-like reaction for organic pollutants degradation. *Appl. Surf. Sci.* **2021**, *542*, 148631. [CrossRef]
67. Ghorai, K.; Panda, A.; Bhattacharjee, M.; Mandal, D.; Hossain, A.; Bera, P.; Seikh, M.M.; Gayen, A. Facile synthesis of $CuCr2O_4/CeO_2$ nanocomposite: A new Fenton like catalyst with domestic LED light assisted improved photocatalytic activity for the degradation of RhB, MB and MO dyes. *Appl. Surf. Sci.* **2021**, *536*, 147604. [CrossRef]
68. Bao, C.; Zhang, H.; Zhou, L.; Shao, Y.; Ma, J.; Wu, Q. Preparation of copper doped magnetic porous carbon for removal of methylene blue by a heterogeneous Fenton-like reaction. *RSC Adv.* **2015**, *5*, 72423–72432. [CrossRef]
69. Yimin, D.; Jiaqi, Z.; Danyang, L.; Lanli, N.; Liling, Z.; Yi, Z.; Xiaohong, Z. Preparation of Congo red functionalized Fe_3O_4@SiO_2 nanoparticle and its application for the removal of methylene blue. *Colloids Surf. A Physicochem. Eng. Asp.* **2018**, *550*, 90–98. [CrossRef]
70. Zhang, L.; Song, X.; Tan, L.; Ma, H.; Guo, D.; Pang, H.; Wang, X. Fabrication of double-shell hollow NiO@N-C nanotubes for a high-performance supercapacitor. *New J. Chem.* **2019**, *43*, 13457–13462. [CrossRef]
71. Yang, B.; Zhou, P.; Cheng, X.; Li, H.; Huo, X.; Zhang, Y. Simultaneous removal of methylene blue and total dissolved copper in zero-valent iron/H_2O_2 Fenton system: Kinetics, mechanism and degradation pathway. *J. Colloid Interface Sci.* **2019**, *555*, 383–393. [CrossRef] [PubMed]
72. Yang, C.; Dong, W.; Cui, G.; Zhao, Y.; Shi, X.; Xia, X.; Tang, B.; Wang, W. Highly efficient photocatalytic degradation of methylene blue by P2ABSA-modified TiO_2 nanocomposite due to the photosensitization synergetic effect of TiO_2 and P2ABSA. *RSC Adv.* **2017**, *7*, 23699–23708. [CrossRef]
73. Zheng, S.; Chen, H.; Tong, X.; Wang, Z.; Crittenden, J.C.; Huang, M. Integration of a Photo-Fenton Reaction and a Membrane Filtration using CS/PAN@FeOOH/g-C_3N_4Electrospun Nanofibers: Synthesis, Characterization, Self-cleaning Performance and Mechanism. *Appl. Catal. B Environ.* **2021**, *281*, 119519. [CrossRef]
74. Wolski, L.; Ziolek, M. Insight into pathways of methylene blue degradation with H_2O_2 over mono and bimetallic Nb, Zn oxides. *Appl. Catal. B Environ.* **2018**, *224*, 634–647. [CrossRef]
75. Wang, Q.; Tian, S.; Ning, P. Degradation Mechanism of Methylene Blue in a Heterogeneous Fenton-like Reaction Catalyzed by Ferrocene. *Ind. Eng. Chem. Res.* **2013**, *53*, 643–649. [CrossRef]

MDPI
St. Alban-Anlage 66
4052 Basel
Switzerland
www.mdpi.com

Nanomaterials Editorial Office
E-mail: nanomaterials@mdpi.com
www.mdpi.com/journal/nanomaterials

www.ingramcontent.com/pod-product-compliance
Lightning Source LLC
Chambersburg PA
CBHW061818210926
51850CB00006B/740
9783725804344